Healthy Ageing and Longevity

Volume 17

Rapidly changing demographics worldwide towards increased proportion of the elderly in the population and increased life-expectancy have brought the issues, such as "why we grow old", "how we grow old", "how long can we live", "how to maintain health", "how to prevent and treat diseases in old age", "what are the future perspectives for healthy ageing and longevity" and so on, in the centre stage of scientific, social, political, and economic arena. Although the descriptive aspects of ageing are now well established at the level of species, populations, individuals, and within an individual at the tissue, cell and molecular levels, the implications of such detailed understanding with respect to the aim of achieving healthy ageing and longevity are ever-changing and challenging issues. This continuing success of gerontology, and especially of biogerontology, is attracting the attention of both the well established academicians and the younger generation of students and researchers in biology, medicine, bioinformatics, bioeconomy, sports science, and nutritional sciences, along with sociologists, psychologists, politicians, public health experts, and health-care industry including cosmeceutical-, food-, and lifestyle-industry. Books in this series will cover the topics related to the issues of healthy ageing and longevity. This series will provide not only the exhaustive reviews of the established body of knowledge, but also will give a critical evaluation of the ongoing research and development with respect to theoretical and evidence-based practical and ethical aspects of interventions towards maintaining, recovering and enhancing health and longevity.

Francesco Marotta
Editor

Gut Microbiota in Aging and Chronic Diseases

Springer

Editor
Francesco Marotta
ReGenera R&D International for Aging
Intervention
Milan, Italy

ISSN 2199-9007 ISSN 2199-9015 (electronic)
Healthy Ageing and Longevity
ISBN 978-3-031-14022-8 ISBN 978-3-031-14023-5 (eBook)
https://doi.org/10.1007/978-3-031-14023-5

This Springer imprint is published by the registered company Springer Nature Switzerland AG
The registered company address is: Gewerbestrasse 11, 6330 Cham, Switzerland

This work is dedicated to my late parents who firmly believed that only education through hard challenges and falls, no matter how far from home, was the worthy one. "To live a life you will remember" as the young Kathy, "the wonderlust", who herself climbs much rougher cliffs to see wider horizons, still offers me as a daily teaching example. Similarly, the younger Carole for reflecting to me her glowing grace, endurance and willpower she proved herself in successfully tackling much harder hurdles. The benevolent tolerance of my wife to get on her shoulder heavier duties while the husband would pop in and disappear around the globe was a vital element too.

Preface

Over 2000 years ago, a Ming dynasty painting would depict a sitting living man with an open abdomen where food would travel along the gut resembling active tiny workers interacting with one another and all within that specific milieu. In modern times, it appeared that the gut contains a massive and hierarchically organized neuronal network and it was labeled as "a second brain". However, due also to the analytical limitations, till only about 18 years ago, the concept of dysbiosis and potential interventions were limited to gastrointestinal diseases. Prompted by an escalating experimental data accumulation, over those years the fundamental systemic implications of gut microbiota pathophysiology especially with chronic inflammation-related disease (metabolic syndrome, obesity, fatty liver, etc.) and extra-colonic cancer came clear. Those were the years laying down a solid basis for the ever-deepening concept of "inflamm-aging" where gut microbiota walked into the stage as a prime actor where New Generation Sequencing technology was a major metagenomic tool to hugely widespread researchers' efforts and investigational creativity. In the last 10 years, over 54.000 papers on gut microbiota have appeared in the indexed literature versus 2600 in the past 45 years. The plasticity and evolutionary dynamics of gut microbiota have been extensively studied from birth to adulthood, while the physiological "aging gut microbiota" per sè has opened a new field in aging hologenomic understanding and potential intervention. Virtually, all age-related diseases have shown specific, albeit not fully unfolded as yet, interrelationships with gut microbiota. So far not much attention has been dedicated to this indissoluble conjuring up cluster, i.e. gut microbiota patterns and aging in the pathogenetic mechanisms of chronic disease. On the other hand, pre- and probiotics technologies are seeing a parallel development and stepping up as novel "biotherapeutics" and "gerobiotics", amenable to become key players for antiaging strategies in healthy persons. In this book, key opinion leaders who enthusiastically adhered to the project offered their fine expertise fully embracing the above "cluster" approach each one within his super-specialty.

Milan, Italy Francesco Marotta

Acknowledgments The punctual and elegantly professional support received by Ms. Marleen Moore and her tireless editorial staff is gladly acknowledged. This work came to an end also thanks to them.

Contents

Part I
The Gut Microbiota in Ageing

Chapter 1
The Gut Microbiota and Its Metabolites Contribute to Ageing and Ageing-Related Diseases

Chuan Chen, Zheng Cao, Hehua Lei, and Limin Zhang

Abstract The human gut is regarded as a metabolic "super organ" populated by millions of microorganisms and functions as a dynamic barrier to the environment with signal integration properties. Increasing evidence has shown that the gut microbiota and its metabolites are highly associated with many physiological processes, including nutrient intake, growth, and development as well as immune and inflammatory status. It is well known that ageing is a biological and psychological process with progressive decline in host functional homeostasis. Furthermore, ageing is one of the biggest risk factors for many metabolic syndromes such as osteoporosis, cancer, and neurodegenerative pathologies, which negatively impact human vitality, healthspan, and lifespan. Thus, it can be seen that the gut microbiota and its metabolites are guaranteed to be responsible for ageing and ageing-related diseases. This chapter highlights the roles of intestinal barrier functions, the gut microbiota, and microbiota-derived metabolites mainly including short chain fatty acids (SCFAs), bile acids (BAs), and tryptophan metabolites in ageing and ageing-related chronic diseases.

Keywords The gut microbiota · Intestinal barrier function · Short chain fatty acids · (SCFAs) · Bile acids (BAs) · Tryptophan · Ageing

---

Chuan Chen and Zheng Cao, these authors contributed to this work equally.

C. Chen · Z. Cao · H. Lei · L. Zhang (✉)
State Key Laboratory of Magnetic Resonance and Atomic and Molecular Physics, National Centre for Magnetic Resonance in Wuhan, Innovation Academy of Precision Measurement Science and Technology, Chinese Academy of Sciences (CAS), Wuhan 430071, P. R. China
e-mail: zhanglm@wipm.ac.cn

C. Chen · Z. Cao · L. Zhang
University of Chinese Academy of Sciences, Beijing 100049, P. R. China

F. Marotta (ed.), *Gut Microbiota in Aging and Chronic Diseases*, Healthy Ageing and Longevity 17, https://doi.org/10.1007/978-3-031-14023-5_1

1.1 Introduction

Ageing is a physiological and pathological process with a functional decline of tissues and organs ultimately contributing to healthspan and lifespan. Ageing is one of the largest risk factors for a variety of ageing-related diseases such as cancer (d'Adda di Fagagna 2008), osteoporosis (Hendrickx et al. 2015), vascular dysfunction (Jia et al. 2019), and neurodegenerative pathologies (Cheng et al. 2022), although causality between them has not been determined. During the past several decades, molecular and biochemical mechanisms underlying individual ageing-related chronic disorders were investigated in laboratories using animal models such as mouse, *C. elegans,* and *Drosophila* (Koyuncu et al. 2021; Ulgherait et al. 2021; Bárcena et al. 2019; Fontana and Partridge 2015). For example, the dietary restriction has been found to extend lifespan and prevent against the development of ageing-related diseases in mice and rats with restriction of caloric intake (Ekmekcioglu 2020). For many years, ageing research mainly focused on identifying host ageing phenotypes and corresponding genetic pathways such as insulin-like signalling pathway, circadian clocks, senescence, mitochondria and oxidative stress, and chronic inflammation, which are involved in the maintenance of host metabolic homeostasis (Campisi et al. 2019). However, less work has been performed on the modulation of the gut microbiota and its metabolites that are closely associated with many metabolic syndromes such as obesity, diabetes, and cancer (Kumar et al. 2019; Venkatachalam et al. 2017).

It is well established that intestinal barrier functions can be attributed to two parts: one is intestinal epithelium involving several types of enterocytes, and another is the gut microbiota colonized in the intestinal cavity and microbiota-derived metabolites. During ageing, intestinal barrier functions are gradually declined, which are mainly reflected in progenitor differentiation and apoptosis of intestinal epithelial cells and disruption of gut microbiota. The intestinal epithelium functions as a dynamic barrier to many environmental factors such as diet, drugs, and toxins (Zhang et al. 2015). It serves as an important physical and chemical barrier for the maintenance of epithelial homeostasis. Previous studies showed that microbial products, such as endotoxin lipopolysaccharide (LPS) and flagellin leading to systemic inflammation, can go through the mucosa and spread to the whole body due to the altered intestinal barrier structure, which induces increased gut permeability and gut leakiness (Li et al. 2016). Intestinal dysbiosis triggers an innate immune response and chronic low-grade inflammation, leading to many age-related degenerative diseases (Clemente et al. 2012; Wendy et al. 2010). Some bacteria such as *Akkermansia*, *Bifidobacterium,* and *Lactobacillus* have the ability to maintain intestinal integrity and permeability for anti-inflammation (Paone and Cani 2020). Meanwhile, gut microbiota-derived metabolites such as short chain fatty acids (SCFAs), bile acids (BAs), and tryptophan metabolites are also beneficial for immunoregulation function through a variety of signalling pathways by interacting with host organs (Fig. 1.1).

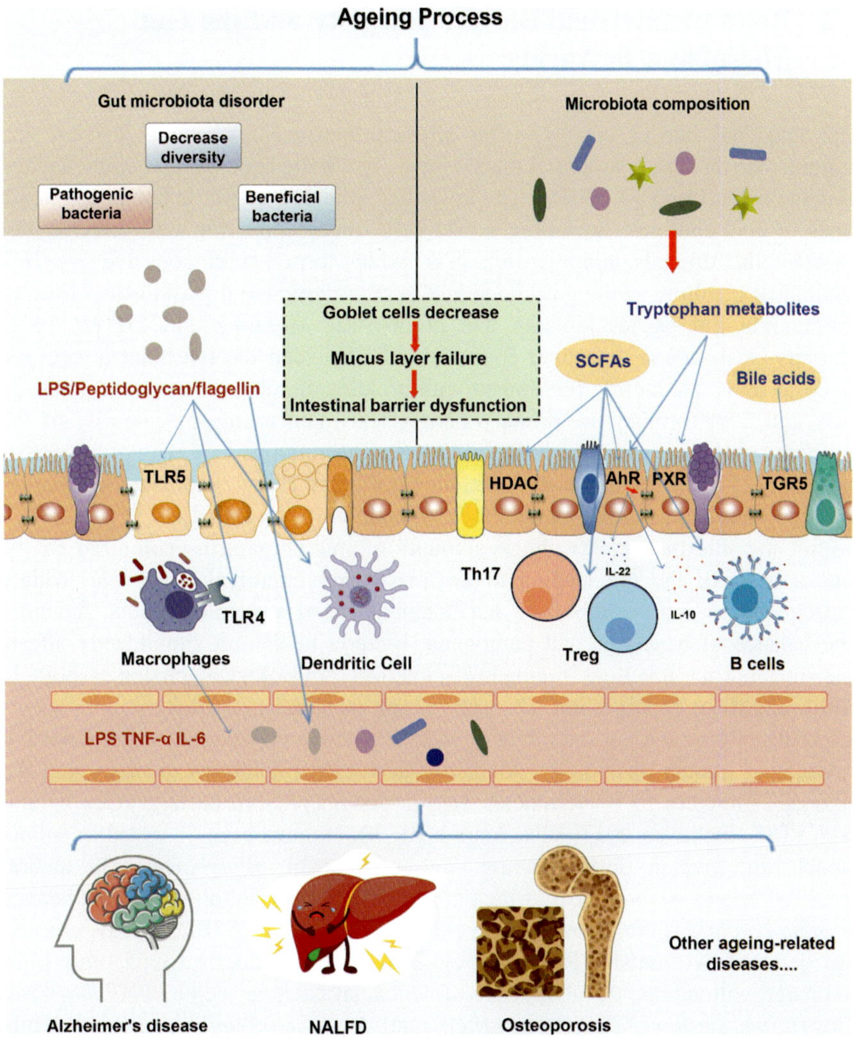

Fig. 1.1 Summary of the molecular signalling transportation between the gut microbiota and host organs with ageing process. During ageing, the gut microbiota and its metabolites are altered which leads to dysfunction of the gut barrier and gut leakiness, systemic inflammation, and ultimately ageing-related diseases

In recent years, intestinal barrier functions, the gut microbiota, and microbiota-derived metabolites have contributed evidence to many of the signalling pathways and processes closely associated with ageing and ageing-related diseases. This review attempts to focus on the roles of intestinal homeostasis, the gut microbiota, and its metabolites in ageing, highlighting recent achievements about the potential consequences of gut-host interactions on the intestinal barrier and health.

1.2 Roles of Intestinal Barrier Integrity and the Gut Microbiota in Ageing

The intestinal barrier consists of a physical barrier, biochemical barrier, and immune barrier with a structural mucus layer, symbiotic bacteria, and epithelial and immune cells (Gao et al. 2018). The intestinal epithelial barrier is composed of six types of epithelial cells including goblet cells, enteroendocrine cells, enterocytes, Paneth cells, tuft cells, and microfold cells, which form a continuous and polarized monolayer resulting in the gut lumen and lamina propria and play distinct roles in maintaining gut barrier function and homeostasis (Beumer and Clevers 2021; Martens et al. 2018). Intestinal epithelial cells between the intestinal lumen and mucosal tissue are closely interlinked with proteins of adhesive and tight junctions (AJs and TJs) forming the primary barrier and maintaining the integrity of the intestinal epithelial barrier (Zihni et al. 2016). The immune barrier is composed of lymphocytes and macrophages in mucosal epithelium excreting antibodies immunoglobulin IgA and immune factors such as IL-22 and IL-10 (Pabst 2012). Within the intestinal cavity, huge amounts of microorganisms colonized in the mucus layer of the gut leading to an interrelated microbial ecosystem, widely participating in host metabolism and maintenance of systemic immune function. The balance of beneficial and pathogenic bacteria in the gut significantly affects intestinal barrier functions regulating numerous physiological processes both in health and disease. Digestion, absorption, and metabolism of dietary components markedly rely on the gut microbiota generating active metabolites or nutrients and modulating human physiology. Markedly decreased microbiota diversity and the ratio of Firmicutes to Bacteroidetes, two dominant phylum bacteria (Gentile and Weir 2018; Donovan 2017; Vaiserman et al. 2017), have been observed in animal models and human during ageing. The number of *Bifidobacteria* in natural breastfed infants is about 90%, which is less than 5% in adulthood. The abundance of *Bifidobacteria* is much lower in the elderly or patients with IBD, allergy, obesity, and degenerative autism (Rivière et al. 2016). In the elderly (> 65 years old), compared with adults, the abundance of beneficial bacteria such as *Bifidobacteria, Clostridium cluster XIV*, *Eubacterium,* and *Verrucomicrobia* decreased while *Bacteroidetes*, *Enterobacteriaceae,* and *Cyanobacteria* increased (Bárcena et al. 2019; Rondanelli et al. 2015), which may be classified as pathogenic bacteria. In addition, mouse models of progeria are characterized by intestinal dysbiosis with alterations of *Proteobacteria*, *Verrucomicrobia,* and *Cyanobacteria* abundance (Bárcena et al. 2019). Oral administration of *Akkermansia* significantly increases healthy lifespan of aged mice (Shin et al. 2021). Supplementation with *Lactobacillus casei Shirota* effectively reduced age-related declines in muscle mass and stress (Chen et al. 2022).

Intestinal barrier dysfunction has been shown to highly associate with systemic inflammatory response resulting in aggravation of pathophysiological status in various gastrointestinal and immune diseases. Chronic inflammation occurs during the aging process, which is accompanied by a decline in immune system function called "immune aging" (Appay and Sauce 2014). A growing number of studies have suggested that the gut microbiota is considered to be the largest immune organ of the human body. In general, older people are considered to be more susceptible to infections and other diseases, including many gastrointestinal diseases due to the worsening inflammatory status and intestinal barrier functions (Wilmanski et al. 2021; He et al. 2020; Sovran et al. 2019). The alterations of ageing in the gut microbiota composition are characterized by declines in microbiota diversity and saccharolytic bacteria and upregulation of proteolytic bacteria. Especially, down-regulation of *Bifidobacteria* and the ratio of dominant Firmicutes to Bacteroides (F/B) was found to be associated with ageing (Perez et al. 2014). A previous study reported that the gut microbiota of people over the age of 65 in Ireland has significant increases in the proportion of *Bacteroides spp.* and *Clostridium* in comparison with younger individuals (Ghosh et al. 2020).

Increasing evidence has shown that intestinal barrier dysfunction due to disruption of AJs and TJs under the conditions of ageing or diseases directly increased intestinal gut permeability leading to significant leakiness. Subsequently, microbial products, LPS, and flagellin secreting by some pathogenic bacteria such as *Bacteroides*, *Proteobacteria*, and other pathogens leak from the intestinal cavity and spread to the whole body consequently triggering systemic inflammation (Schiffrin et al. 2010). In ovariectomy-induced osteoporosis models, intestinal T helper cells and TNF + T cells were found to expand from lamina propria of the intestine to bone marrow due to the disruption of the intestinal barrier and gut leakage induced by TLR5-flagellin interaction, leading to osteoclastogenesis and osteoporosis (Yu et al. 2021). On the basis of these findings, homeostasis maintenance of the gut microbiota and intestinal barrier integrity are the key aspects of health and diseases. In recent years, supplementation with some probiotics such as *Lactobacillus*, *Bifidobacterium,* and *Akkermansia* via faecal microbiota transplantation (FMT) or dietary prebiotics has been widely used to ameliorate many metabolic diseases such as obesity (Wu et al. 2019) and ageing and ageing-related diseases (osteoporosis and Alzheimer's disease) (Ou et al. 2020) by restoring intestinal barrier functions and improving systemic immunity. Interestingly, supplementation of microbial metabolites such as short chain fatty acids, bile acids, and tryptophan metabolites is being applied to prevent against numerous disorders in animal models, since they exhibit a strong capacity in improving intestinal barrier function and host cellular homeostasis.

1.3 Roles of Short Chain Fatty Acids in Ageing

Short chain fatty acids (SCFAs), also known as volatile fatty acids, mainly include acetic acid, propionic acid, isobutyric acid, butyric acid, isovaleric acid, and valeric acid, which are primarily produced by anaerobic bacteria fermentation of dietary fibres and non-digestible sugars (non-digestible polysaccharides and resistant starch) in the gastrointestinal tract (Makki et al. 2018). Isobutyric acid and isovaleric acid are also produced by the catabolism of branched chain amino acids (BCAAs: valine, leucine, and isoleucine). Specifically, butyric acid is produced from Bacteroidetes and Firmicutes, the dominant bacteria producing SCFAs in the intestine (Koh et al. 2016). Acetic acid and propionic acid are originated from *Veillonellaceae* and *Bacteroides* by converting lactic acid and succinic acid (Koh et al. 2016). In the colon, SCFAs are rapidly absorbed through interaction with G-protein-coupled receptors (GPR41, GPR43, and GPR109A) and regulate the expression of downstream genes by inhibiting histone deacetylases (HDACs) and activating downstream signalling cascades, including phospholipase C (PLC), mitogen-activated protein kinase (MAPK), phospholipid A2 (PLA2), and NF-kB pathways (Makki et al. 2018). After absorption, SCFAs provide the main energy and nutrients source of intestinal epithelial cells commonly involved in host growth and energy metabolism and serve as cellular mediators regulating cell differentiation and development (Makki et al. 2018).

As one kind of the most key products of the gut microbiota, SCFAs play important roles in maintaining intestinal barrier functions and the morphology and structure of colonic epithelial cells (Aho et al. 2021). In particular, butyric acid has been shown as a mediator of barrier function by promoting the proliferation of epithelial cells and enhancing the expression of tight junction protein in the intestine (van der Hee and Wells 2021). Acetate exhibits the ability in protecting cognitive function impairment through gut-brain interaction via vagus nerve stimulation. FMT or supplementation with acetate significantly recovered hippocampal reduction of synaptophysin (SYP) in diabetic mice (Zheng et al. 2021a, b). Numerous studies have suggested that SCFAs, as a potent class of immune-modulatory compounds, owns the capacity to modulate host immune responses and affect Treg, T helper 17 (Th17) cells, and macrophage differentiation and development through membrane protein in the gut and periphery, which can establish an anti-inflammation environment (Wang et al. 2020; Shen et al. 2019). A common feature of physical sub-health and aging is the changes of the gut microbiota and microbial metabolites related to nutrition acquisition, as well as inflammatory status and intestinal integrity destruction. Given that the whole aging process is accompanied by the decline of intestinal barrier function, interventions with dietary fibre rich "Mediterranean" diet modulating the gut microbiota and producing SCFAs for older people can improve health outcomes and reduce all-cause mortality by up to 50% (Wang et al. 2021). A randomized clinical study reported that dietary fibres producing SCFAs from carbohydrate fermentation by the gut microbiota effectively ameliorate patients with T2DM, thus providing a new approach with micro-ecological regulation for managing T2DM (Zhao et al. 2021). Some recent studies have shown that transplantation of faecal bacteria (*Bifidobacterium longum*, *Clostridium faecalis*, *Akkermansia muciniphila* (*Akk*), and *Lactobacillus fermentum*) producing SCFAs significantly reduced neurological defects and inflammation in elderly stroke mice (Lee et al. 2020). It is known that osteoporosis and arthritis are two typical ageing-related bone degenerative diseases. Oral supplementation of *Akk*-abundant gut microbiota of children or probiotics such as *Akk*, *Lactobacillus,* and *Bifidobacterium* has also been used to ameliorate ovx-induced osteoporosis in mice (Liu et al. 2021). Furthermore, direct supplementation with SCFAs especially propionate and butyrate was found to protect against postmenopausal and mitigating arthritis via suppression of osteoclast differentiation and inhibition of bone resorption (Lucas et al. 2018). Elizabeth C. R. et al. recently reported that butyrate supplementation reduced arthritis in a B_{reg}-dependent manner by increasing the level of the serotonin-derived metabolite 5-Hydroxyindole-3-acetic acid (5-HIAA), which activates the aryl-hydrocarbon receptor (AhR) (Rosser et al. 2020). Collectively, these achievements provided sufficient evidence that SCFAs or/and bacteria producing SCFAs can effectively improve ageing-related diseases via maintenance of intestinal barrier function and enhancement of systemic immunity.

1.4 Roles of Tryptophan Metabolites in Ageing

Tryptophan (Trp) is an essential aromatic amino acid with the lowest abundance in proteins and cells. It is also a biosynthetic precursor of many microbial metabolites interacting with the host physiology. Dietary intake of Trp from common natural food is the main exogenous source since it is not able to be produced by animal cells. In the gastrointestinal tract, microbiota-derived Trp metabolism yields potent bioactive metabolites involved in three major metabolic pathways, one of which is the direct transformation of Trp by the gut microbiota into indole and its derivatives (Agus et al. 2018). Trp-microbiota metabolizing pathway has been identified in some animal bacteria. For example, *Clostridium sporogen* converts Trp to tryptamine, one of the neurotransmitters, and another two metabolites indole acetic acid (ILA) and indole-3-propionic acid (IPA) affecting intestinal barrier permeability and host immunity (Dodd et al. 2017; Williams et al. 2014). Similarly, digestive streptococci, including Roche, anaerobic, and stomatitis bacilli, are known to convert Trp to indoleacrylic acid (IA) and IPA by *Clostridium sporogenes, Clostridium suis, Clostridium botulinum,* and *Streptococcus* with the presence of phenyllactate hydratase gene cluster on the chromosomes of these species (Wlodarska et al. 2017). In addition, *Lactobacillus* and *Bifidobacteria* can also metabolize Trp into IALD and ILA through aromatic amino acid aminotransferase (ARAT) and indole acetate dehydrogenase (ILDH) (Cervantes-Barragan et al. 2017; Wilck et al. 2017). Indole and most of the indole derivatives, such as indole-3-aldehyde (IAld), indole-3-acid-acetic (IAA), IPA, IA, and indole-3-acetaldehyde (IAAld), are endogenous ligands of the aryl-hydrocarbon receptor (AhR), which is extensively expressed in the intestinal epithelial cells and especially immune cells such as lymphocytes, T_H17, T_{reg}, and B_{reg} cells (Stockinger et al. 2021; Rothhammer and Quintana 2019; Agus et al. 2018). AhR signalling has crucial roles in maintaining host homeostasis by immunoregulatory networks through crosstalk between the host and the gut microbiota. In particular, the intestinal immune is highly dependent on intestinal epithelial barrier immunosurveillance in the prevention of systemic stimuli and microbial infection. The declined transformation ability of the gut microbiota markedly results in defective intestinal barrier integrity and impaired AhR agonist activity, which ultimately accelerates the occurrence and development of metabolic syndromes including high blood pressure, diabetes, obesity, and cancer.

Given that the AhR signalling plays a pivotal and beneficial role in the intestinal function and host metabolic homeostasis, dietary interventions or direct supplementation with AhR ligands to increase the availability of AhR activity could be an effective strategy to maintain homeostatic conditions of intestinal microecology and host health. Many studies have highlighted the roles of the gut microbiota and diet in the production of AhR ligands, and direct or indirect supplementation with AhR ligands has also received comprehensive attention in the prevention of ageing and ageing-related diseases in the past few years. For example, a previous study showed that colonization of bacteria that secrete indoles or administration of the indole derivative indole-3 aldehyde (ICA) promotes epithelial cells proliferation via the AhR and IL-10 signalling, thus exhibiting an effect of aging (Powell et al. 2020). Multiple studies have suggested that indoles from commensal microbiota extend the healthspan of diverse organisms, including *C. elegans*, *Drosophila melanogaster*, and mice in an AhR-dependent manner (Sonowal et al. 2017). It is known that ageing is an inflammaging process with the intestinal barrier becoming more permeable, and bacterial antigens and products such as LPS leak out from the lumen to the whole body precipitating a systemic inflammatory response and leading to various inflammatory diseases. Three indole metabolites derived from gut bacterial metabolism of Trp including indole-3-ethanol, indole-3-pyruvate (IPyA), and IAld have been shown to ameliorate dextran sodium sulphate (DSS)-induced colitis of mice by protecting against increased gut permeability and regulating gut barrier function via the AhR activation (Scott et al. 2020). Furthermore, these indole metabolites exhibit anti-inflammation through modulation of immune cells such as CD103 + DCs, T regulatory 1 ($T_{reg}1$) cells, and Nkp46 + ILCs (Aoki et al. 2018).

In addition to indole metabolism, kynurenine (Kyn) and serotonin (5-HT) pathways in the gut are also mediated by the gut microbiota in stimulating indoleamine 2,3-dioxygenase 1 (IDO1) and Trp hydroxylase 1 enzyme (TpH1), which are close with host physiological processes involving inflammatory status and immune response. Two previous independent studies indicated that the ratio of Kyn to Trp reflecting the degradation rate of Trp markedly increased in the elderly, suggesting that ageing is in line with accelerated degradation of Trp through the Kyn pathway (Pertovaara et al. 2006; Frick et al. 2004). In addition, the higher urinary Trp ratio can be regarded as a biomarker for predicting higher mortality in a group of people in their 90 s (Pertovaara et al. 2006). Mice with intestinal $IDO1^{-/-}$ are more susceptible to colitis and IBD, suggesting that IDO1 negatively regulates intestinal inflammation. Increased IDO 1 activity enhancing Trp degradation has been shown to contribute to the incidence of pathologies of IBD or cancer patients with ageing (Chen et al. 2021). Most of 5-HT, a key intestinal signalling molecule neurotransmitter originated from the intestinal epithelial cell in the gut, especially enterochromaffin cells (ECs), mediated by TpH1 (Yadav et al. 2008). Although 5-HT cannot directly go through the blood–brain barrier upon physiological conditions, it delivers signals from the gut to neurons and influences intestinal motility, secretion, and nutrient absorption (Stockinger et al. 2021). High blood level and low intestinal level of 5-HT are observed in patients with autism spectrum disorder (ASD) and Alzheimer's disease (AD) attributing to the gut-brain signalling axis (Wiley et al. 2021). Moreover, 5-HT-derived metabolite 5-Hydroxyindole-3-acetic acid (5-HIAA) was found to ameliorate ageing-related rheumatoid arthritis by activating (AhR) in Regulatory B Cells through the gut-bone axis (Rosser et al. 2020). Taken together, the Trp metabolism involved in these pathways is directly or indirectly mediated by the gut microbiota. Manipulation of the IDO1, 5-HT, and AhR pathways in the gut based on microbiota-host interaction ought to be useful therapeutic strategies for ageing-related diseases, which guarantee further investigation.

1.5 Roles of Bile Acids in Ageing

Bile acids (BAs) are synthetized in hepatocytes from hepatic cholesterol mediated by more than 17 different enzymes (de Aguiar Vallim et al. 2013; Russell 2003) through two different pathways, the classical pathway accounting for 75% of primary BAs and alternative pathways for the rest. Cholesterol 7α-hydroxylase (CYP7A1) (Thomas et al. 2008) and sterol 27-hydroxylase (CYP27A1) (Sayin et al. 2013) are respective key rate-limiting enzymes. Cholic acid (CA) and chenodeoxy-cholic acid (CDCA) are the main primary BAs and then catalyzed by the bacteria in the gut forming secondary BAs including deoxycholic acid (DCA), ursodesoxycholic acid (UDCA), ω-muricholic acid (ω-MCA), and lithocholic acid (LCA). Some BAs can also be conjugated with taurine or glycine generating conjugated BAs, which are stored in the gallbladder by a bile salt export pump (BSEP, also known as ABCB11) (Thomas et al. 2008) and are hydrolyzed by bile salt hydrolase (BSH) produced from some gut bacteria such as *Bifidobacterium* (Kim et al. 2004), *Lactobacillus* (Elkins et al. 2001; Corzo and Gilliland 1999), *Clostridium* (Coleman and Hudson 1995), and *Bifidobacterium* spp. after releasing into the intestine. Recycling of BAs 6–10 times per day, and around 95% of BAs pool is recycled by enterohepatic circulation, the remaining excreted into the faeces (Russell 2003).

Within enterohepatic circulation, BAs need to undergo several processes including transportation, absorption, and metabolism, which are mainly mediated by the farnesoid X receptor (FXR) and one G protein-coupled receptor (TGR5) in the intestine (Perino et al. 2021). FXR and TGR5 also mediated the different roles of some BAs in health and various diseases, which are classified into beneficial and pathogenic BAs primarily summarized here (Table 1.1). Numerous studies have shown that interaction between BAs and the gut microbiota contributes to intestinal barrier functions and thus affects host immune homeostasis. For example, the proportion of UDCA to DCA is regarded as one of the indicators associated with impaired gut barrier function of high fat diet-induced obesity in mice (Shi et al.

2021). Compared with mice fed a chow diet, mice with dietary rich in DCA exhibited markedly increased gut permeability (>1.5-fold). Furthermore, DCA and CDCA were found to exacerbate LPS-induced barrier disruption with increased intestinal permeability and decreased TJs levels in the colitis mouse model (Zheng et al. 2021a, b). However, CDCA administration to prematurely weaned piglets ameliorates the gut barrier disruption by increasing the ZO-1 level and reducing TNF-a, IL-6, and IL-10 (Su et al. 2013). LCA is regarded as a modulator of T_H17 and T_{reg} differentiation in the regulation of host adaptive immunity (Hang et al. 2019). Recently, the secondary bile acid 3β-hydroxydeoxycholic acid (isoDCA) increases the induction of Foxp3 by acting on dendritic cells (DCs) to diminish their immunostimulatory properties (Campbell et al. 2020). The levels of both 3-oxoLCA and isoLCA and the 3α-hydroxysteroid dehydrogenase genes were significantly reduced in patients with IBD in comparison with health (Paik et al. 2022). TUDCA was reported to ameliorate HFD-induced obesity of mice through the improvement of intestinal barrier function with increasing TJs levels and modulation of gut dysbiosis by protecting ER stress (Wang et al. 2018). Notably, in progeroid mouse models, FMT either from wild-type mice or direct supplementation with *Akkermansia muciniphila* has been found to markedly prolong healthspan and lifespan through the restoration of secondary BAs reestablishing a healthy microbiome (Bárcena et al. 2019). Some drugs have been also shown to intervene or ameliorate some chronic metabolic diseases. For example, obeticholic acid (INT 747), a typical FXR agonist, attenuates intestinal inflammation and gut barrier dysfunction in cholestatic rats, DSS-induced IBD mice, and patients with chronic liver diseases via FXR activation (Lajczak-McGinley et al. 2020). Ginsenoside compound K (CK) can be used to ameliorate obesity and type 2 diabetes mellitus by modulating gut microbiota, especially *Ruminococcaceae family*, which positively upregulates LCA and DCA via activation of TGR5 (Xu et al. 2017). However, imidacloprid disturbs the gut barrier function by interfering with bile acid metabolism in mice (Yang et al. 2020). Methionine restriction (MR) and dietary CA were found to extend the healthspan and lifespan of progeroid mice by modulating the BAs pool (Bárcena et al. 2018). The gut microbiota remodelling reverses ageing-associated inflammation and dysregulation of systemic BAs homeostasis in mice sex-specifically (Ma et al. 2020). For instance, antioxidant tempol increases TβMCA levels by remodelling of the gut microbiota, resulting in an inhibition of intestinal FXR signalling and an improvement in obesity (Li et al. 2013). Collectively, BAs are necessary for the maintenance of intestinal barrier integrity and host immune homeostasis.

Table 1.1 Summary of beneficial and pathogenic bile acids (BAs) participating in various diseases in clinical trials and animal models studies

Species	Beneficial BAs	Roles or functions	References
vitro	UDCA	Controls *Clostridium difficile* germination and growth	Weingarden et al. (2016a, b)
		Inhibits *Clostridium difficile* spore germination and vegetative growth, and prevents the recurrence of ileac Pouchitis after infection	Weingarden et al. (2016a, b)
Mouse	CA	Hypercholesterolemia and gallstones in diabetic mice	Wang et al. (2006)
	TUDCA	Inhibits intestinal inflammation and barrier disruption in HFD mice with non-alcoholic fatty liver disease	Wang et al. (2018)
	TCDCA or TUDCA	Treating HFD-induced obesity and hypercholesterolemia	Huang et al. (2019)
	UDCA	Suppressed YAP signalling by activating the membrane G-protein-coupled	Zhang et al. (2021)

(continued)

Table 1.1 (continued)

Species	Beneficial BAs	Roles or functions	References
		bile acid receptor (TGR5)	
	UDCA	Retinal detachment and protects against retinal degeneration	Daruich et al. (2021)
	LCA	LCA acts by dampening DSS-induced inflammatory response in the colon	Lajczak-McGinley et al. (2020)
Human	UDCA	Improves digestive disease	Solá et al. (2006)
	UDCA	For dissolution of cholesterol gallstones or treatment of primary biliary cirrhosis	Solá et al. (2006)
	CDCA	Improves BAs-deficient	Li and Chiang (2014)
	CDCA, CA, and UDCA	For gallstone dissolution	Li and Chiang (2014)
	CDCA	Improves diarrhoea	Bazzoli et al. (1983)
	UDCA	For treatment for primary biliary cholangitis	Floreani and Mangini (2018)
	UDCA	Clinical use in the treatment of primary sclerosing cholangitis	Mousa et al. (2021)

(continued)

Table 1.1 (continued)

Species	Beneficial BAs	Roles or functions	References
	UDCA,UDCA, TUDCA	Improves amyotrophic lateral sclerosis	Daruich et al. (2019)
	OCA	Improves T2DM and NAFLD	Mudaliar et al. (2013)
	OCA	Improves NAFLD and NASH	Hameed et al. (2018), Neuschwander-Tetri et al. (2015), Siddiqui et al. (2020)
	OCA	Improves primary biliary cholangitis (PBC)	Bowlus et al. (2019)
	OCA	Improves BAs malabsorption and chronic diarrhoea	Walters et al. (2015)
	OCA	Improves obesity and gallstones	Al-Dury et al. (2019)
	OCA	Improves liver cirrhosis	Edwards et al. (2016)
	OCA	Improves primary sclerosing cholangitis (PSC)	Kowdley et al. (2020)
	Secondary Bile Acid	Anti-inflammatory in colitis	Sinha et al. (2020)

(continued)

Table 1.1 (continued)

Species	Beneficial BAs	Roles or functions	References
Detrimental		**BAs**	
Mouse	DCA	Intestinal inflammation	Xu et al. (2021a, b)
	DCA	Intestinal inflammation	Xu et al. (2021a, b)
Human	CA converted to DCA	a colon cancer promoter	Li and Chiang (2014)
	CA	Gallstones	Wang et al. (2006)
	OCA	Increased risk of acute liver decompensation, and increased LDL cholesterol levels	Fiorucci et al. (2019)

1.6 Conclusions

This review provides a brief and general summary of current achievements of the alterations of the gut microbiota and its metabolites in ageing and ageing-related diseases. We outline the importance of intestinal barrier integrity and the roles of microbiota-derived metabolites including SCFAs, BAs, and Trp metabolites in the maintenance of host metabolic homeostasis. Crosstalk between the gut microbiota and host organs which can be called "Gut-X" signalling axis, such as gut-liver, gut-bone, and gut-brain axes, highly contributes to health and disease (Fig. 1.1). Thus, potential strategies targeting intestinal microbial homeostasis is being emerging topics of aging and age-related pathologies. Although ageing and ageing-related diseases are closely associated with the gut microbiota and its metabolites, their causal relationship remains largely unknown. Furthermore, many studies of the molecular mechanisms of ageing and ageing-related diseases are limited to animal models in laboratories lacking clinical studies. Finally, single cell or single bacterial omics techniques could be powerful tools for further investigations to figure out the functions of specific cells or strains in ageing and ageing-related diseases.

Acknowledgements This work was financially supported by the National Key Research and Development Project of China (2018YFE0110800) and Foshan Core Technology Tackling Key Project (1920001000262).

Competing Interests The authors declare no competing interests.

References

Agus A et al (2018) Gut microbiota regulation of tryptophan metabolism in health and disease. Cell Host & Microbes 23:716–724

Aho VTE et al. (2021) Relationships of gut microbiota, short-chain fatty acids, inflammation, and the gut barrier in Parkinson's disease. Mol Neurodegener 16.6

Al-dury S et al (2019) Obeticholic acid may increase the risk of gallstone formation in susceptible patients. J Hepatol 71:986–991

Aoki R et al (2018) Indole-3-Pyruvic acid, an Aryl hydrocarbon receptor activator, suppresses experimental colitis in mice. J Immunol 201:3683–3693

Appay V, Sauce D (2014) Naive T cells: the crux of cellular immune aging? Exp Gerontol 54:90–93

Bárcena C et al (2018) Methionine restriction extends lifespan in progeroid mice and alters lipid and bile acid metabolism. Cell Rep 24:2392–2403

Bárcena C et al (2019) Healthspan and lifespan extension by fecal microbiota transplantation into progeroid mice. Nat Med 25:1234–1242

Bazzoli F et al (1983) Treatment of constipation with chenodeoxycholic acid. J Int Med Res 11:120–123

Beumer J, Clevers H (2021) Cell fate specification and differentiation in the adult mammalian intestine. Nat Rev Mol Cell Biol 22:39–53

Bowlus CL et al (2020) Long-term obeticholic acid therapy improves histological endpoints in patients with primary biliary cholangitis. Clin Gastroenterol Hepatol 18:1170-1178.e6
Campbell C et al (2020) Bacterial metabolism of bile acids promotes generation of peripheral regulatory T cells. Nature 581:475–479
Campisi J et al (2019) From discoveries in ageing research to therapeutics for healthy ageing. Nature 571:183–192
Cervantes-barragan L et al (2017) Lactobacillus reuteri induces gut intraepithelial CD4(+)CD8αα (+) T cells. Science 357:806–810
Chen LM et al (2021) Tryptophan-kynurenine metabolism: a link between the gut and brain for depression in inflammatory bowel disease. J Neuroinflammation 18:135
Chen LH et al (2022) Probiotic supplementation attenuates age-related sarcopenia via the gut-muscle axis in SAMP8 mice. J Cachexia Sarcopenia Muscle 13:515–531
Cheng WY et al (2022) Linking circadian rhythms to microbiome-gut-brain axis in aging-associated neurodegenerative diseases. Ageing Res Rev 78:101620
Clemente JC et al (2012) The impact of the gut microbiota on human health: an integrative view. Cell 148:1258–1270
Coleman JP, Hudson LL (1995) Cloning and characterization of a conjugated bile acid hydrolase gene from Clostridium perfringens. Appl Environ Microbiol 61:2514–2520
Corzo G, Gilliland SE (1999) Bile salt hydrolase activity of three strains of Lactobacillus acidophilus. J Dairy Sci 82:472–480
D'adda di fagagna F (2008) Living on a break: cellular senescence as a DNA-damage response. Nat Rev Cancer 8:512–522
Daruich A et al (2019) Review: the bile acids urso- and tauroursodeoxycholic acid as neuroprotective therapies in retinal disease. Mol vis 25:610–624
Daruich A et al (2021) Oral ursodeoxycholic acid crosses the blood retinal barrier in patients with retinal detachment and protects against retinal degeneration in an Ex Vivo model. Neurotherapeutics 18:1325–1338
De aguiar vallim TQ et al. (2013) Pleiotropic roles of bile acids in metabolism. Cell Metab 17:657–669
Dodd D et al (2017) A gut bacterial pathway metabolizes aromatic amino acids into nine circulating metabolites. Nature 551:648–652
Donovan SM (2017) Introduction to the special focus issue on the impact of diet on gut microbiota composition and function and future opportunities for nutritional modulation of the gut microbiome to improve human health. Gut Microbes 8:75–81
Edwards JE et al (2016) Modeling and experimental studies of obeticholic acid exposure and the impact of cirrhosis stage. Clin Transl Sci 9:328–336
Ekmekcioglu C (2020) Nutrition and longevity - from mechanisms to uncertainties. Crit Rev Food Sci Nutr 60:3063–3082
Elkins CA et al (2001) Genes encoding bile salt hydrolases and conjugated bile salt transporters in Lactobacillus johnsonii 100–100 and other Lactobacillus species. Microbiology 147:3403–3412
Fiorucci S et al (2019) Obeticholic acid: an update of its pharmacological activities in liver disorders. Handb Exp Pharmacol 256:283–295
Frick B et al (2004) Increasing production of homocysteine and neopterin and degradation of tryptophan with older age. Clin Biochem 37:684–687
Fontana L, Partridge L (2015) Promoting health and longevity through diet: from model organisms to humans. Cell 161:106–118
Gao X et al (2018) Effect of Montmorillonite powder on intestinal mucosal barrier in children with abdominal Henoch-Schonlein purpura: a randomized controlled study. Medicine 97:e12577
Gentile CL, Weir TL (2018) The gut microbiota at the intersection of diet and human health. Science 362:776–780
Ghosh TS et al (2020) Mediterranean diet intervention alters the gut microbiome in older people reducing frailty and improving health status: the NU-AGE 1-year dietary intervention across five European countries. Gut 69:1218–1228

Hameed B et al (2018) Clinical and metabolic effects associated with weight changes and obeticholic acid in non-alcoholic steatohepatitis. Aliment Pharmacol Ther 47:645–656
Hang S et al (2019) Bile acid metabolites control T(H)17 and T(reg) cell differentiation. Nature 576:143–148
He D et al (2020) Gut stem cell aging is driven by mTORC1 via a p38 MAPK-p53 pathway. Nat Commun 11:37
Hendrickx G et al (2015) A look behind the scenes: the risk and pathogenesis of primary osteoporosis. Nat Rev Rheumatol 11:462–474
Huang F et al (2019) Theabrownin from Pu-erh tea attenuates hypercholesterolemia via modulation of gut microbiota and bile acid metabolism. Nat Commun 10:4971
Jia G et al (2019) Endothelial cell senescence in aging-related vascular dysfunction. Biochim Biophys Acta Mol Basis Dis 1865:1802–1809
Kim GB et al (2004) Purification and characterization of three different types of bile salt hydrolases from Bifidobacterium strains. J Dairy Sci 87:258–266
Koh A et al (2016) From dietary fiber to host physiology: short-chain fatty acids as key bacterial metabolites. Cell 165:1332–1345
Kowdley KV et al (2020) A randomized, placebo-controlled, phase II study of obeticholic acid for primary sclerosing cholangitis. J Hepatol 73:94–101
Koyuncu S et al. (2021) Rewiring of the ubiquitinated proteome determines ageing in C. elegans. Nature 596:285–290
Kumar R et al (2019) Epigallocatechin gallate suppresses premature senescence of preadipocytes by inhibition of PI3K/Akt/mTOR pathway and induces senescent cell death by regulation of Bax/Bcl-2 pathway. Biogerontology 20:171–189
Lajczak-mcginley NK et al (2020) The secondary bile acids, ursodeoxycholic acid and lithocholic acid, protect against intestinal inflammation by inhibition of epithelial apoptosis. Physiol Rep 8:e14456
Lee J et al (2020) Gut microbiota-derived short-chain fatty acids promote post-stroke recovery in aged mice. Circ Res 127:4
Li T, Chiang JY (2014) Bile acid signaling in metabolic disease and drug therapy. Pharmacol Rev 66:948–983
Li F et al (2013) Microbiome remodelling leads to inhibition of intestinal farnesoid X receptor signalling and decreased obesity. Nat Commun 4:2384
Li JY et al (2016) Sex steroid deficiency-associated bone loss is microbiota dependent and prevented by probiotics. J Clin Invest 126:2049–2063
Liu JH et al. (2021) Extracellular vesicles from child gut microbiota enter into bone to preserve bone mass and strength. Adv Sci (Weinh) 8:2004831
Lucas S et al (2018) Short-chain fatty acids regulate systemic bone mass and protect from pathological bone loss. Nat Commun 9:55
Ma J et al (2020) Gut microbiota remodeling reverses aging-associated inflammation and dysregulation of systemic bile acid homeostasis in mice sex-specifically. Gut Microbes 11:1450–1474
Makki K et al. (2018) The impact of dietary fiber on gut microbiota in host health and disease. Cell Host Microbes 23:705–715
Martens EC et al (2018) Interactions of commensal and pathogenic microorganisms with the intestinal mucosal barrier. Nat Rev Microbiol 16:457–470
Mousa OY et al (2021) Bile acid profiles in primary sclerosing cholangitis and their ability to predict hepatic decompensation. Hepatology 74:281–295
Mudaliar S et al (2013) Efficacy and safety of the farnesoid X receptor agonist obeticholic acid in patients with type 2 diabetes and nonalcoholic fatty liver disease. Gastroenterology 145:574–82.e1
Neuschwander-tetri BA et al (2015) Farnesoid X nuclear receptor ligand obeticholic acid for non-cirrhotic, non-alcoholic steatohepatitis (FLINT): a multicentre, randomised, placebo-controlled trial. Lancet 385:956–965

Floreani A, MANGINI C (2018) Primary biliary cholangitis: Old and novel therapy. Eur J Intern Med 47:1-5

Ou Z et al (2020) Protective effects of Akkermansia muciniphila on cognitive deficits and amyloid pathology in a mouse model of Alzheimer's disease. Nutr Diabetes 10:12

Pabst O (2012) New concepts in the generation and functions of IgA. Nat Rev Immunol 12:821–832

Paik D et al (2022) Human gut bacteria produce T(H)17-modulating bile acid metabolites. Nature 603:907–912

Paone P, CANI PD (2020) Mucus barrier, mucins and gut microbiota: the expected slimy partners? Gut 69:2232–2243

Perino A et al (2021) Molecular physiology of bile acid signaling in health, disease, and aging. Physiol Rev 101:683–731

Pertovaara M et al (2006) Indoleamine 2,3-dioxygenase activity in nonagenarians is markedly increased and predicts mortality. Mech Ageing Dev 127:497–499

Powell DN et al. (2020) Indoles from the commensal microbiota act via the AHR and IL-10 to tune the cellular composition of the colonic epithelium during aging. Proc Natl Acad Sci USA 117:21519–21526

Rivière A et al (2016) Bifidobacteria and butyrate-producing colon bacteria: importance and strategies for their stimulation in the human gut. Front Microbiol 7:979

Rondanelli M et al (2015) Review on microbiota and effectiveness of probiotics use in older. World J Clin Cases 3:156–162

Rosser EC et al (2020) Microbiota-derived metabolites suppress arthritis by amplifying Aryl-hydrocarbon receptor activation in regulatory B cells. Cell Metab 31:837-851.e10

Rothhammer V, Quintana FJ (2019) The aryl hydrocarbon receptor: an environmental sensor integrating immune responses in health and disease. Nat Rev Immunol 19:184–197

Russell DW (2003) The enzymes, regulation, and genetics of bile acid synthesis. Annu Rev Biochem 72:137–174

Sayin SI et al (2013) Gut microbiota regulates bile acid metabolism by reducing the levels of tauro-beta-muricholic acid, a naturally occurring FXR antagonist. Cell Metab 17:225–235

Schiffrin EJ et al (2010) The inflammatory status of the elderly: the intestinal contribution. Mutat Res 690:50–56

Scott SA et al. (2020) Microbial tryptophan metabolites regulate gut barrier function via the aryl hydrocarbon receptor. Proc Natl Acad Sci USA 117:19376–19387

Shen B et al (2019) Antibiotics exacerbated colitis by affecting the microbiota, Treg cells and SCFAs in IL10-deficient mice. Biomed Pharmacother 114:108849

Shi Z et al (2021) Gut microbiota and its metabolite deoxycholic acid contribute to sucralose consumption-induced nonalcoholic fatty liver disease. J Agric Food Chem 69:3982–3991

Shin J et al (2021) Ageing and rejuvenation models reveal changes in key microbial communities associated with healthy ageing. Microbiome 9:240

Siddiqui MS et al (2020) Impact of obeticholic acid on the lipoprotein profile in patients with non-alcoholic steatohepatitis. J Hepatol 72:25–33

Sinha SR et al (2020) Dysbiosis-induced secondary bile acid deficiency promotes intestinal inflammation. Cell Host & Microbes 27:659-670.e5

Solá S et al (2006) Modulation of hepatocyte apoptosis: cross-talk between bile acids and nuclear steroid receptors. Curr Med Chem 13:3039–3051

Sonowal R et al. (2017) Indoles from commensal bacteria extend healthspan. Proc Natl Acad Sci USA 114:E7506-e7515

Sovran B et al (2019) Age-associated impairment of the mucus barrier function is associated with profound changes in microbiota and immunity. Sci Rep 9:1437

Stockinger B et al (2021) AHR in the intestinal microenvironment: safeguarding barrier function. Nat Rev Gastroenterol Hepatol 18:559–570

Su KC et al (2013) Bile acids increase alveolar epithelial permeability via mitogen-activated protein kinase, cytosolic phospholipase A2, cyclooxygenase-2, prostaglandin E2 and junctional proteins. Respirology 18:848–856

Thomas C et al (2008) Targeting bile-acid signalling for metabolic diseases. Nat Rev Drug Discov 7:678–693

Ulgherait M et al (2021) Circadian autophagy drives iTRF-mediated longevity. Nature 598:353–358

Vaiserman AM et al (2017) Gut microbiota: a player in aging and a target for anti-aging intervention. Ageing Res Rev 35:36–45

Van der hee B, Wells JM (2021) Microbial regulation of host physiology by short-chain fatty acids. Trends Microbiol 29:700–712

Venkatachalam G et al (2017) Replication stress-induced endogenous DNA damage drives cellular senescence induced by a sub-lethal oxidative stress. Nucleic Acids Res 45:10564–10582

Walters JR et al (2015) The response of patients with bile acid diarrhoea to the farnesoid X receptor agonist obeticholic acid. Aliment Pharmacol Ther 41:54–64

Wang J et al (2006) Critical role of cholic acid for development of hypercholesterolemia and gallstones in diabetic mice. Biochem Biophys Res Commun 342:1382–1388

Wang W et al (2018) Tauroursodeoxycholic acid inhibits intestinal inflammation and barrier disruption in mice with non-alcoholic fatty liver disease. Br J Pharmacol 175:469–484

Wang Z et al (2020) Inulin alleviates inflammation of alcoholic liver disease via SCFAs-inducing suppression of M1 and facilitation of M2 macrophages in mice. Int Immunopharmacol 78:106062

Wang DD et al (2021) The gut microbiome modulates the protective association between a Mediterranean diet and cardiometabolic disease risk. Nat Med 27:333–343

Weingarden AR et al (2016a) Ursodeoxycholic acid inhibits clostridium difficile spore germination and vegetative growth, and prevents the recurrence of ileal pouchitis associated with the infection. J Clin Gastroenterol 50:624–630

Weingarden AR et al (2016b) Changes in colonic bile acid composition following fecal microbiota transplantation are sufficient to control clostridium difficile germination and growth. PLoS ONE 11:e0147210

Wendy et al. (2010) Homeostasis and inflammation in the intestine. Cell 140:859–870

Wilck N et al (2017) Salt-responsive gut commensal modulates T(H)17 axis and disease. Nature 551:585–589

Wiley NC et al (2021) Production of psychoactive metabolites by gut bacteria. Mod Trends Psychiatry 32:74–99

Williams BB et al (2014) Discovery and characterization of gut microbiota decarboxylases that can produce the neurotransmitter tryptamine. Cell Host & Microbes 16:495–503

Wilmanski T et al (2021) Gut microbiome pattern reflects healthy ageing and predicts survival in humans. Nat Metab 3:274–286

Wlodarska M et al (2017) Indoleacrylic acid produced by commensal peptostreptococcus species suppresses inflammation. Cell Host & Microbes 22:25-37.e6

Wu TR et al (2019) Gut commensal Parabacteroides goldsteinii plays a predominant role in the anti-obesity effects of polysaccharides isolated from Hirsutella sinensis. Gut 68:248–262

Xu J et al (2017) Understanding the molecular mechanisms of the interplay between herbal medicines and gut microbiota. Med Res Rev 37:1140–1185

Xu M et al (2021a) Deoxycholic acid-induced gut dysbiosis disrupts bile acid enterohepatic circulation and promotes intestinal inflammation. Dig Dis Sci 66:568–576

Xu M et al (2021b) Modulation of the gut microbiota-farnesoid X receptor Axis improves deoxycholic acid-induced intestinal inflammation in Mice. J Crohns Colitis 15:1197–1210

Yadav VK et al (2008) Lrp5 controls bone formation by inhibiting serotonin synthesis in the duodenum. Cell 135:825–837

Yang G et al (2020) Imidacloprid disturbed the gut barrier function and interfered with bile acids metabolism in mice. Environ Pollut 266:115290

Yu M et al. (2021) Ovariectomy induces bone loss via microbial-dependent trafficking of intestinal TNF+ T cells and Th17 cells. J Clin Invest 131

Zhang L et al (2015) Persistent organic pollutants modify gut microbiota-host metabolic homeostasis in mice through aryl hydrocarbon receptor activation. Environ Health Perspect 123:679–688

Zhang H et al (2021) Ursodeoxycholic acid suppresses the malignant progression of colorectal cancer through TGR5-YAP axis. Cell Death Discov 7:207

Zhao LP et al (2018) Gut bacteria selectively promoted by dietary fibers alleviate type 2 diabetes. Science 359:1151–1156

Zheng H et al (2021a) Depletion of acetate-producing bacteria from the gut microbiota facilitates cognitive impairment through the gutbrain neural mechanism in diabetic mice. Microbiome 9:2–19

Zheng X et al (2021b) Hyocholic acid species improve glucose homeostasis through a distinct TGR5 and FXR signaling mechanism. Cell Metab 33:791-803.e7

Zihni C et al (2016) Tight junctions: from simple barriers to multifunctional molecular gates. Nat Rev Mol Cell Biol 17:564–580

Chapter 2
Ageing and Human Gut Microbiome: The Taxonomic and Functional Transition Towards an Elderly-Type Microbiome

Simone Rampelli, Silvia Turroni, and Marco Candela

Abstract The gut microbiome is recognized as an important component in determining the health status of ageing individuals due to its role in modulating digestive functions, immunity and resistance to pathogen infections, and cognitive functions. Microbial data from long-lived individuals showed the progressive transition to a peculiar elderly-type microbiome, often related to immunosenescence, inflammageing, and frailty, along with the emergence of some possibly ageing-supportive features. Here, we extensively describe the characteristics of such a microbiome configuration, with a focus on the concomitant ageing-associated unavoidable physiological modifications, in order to bring out the close connections between the human host and its microbial counterpart.

Keywords Microbiome · Aging · Extreme longevity · Metagenome

2.1 The Human Gut Microbiome

It is indisputable that the gut microbiome, *i.e.*, the community of over 10 trillion microbes that inhabit our intestine, is a leading player in human physiology, capable of providing a varied panel of ecosystem services throughout our entire life (Kundu et al. 2017; Turroni et al. 2018). Our microbial counterpart is in fact essential for multiple functions, such as energy balance (through fermentation of otherwise indigestible polysaccharides), metabolic regulation (e.g., through the synthesis of B vitamins), the barrier effect against potential enteropathogens, development, education, and regulation of (innate and adaptive) immune system, and modulation of

S. Rampelli (✉) · S. Turroni · M. Candela
Unit of Microbiome Science and Biotechnology, Department of Pharmacy and Biotechnology, University of Bologna, Bologna, Italy
e-mail: simone.rampelli@unibo.it

F. Marotta (ed.), *Gut Microbiota in Aging and Chronic Diseases*, Healthy Ageing and Longevity 17, https://doi.org/10.1007/978-3-031-14023-5_2

the central nervous system (through the bidirectional gut-brain axis) (de Vos et al. 2022). These functions are mainly exercised by bacteria, which are certainly the dominant component of the gut microbiome and the one most studied to date (to which all statements in the following chapters refer). Fungi and viruses are also usually present but their contribution to health conditions (*i.e.*, eubiosis) is still questionable (Huseyin et al. 2017; Shkoporov et al. 2019).

The immense amount of microorganisms in our gut has an undoubted potential to produce a myriad of different bioactive molecules, which can enter the bloodstream and reach extraintestinal organs, for effects on the whole human body (Turroni et al. 2018). Among these, we must mention short-chain fatty acids (SCFAs), deriving from the fermentation of polysaccharides, whose signalling, through the binding to G protein-coupled receptors or the inhibition of histone deacetylases, allows the maintenance of metabolic and immunological homeostasis (Koh et al. 2016). SCFAs can in fact influence, among others, satiety, intestinal gluconeogenesis, adipose tissue, and liver function, as well as exert anti-inflammatory and immunomodulatory activities, which make them key metabolites in multiple disease contexts. Phenolic compounds, resulting from aromatic amino acid metabolism, are instead normally associated with poor cardiometabolic and cognitive outcomes and generally increased with advancing age (Collino et al. 2013). The gut microbiome is also known to produce trimethylamine from dietary choline and carnitine, which is then oxidized by host enzymes into trimethylamine-N-oxide (TMAO), a risk factor for cardiovascular diseases, and is involved in metabolism (by deconjugation of the primary bile acids and production of secondary ones) and consequently in the signalling of bile acids (Barone et al. 2021). In this regard, it has recently been shown that centenarians are enriched in microorganisms capable of generating unique secondary bile acids, including various isoforms of lithocholic acid, which show a potent antimicrobial effect against multidrug-resistant Gram-positive pathogens, such as *Clostridioides difficile* and *Enterococcus faecium*, potentially contributing to the maintenance of intestinal homeostasis (Sato et al. 2021).

In addition to showing diversity, index of functioning, stability, and productivity of the population (Fassarella et al. 2021), the gut microbiome boasts a significant degree of plasticity that allows it to adapt to changes in host-related variables, while maintaining host service provider function. Major microbiome variation factors are recognized as exogenous in nature and include diet, lifestyle, host location (or geographic effect) (He et al. 2018), or, more generally, the exposome (Vujkovic-Cvijin et al. 2020). As expected, age, which encompasses the above variables, has also been identified as a major microbiome-associated confounder. However, under certain conditions, the dynamic balance of the gut microbiome can be severely compromised, resulting in a reduction in diversity, a decrease in health-associated microorganisms (mainly SCFA producers), and an increase in pathogens or pathobionts (Duvallet et al. 2017). The establishment of these altered or dysbiotic states has been associated with various disorders, including intestinal (e.g., inflammatory bowel disease, irritable bowel syndrome, etc.), hepatic, metabolic, respiratory, cardiovascular, neurological, and oncological (Lynch and

Pedersen 2016). According to recent research, some microbiome profiles could even predict the onset and progression of diseases, including response to treatment (Helmink et al. 2019; Integrative HMP (iHMP) Research Network Consortium 2019), which stresses the inclusion of microbiome data in current models of prediction, diagnosis, and prognosis. Precise identification of microbiome signatures is highly advocated in multiple contexts, to guide the development of personalized, precision, integrated, preventive, and therapeutic strategies (e.g., based on prebiotics, probiotics, postbiotics, or even faecal microbiota transplantation) also aimed at modulating the gut microbiome to promote long-term health.

2.2 The Transition from an Adult-Type Gut Microbiome to an Elderly-Type Gut Microbiome

In the last years, several studies have sought to shed light on differences in the gut microbiome structure between people of different ages, such as adults and the elderly, also including different health conditions, nationalities, or medication burdens (Bartosch et al. 2004; van Tongeren et al. 2005; Mueller et al. 2006; Biagi et al. 2010; Claesson et al. 2010; Claesson et al. 2012; Wang et al. 2015; Jeffery et al. 2016; Biagi et al. 2016; Qato et al. 2016; Ghosh et al. 2020; Tavella et al. 2021a). In brief, studies have mostly been in line in reporting an age-related reduction in biodiversity in the gastrointestinal tract of elderly people, as well as an increased colonization by pathobionts, such as *Enterobacteriaceae* members, with a concomitant reduction in species involved in the production of SCFAs. This ecosystem disruption was even more pronounced in centenarians and supercentenarians (aged over 105 years), *i.e.*, in extreme ageing, parallel to the profound modification and impairment of most of the physiological functions of the human host. Indeed, for several microbiome modifications, an intuitive connection with the physiological and lifestyle changes associated with the ageing process is evident: for example, the loss of teeth or the alteration of taste and smell are linked to changes in eating habits, with a lower consumption of fibre-rich foods, and consequent reduction of fibre-degrading bacteria within the gut microbiome. Furthermore, the reduction in physical activity often characteristic of aged people contributes to the age-associated decrease in intestinal motility, with a consequent reduction in bacterial excretion (*i.e.*, lower number of bowel movements per week) and greater chances of proliferation for opportunistic bacteria. Also, some characteristics of the elderly-type gut microbiome have been proposed to be nurtured by the chronic low-grade inflammation that is peculiar to elderly people, *i.e.*, inflammageing, and in turn promote the inflammatory process (Guigoz et al. 2008). Finally, some elderly-type gut microbiome features have been hypothesized to be associated with the frailty degree (Jeffery et al. 2016), the increase in intestinal permeability (Nicoletti 2015), the reduction of bone health (Villa et al. 2017), as well as the loss of cognitive health (Leung and Thuret 2015), although, as expected, causality cannot be established.

Collectively, the transition to an elderly-type gut microbiome can be described as a progressive, multi-step process that begins in the elderly after a subject-specific "age threshold", which depends on individual characteristics such as diet, lifestyle, country, or, more generally, the personal exposome (Biagi et al. 2012). This transition profoundly affects not only the gut microbiome composition and function, but also the microbial trophic web, as highlighted by changes in positive and negative correlations between microorganisms in a network analysis (Fig. 2.1). Such analysis, conducted using the microbiome data produced by Biagi and colleagues (2016) on the Italian population, showed the emergence of different co-abundance groups of correlated microorganisms, which were found to be exclusively associated with the most important "age stages" (*i.e.*, adulthood, seniority, and extreme ageing). In particular, the microbiome of the adult group was characterized by a *Faecalibacterium prausnitzii*-centred network, with several co-abundant saccharolytic butyrate-producing taxa, such as *Blautia*, *Roseburia intestinalis*, *Roseburia inulinivorans*, *Eubacterium ventrosium*, *Agathobacter faecis,* and *Coprococcus comes* that clustered together. Two other auxiliary networks were closely correlated to these bacteria and included *Coprococcus eutactus*, *Coprococcus catus*, *Eubacterium coprostanoligenes*, *Bacteroides stercoris,* and *Lachnospira*, which are SCFA producers as well. Conversely, the elderly-type gut microbiome was found to be centred around the *Christensenella* network, which included longevity-related *Christensenella* species (Waters and Ley 2019), along with some pathobionts, such as *Escherichia coli* and *Bacteroides fragilis*, the mucus degrader *Akkermansia muciniphila,* and the SCFA producers *Clostridium leptum* and *Ruminoclostridium*. Together, such microorganisms constitute a very eclectic cluster, potentially capable of expressing disparate functionalities and therefore providing a varied panel of physiological services to the ageing host. Unlike the previous ones, the *Bacteroides vulgatus* cluster did not show appreciable variations between age groups, but only some significant correlations with members of other groups, as if its abundance was not closely associated with the ageing process. Finally, we identified a small cluster containing *Bifidobacterium* and *Ruminococcus bromii*, which does not seem to represent a trophic network, as it was present when the *B. vulgatus* cluster and some ancillary cluster taxa were absent, in a sort of dichotomous alternative. Probably the most relevant finding of the network analysis is the switch of keystone species from *F. prausnitzii* and *Blautia* to *Christensenella*, in the transition from the adult- to the elderly-type microbiome. This supports the emerging interest in the *Christensenella* genus, which has recently been proposed as a next-generation probiotic candidate, also due to its potent immunomodulatory properties (Tavella et al. 2021b; Wilmanski et al. 2021; Kropp et al. 2021). Indeed, it has been recently demonstrated that centenarians are enriched in microorganisms that are capable of generating unique secondary bile acids, including various isoforms of lithocholic acid, which show a potent antimicrobial effect against Gram-positive multidrug-resistance pathogens such as *Clostridioides difficile* and *Enterococcus faecium* (Sato et al. 2021). All these findings suggest to the authors that the centenarian-specific members of the gut microbiome (*i.e.*, the members of the *Christensenella* network), rather than increasing their abundance as a mere consequence of ageing, might also actively contribute to resistance against pathogenic infection and to the maintenance of the intestinal homeostasis in old age.

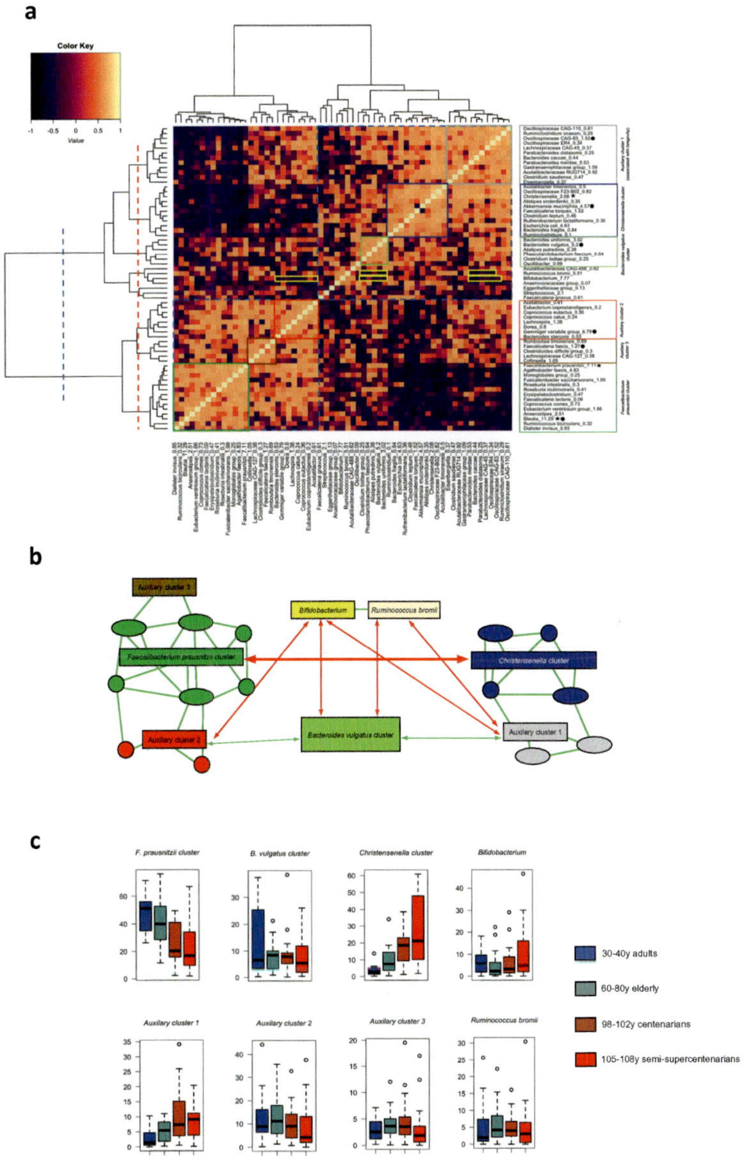

◂**Fig. 2.1** The gut microbiome network along the human lifespan. To build these graphs, we used gut microbiome data from adult, elderly, centenarian, and semi-supercentenarian individuals from Biagi and colleagues (2016). 16S rRNA gene sequences were subjected to bioinformatic analysis, including quality trimming, OTU (operational taxonomic unit) clustering and taxonomic assignment using QIIME 2 (Bolyen et al. 2019), oligotyping (Eren et al. 2014), and the global earth microbiome database (Nayfach et al. 2020), respectively. **a** A network heatmap based on Kendall's correlation coefficient and microbiome data was generated using the most abundant taxa at the genus or species level across all samples. The most dominant clusters identified are highlighted by different coloured boxes and were confirmed by permutation test with pseudo-F ratio ($P < 0.05$, "adonis" of the R package "vegan"). Two different settings were utilized for cluster analysis (blue and red dashed lines), which identified two and eight clusters, respectively. The *Faecalibacterium prausnitzii* cluster is highlighted in green, the *Bacteroides vulgatus* cluster in light green, the *Christensenella* cluster in blue, and the three auxiliary groups in grey, red, and brown, respectively. Yellow boxes denote a negative correlation between *Bifidobacterium* or *Ruminococcus bromii* and other taxa. The main representative taxa of each cluster are marked with a star. The most important taxa for the network structure, as highlighted by the network analysis of betweenness centrality, closeness centrality, and degree, are marked with a dot. The mean relative abundances for each taxon in the overall cohort are reported next to the taxon name. **b** Bacterial network scheme. Only significant Kendall's tau correlation coefficients were considered. The leading taxa in each network are highlighted. A positive correlation is shown with a green line and a negative correlation with a red line. **c** Relative abundance of the different groups of taxa among the age groups.

2.3 A Functional Perspective: Age-Associated Changes in Saccharolytic and Proteolytic Potential and an Increased Propensity for Xenobiotics Degradation

The functional peculiarities of the ageing gut microbiome layout were first described in a pilot comparative metagenomic study, in which gut metagenomes from 3 centenarians, 5 70-year-old elderly subjects, and one 30-year-old young adult were deep-sequenced (Rampelli et al. 2013). According to this study, the centenarian gut microbiome shows a proteolytic propensity, with a higher proportion of genes involved in the metabolism of aromatic amino acids (tryptophan, tyrosine, and phenylalanine), as well as valine and lysine, while a lower amount of functions involved in carbohydrate metabolism, compared to the gut microbiome of younger people. These findings were subsequently confirmed by other studies investigating the extreme part of the microbiome-host trajectory across the human lifespan (Wu et al. 2019; Rampelli et al. 2020; Tavella 2021b), and integrated with other observations, especially concerning the lipid and xenobiotic metabolism. In particular, the elderly-type gut microbiome was enriched with genes for the degradation of various xenobiotics, such as toluene, ethylbenzene, caprolactam, chlorocyclohexane, and chlorobenzene (Rampelli et al. 2020). At the same time, it showed a higher propensity for the metabolism of alpha-linoleic acid and glycerolipids probably derived from plant foods (Stark et al. 2008). While changes in saccharolytic, proteolytic, and lipid metabolism may be related to a recent change in eating habits, the intriguing increase in genes for xenobiotic metabolism may reflect longer or even lifelong lifestyle habits. Ethylbenzene, chlorobenzene, chlorocyclohexane, and toluene are pervasive chemicals with toxic effects, the release of which into the atmosphere is under monitoring all over the world as part of the Paris Agreement, the legally binding international treaty on climate change signed in 2016 (https://ec.europa.eu/clima/eu-action/international-action-climate-change/climate-negotiations/paris-agreement_en). The major sources of these molecules are motor and exhaust vehicle emissions, along with cigarette smoke (Bruno et al. 2008a; Buczynska et al. 2009). Furthermore, they are generated during the processing or production of refined petroleum products, such as plastics, and contained in common consumer products, such as paints and lacquers, thinners, and rubber products (Leusch and Bartkow 2010). Among all, toluene is always present in indoor air at detectable levels caused by indoor source emissions from paint organic solvent and household cleaning chemicals, and, at higher levels, in workplaces such as photocopy centres and automotive paint shops (Bruno et al. 2008b; Song and Chun 2021). As for caprolactam, it is the raw material of nylon, used for the production of many indoor products, such as synthetic fibres, resins, synthetic leather, and plasticizers. Other previous studies have demonstrated the higher indoor load of these molecules compared to the outdoor environment, emphasizing the exceptional importance of indoor exposure for human health (Massolo et al. 2010; Esplugues et al. 2010). Such an increasing burden of catabolic potential towards xenobiotic chemicals could therefore be intuitively linked to lifestyle changes associated with the ageing process. Indeed, elderly individuals, especially long-lived ones, such as centenarians and semi-supercentenarians, have reduced mobility and tend to spend a lot of time in their own houses, resulting in greater exposure to indoor pollutants than younger people. Additionally, they may boast an important history of exposure to xenobiotic stressors, creating the perfect environment for selecting bacteria with a higher propensity for their degradation. Another intriguing aspect concerns the taxonomic assignment of these functions, which were found to belong mostly to core gut microbiome taxa, such as members of the *Ruminococcaceae*, *Lachnospiraceae*, *Bacteroidaceae*, and *Bifidobacteriaceae* families, underlining the adaptive nature of microbiome remodelling during ageing (Rampelli et al. 2020). Although questions remain open about the biological mechanisms that lead to the enrichment of xenobiotic-degrading propensity in elderly-type microbiomes, the most consolidated opinion is that it is a top-down selection process related to the lifestyle habits of elderly people, *i.e.*, stable and constant living settings within their homes, together with longer exposure and possibly the accumulation of these chemicals in host tissues due to their longer life.

2.4 Age-Related Changes in the Human Gut Resistome

Intestinal microbes with antibiotic resistances constitute a long-lasting problem for human health, above all for what concern compromised individuals, such as preterm infants receiving early-life antibiotics (Gasparrini et al. 2019), elderly people with a compromised immune system (Araos et al. 2019), and people with autoimmune disorders or cancers (D'Amico et al. 2019). For investigating the variation of the human intestinal resistome during ageing, Tavella and colleagues (Tavella et al. 2021a) recently characterized the human gut resistome of an exceptional cohort of semi-supercentenarians (>104y), centenarians, seventy-year-old elderly, and young adults. Intriguingly, the taxonomic composition of the resistome largely matches that of the microbiome, as if antibiotic resistance (AR)-coding genes were mainly part of the dominant microbial families composing the gut ecosystem, such as Lachnospiraceae, Ruminococcaceae, and Bifidobacteriaceae. This could be caused by an intense microorganism-microorganism cross-talk with the propagation of AR genes via lateral gene transmission, potentially nurturing by antibiotic consumption and/or exposure. Conversely, the beneficial short-chain fatty acid-producing taxa (i.e., *Faecalibacterium*, *Roseburia*, and *Ruminococcus*) (Koh et al. 2016), of long-lived people, did not encode for any AR, but the majority of the AR reads were assigned to known pathobionts of the family Enterobacteriaceae, such as *Escherichia coli*, and Eggerthella. These associations were also confirmed considering age as a continuous variable and are associated with a progressive increase with ageing of genes for lipopolysaccharide biosynthesis (ko00540, KEGG database, Rampelli et al. 2020), together with the concomitant chronic inflammation in elderly people (i.e., inflammageing).

Based on the increased vulnerability of old people to infectious diseases (Liang et al. 2007), the increase of resistant bacteria along with ageing could pose a serious problem for health, as well as urging the need for resistome characterization in clinical practice, as it has recently been highlighted (D'Amico et al. 2019). For what concerns AR mechanisms, the gut resistome is mainly encoded for genes for antibiotic efflux pumps, or for microbial enzymes that directly modify the antibiotic target. In particular, antibiotic efflux pumps were similarly represented in all subjects regardless of age. On the other hand, ageing was associated with a higher abundance of genes for multidrug and sulfonamide resistance (Tavella et al. 2021a).

It is important to note that the modification of the resistome composition along with ageing may derive from an adaptive response to lifelong exposure to antibiotics, including those utilized in the food chain production and for medical practice. While being a model of healthy ageing, centenarians and semi-supercentenarians have been probably more exposed to antibiotics, also due to ageing-related physiological modifications, such as immunosenescence, which contributes to nurturing susceptibility to infections, potentially implying a greater usage of antibiotics (Giarratano et al. 2018; Franceschi et al. 2018). On the other hand, AR genes are well known to be widely present in any environment, including soil, water, air, and even household dust (Li et al. 2018; Maamar et al. 2020). In particular, built environments have recently been identified as a conspicuous reservoir for AR genes, with the accumulation of resistant microorganisms, loss of microbial diversity, and higher abundance of virulence genes (Hartmann et al. 2016; Mahnert et al. 2019). Based on this evidence, some speculations connecting the constant living settings within homes to AR accumulation in extreme longevity have been advanced, as if the greater abundance of AR genes in the older individuals, i.e., people with reduced mobility who spend more time at home (Rampelli et al. 2020), was the result of a long process associated not only to the short-term medical history, but also to lifestyle habits, which involved longer and more extensive exposure to various chemicals. On the other hand, the microbiome of younger individuals was characterized by a higher abundance of AR genes for beta-lactam antibiotics when compared to centenarians and semi-supercentenarians. Such genes were mainly encoded by Bacteroidetes members, as frequently found also in other studies (Forslund et al. 2013; Hu et al. 2013). Interestingly, these AR genes against beta-lactam are not transferred to opportunistic pathogens, such as Enterobacteriaceae (Sommer et al. 2009), possibly explaining their poor representation in the intestinal resistome of older people.

2.5 Christensenella-Enriched Elderly-Type Microbiome is Related to Different Quantities of Visceral Adipose Tissue and a Healthier Metabolic Profile

In a very recent study, conducted by Tavella and colleagues (Tavella et al. 2021b), the authors found three significantly different groups (G1 to G3) of elderly individuals showing three different intestinal microbiome configurations. Such structures were associated with body composition and other health-related parameters. In particular, the G1 group was characterized by a higher abundance of Lachnospiraceae; the G2 group showed a higher abundance of Christensenellaceae, Porphyromonadaceae, and Rikenellaceae; and G3 was enriched in Ruminococcaceae. Additionally, G2 and G3 were also characterized by a higher level of biodiversity with respect to G1 configuration. It is important to note that the G2 microbiome cluster showed the lowest value of visceral adipose tissue (VAT), a specific measure of abdominal obesity. Finally, by using DXA data for android and gynoid fat, the authors found that the microbial communities defining the three elderly groups were neither sex-related nor age-driven.

The authors speculated that the G2 microbiome (i.e., the higher abundance of Christensenellaceae, Porphyromonadaceae, and Rikenellaceae) might contribute to the reduced amount of VAT. In support of this opinion, the family Christensenellaceae has been reported as a marker of lean phenotype and also negatively correlated to a higher amount of VAT (Beaumont et al. 2016). In addition, Porphyromonadaceae and Rikenellaceae (families of the Bacteroidetes phylum) might be involved in the modulation of adiposity by acetate and propionate production (Lu et al. 2016). Expressly, acetate contributes to adiposity reduction by stimulating fatty acid oxidation in the liver (Kondo et al. 2009). Furthermore, the contributions of Christensenellaceae and Rikenellaceae have recently been found to be significantly lower in obese than lean individuals (Oki et al. 2016). Quite the opposite, the Lachnospiraceae family, found to be characteristic of the higher VAT-related and low-diverse G1 microbiome, has been associated with dietary lipid metabolism and a higher level of VAT (Beaumont et al. 2016). Another Clostridiales family, Ruminococcaceae (a microbiome component enriched in the high-diverse but VAT-related G3 group) showed conflicting connections to visceral fat. In particular, *Faecalibacterium* has sometimes been associated with higher levels of VAT in obese individuals, despite its immunomodulating and anti-inflammatory characteristics, probably due to its ability to metabolite unabsorbable polysaccharides (Balamurugan et al. 2010). On the other hand, the genera Ruminococcaceae UCG 014 and Ruminococcaceae UCG 005 have both been negatively connected to lipid metabolism (Whutti-In et al. 2020), together with *Subdoligranulum*, which has recently been identified as a key species for microbiome-host metabolic interplay (Visconti et al. 2019).

Intriguingly, in the same study (Tavella et al. 2021b), the elderly of the G2 group, compared to G1 and G3, showed a significantly lower level of distinct metabolic, anthropometric, renal, and cardiovascular risk parameters, such as BMI, hip and waist circumference, diastolic pressure, creatinine and uric acid, and higher values of adiponectin, an adipose-related cytokine with anti-inflammatory effects. These results were of particular interest and strengthen the role of Christensenellaceae, as a potential marker of healthy ageing and longevity in the early old age (60–70 years, Biagi et al. 2016; Waters et al. 2019). Furthermore, the G2 group showed also lower serum concentrations of BCAAs, which are connected to insulin-deficient and correlated positively with VAT (Holecek 2018). The same elderly group showed also a lower circulating abundance of methyl ester fatty acids and myristic acid, for which an inverse correlation with HDL cholesterol has been demonstrated in Italian individuals following a Mediterranean diet (Noto et al. 2015). Elderly from the G2 group showed also higher propension for secondary bile acids production (Tavella et al. 2021b), with the G2 discriminating taxa, especially Bacteroidetes members, that are probably responsible for the deconjugation and metabolism of primary bile acids into secondary ones (Yao et al. 2018). Further highlighting this propensity, a positive correlation between secondary bile acid metabolism and the Christensenellaceae family has recently been found (Aleman et al. 2018). Based on an additional search in distinct databases, Christensenellaceae species actually encode for both bile salt hydrolase (EC 3.5.1.24) and bile acid 7-alpha-dehydratase (EC 4.2.1.106), participating in the 7-dehydroxylation process connected to bile acid degradation (Tavella et al. 2021b). Taking together such results suggests the relevance of high amounts of Christensenellaceae as protective of cardiovascular and metabolic diseases related to visceral fat and, thus, potential markers of healthy ageing and, possibly, longevity.

2.6 Conclusions

The elderly-type gut microbiome is generated progressively, after a subject-specific "age threshold" that depends on individual characteristics, such as dietary habits, lifestyle, geographical location, history of diseases, and, in general, personal exposures. In summary, elderly individuals possess a distinct gut microbiome enriched in microorganisms that show a higher propensity for proteolytic metabolism, a lower abundance of genes involved in carbohydrate metabolism, and an increase in genes for mechanisms of resistance against antibiotics, reflecting progressive growth with ageing of pathobionts such as *Enterobacteriaceae*. Furthermore, the hypothesis has been advanced that some members of the elderly-type gut microbiome could actively contribute to the maintenance of microbiome-host homeostasis in old age, such as *Christensenella*, which also for this reason has been proposed as a next-generation probiotic candidate. Notably, ageing in Western urban environments progressively selects for microbiome commensals with metabolic abilities towards specific xenobiotics and specific antibiotics. In our opinion, this could represent an adaptive response of the human holobiont to the increased exposure to, and accumulation of, xenobiotic substances along the ageing process. In this scenario, future studies should aim to better depict the interplay between xenobiotic exposure and the human gut microbiome. The individual gut microbiome configuration and its metagenome will have to be matched to the personal exposure level, by directly monitoring xenobiotics in faeces and body fluids, possibly also conceiving very long-term longitudinal studies.

References

Alemán JO, Bokulich NA, Swann JR, Walker JM, De Rosa JC, Battaglia T, Costabile A, Pechlivanis A, Liang Y, Breslow JL, Blaser MJ, Holt PR (2018) Fecal microbiota and bile acid interactions with systemic and adipose tissue metabolism in diet-induced weight loss of obese postmenopausal women. J Transl Med 16(1):244. https://doi.org/10.1186/s12967-018-1619-z. PMID:30176893;PMCID:PMC6122649

Araos R, Battaglia T, Ugalde JA, Rojas-Herrera M, Blaser MJ, D'Agata EMC (2019) Fecal microbiome characteristics and the resistome associated with acquisition of multidrug-resistant organisms among elderly subjects. Front Microbiol 10:2260. https://doi.org/10.3389/fmicb.2019.02260

Balamurugan R, George G, Kabeerdoss J, Hepsiba J, Chandragunasekaran AM, Ramakrishna BS (2010) Quantitative differences in intestinal Faecalibacterium prausnitzii in obese Indian children. Br J Nutr 103(3):335–338. https://doi.org/10.1017/S0007114509992182 Epub 2009 Oct 23 PMID: 19849869

Barone M, D'Amico F, Fabbrini M, Rampelli S, Brigidi P, Turroni S (2021) Over-feeding the gut microbiome: a scoping review on health implications and therapeutic perspectives. World J Gastroenterol 27(41):7041–7064. https://doi.org/10.3748/wjg.v27.i41.7041.PMID:34887627; PMCID:PMC8613651

Bartosch S, Fite A, Macfarlane GT, McMurdo ME (2004) Characterization of bacterial communities in feces from healthy elderly volunteers and hospitalized elderly patients by using real-time PCR and effects of antibiotic treatment on the fecal microbiota. Appl Environ Microbiol 70(6):3575–3581. https://doi.org/10.1128/AEM.70.6.3575-3581.2004.PMID: 15184159;PMCID:PMC427772

Beaumont M, Goodrich JK, Jackson MA, Yet I, Davenport ER, Vieira-Silva S, Debelius J, Pallister T, Mangino M, Raes J, Knight R, Clark AG, Ley RE, Spector TD, Bell JT (2016) Heritable components of the human fecal microbiome are associated with visceral fat. Genome Biol 17(1):189. https://doi.org/10.1186/s13059-016-1052-7.PMID:27666579;PMCID: PMC5036307

Biagi E, Nylund L, Candela M, Ostan R, Bucci L, Pini E, Nikkïla J, Monti D, Satokari R, Franceschi C, Brigidi P, De Vos W (2010) Through ageing, and beyond: gut microbiota and inflammatory status in seniors and centenarians. PLoS ONE 5(5):e10667. https://doi.org/10.1371/journal.pone.0010667.Erratum.In:PLoSOne.2010;5(6).doi:10.1371/annotation/df45912f-d15c-44ab-8312-e7ec0607604d.PMID:20498852;PMCID:PMC2871786

Biagi E, Candela M, Fairweather-Tait S, Franceschi C, Brigidi P (2012) Aging of the human metaorganism: the microbial counterpart. Age (dordr) 34(1):247–267. https://doi.org/10.1007/s11357-011-9217-5

Biagi E, Franceschi C, Rampelli S, Severgnini M, Ostan R, Turroni S, Consolandi C, Quercia S, Scurti M, Monti D, Capri M, Brigidi P, Candela M (2016) Gut Microbiota and Extreme Longevity. Curr Biol 26(11):1480–1485. https://doi.org/10.1016/j.cub.2016.04.016 Epub 2016 May 12 PMID: 27185560

Bolyen E, Rideout JR, Dillon MR, Bokulich NA, Abnet CC, Al-Ghalith GA, Alexander H, Alm EJ, Arumugam M, Asnicar F, Bai Y, Bisanz JE, Bittinger K, Brejnrod A, Brislawn CJ, Brown CT, Callahan BJ, Caraballo-Rodríguez AM, Chase J, Cope EK, Da Silva R, Diener C, Dorrestein PC, Douglas GM, Durall DM, Duvallet C, Edwardson CF, Ernst M, Estaki M, Fouquier J, Gauglitz JM, Gibbons SM, Gibson DL, Gonzalez A, Gorlick K, Guo J, Hillmann B, Holmes S, Holste H, Huttenhower C, Huttley GA, Janssen S, Jarmusch AK, Jiang L, Kaehler BD, Kang KB, Keefe CR, Keim P, Kelley ST, Knights D, Koester I, Kosciolek T, Kreps J, Langille MGI, Lee J, Ley R, Liu YX, Loftfield E, Lozupone C, Maher M, Marotz C, Martin BD, McDonald D, McIver LJ, Melnik AV, Metcalf JL, Morgan SC, Morton JT, Naimey AT, Navas-Molina JA, Nothias LF, Orchanian SB, Pearson T, Peoples SL, Petras D, Preuss ML, Pruesse E, Rasmussen LB, Rivers A, Robeson MS 2nd, Rosenthal P, Segata N, Shaffer M, Shiffer A, Sinha R, Song SJ, Spear JR, Swafford AD, Thompson LR, Torres PJ, Trinh P, Tripathi A, Turnbaugh PJ, Ul-Hasan S, van der Hooft JJJ, Vargas F, Vázquez-Baeza Y, Vogtmann E, von Hippel M, Walters W, Wan Y, Wang M, Warren J, Weber KC, Williamson CHD, Willis AD, Xu ZZ, Zaneveld JR, Zhang Y, Zhu Q, Knight R, Caporaso JG (2019 Aug) Reproducible, interactive, scalable and extensible microbiome data science using QIIME 2. Nat Biotechnol 37(8):852–857. https://doi.org/10.1038/s41587-019-0209-9.Erratum.In:NatBiotechnol.2019Sep;37(9):1091.PMID:31341288; PMCID:PMC7015180

Bruno P, Caselli M, de Gennaro G, Iacobellis S, Tutino M (2008) Monitoring of volatile organic compounds in non-residential indoor environments. Indoor Air 18(3):250–256. https://doi.org/10.1111/j.1600-0668.2008.00528.x Epub 2008 Apr 21 PMID: 18429995

Bruno P, Caselli M, de Gennaro G, Scolletta L, Trizio L, Tutino M (2008) Assessment of the impact produced by the traffic source on VOC level in the urban area of Canosa di Puglia (Italy). Water Air Soil Pollut 193:37–50. https://doi.org/10.1007/s11270-008-9666-3

Buczynska AJ, Krata A, Stranger M, Locateli Godoi AF, Kontozova-Deutsch V, Bencs L, Naveau I, Roekens E, Van Grieken R (2009) Atmospheric BTEX-concentrations in an area with intensive street traffic. Atmos Environ 43:311–318. https://doi.org/10.1016/j.atmosenv.2008.09.071

Claesson MJ, Jeffery IB, Conde S, Power SE, O'Connor EM, Cusack S, Harris HM, Coakley M, Lakshminarayanan B, O'Sullivan O, Fitzgerald GF, Deane J, O'Connor M, Harnedy N, O'Connor K, O'Mahony D, van Sinderen D, Wallace M, Brennan L, Stanton C, Marchesi JR, Fitzgerald AP, Shanahan F, Hill C, Ross RP, O'Toole PW (2012) Gut microbiota composition correlates with diet and health in the elderly. Nature 488(7410):178–184. https://doi.org/10.1038/nature11319 PMID: 22797518

Claesson MJ, Wang Q, O'Sullivan O, Greene-Diniz R, Cole JR, Ross RP, O'Toole PW (2010) Comparison of two next-generation sequencing technologies for resolving highly complex microbiota composition using tandem variable 16S rRNA gene regions. Nucleic Acids Res 38 (22):e200. https://doi.org/10.1093/nar/gkq873. Epub 2010 Sep 29. PMID: 20880993; PMCID: PMC3001100

Collino S, Montoliu I, Martin FP, Scherer M, Mari D, Salvioli S, Bucci L, Ostan R, Monti D, Biagi E, Brigidi P, Franceschi C, Rezzi S (2013) Metabolic signatures of extreme longevity in northern Italian centenarians reveal a complex remodeling of lipids, amino acids, and gut microbiota metabolism. PLoS One 8(3):e56564. https://doi.org/10.1371/journal.pone.0056564. Epub 2013 Mar 6. Erratum in: PLoS One. 2013;8(8). https://doi.org/10.1371/annotation/5fb9fa6f-4889-4407-8430-6dfc7ecdfbdd. PMID: 23483888; PMCID: PMC3590212

D'Amico F, Soverini M, Zama D, Consolandi C, Severgnini M, Prete A, Pession A, Barone M, Turroni S, Biagi E, Brigidi P, Masetti R, Rampelli S, Candela M (2019) Gut resistome plasticity in pediatric patients undergoing hematopoietic stem cell transplantation. Sci Rep 9:5649. https://doi.org/10.1038/s41598-019-42222-w

Duvallet C, Gibbons SM, Gurry T, Irizarry RA, Alm EJ (2017) Meta-analysis of gut microbiome studies identifies disease-specific and shared responses. Nat Commun 8(1):1784. https://doi.org/10.1038/s41467-017-01973-8.PMID:29209090;PMCID:PMC5716994

Eren AM, Borisy GG, Huse SM, Mark Welch JL. Oligotyping analysis of the human oral microbiome. Proc Natl Acad Sci USA 111(28):E2875–84. https://doi.org/10.1073/pnas.1409644111. Epub 2014 Jun 25. PMID: 24965363; PMCID: PMC4104879

Esplugues A, Ballester F, Estarlich M, Llop S, Fuentes-Leonarte V, Mantilla E, Iñiguez C (2010) Indoor and outdoor air concentrations of BTEX and determinants in a cohort of one-year old children in Valencia, Spain. Sci Total Environ 409(1):63–69. https://doi.org/10.1016/j.scitotenv.2010.09.039 Epub 2010 Oct 20 PMID: 20965548

Fassarella M, Blaak EE, Penders J, Nauta A, Smidt H, Zoetendal EG (2021) Gut microbiome stability and resilience: elucidating the response to perturbations in order to modulate gut health. Gut 70(3):595–605. https://doi.org/10.1136/gutjnl 2020-321747 Epub 2020 Oct 13 PMID: 33051190

Forslund K, Sunagawa S, Kultima JR, Mende DR, Arumugam M, Typas A, Bork P (2013) Country-specific antibiotic use practices impact the human gut resistome. Genome Res 23 (7):1163–9. https://doi.org/10.1101/gr.155465.113. Epub 2013 Apr 8. PMID: 23568836; PMCID: PMC3698509

Franceschi C, Garagnani P, Parini P, Giuliani C, Santoro A (2018) Inflammaging: a new immune-metabolic viewpoint for age-related diseases. Nat Rev Endocrinol 14(10):576–590. https://doi.org/10.1038/s41574-018-0059-4 PMID: 30046148

Gasparrini AJ, Wang B, Sun X, Kennedy EA, Hernandez-Leyva A, Ndao IM, Tarr PI, Warner BB, Dantas G (2019) Persistent metagenomic signatures of early-life hospitalization and antibiotic treatment in the infant gut microbiota and resistome. Nat Microbiol 4:2285–2297. https://doi.org/10.1038/s41564-019-0550-2

Ghosh TS, Rampelli S, Jeffery IB, Santoro A, Neto M, Capri M, Giampieri E, Jennings A, Candela M, Turroni S, Zoetendal EG, Hermes GDA, Elodie C, Meunier N, Brugere CM, Pujos-Guillot E, Berendsen AM, De Groot LCPGM, Feskins EJM, Kaluza J, Pietruszka B, Bielak MJ, Comte B, Maijo-Ferre M, Nicoletti C, De Vos WM, Fairweather-Tait S, Cassidy A, Brigidi P, Franceschi C, O'Toole PW (2020) Mediterranean diet intervention alters the gut microbiome in older people reducing frailty and improving health status: the NU-AGE 1-year dietary intervention across five European countries. Gut 69(7):1218–1228. https://doi.org/10.1136/gutjnl-2019-319654. Epub 2020 Feb 17. PMID: 32066625; PMCID: PMC7306987

Giarratano A, Green SE, Nicolau DP (2018) Review of antimicrobial use and considerations in the elderly population. Clin Interv Aging 17(13):657–667. https://doi.org/10.2147/CIA.S133640. PMID:29713150;PMCID:PMC5909780

Guigoz Y, Doré J, Schiffrin EJ (2008) The inflammatory status of old age can be nurtured from the intestinal environment. Curr Opin Clin Nutr Metab Care 11(1):13–20. https://doi.org/10.1097/MCO.0b013e3282f2bfdf PMID: 18090652

Hartmann EM, Hickey R, Hsu T, Betancourt Román CM, Chen J, Schwager R, Kline J, Brown GZ, Halden RU, Huttenhower C, Green JL (2016) Antimicrobial chemicals are associated with elevated antibiotic resistance genes in the indoor dust microbiome. Environ Sci Technol 50(18):9807–15. https://doi.org/10.1021/acs.est.6b00262. Epub 2016 Sep 7. PMID: 27599587; PMCID: PMC5032049

He Y, Wu W, Zheng HM, Li P, McDonald D, Sheng HF, Chen MX, Chen ZH, Ji GY, Zheng ZD, Mujagond P, Chen XJ, Rong ZH, Chen P, Lyu LY, Wang X, Wu CB, Yu N, Xu YJ, Yin J, Raes J, Knight R, Ma WJ, Zhou HW (2018) Author correction: regional variation limits applications of healthy gut microbiome reference ranges and disease models. Nat Med 24 (12):1940. https://doi.org/10.1038/s41591-018-0219-z.Erratumfor:NatMed.2018Oct;24(10): 1532-1535 PMID: 30250144

Helmink BA, Khan MAW, Hermann A, Gopalakrishnan V, Wargo JA (2019 Mar) The microbiome, cancer, and cancer therapy. Nat Med 25(3):377–388. https://doi.org/10.1038/s41591-019-0377-7 Epub 2019 Mar 6 PMID: 30842679

Holeček M (2018 May) Branched-chain amino acids in health and disease: metabolism, alterations in blood plasma, and as supplements. Nutr Metab (lond) 3(15):33. https://doi.org/10.1186/s12986-018-0271-1.PMID:29755574;PMCID:PMC5934885

Hu Y, Yang X, Qin J, Lu N, Cheng G, Wu N, Pan Y, Li J, Zhu L, Wang X, Meng Z, Zhao F, Liu D, Ma J, Qin N, Xiang C, Xiao Y, Li L, Yang H, Wang J, Yang R, Gao GF, Wang J, Zhu B (2013) Metagenome-wide analysis of antibiotic resistance genes in a large cohort of human gut microbiota. Nat Commun 4:2151. https://doi.org/10.1038/ncomms3151 PMID: 23877117

Huseyin CE, O'Toole PW, Cotter PD, Scanlan PD (2017) Forgotten fungi-the gut mycobiome in human health and disease. FEMS Microbiol Rev 41(4):479–511. https://doi.org/10.1093/femsre/fuw047 PMID: 28430946

Integrative HMP (iHMP) Research Network Consortium (2019) The integrative human microbiome project. Nature 569(7758):641–648. https://doi.org/10.1038/s41586-019-1238-8. Epub 2019 May 29. PMID: 31142853; PMCID: PMC6784865

Jeffery IB, Lynch DB, O'Toole PW (2016) Composition and temporal stability of the gut microbiota in older persons. ISME J 10(1):170–82. https://doi.org/10.1038/ismej.2015.88. Epub 2015 Jun 19. PMID: 26090993; PMCID: PMC4681863

Koh A, De Vadder F, Kovatcheva-Datchary P, Bäckhed F (2016) From dietary fiber to host physiology: short-chain fatty acids as key bacterial metabolites. Cell 165(6):1332–1345. https://doi.org/10.1016/j.cell.2016.05.041 PMID: 27259147

Kondo T, Kishi M, Fushimi T, Kaga T (2009) Acetic acid upregulates the expression of genes for fatty acid oxidation enzymes in liver to suppress body fat accumulation. J Agric Food Chem 57 (13):5982–5986. https://doi.org/10.1021/jf900470c PMID: 19469536

Kropp C, Le Corf K, Relizani K, Tambosco K, Martinez C, Chain F, Rawadi G, Langella P, Claus SP, Martin R (2021) The Keystone commensal bacterium Christensenella minuta DSM 22607 displays anti-inflammatory properties both in vitro and in vivo. Sci Rep 11(1):11494. https://doi.org/10.1038/s41598-021-90885-1.PMID:34075098;PMCID:PMC8169850

Kundu P, Blacher E, Elinav E, Pettersson S (2017) Our gut microbiome: the evolving inner self. Cell 171(7):1481–1493. https://doi.org/10.1016/j.cell.2017.11.024 PMID: 29245010

Leung K, Thuret S (2015) Gut microbiota: a modulator of brain plasticity and cognitive function in ageing. Healthcare (basel) 3(4):898–916. https://doi.org/10.3390/healthcare3040898.PMID: 27417803;PMCID:PMC4934620

Leusch F, Bartkow M (2010). A short primer on benzene, toluene, ethylbenzene and xylenes (BTEX) in the environment and hydraulic fracturing fluids. Smart Water Research Centre, Griffith University, Queensland, Australia. https://environment.des.qld.gov.au/management/coal-seam-gas/pdf/btex-report.pdf

Li J, Cao J, Zhu YG, Chen QL, Shen F, Wu Y, Xu S, Fan H, Da G, Huang RJ, Wang J, de Jesus AL, Morawska L, Chan CK, Peccia J, Yao M (2018) Global survey of antibiotic resistance genes in air. Environ Sci Technol 52(19):10975–10984. https://doi.org/10.1021/acs.est.8b02204 Epub 2018 Jul 25 PMID: 30043612

Liang SY, Mackowiak PA (2007) Infections in the elderly. Clin Geriatr Med 23(2):441–456, viii. https://doi.org/10.1016/j.cger.2007.01.010. PMID: 17462528

Lu Y, Fan C, Li P, Lu Y, Chang X, Qi K (2016) Short Chain fatty acids prevent high-fat-diet-induced obesity in mice by regulating G protein-coupled receptors and gut microbiota. Sci Rep 28(6):37589. https://doi.org/10.1038/srep37589.PMID:27892486;PMCID: PMC5124860

Lynch SV, Pedersen O (2016) The human intestinal microbiome in health and disease. N Engl J Med 375(24):2369–2379. https://doi.org/10.1056/NEJMra1600266 PMID: 27974040

Mahnert A, Moissl-Eichinger C, Zojer M, Bogumil D, Mizrahi I, Rattei T, Martinez JL, Berg G (2019) Man-made microbial resistances in built environments. Nat Commun 10(1):968. https://doi.org/10.1038/s41467-019-08864-0.PMID:30814504;PMCID:PMC6393488

Massolo L, Rehwagen M, Porta A, Ronco A, Herbarth O, Mueller A (2010) Indoor-outdoor distribution and risk assessment of volatile organic compounds in the atmosphere of industrial and urban areas. Environ Toxicol 25(4):339–349. https://doi.org/10.1002/tox.20504 PMID: 19449388

Mueller S, Saunier K, Hanisch C, Norin E, Alm L, Midtvedt T, Cresci A, Silvi S, Orpianesi C, Verdenelli MC, Clavel T, Koebnick C, Zunft HJ, Doré J, Blaut M (2006) Differences in fecal microbiota in different European study populations in relation to age, gender, and country: a cross-sectional study. Appl Environ Microbiol 72(2):1027–1033. https://doi.org/10.1128/AEM.72.2.1027-1033.2006.PMID:16461645;PMCID:PMC1392899

Nayfach S, Roux S, Seshadri R, Udwary D, Varghese N, Schulz F, Wu D, Paez-Espino D, Chen IM, Huntemann M, Palaniappan K, Ladau J, Mukherjee S, Reddy TBK, Nielsen T, Kirton E, Faria JP, Edirisinghe JN, Henry CS, Jungbluth SP, Chivian D, Dehal P, Wood-Charlson EM, Arkin AP, Tringe SG, Visel A; IMG/M Data Consortium, Woyke T, Mouncey NJ, Ivanova NN, Kyrpides NC, Eloe-Fadrosh EA (2021) A genomic catalog of Earth's microbiomes. Nat Biotechnol 39(4):499–509. https://doi.org/10.1038/s41587-020-0718-6. Epub 2020 Nov 9. Erratum in: Nat Biotechnol. 2020 Nov 18;: Erratum in: Nat Biotechnol. 2021 Apr 1;: PMID: 33169036; PMCID: PMC8041624

Nicoletti C (2015) Age-associated changes of the intestinal epithelial barrier: local and systemic implications. Expert Rev Gastroenterol Hepatol 9(12):1467–1469. https://doi.org/10.1586/17474124.2015.1092872 Epub 2015 Sep 22 PMID: 26395417

Noto D, Fayer F, Cefalù AB, Altieri I, Palesano O, Spina R, Valenti V, Pitrone M, Pizzolanti G, Barbagallo CM, Giordano C, Averna MR (2016) Myristic acid is associated to low plasma HDL cholesterol levels in a Mediterranean population and increases HDL catabolism by enhancing HDL particles trapping to cell surface proteoglycans in a liver hepatoma cell model. Atherosclerosis 246:50–56. https://doi.org/10.1016/j.atherosclerosis.2015.12.036 Epub 2015 Dec 28 PMID: 26756970

Oki K, Toyama M, Banno T, Chonan O, Benno Y, Watanabe K (2016) Comprehensive analysis of the fecal microbiota of healthy Japanese adults reveals a new bacterial lineage associated with a phenotype characterized by a high frequency of bowel movements and a lean body type. BMC Microbiol 16(1):284. https://doi.org/10.1186/s12866-016-0898-x.PMID:27894251;PMCID: PMC5127096

Qato DM, Wilder J, Schumm LP, Gillet V, Alexander GC (2016) Changes in prescription and over-the-counter medication and dietary supplement use among older adults in the United States, 2005 vs 2011. JAMA Intern Med 176(4):473–482. https://doi.org/10.1001/jamainternmed.2015.8581.PMID:26998708;PMCID:PMC5024734

Rampelli S, Soverini M, D'Amico F, Barone M, Tavella T, Monti D, Capri M, Astolfi A, Brigidi P, Biagi E, Franceschi C, Turroni S, Candela M (2020) Shotgun metagenomics of gut microbiota in humans with up to extreme longevity and the increasing role of xenobiotic degradation.

mSystems 5(2):e00124–20. https://doi.org/10.1128/mSystems.00124-20. PMID: 32209716; PMCID: PMC7093822

Sato Y, Atarashi K, Plichta DR, Arai Y, Sasajima S, Kearney SM, Suda W, Takeshita K, Sasaki T, Okamoto S, Skelly AN, Okamura Y, Vlamakis H, Li Y, Tanoue T, Takei H, Nittono H, Narushima S, Irie J, Itoh H, Moriya K, Sugiura Y, Suematsu M, Moritoki N, Shibata S, Littman DR, Fischbach MA, Uwamino Y, Inoue T, Honda A, Hattori M, Murai T, Xavier RJ, Hirose N, Honda K (2021) Novel bile acid biosynthetic pathways are enriched in the microbiome of centenarians. Nature 599(7885):458–464. https://doi.org/10.1038/s41586-021-03832-5 Epub 2021 Jul 29 PMID: 34325466

Shkoporov AN, Clooney AG, Sutton TDS, Ryan FJ, Daly KM, Nolan JA, McDonnell SA, Khokhlova EV, Draper LA, Forde A, Guerin E, Velayudhan V, Ross RP, Hill C (2019) The human gut virome is highly diverse, stable, and individual specific. Cell Host Microbe 26 (4):527-541.e5. https://doi.org/10.1016/j.chom.2019.09.009 PMID: 31600503

Sommer MOA, Dantas G, Church GM (2009) Functional characterization of the antibiotic resistance reservoir in the human microflora. Science 325(5944):1128–1131. https://doi.org/10.1126/science.1176950.PMID:19713526;PMCID:PMC4720503

Song MY, Chun H (2021) Species and characteristics of volatile organic compounds emitted from an auto-repair painting workshop. Sci Rep 11(1):16586. https://doi.org/10.1038/s41598-021-96163-4.PMID:34400724;PMCID:PMC8368184

Stark AH, Crawford MA, Reifen R (2008) Update on alpha-linolenic acid. Nutr Rev 66(6):326–332. https://doi.org/10.1111/j.1753-4887.2008.00040.x PMID: 18522621

Tavella T, Turroni S, Brigidi P, Candela M, Rampelli S (2021a) The human gut resistome up to extreme longevity. mSphere 6(5):e0069121. https://doi.org/10.1128/mSphere.00691-21. Epub 2021a Sep 8. PMID: 34494880; PMCID: PMC8550338

Tavella T, Rampelli S, Guidarelli G, Bazzocchi A, Gasperini C, Pujos-Guillot E, Comte B, Barone M, Biagi E, Candela M, Nicoletti C, Kadi F, Battista G, Salvioli S, O'Toole PW, Franceschi C, Brigidi P, Turroni S, Santoro A (2021b) Elevated gut microbiome abundance of Christensenellaceae, Porphyromonadaceae and Rikenellaceae is associated with reduced visceral adipose tissue and healthier metabolic profile in Italian elderly. Gut Microbes 13(1):1–19. https://doi.org/10.1080/19490976.2021b.1880221.PMID: 33557667; PMCID: PMC7889099

Turroni S, Brigidi P, Cavalli A, Candela M (2018) Microbiota-host transgenomic metabolism, bioactive molecules from the inside. J Med Chem 61(1):47–61. https://doi.org/10.1021/acs.jmedchem.7b00244 Epub 2017 Aug 3 PMID: 28745893

van Tongeren SP, Slaets JP, Harmsen HJ, Welling GW (2005) Fecal microbiota composition and frailty. Appl Environ Microbiol 71(10):6438–6442. https://doi.org/10.1128/AEM.71.10.6438-6442.2005.PMID:16204576;PMCID:PMC1265947

Villa CR, Ward WE, Comelli EM (2017) Gut microbiota-bone axis. Crit Rev Food Sci Nutr 57 (8):1664–1672. https://doi.org/10.1080/10408398.2015.1010034 PMID: 26462599

Visconti A, Le Roy CI, Rosa F, Rossi N, Martin TC, Mohney RP, Li W, de Rinaldis E, Bell JT, Venter JC, Nelson KE, Spector TD, Falchi M (2019) Interplay between the human gut microbiome and host metabolism. Nat Commun 10(1):4505. https://doi.org/10.1038/s41467-019-12476-z.PMID:31582752;PMCID:PMC6776654

de Vos WM, Tilg H, Van Hul M, Cani PD (2022) Gut microbiome and health: mechanistic insights. Gut gutjnl-2021–326789. https://doi.org/10.1136/gutjnl-2021-326789. Epub ahead of print. PMID: 35105664

Vujkovic-Cvijin I, Sklar J, Jiang L, Natarajan L, Knight R, Belkaid Y (2020) Host variables confound gut microbiota studies of human disease. Nature 587(7834):448–454. https://doi.org/10.1038/s41586-020-2881-9. Epub 2020 Nov 4. PMID: 33149306; PMCID: PMC7677204

Waters JL, Ley RE (2019 Oct 28) The human gut bacteria Christensenellaceae are widespread, heritable, and associated with health. BMC Biol 17(1):83. https://doi.org/10.1186/s12915-019-0699-4.PMID:31660948;PMCID:PMC6819567

Wilmanski T, Diener C, Rappaport N, Patwardhan S, Wiedrick J, Lapidus J, Earls JC, Zimmer A, Glusman G, Robinson M, Yurkovich JT, Kado DM, Cauley JA, Zmuda J, Lane NE, Magis AT,

Lovejoy JC, Hood L, Gibbons SM, Orwoll ES, Price ND (2021) Gut microbiome pattern reflects healthy ageing and predicts survival in humans. Nat Metab 3(2):274–286. https://doi.org/10.1038/s42255-021-00348-0. Epub 2021 Feb 18. Erratum in: Nat Metab. 2021 Apr;3(4): 586. PMID: 33619379; PMCID: PMC8169080.Copy

Wu L, Zeng T, Zinellu A, Rubino S, Kelvin DJ, Carru C (2019) A cross-sectional study of compositional and functional profiles of gut microbiota in sardinian centenarians. mSystems 4 (4):e00325–19. Published 2019 Jul 9. doi:https://doi.org/10.1128/mSystems.00325-19

Wutthi-In M, Cheevadhanarak S, Yasom S, Kerdphoo S, Thiennimitr P, Phrommintikul A, Chattipakorn N, Kittichotirat W, Chattipakorn S (2020) Gut microbiota profiles of treated metabolic syndrome patients and their relationship with metabolic health. Sci Rep 10(1):10085. https://doi.org/10.1038/s41598-020-67078-3.PMID:32572149;PMCID:PMC7308281

Yao L, Seaton SC, Ndousse-Fetter S, Adhikari AA, DiBenedetto N, Mina AI, Banks AS, Bry L, Devlin AS (2018) A selective gut bacterial bile salt hydrolase alters host metabolism. Elife 17 (7):e37182. https://doi.org/10.7554/eLife.37182.PMID:30014852;PMCID:PMC6078496

Chapter 3
Ageing of the Gut Microbiome and Its Potential Contribution Towards Immunesenescence and Inflammaging

Jessica Conway, Nia Paddison Rees, and Niharika A. Duggal

Abstract The gut microbiome plays an essential role in host health, mediated via its interaction with the immune system. Advancing age is accompanied by alterations in gut microbiome composition and their functional features. In this chapter, we explore the impact of these age-related microbiome changes in contributing towards a state of immune dysfunction (immunesenescence) and low-grade basal inflammation (inflammaging) in older adults. We propose that modulating the ageing gut microbiome via probiotic/prebiotic supplementation, dietary interventions, fecal microbiota transplantation and therapeutic administration of microbiome metabolites may serve as effective approaches to boost immune responses in older adults. However, future studies focused on developing a deeper understanding of the impact of aged microbiota on immunity is crucial for identifying novel microbiome targets that can serve as anti-immunesenescence medicines, which is valuable for healthier ageing.

Keywords Immunesenescence · Inflammaging · Microbiome dysbiosis · Intestinal barrier leakage · Microbiome-based interventions

3.1 Introduction

Human life expectancy has risen drastically over the last century. During the period 2016–2018, the life expectancy increased by 0.8 and 0.6 years for males and females, respectively. Unfortunately, healthspan has not kept pace with increases in lifespan in recent years and older adults are spending these additional years of life in ill health, healthy life expectancy for males increased by 0.4 years and for females by only 0.2 years during this period (House of Lords Report 2021). Advancing ageing is accompanied by functional deterioration across multiple

J. Conway · N. P. Rees · N. A. Duggal (✉)
MRC-Versus Arthritis Centre for Musculoskeletal Ageing Research, Institute of Inflammation and Ageing, University of Birmingham, Birmingham, UK
e-mail: n.arora@bham.ac.uk

F. Marotta (ed.), *Gut Microbiota in Aging and Chronic Diseases*, Healthy Ageing and Longevity 17, https://doi.org/10.1007/978-3-031-14023-5_3

systems that culminates in an increased susceptibility to infections, cancer, autoimmunity, atherosclerosis, metabolic diseases, chronic illnesses and all-cause mortality and increasing prevalence of multimorbidity (Carmona and Michan 2016; Guerra et al. 2019); together making older adults a vulnerable population. Unravelling what promotes healthy ageing remains an important research question, which may translate into significant improvements in the quality of life for the global ageing population.

The gut microbiota consists of a highly diverse community of bacterial, fungal, parasitic and viral organisms that exist in a homeostatic balance, to promote host health. The gut microbiome has grabbed the attention of scientists across the globe due to its critically important role in host well-being throughout life (Goulet 2015). Notably, the microbiome impacts a range of functions ranging from host metabolism, biosynthesis of vitamins and amino acids, carbohydrate fermentation, and development of the central nervous system (Lloyd-Price et al. 2016; Mohajeri et al. 2018). Advancing age is accompanied by a shift in the microbial composition profile, a loss of microbial diversity (Biagi et al. 2010) and modified intestinal barrier integrity (Wilson et al. 2018).

In recent years, multiple studies have begun to untangle the profound impact of the gut microbiome in developing and maintaining the immune system (Zheng et al. 2020a, b). However, despite progress in our understanding of how the microbiome and immune system influence one another, still very little is known about how age-associated changes in the gut microbiome influence the trajectory of the host immune system, which will be addressed in this chapter. We first outline how the microbiome and immune system change with age and then delve into how the microbiome and immune system are intimately interlinked, how these links change with age and the implications on host health. Importantly, identifying therapeutic targets within the gut microbiome that could combat immune ageing will have major clinical implications for the growing population of older adults across the world, and will be discussed towards the end of this chapter in which we highlight microbiome-based strategies to reverse the immune ageing clock.

3.2 Age-Associated Compositional and Functional Changes in the Microbiome

It is well established that the gut microbiome composition changes with advancing age, despite remaining relatively stable throughout adult life; characterised by reduced species richness and loss of diversity, termed microbiome dysbiosis (Biagi et al. 2016). The abundance of core commensals (e.g. *Bacteroides animalis*, *Bifidobacteria bifidium* and *Lactobacilli casei*) are reduced in the older population, which is accompanied by an expansion of opportunist microbial communities such as Enterobacteria and Proteobacteria (e.g. *E. coli*, *Haemophilus*, *Pseudomonas* and *Clostridium difficile* which are usually present at low abundance in the intestine of healthy young individuals (Salazar et al. 2019). Additionally, the microbiome of older adults displays a decrease in the Firmicutes/Bacteroidetes ratio in individuals (Mariat et al. 2009). Furthermore, studies on the gut microbiome composition of centenarians have provided further insight into the gut microbiome trajectory associated with healthy ageing with centenarians displaying a decreased abundance of *Coprococcus, Roseburia* and increased abundance of *Akkermansia.* Interestingly a recent study done on a large cohort of 9,000 individuals has highlighted that an age-associated increase in the uniqueness of gut microbiome composition is reflective of healthy ageing and is associated with a decreased survival during the course of a 4-year follow up (Wilmanski et al. 2021). Multiple factors have been recognised to contribute to age-associated microbial dysbiosis including; altered dietary and lifestyle schedules, physical inactivity, altered gut morphology, recurrent infections, repeated hospitalisation and overuse of antibiotics (Yatsunenko et al. 2012). Furthermore, what is increasingly becoming clear is the potential of a healthy diet (rich in dietary fibres and protein) in contributing toward a more diverse gut microbiota and thus promoting healthier ageing (Claesson et al. 2012).

The commensal gut bacteria produce metabolites, such as short-chain fatty acids SCFA (butyrate, acetate, propionate), that are end-products of bacterial fermentation of non-digestible dietary fibres, and are known regulatory of the host immune system. An age-associated decline in the gut commensal SCFA-producing bacteria, such as *Clostridium leptum* and *Eubacterium rectale* (Biagi et al. 2010) and changes in dietary patterns; result in a decline in stool and circulating SCFA levels (Rampelli et al. 2013); driving a shift in microbial metabolic profile. Another key product of microbial metabolism of dietary proteins (e.g. tryptophan) is trimethylamine-N-oxide (TMAO), which impacts a range of host's physiological processes; including intestinal homeostasis and also impacts the immune system (Roager and Licht 2018). An increase in plasma levels of TMAO has been reported in both aged humans and mice (Li et al. 2018).

Age-associated alterations in gut microbiota have profound effects on human health. Evidence from mice models has highlighted that a loss of microbiome diversity is associated with an increased risk of respiratory viral infection (Grayson et al. 2018). Furthermore, microbiome dysbiosis promotes muscle atrophy in older adults by altering skeletal muscle metabolism and thereby contributing to age-associated increased frailty (Picca et al. 2020). Microbial composition changes have also been linked with vascular ageing and an increased risk of atherosclerosis; driven by increasing TMAO production, which leads to the accumulation of macrophage foam cells and endothelial dysfunction (Wang et al. 2011). On the other hand, SCFAs have been reported to regulate blood pressure in animal models (Wang et al. 2017), and age-associated decline in SCFA levels could be a contributing factor toward age-associated increased risk of hypertension and poor cardiovascular health (Chambers et al. 2018). Microbial dysbiosis-induced inflammation has also been recognised as a contributing factor driving age-related neurodegeneration and protein aggregation, together contributing to an age-associated predisposition towards cognitive impairment (Pan et al. 2021) and

Alzheimer's disease (Zhan et al. 2016); driven by elevated TMAO levels (Chelakkot et al. 2018). Overall, our understanding of the gut microbiome and its impact on the ageing trajectory is still in its infancy.

3.3 Ageing and Intestinal Barrier Dysfunction

The intestinal epithelial barrier functions as the primary barrier that regulates the transportation of nutrients, essential ions and microbial components between the gut lumen and systemic circulation. The intestinal barrier is comprised of an overlying mucus layer, produced by goblet cells, that overlays the one-cell-thick intestinal epithelium layer made up of intestinal epithelial cells (IECs) joined together by junctional complexes such as tight junction proteins (zonulin and occludin) and adherent junctions (Chelakkot 2016). Below the epithelium layer is the lamina propria, a thin layer of connective tissue that hosts both adaptive and innate immune cells such as T cells, B cells and macrophages. Commensal bacteria residing in the gut positively impact the intestinal barrier integrity through the production of metabolites such as short-chain fatty acids (SCFAs) and secondary bile acids. Advancing age is accompanied by remodelling of intestinal epithelial tight junction proteins that leads to increased intestinal permeability in non-human primates, also known as 'leaky gut' (Tran and Greenwood-van Meerland 2013). Loss of intestinal barrier integrity can lead to leakage of microbes and microbe-associated molecular patterns (MAMPs), such as lipopolysaccharide (LPS), bacterial flagellum and CpG DNA, through the gut barrier into the circulation, that possess inflammagenic potency and can exert numerous pathological implications such as the increased risk of inflammatory diseases (such as rheumatoid arthritis) and metabolic disorders in the older adults (Kell and Pretorious 2015; Mu et al. 2017). Circulating microbial products also accumulate in the fatty acids of arterial walls causing degradation of vascular endothelium, thus increasing age-related susceptibility to cardiovascular disease. Whilst numerous studies have reported increased intestinal barrier dysfunction with age, the mechanisms behind these processes and potential implications on health are yet to be fully understood.

3.4 Ageing and the Immune System and Inflammaging

Advancing age is accompanied by profound remodelling of the innate and adaptive arms of the immune system and results in an impaired immune response to an antigenic challenge, termed immune senescence (Duggal 2018). This is a many-sided process in which some functions decline sharply with ageing, and others are maintained or even increased, to varying degrees in the different subjects (Fig. 3.1). Neutrophils are the first innate immune cell that migrates toward the site of infection and these cells are equipped with several defence strategies to eliminate

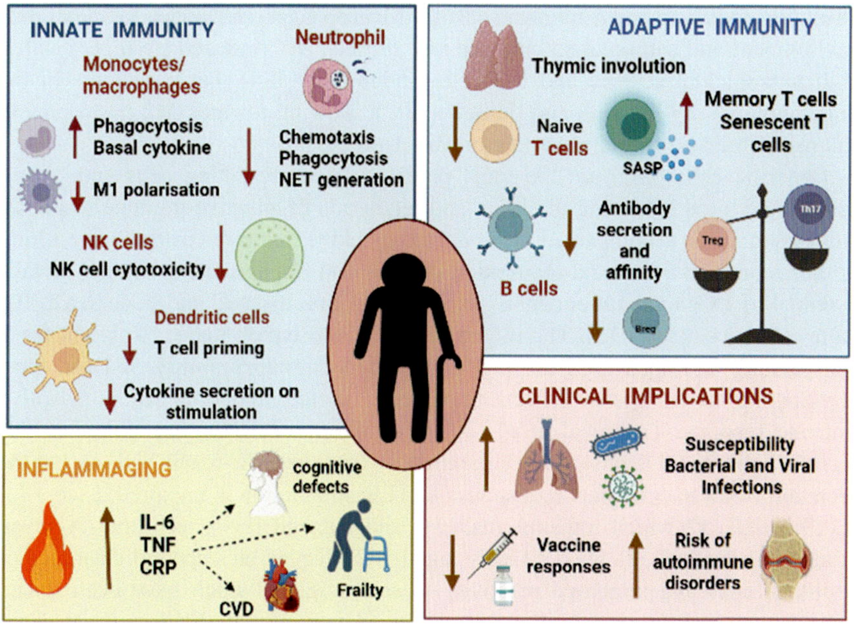

Fig. 3.1 Remodelling of the innate and adaptive immune system with advancing age. Advancing age is accompanied by remodelling of the immune system, known as immunesenescence. Key hallmarks of an adaptive immune ageing which has implications on health of older adults. Another universal feature of physiological ageing is a low-grade increase in systemic levels of pro-inflammatory cytokines, termed Inflammaging. which has been associated with frailty, cognitive impairment, increased risk of cardiovascular diseases, together contributing towards elevated mortality in older adults

bacterial and fungal pathogens. Key features of neutrophil ageing include; compromised neutrophil chemotaxis (Sapey et al. 2014), phagocytosis (ingestion) of the pathogen (Butcher et al. 2001) to form phagosome into which reactive oxygen species are released to kill the engulfed bacteria and reduced ability to extrude neutrophil extracellular traps to entrap bacteria extracellularly (Hazeldine et al. 2014); contributing towards the vulnerability towards bacterial infections and elevated mortality in older adults (Panda et al. 2010).

Monocytes/macrophages are specialised innate phagocytes that also secrete a range of inflammatory mediators; mediating a critical role in early immune responses. Although ageing does not impact the total number of circulating monocytes, an age dependant redistribution of monocyte subsets has been observed; driven by an increase in non-classical ($CD14^{+ve}CD16^{++ve}$) monocytes that exhibit a senescence-associated secretory phenotype (SASP) (Ong et al. 2018). Furthermore, age-associated alterations in monocyte functional capacity include; an increase in pro-inflammatory cytokine production in basal conditions, decline in bactericidal properties, delayed clearance of apoptotic cells and impaired wound healing; which together contribute to age-associated impairment of immune resolution (Maeyer and Chambers 2021). Macrophages can adapt to their local environment and can polarise into two key subsets; M1 (induced by IFNγ) with a high microbicidal activity and M2 cells (induced by IL4) that participate in the immunoregulatory function and tissue repair; a skewing towards M2 macrophages has been observed in old mice (Jackaman et al. 2013).

Dendritic cells (DC) are the most potent antigen-presenting cells and play a central role in orchestrating the onset and regulation of adaptive immune response. Advancing age is accompanied by an altered cytokine micro-environment; resulting in premature activation and maturation of DCs and impairment of antigen uptake potential of DCs and subsequent T cell priming (Bruunsgaard and Pedersen 2003; Agrawal and Gupta 2011). The migration of DCs to lymph nodes is pivotal for T cell priming and aged DCs display an impaired migratory ability of DCs; thus, contributing to an age-associated impairment in the initiation of the adaptive immune response (Agrawal et al. 2007). Furthermore, DCs from older adults secrete higher basal levels of pro-inflammatory cytokines (IL-6 and TNF-a), but the secretion of these cytokines in response to TLR stimulation is impaired, except for IL-10 and IL-12 which remains unaltered and reduced IFNγ secretion (Agrawal et al. 2007; Della-Bella 2007). Lastly, aged DCs display an impaired clearance of apoptotic cells and increased reactivity to self-antigens, which have been linked with the age-associated risk of autoimmunity (Agrawal et al. 2012).

NK cells are innate cytotoxic lymphocytes that play a crucial role in the recognition and host defence against viral pathogens and tumours. Aged individuals display an increase in systemic numbers of NK cells, driven by an accumulation of senescent CD57 expressing NK cells (Garff-Tavernier et al. 2010). On assessing NK cell functioning, a reduction in per-cell cytotoxicity (Hazeldine et al. 2012) and impaired secretion of IFN-γ and IL-8 but not TNF-α have been observed by aged NK cells on target cell stimulation (Hazeldine 2015); together resulting in an age-associated impaired response to viral challenges. Recent evidence suggests that NK cells play a key role in the resolution of inflammation via clearance of senescent cells, although unexplored it can be postulated that age-associated impairments in NK cell clearance of senescent cells, contribute towards the age-associated accumulation of senescent cells, which has been associated with the emergence of pathologies, such as sarcopenia (Baker 2011).

The thymus plays a vital role in the maintenance of the peripheral adaptive immune system as it is devoted to T cell development and maturation. In humans, advancing age is accompanied by thymic involution; involving a loss of tissue architecture and cellularity, accompanied by infiltration of adipocytes and an altered thymic microenvironment (Lynch et al. 2009), resulting in a decline in the net thymic output of naïve antigen-inexperienced T cells (Haines et al. 2009). This contributes to an age-associated increased vulnerability of older adults towards novel pathogens such as severe acute respiratory syndrome (SARS)-CoV-2 virus (Cunha et al. 2020). The diminution of naive T cells is accompanied by an accumulation of antigen-experienced memory T cells (Dorshkind et al. 2009). Repeated antigenic encounters in aged individuals drive further differentiation of memory T cells into a state of replicative senescence that exhibits a SASP phenotype (Goronzy and Weyand 2019). Importantly, age-associated defects in CD4 T cell helper functions exert an impact on the humoral responses of an aged host (Eaton et al. 2004). Another hallmark of T cell ageing is a skewing of T-cell responses toward a pro-inflammatory Th17 cell polarisation (Ouyang et al. 2011). Regulatory T cells (T_{reg}) play a pivotal role in maintaining immune homeostasis. An expansion of circulating memory T_{regs} has been observed in aged mice and humans, but these cells display an impaired suppressive functional capacity (Churov et al. 2020); shifting the Th17/T_{reg} balance towards a pro-inflammatory environment with age which has been associated with an increased risk of autoimmunity (Prelog 2006).

Ageing is accompanied by impairments in B cell haematopoiesis in the bone marrow and a skewing towards 'antigen-experienced' B cells and an accumulation of age-associated B cells (De Mol et al. 2021). Functional impairments in antibody production and secretion of antibodies with poor affinity by aged B cells; result in poor vaccination efficacy in older people, making protecting the aged population from infectious diseases even more challenging (Frasca and Blomberg 2020). Thus, it's no surprise that the World Health Organisation has included the development of vaccines targeting older adults as a future research priority (Thomas-Crusells et al. 2012). Furthermore, dysfunctional B cell responses in older adults, such as an accumulation of Age-Associated B cells (ABC) that secrete pro-inflammatory cytokines (TNFα) and autoantibodies secretion (Riley et al. 2017) and numerical and functional loss in immunoregulatory B cells (B_{regs}); a potential contributor towards the increased risk for autoimmune diseases (Duggal et al. 2013).

Another universal feature of ageing is chronic, elevated basal levels of peripheral pro-inflammatory cytokines (IL6, TNFα, CRP) termed inflammaging (Franceschi and Campisi 2014). There is mounting evidence suggesting a role of inflammaging in older adults as a risk factor for the prevalence of cardiovascular diseases (CVD) (Cesari et al. 2003), loss of muscle mass and strength (Visser et al. 2002), poor physical performance (Peterson 2014), which together contribute towards age-associated frailty (Van Epps et al. 2016) and also central role in driving the development of neurodegenerative diseases, cognitive defects and impaired memory with advancing age (Mészáros 2020). Importantly, inflammaging prior to infection has recently been recognised as a potential contributor that predisposes older adults to severe COVID-19 infections (Zheng et al. 2020a, b) and other viral infections (Goldstein 2010). Together these studies provide strong evidence of the role of inflammaging in the development of multiple age-related conditions and thus it's no surprise that it is a powerful predictor of mortality and morbidity in older adults (Sayed et al. 2021). The dynamic interaction of multifactorial processes, such as a lifelong exposure to antigen stressors, immunesenescence, oxidative stress, the accumulation of senescent cells, gut microbiota dysbiosis, an unbalanced diet, increased central adiposity and physical inactivity, together culminate in inflammaging (Franceschi et al. 2000). Together, making it clear that the immune system does not operate in isolation and can be modified by a broad range of signals, we now need to consider how we boost the reduced immune responses of older adults.

3.5 Intestinal Microbiome: Shaping Mucosal and Systemic Immune Responses in Homeostasis

In recent years, the gut microbiota has gained considerable interest for its role in the development and maintenance of the immune system (summarised in Fig. 3.2). Although several reports indicate that the immune system and microbiome communicate and influence one another, the complex relationship between the immune system and the microbiome remains incompletely understood (Molloy et al. 2012). In this section, we highlight some of the existing mechanistic dissections of multifaceted microbiome-immunity dialogues. Recognition of microbial molecules, termed microbe-associated molecular patterns (MAMPs), by pattern recognition receptors (PRRs) expressed on the surface of immune cells is critical for coordinating effective inflammatory responses to infectious agents (Chu and Mazmanian

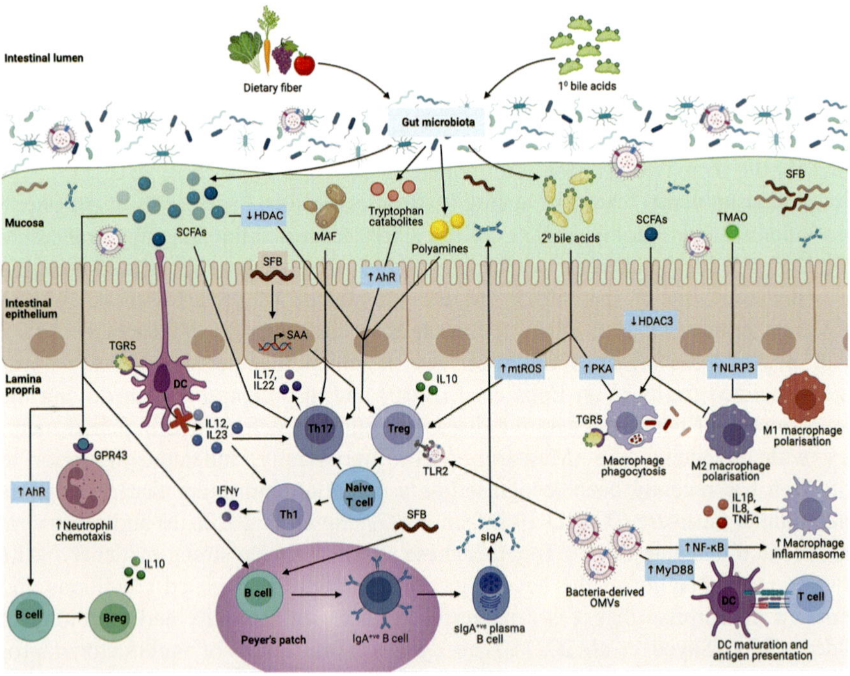

Fig. 3.2 Dissecting interactions between gut microbiota and the immune system. Indigestible dietary fibres are fermented by the gut microbiota to produce short-chain fatty acids (SCFAs) that possess multiple immunomodulatory properties. Additionally, primary (1^0) bile acids synthesised by the liver can also be metabolised by the microbiota to generate secondary (2^0) bile acids that also impacts Treg/Th17 balance. Other immunomodulatory microbial metabolites include polyamines, tryptophan catabolites and trimethylamine N-oxide (TMAO). Moreover, recently bacteria-derived outer membrane vesicles (OMVs) have been reported exert immunoregulatory effects on macrophages and dendritic cells

2013). For example, PRR signalling induced in response to MAMPs (e.g. bacterial peptidoglycan, flagellin and lipoproteins) leads to the production of antimicrobial peptides and mucin by secretory epithelial cells (ECs). This is especially important for maintaining intestinal homeostasis and regulating microbiota composition (Birchenough et al. 2016; Wlodarska et al. 2014). In addition to direct stimulation, commensal bacteria can indirectly regulate immune responses through interactions with the intestinal microenvironment. For example, adherence of segmented filamentous bacterium (SFB) to intestinal ECs promotes T helper 17 (Th17) differentiation, resulting in the recruitment of effector immune cells for enhanced clearance of opportunists (Atarashi et al. 2015; Ivanov et al. 2008). SFB are also involved in the development of Peyer's patch germinal centre responses for the generation of secretory IgA-producing plasma B cells (Lécuyer et al.2014). Secretory IgA antibodies serve as the first line of defence against opportunistic bacteria and are therefore essential for mucosal immunity (Boullier et al. 2009; Forbes et al. 2008).

Emerging evidence suggests that SCFAs (namely butyrate, acetate propionate), the by-products of microbial fermentation of non-digestible fibre exert multiple effects on a range of immune cells. For instance, butyrate enhances macrophage antimicrobial function via histone deacetylase 3 (HDAC3) inhibition (Schulthess et al. 2019) and inhibits inflammatory cytokine production (Usami et al. 2008). Additionally, SCFAs can partially suppress M2 polarisation in alveolar macrophages by HDAC3 inhibition (Huang et al. 2022). SCFAs can act as chemoattractants for neutrophils and play a critical role in G protein-coupled receptor (GPR) 43-mediated neutrophil chemotaxis during intestinal inflammation (Sina et al. 2009). On the other hand, butyrate and propionate inhibit IL-12 and IL-23 secretion by dendritic cells (DCs), thus modulating the polarisation of Th1 and Th17 cells (Nastasi et al. 2017). Furthermore, SCFAs, in particular butyrate, promote the differentiation of naïve T cells to regulatory T cells (Treg) by HDAC inhibition at the *Foxp3* locus (Furusawa et al. 2013), enhancing regulatory B cell differentiation (Rosser et al. 2020), upregulating anti-inflammatory cytokine IL-10 and suppress Th17 polarisation (Zhang et al. 2016). SCFAs also control mucosal and systemic immunity by promoting plasma B cell differentiation and IgA and IgG antibody production (Kim et al. 2016).

Cholesterol-derived bile acids are synthesised in the liver and secreted into the small intestines to aid lipid digestion. However, approximately 5% of these molecules escape reabsorption and are converted into secondary bile acids by the colonic microbiota (Guzior and Quinn 2021). Secondary bile acids, such as deoxycholic acid (DCA) and lithocholic acid (LCA), can dampen gut inflammation (Sinha et al. 2020; Guo et al. 2016), protect against intestinal barrier dysfunction (Lajczak-McGinley et al. 2020; Golden et al. 2018) and confer gut mucosal protection through interactions with the immune system. In vitro studies have shown that secondary bile acids can inhibit pro-inflammatory cytokine production (IL-6, IL-12 and TNFα) and reduce the basal phagocytic activity of macrophages via induction of the G protein-coupled bile acid receptor TGR5 and PKA activation (Wammers et al. 2018). Conversely, elevated levels of DCA induced by a high-fat diet promote a pro-inflammatory M1 phenotype (classical activated) in murine macrophages (Wang et al. 2020), suggesting a dual role of secondary bile acids that could be dependent on circulating levels. Moreover, secondary bile acid-induced TGR5 signalling suppresses classical DC activation and promotes the differentiation of IL12 hypo-producing DCs through activation of the cAMP-PKA signalling pathway and NF-kB inhibition (Ichikawa 2016; Hu et al. 2021). With regards to the adaptive immune system, LCA derivatives can increase the Treg/Th17 ratio through enhanced mitochondrial reactive oxygen species production (Campbell et al. 2020). Furthermore, DCA and LCA have been shown to decrease Th1 development and IFNy secretion in mice (Hu et al. 2021).

Other microbial-derived metabolites include polyamines, tryptophan catabolites and trimethylamine N-oxide (TMAO), though their effect on immune cells at the molecular level is less known. Polyamines, such as spermidine display immunomodulatory properties, governing the ability of $CD4^{+ve}$ T cells to differentiate into helper T cell subsets while preferentially promoting the generation of Tregs (Carriche et al. 2021) and spermidine has also been reported to restore B cell responses and boost vaccine response in aged mice (Zhang et al. 2019). Microbial tryptophan catabolites (e.g. indole, indoleacetic acid and indolepropionic acid) regulate immune responses through the activation of the aryl hydrocarbon receptor (AhR), which has a crucial role in Treg gut homing and immunosuppressive function (Ye et al. 2017). On the other hand, TMAO can drive M1 macrophage polarisation through activation of the NLRP3 inflammasome (Wu et al. 2020) and modulate inflammatory response in macrophages (Krishnan et al. 2018).

Microbiota-released outer membrane vesicles (OMVs) carry bioactive cargo (e.g. DNA, RNA, proteins and lipids) that are immunomodulatory, triggering immune activation. Bacteria-derived OMVs (20–400 nm) induce DC maturation via toll-like receptor (TLR) activation and subsequent MyD88 signalling as well as the activation of macrophage inflammasomes and proinflammatory cytokine production (IL1ß, IL8 and TNF) (Cecil et al. 2017; Schetters et al. 2019). Moreover, *Bacteroides fragilis*-released OMVs containing polysaccharide A can dampen intestinal inflammation by promoting the generation of IL10-producing $Foxp3^{+ve}$ Tregs through the activation of tolerogenic DCs (Shen et al. 2012).

3.6 Ageing Microbiome: A Potential Driver of Immunesenescence and Inflammaging

The immune system does not operate in isolation and responds to signals from the environment. We have discussed in the section above that immune fitness declines with age and in this section discusses the potential influence of age-associated host-microbiome changes on host immunity and the ability to combat pathogen invasion. The first evidence supporting the hypothesis that the age-associated changes in microbiota drives inflammaging comes from, a study done in aged germ-free mice has recently shown that this translocation of gut microbes and microbial products into the bloodstream contributes to the low-grade chronic inflammation which occurs with age, termed inflammaging, and macrophage dysfunction (Thevaranjan et al. 2017). Notably, microbial transfer via co-housing conventionally aged mice with aged germ-free mice, resulted in an increased circulating level of TNFα; which was not replicated by co-housing with young mice (Thevaranjan et al. 2017). Further evidence of this causal relationship between microbial dysbiosis and inflammation comes from a study reporting that a fecal microbiota transplant (FMT) from an aged mouse into young GF recipients, results in an 'aged' gut microbiota that elicits intestinal inflammation, elevated levels of circulating microbiota-derived inflammatory components and induced T cell activation study (Fransen et al. 2017). Further evidence of microbiome modulation associated impact on the adaptive immune system comes from a latest study in rhesus monkey reporting an association between age-associated changes in microbiome profile and a skewing of the naïve: memory CD4 T cell ratio (Pallikuth et al. 2021). So far, the only evidence of a direct impact of a state of microbial dysbiosis in contributing toward a state of immunesenescence in humans comes from a recent study reporting that fecal microbiota transplant (FMT) in aged *C. difficile* infection patients, results in a decline in systemic senescent T cells and an expansion of ant-inflammatory cytokine IL10 producing regulatory T and B cells post-FMT, and directly have an anti-immunosenescent effect (Monaghan et al. 2021) providing evidence for the potential of microbiome manipulation to restore immunity in aged individuals. To summarise, it is not unreasonable to suggest that this age-associated increased intestinal permeability and translocation of pathobionts and microbial products, and a state of microbial dysbiosis and reduction in SCFAs, which are known to play a key role in maintaining immune homeostasis may be involved with disrupted immune homeostasis and contribute towards inflammaging (Fig. 3.3) and the development of an in-depth understanding of the long-term impacts of microbial dysbiosis may bear an impact on understanding and potential intervention of immune ageing.

3.7 Taking Action: Microbiome-Based Interventions to Combat Immune Ageing

In the previous section, we discussed evidence suggesting a potential link between age-associated altered gut microenvironment and immunosenescence/ inflammaging. Given that ageing is a malleable process, this section will explore the immune-stimulatory, anti-inflammatory and healthspan extension effects of microbiome-based intervention strategies.

Probiotics are live microorganisms of healthy gut flora (e.g. lactobacilli and bifidobacteria) that exert a beneficial effect on the host upon adequate ingestion. In contrast, prebiotics are typically indigestible fibre compounds (e.g. inulin,

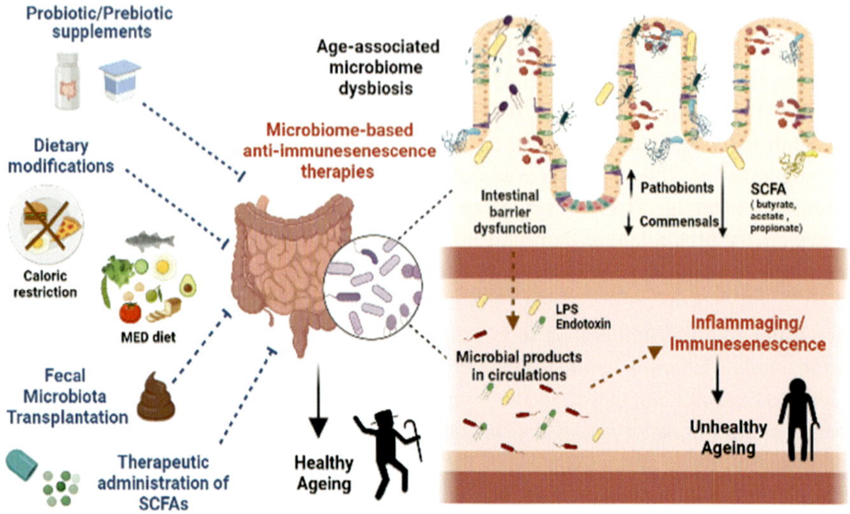

Fig. 3.3 The effects of age associated microbiome dysbiosis and intestinal barrier impairments in instigating aging-related inflammaging and immunesenescence. The gut microbial composition changes drastically with advancing age, termed microbial dysbiosis; characterised by a loss of beneficial commensals and an expansion of opportunistic microbes, contributing towards an increased intestinal barrier permeability, which allows translocation of microbes and their products (toxins, antigens) into the circulation, increasingly being recognised as a contributing factor towards inflammaging and immunesenescence. Given that age-associated microbiome alterations are modifiable, we propose that microbiome-based interventions that cultivate a healthier microbiota in aged individuals can serve as novel intervention strategy to reverse the immune ageing clock

galacto-oligosaccharides) which promote the growth of probiotic gut bacteria and also act as a substrate for fermentation and production of SCFAs.

Evidence from clinical trials has shown that prebiotics/probiotics can restore gut microbiome homeostasis (Cheng et al. 2018) and maintenance of intestinal membrane integrity (Wilms et al. 2019) and contribute toward healthy ageing (Tiihonen et al. 2010). Additionally, probiotics (*Lactobacillus*, *Bifidobacterium* strains) possess immunomodulatory properties and have been shown to enhance NK cell cytotoxicity (Gill et al. 2001), granulocyte phagocytosis (Maneerat et al. 2014), induce T cell subset redistribution (Finamore 2019), expansion of regulatory T cells (Mane 2011) and an overall improved immunity in aged hosts (Finamore 2019; Mane 2011). Additionally, probiotic consumption exerts an anti-inflammaging effect via downregulation of anti-inflammatory cytokine secretion and reduced systemic inflammation (Dong 2013; Matsumoto 2006) and improved vaccine responses in older adults (Boge 2009) and reduced susceptibility towards infections (Eguchi 2019).

Prebiotics have recently gained attention due to their gut health enhancing, particularly in patients with inflammatory intestinal disorders (Akutko 2021); driven by induction of SCFA production. The benefit of co-administrating prebiotics with probiotics referred to as synbiotics has also been reported in aged hosts. For instance, the prebiotic composition of inulin and oligofructose decreases cecal inflammation (IL1β) in rats (Hoentjen et al. 2005). Results from human studies, investigating the impact of supplementation with galactans reducing systemic levels of pro-inflammatory cytokines (IL-1, IL-6 and TNF-α) in older adults (Shokryazdan et al. 2017), but combinational symbiotic therapy has been reported to exert a greater anti-inflammatory effect than probiotics alone (Fujimori et al. 2009), but the other immunomodulatory effects of prebiotics need further investigation. Although manipulating the gut microbiome via dietary nutraceuticals (probiotics, prebiotics and synbiotics) represents a relatively simple and cost-effective prospect if targeted appropriately, it's important to remember that these health-promoting benefits are short-lived, and the gut microbial profile returns to its pre-supplementation condition within weeks of supplement discontinuation. Thus, presenting a major challenge in translating the initial promising findings of these microbiome-based strategies in boosting immune health in aged hosts.

The microbiota is dependent on food for survival and fermentation of dietary products is key for regulating their functions. Another method of altering the microbiome is via modulation of dietary patterns. Caloric restriction (CR) involving a reduction in calorie intake of diet without incurring deprivation of essential nutrients, which has been shown to extend lifespan and delay the onset of age-related diseases in a variety of animal species including primates (Mattison et al. 2017) Furthermore, animal studies have shown that short-term caloric restriction can result in a shift of a dysbiotic microbiota towards a healthier state, via expansion of beneficial gut lactobacilli counts and an upregulating of SCFAs (Fabbiano et al. 2018; Tanca et al. 2018). Importantly, another exploratory study in aged mice reported highlighted the impact of CR on fecal microbiome and its relationship with influenza vaccine responses (Bartley et al. 2017). Additionally, a 4-week caloric restriction diet was reported to improve gut barrier integrity, thereby reducing both local and systemic inflammation in obese women (Ott et al. 2017). However, the beneficial effects of CR on microbiome composition and intestinal barrier integrity in older adults need to be investigated in future studies.

The Mediterranean diet (MED-diet) is a well-balanced diet that features high intake of fruits, vegetables, nuts, legumes, whole grains, fish and olive oil, and low consumption of meat, refined carbohydrates and sweets. Importantly, adherence to the MED diet has been associated with increased longevity, reduced mortality, risk of frailty and cardiovascular diseases and improved cognitive function (Ghosh et al. 2020; Estruch et al. 2018; Valls-Pedret et al. 2015). Adopting a Mediterranean diet has also been associated with a lower Firmicutes/Bacteroidetes ratio, induce expansion of *Bifidobacterium* counts, a greater abundance of polysaccharide-degrading microbes and high faecal SCFA levels (Garcia-Mantrana et al. 2018). Unfortunately, older adults across the globe are known to consume a restricted diet, a contributing factor to a low diversity gut microbiome and these studies highlight the potential of a balanced diet rich in dietary fibres in restoring microbiome homeostasis in aged hosts.

A novel approach involving the transfer of healthy intestinal microorganisms called faecal microbiome transplantation (FMT), has gained considerable attention in recent years due to its ability to restore microbiome homeostasis in patients suffering from intestinal pathologies (Benech and Sokol 2020). FMT administration in animals not only re-establishes microbial communities, expands SCFAs and SBAs and also reduces intestinal permeability (Li et al. 2015), but has also been reported to exert an immune-modulatory; increasing IL-10 production by leukocytes and decreasing T cell responses by inhibiting antigen presentation (Burrello et al. 2018) and similar anti-inflammatory and anti-immunesenescence effects have also been reported in aged *C.Difficile* infection patients (Monaghan et al. 2021). Interestingly, a study in aged mice demonstrated that FMT corrects age-associated defects in germinal centre reactions, suggesting the possibility of FMT as a microbiome-based intervention that could restore age-associated impaired immune responses (Stebegg et al. 2019). Although the current evidence points toward FMT being an effective therapy and unlike the consumption of probiotics/prebiotics, the beneficial effect of FMT on microbiome composition and microbiome metabolite profile persists up to a year post successful stool transfer. However, FMT treatment has several limitations including donor microbiome variability, risk of transferring pathobionts and unknown long-term efficacy. Future studies need to focus on creating a standardised multi-species cocktail of bacteria that specifically boosts immune responses in older adults while diminishing the side effects of FMT.

Metabolites derived from gut commensal microbiota (such as SCFA) have the potential to exert an anti-inflammatory effect and improve immune homeostasis. In recent years, there has been an expansion in the number of animal studies reporting the anti-inflammatory potential of SCFAs in treating a range of conditions; including lipopolysaccharide-induced lung injury (Ni et al. 2010), preventing lethality of sepsis (Zhang et al. 2007) and atherosclerosis (Aguilar et al. 2014). So far there have only been two studies of butyrate administration via enema in humans, one study reporting successfully ameliorates colonic inflammation in IBD (Scheppach et al. 1992) and another study in older adults undergoing upper abdominal surgery observed an increased butyrate concentration in portal vein blood (Van der Beek et al. 2015). Together these studies confirm the safety of therapeutic administration of SCFAs in humans. However, key questions regarding the dose of butyrate and the form of safe delivery in older adults remain unanswered and further investigation in future studies are required before exploiting the potential of SCFAs as an anti-immunesenescence intervention.

3.8 Conclusion Remarks

The global expansion of the ageing population has highlighted the urgent need to identify novel targets that help increase healthspan in older adults and relieve the burden of age-related chronic diseases and infections, potentially via boosting the aged immune system. We are only beginning to understand the importance of the gut microbiome in modulating immune responses and have proposed a causal relationship between microbiome dysbiosis, intestinal barrier dysfunction and inflammaging/immunesenescene based on evidence from studies in aged germ-free mice. However, this hypothesis needs to be strengthened and there is room for future well planned studies in humans to help increase the translational potential of these findings. If the hypothesis suggesting a causal link between the age-associated microbiome and immune changes with advancing age in humans is proven, this would provide evidence for benefits of microbiome-based therapies as a novel strategy for ameliorating immunesenescence. In conclusion, we propose exploitation of the modifiable nature of the microbiome and development of 'personalised' nutritional strategies that use gut microbes as agents to improve immune health and exert a positive impact on health of the ageing population.

References

Agrawal A, Agrawal S, Cao JN, Su H, Osann K, Gupta S (2007) Altered innate immune functioning of dendritic cells in elderly humans, a role of phosphoinositide 3-kinase-signaling pathway. J Immunol 178:6912–6922. https://doi.org/10.4049/jimmunol.178.11.6912

Agrawal A, Gupta S (2011) Impact of aging on dendritic cell functions in humans. Ageing Res Rev 336–345. https://doi.org/10.1016/j.arr.2010.06.004

Agrawal A, Sridharan A, Prakash S, Agrawal H (2012) Dendritic cells and ageing: consequences for autoimmunity. Expert Rev Clin Immunol 8:73–80. https://doi.org/10.1586/eci.11.77

Aguilar EC, Leonel AJ, Teixeira LG, Silva AR, Silva JF, Pelaez JM, Capettini LS, Lemos VS, Santos RA, Alvarez-Leite JI (2014) Butyrate impairs atherogenesis by reducing plaque inflammation and vulnerability and decreasing NFκB activation. Nutr Metab Cardiovasc Dis 24:606–613. https://doi.org/10.1016/j.numecd.2014.01.002

Atarashi K, Tanoue T, Ando M, Kamada N, Nagano Y, Narushima S, Suda W, Imaoka A, Setoyama H, Nagamori T, Ishikawa E, Shima T, Hara T, Kado S, Jinnohara T, Ohno H, Kondo T, Toyooka K, Watanabe E, Yokoyama S, Tokoro S, Mori H, Noguchi Y, Morita H, Ivanov II, Sugiyama T, Nuñez G, Camp JG, Hattori M, Umesaki Y, Honda K (2105) Th17 cell induction by adhesion of microbes to intestinal epithelial cells. Cells 163:367–80. https://doi.org/10.1016/j.cell.2015.08.058

Bartley JM, Zhou X, Kuchel GA, Weinstock GM, Haynes L (2017) Impact of age, caloric restriction and influenza infection on mouse gut microbiome: an exploratory study of the role of age-related microbiome changes on influenza responses. Front Immunol 8:1164. https://doi.org/10.3389/fimmu.2017.01164

Benech N, Sokol H (2020) Fecal microbiota transplantation in gastrointestinal disorders: time for precision medicine. Genome Med 12:58. https://doi.org/10.1186/s13073-020-00757-y

Biagi E, Nylund L, Candela M, Ostan R, Bucci L, Pini E, Nikkila J, Monti D, Satokari R, Franceschi C, Brigidi P, De Vos W (2010) Through ageing, and beyond: gut microbiota and inflammatory status in seniors and centenarians. PLoS One e10667

Biagi E, Franceschi C, Rampelli S, Severgnini M, Ostan R, Turroni S, Consolandi C, Quercia S, Scurti M, Monti D, Capri M, Brigidi P, Candela M (2016) Gut microbiota and extreme longevity. Curr Biol 26:1480–1485. https://doi.org/10.1016/j.cub.2016.04.016

Birchenough GMH, Nyström EEL, Johansson MEV, Hansson GC (2016) A sentinel goblet cell guards the colonic crypt by triggering Nlrp6-dependent Muc2 secretion. Science 352:1535–1542. https://doi.org/10.1126/science.aaf7419

Boullier S, Tanguy M, Kadaoui KA, Caubet C, Sansonetti P, Corthésy B, Phalipon A (2009) Secretory IgA-mediated neutralization of Shigella flexneri prevents intestinal tissue destruction by down-regulating inflammatory circuits. J Immunol 183:5879–5885. https://doi.org/10.4049/jimmunol.0901838

Bruunsgaard H, Pedersen BK (2003) Age-related inflammatory cytokines and disease. Immunol Allergy Clin North Am 23:15–39. https://doi.org/10.1016/s0889-8561(02)00056-5

Burrello C, Garavaglia F, Cribiù FM, Ercoli G, Lopez G, Troisi J, Colucci A, Guglietta S, Carloni S, Guglielmetti S, Tavernii V, Nizzoli G, Bosari S, Caprioli F, Rescigno M, Facciotti C (2018) Therapeutic faecal microbiota transplantation controls intestinal inflammation through IL10 secretion by immune cells. Nat Commun 9:5184. https://doi.org/10.1038/s41467-018-07359-8

Butcher SK, Chahal H, Nayak L, Sinclair A, Henriquez NV, Sapey E, D'O Mahony, Lord J M, (2001) Senescence in innate immune responses: reduced neutrophil phagocytotic capacity and CD16 expression in elderly humans. J Leukoc Biol 70:881–886. https://doi.org/10.1189/jlb.70.6.881

Campbell C, McKenney PT, Konstantinovsky D, Isaeva O, Schizas M, Verter J, Mai C, Jin WB, Guo CJ, Violante S, Ramos RJ, Cross JR, Kadaveru K, Hambor J, Rudensky AY (2020) Bacterial metabolism of bile acids promotes generation of peripheral regulatory T cells. Nature 581:475–479. https://doi.org/10.1038/s41586-020-2193-0

Carmona JJ, Michan S (2016) Biology of health aging and longevity. Rev Inves Clin 68:7–16

Carriche GM, Almeida L, Stüve P, Velasquez L, Dhillon-LaBrooy A, Roy U, Lindenberg M, Strowig T, Plaza-Sirvent C, Schmitz I, Lochner M, Katharina Simon A, Sparwasser T (2021) Regulating T-cell differentiation through polyamine spermidine. J Allergy Clin Immunol 147:335-348.e11. https://doi.org/10.1016/j.jaci.2020.04.037

Cecil JD, O'Brien-Simpson NM, Lenzo JC, Holden JA, Singleton W, Perez-Gonzalez A, Mansell A, Reynolds EC (2017) Outer membrane vesicles prime and activate macrophage inflammasomes and cytokine secretion in vitro and in vivo. Front Immunol 8:1017. https://doi.org/10.3389%2Ffimmu.2017.01017

Cesari M, Penninx BW, Newman AB, Kritchevsky SB, Nicklas BJ, Sutton-Tyrrell K, Tracy RP, Rubin SM, Harris TB, Pahor M (2003) Inflammatory markers and cardiovascular disease (The health, aging and body composition [Health ABC] study). Am J Cardiol 92:522–528. https://doi.org/10.1161/01.CIR.0000097109.90783.FC

Claesson ML, Jeffery IB, Conde S, Power SE, O'Connor EM, Cusack S, Harris HB, Coakley M, Lakshminarayanan B, O'Sullivan O, Fitzgerald GF, Deane J, O'Connor M, Harnedy N, O'Connor K, O'Mahony D, Sinderen D, Wallace M, Brennan L, Stanton C, Marchesi JR, Fitzgerald AP, Shanahan F, Hill C, Ross PR, O'Toole PW (2012) Gut microbiota composition correlates with diet and health in the elderly. Nature 488:178–184. https://doi.org/10.1038/nature11319

Chambers ES, Preston T, Frost G, Morrison DJ (2018) Role of gut-microbiota generated short chain fatty acids in metabolic and cardiovascular health. Curr Nutr Rep 7:198–206. https://doi.org/10.1007%2Fs13668-018-0248-8

Chambers ES, Preston T, Frost G, Morrison DJ (2018) Role of gut-microbiota generated short chain fatty acids in metabolic and cardiovascular health. Curr Nutr Rep 7:198–206. https://doi.org/10.1007%2Fs13668-018-0248-8

Cheng W, Lu J, Lin W, Wei X, Li H, Zhao X, Jiang A, Yuan J (2018) Effects of a galacto-oligosaccharide-rich diet on fecal microbiota and metabolite profiles in mice. Food Funct 9:1612–1620. https://doi.org/10.1039/C7FO01720K

Chu H, Mazmanian SK (2013) Innate immune recognition of the microbiota promotes host-microbial symbiosis. Nat Immunol 14:668–675. https://doi.org/10.1038/ni.2635

Churov AV, Mamashov KY, Novitskaia AV (2020) Homeostasis and the functional role of CD4+ Treg cells in aging. Immunol Letters 26:83–89. https://doi.org/10.1016/j.imlet.2020.07.004

Cunha LL, Perazzio SF, Azzi J, Cravedi P, Riella LV (2020) Remodelling of the immune response with aging: immunosenescence and its potential impact on COVID-19 immune response. Front Immunol 11:01748. https://doi.org/10.3389/fimmu.2020.01748

De Mol J, Kuiper J, Tsiantoulas D, Foks AC (2021) The dynamics of B cell aging in health and disease. Frontiers of Immunol 12:733566. https://doi.org/10.3389/fimmu.2021.733566
Dorshkind K, Montecino-Rodriguez E, Signer RA (2009) The ageing immune system: is it ever too old to become young again? Nat Rev Immunol 9:57–62. https://doi.org/10.1038/nri2471
Duggal NA, Upton J, Phillips AC, Sapey E, Lord JM (2013) An age-related numerical and functional deficit in $CD19^{+}CD24^{hi}CD38^{hi}$ B cells is associated with an increase in systemic autoimmunity. Ageing Cell 12:873–881. https://doi.org/10.1111/acel.12114
Duggal NA (2018) Reversing the immune aging clock: lifestyle modifications and pharmacological interventions. Biogerontol 19:481–496. https://doi.org/10.1007/s10522-018-9771-7
Eaton SM, Burns EM, Kusser K, Randall TD, Haynes L (2004) Age-related defects in CD4 T cell cognate helper function lead to reductions in humoral responses. J Exp Med 200:1613–1622. https://doi.org/10.1084/jem.20041395
Estruch R, Ros E, Salas-Salvadó J, Covas M, Corella D, Arós F, Gómez-Gracia E, Ruiz-Gutiérrez V, Fiol M, Lapetra J, Lamuela-Raventos RM, Serra-Majem L, Pintó X, Basora J, Muñoz MA, Zorlí JV, Martínez JA, Fitó M, Gea A, Hernán MA, Martínez-González MA (2018) PREDIMED Study Investigators. Primary prevention of cardiovascular disease with a Mediterranean diet supplemented with extra-virgin olive oil or nuts. N Engl J Med 378:e34. https://doi.org/10.1056/nejmoa1800389
Fabbiano S, Suárez-Zamorano N, Chevalier C, Lazarevíc V, Kieser S, Rigo D, Leo S, Veyrat-Durebex C, Gaïa N, Maresca M, Merkler D, de Agüero MG, Macpherson A, Schrenzel J, Trajkovski M (2018) Functional gut microbiota remodeling contributes to the caloric restriction-induced metabolic improvements. Cell Metab 28:907–21.e7. https://doi.org/10.1016/j.cmet.2018.08.005
Forbes SJ, Eschmann M, Mantis NJ (2008) Inhibition of Salmonella enterica serovar typhimurium motility and entry into epithelial cells by a protective antilipopolysaccharide monoclonal immunoglobulin A antibody. Infect Immun 76:4137–4144. https://doi.org/10.1128/iai.00416-08
Fransen F, van Beek AA, Borghuis T, Aidy SE, Hugenholtz F, van der Gaast-de Jongh C, Savelkoul HFJ, De Longe MI, Boekschoten MV, Smidt H , Faas MM , de Vos P (2017) Aged gut microbiota contributes to systemical Inflammaging after transfer to germ-free mice. Front Immunol 8:1385. https://doi.org/10.3389/fimmu.2017.01385
Franceschi C, Bonafe M, Valensin S, Olivieri F, Luca M, Ottaviani E, De Benedictis G (2000) Inflamm-aging. An evolutionary perspective on immunosenescence. Ann NY Acad Sci 908:244–254. https://doi.org/10.1111/j.1749-6632.2000.tb06651.x
Franceschi C, Campisi J (2014) Chronic inflammation (inflammaging) and its potential contribution to age-associated diseases. J Gerontol 69:4–9. https://doi.org/10.1093/gerona/glu057
Frasca D, Blomberg BB (2020) Aging induces B cell defects and decreased antibody responses to influenza infection and vaccination. Immun Ageing 17:37. https://doi.org/10.1186/s12979-020-00210-z
Fujimori S, Gudis K, Mitsui K, Seo T, Yonezawa M, Tanaka S, Tatsuguchi A, Sakamoto C (2009) A randomised controlled trial on efficacy of symbiotic versus probiotic or prebiotic treatment to improve the quality of life in patients with ulcerative colitis. Nutrition 25:520–525. https://doi.org/10.1016/j.nut.2008.11.017
Furusawa Y, Obata Y, Fukada S, Endo TA, Nakato G, Takahashi D, Nakanishi Y, Uetake C, Kato K, Kato T, Takahashi M, Fukuda NN, Murakami S, Miyauchi E, Hino S, Atarashi K, Onawa S, Fujimura Y, Lockett T, Clarke JM, Topping DL, Tomita M, Hori S, Ohara O, Morita T, Koseki H, Kikuchi J, Honda K, Hase K, Ohno H (2013) Commensal microbe-derived butyrate induces differentiation of colonic regulator T cells. Nature 504:446–450. https://doi.org/10.1038/nature12721
Garff-Tavernier L, Beziat V, Decocq J, Siguret V, Gandjbakhch F, Pautas E, Debre P, Merle-Beral H, Vieillard V (2010) Human NK cells display major phenotypic and functional changes over the lifespan. Ageing Cell 9:527–535. https://doi.org/10.1111/j.1474-9726.2010.00584.x

Garcia-Mantrana I, Selma-Royo M, Alcantara C, Collado MC (2018) Shifts on gut microbiota associated to Mediterranean diet adherence and specific dietary intakes on general adult population. Front Microbiol 9:890. https://doi.org/10.3389/fmicb.2018.00890

Ghosh TS, Rampelli S, Jeffery IB, Santoro A, Neto M, Capri M, Giampieri E, Jennings A, Candela M, Turroni S, Zoetendal EG, Hermes GDA, Elodie C, Meunier N, Brugere CM, Pujos-Guillot E, Berendsen AM, De Groot LCP, Feskins EJM, Kaluza J, Pietruszka B, Bielak MJ, Comte B, Maijo-Ferre M, Nicoletti C, De Vos WM, Fairweather-Tait S, Cassidy A, Brigidi P, Franceschi C, O'Toole PW (2020) Mediterranean diet intervention alters the gut microbiome in older people reducing frailty and improving health status: the NU-AGE 1-year dietary intervention across five European countries. Gut 69:1218–1228. https://doi.org/10.1136/gutjnl-2019-319654

Gill HS, Rutherfurd KJ, Cross ML (2001) Dietary probiotic supplementation enhances natural killer cells activity in the elderly: an investigation of age-related immunological changes. J Clin Immunol 21:264–271. https://doi.org/10.1023/a:1010979225018

Golden JM, Escobar OH, Nguyen MVL, Mallicote MU, Kavarian P, Frey MR, Gayer CP (2018) Ursodeoxycholic acid protects against intestinal barrier breakdown by promoting enterocyte migration via EGFR- and COX-2 dependent mechanisms. Am J Physiol Gastrointest Liver Physiol 315:G259–G271. https://doi.org/10.1152/ajpgi.00354.2017

Goldstein DR (2010) Aging, imbalanced inflammation and viral infection. Virulence 1:295–298. https://doi.org/10.4161/viru.1.4.12009

Goronzy JJ, Weyand CM (2019) Mechanisms underlying T cell ageing. Nat Rev Immunol 19:573–583. https://doi.org/10.1038/s41577-019-0180-1

Goulet, O (2015) Potential role of the intestinal microbiota in programming health and disease. Nutr Rev 73:32–40. https://doi.org/10.1093/nutrit/nuv039

Grayson MH, Camarda LE, Hussain SA, Zemple SA, Hayward M, Lam V, Hunter DA, Santoro JL, Rohlfing M, Cheung DS, Salzman NH (2018) Intestinal microbiota disruption reduces regulatory T cells and increases respiratory viral infection mortality though increased IFNγ production. Front Immunol 9:1587. https://doi.org/10.3389/fimmu.2018.01587

Guerra SS, Berbiche D, Vasiliadis HM (2019) Measuring multimorbidity in older adults: comparing different data sources. BMC Geriatr 19:166. https://doi.org/10.1186/s12877-019-1173-4

Guo C, Xie S, Chi Z, Zhang J, Liu Y, Zhang L, Zheng M, Zhang X, Xia D, Ke Y, Lu L, Wang D (2016) Bile acids control inflammation and metabolic disorder through inhibition of NLRP3 inflammasome. Immunity 45:802–816. https://doi.org/10.1016/j.immuni.2016.09.008

Guzior DV, Quinn RA (2021) Review: microbial transformation of human bile acids. Microbiome 9:140. https://doi.org/10.1186/s40168-021-01101-1

Haines CJ, Giffon TD, Lu LS, Lu X, Tessier-Lavigne M, Ross DT, Lewis DB (2009) Human CD4 T cell recent thymic emigrants are identified by protein tyrosine kinase 7 and have reduced immune function. J Exp Med 206:275–285. https://doi.org/10.1084/jem.20080996

Hazeldine J, Lord JM (2013) The impact of ageing on natural killer cell function and potential consequences for health in older adults. Ageing Res Rev 12:1069–1078. https://doi.org/10.1016/j.arr.2013.04.003

Hazeldine J, Hampson P, Lord JM (2012) Reduced release and binding of perforin at the immunological synapse underlies the age-related decline in natural killer cell cytotoxicity. Aging Cell 11:751–759. https://doi.org/10.1111/j.1474-9726.2012.00839.x

Hazeldine J, Harris P, Chapple IL, Grant M, Greenwood H, Livesey A, Sapey E, Lord JM (2014) Impaired neutrophil extracellular trap formation: a novel defect in the innate immune system of aged individuals. Ageing Cell 13:690–698. https://doi.org/10.1111/acel.12222

Hoentjen F, Welling GW, Harmsen JM, Zhang X, Snart J, Tannock GW, Lien K, Churchill TA, Lupicki M, Dieleman LA (2005) Reduction of colitis by prebiotics in HLA-B27 transgenic rats is associated with microflora changes and immunomodulation. Inflamm Bowel Dis 11:977–985. https://doi.org/10.1097/01.MIB.0000183421.02316.d5

Hu J, Wang C, Huang X, Yi S, Pan S, Zhang Y, Yuan G, Cao Q, Ye X, Li H (2021) Gut microbiota-mediated secondary bile acids regulate dendritic cells to attenuate autoimmune uveitis through TGR5 signaling. Cell Rep 36:109726

Huang C, Du W, Ni Y, Lan G, Shi G (2022) The effect of short-chain fatty acids on M2 macrophages polarization in vitro and in vivo. Clin Exp Immunol 207:53–64. https://doi.org/10.1093/cei/uxab028

Ichikawa R, Takayama T, Yoneno K, Kamada N, Kitazume MT, Higuchi H, Matsuoka K, Watanabe M, Itoh H, Kanai T, Hisamatsu T, Hibi T (2012) Bile acids induce monocyte differentiation toward interleukin-12 hypo-producing dendritic cells via a TGR5-dependent pathway. Immunology 136:153–162. https://doi.org/10.1111/j.1365-2567.2012.03554.x

Ivanov II, Frutos Rde L, Manel N, Yoshinagga K, Rifkin DB, Sartor RB, Finlay BB, Littman DR (2008) Specific microbiota direct the differentiation of IL-17-producing T-helper cells in the mucosa of the small intestine. Cell Host Microbe 4:337–349. https://doi.org/10.1016/j.chom.2008.09.009

Jackaman C, Radley-Crabb HG, Soffe Z, Shavlakadze T, Grounds MD, Nelson DJ (2013) Targeting macrophages rescues age-related immune deficiencies in C57BL/6J geriatric mice. Aging Cell 12:345–357. https://doi.org/10.1111/acel.12062

Kell DB, Pretorious E (2015) On the translocation of bacteria and their lipopolysaccharides between blood and peripheral locations in chronic, inflammatory diseases: the central roles of LPS and LPS-induced cell death. Integrative Biol 7:1339–1377. https://doi.org/10.1039/C5IB00158G

Kim M, Qie Y, Park J, Kim CH (2016) Gut microbial metabolites fuel host antibody responses. Cell Host Microbe 20:202–214. https://doi.org/10.1016/j.chom.2016.07.001

Krishnan S, Ding Y, Saeidi N, Choi M, Sridharan GV, Sherr DH, Yarmush ML, Alaniz RC, Jayaraman A, Lee K (2018) Gut-microbiota derived tryptophan metabolites modulates inflammatory response in hepatocytes and macrophages. Cell Rep 23:1099–1111. https://doi.org/10.1016/j.celrep.2019.08.080

Lajczak-McGinley NK, Porru E, Fallon CM, Smyth J, Curley C, McCarron PA, Tambuwala MM, Roda A, Keely SJ (2020) The secondary bile acids, ursodeoxycholic acid and lithocholic acid, protect against intestinal inflammation by inhibition of epithelial apoptosis. Physiol Rep 8:e14456. https://doi.org/10.14814/phy2.14456

Lécuyer E, Rakotobe S, Lengliné-Garnier H, Lebreton C, Picard M, Juste C, Fritzen R, Eberl G, McCoy KD, Macpherson AJ, Reynaud C, Cerf-Bensussan N, Gaboriau-Routhiau V (2014) Segmented filamentous bacterium uses secondary and tertiary lymphoid tissues to induce gut IgA and specific T helper 17 cell responses. Immunity 40:608–620. https://doi.org/10.1016/j.immuni.2014.03.009

Li D, Ke Y, Zhan R, Liu C, Zhao M, Zeng A, Shi X, Ji L, Cheng S, Pan B, Zheng L, Hong H (2018) Trimethylamine-N-oxide promotes brain aging and cognitive impairment in mice. Ageing Cell 17:e12768. https://doi.org/10.1111/acel.12768

Li M, Liang P, Li Z, Wang Y, Zhang G, Gao H, Wen S, Tang L (2015) Fecal microbiota transplantation and bacterial consortium transplantation have comparable effects on the re-establishment of mucosal barrier function in mice with intestinal dysbiosis. Front Microbiol 6:692. https://doi.org/10.3389/fmicb.2015.00692

Lloyd-Price J, Abu-Ali G, Huttenhower C (2016) The healthy human microbiome. Genome Med 8:1–11. https://doi.org/10.1186/s13073-016-0307-y

Lynch HE, Goldberg GL, Chidgey A, Van den Brink MR, Boyd R, Sempowski GD (2009) Thymic involution and immune reconstitution. Trends Immunol. 30:366–373. https://doi.org/10.1016%2Fj.it.2009.04.003

Maneerat S, Lehtinen MJ, Childs CE, Forssten SD, Alhoniemi E, Tiphaine M, Yaqoob P, Ouwehand AC, Rastall RA (2014) Consumption of Bifidobacterium lactis Bi-07 by healthy elderly adults enhances phagocytic activity of monocytes and granulocytes. J Nutr Sci 2:44. https://doi.org/10.1017/jns.2013.31

Mariat D, Firmesse O, Levenez F, Guimaraes V, Sokol H, Dore J, Corthier G, Furet JP (2009) The Firmicutes/Bacteroidetes ratio of the human microbiota changes with age. BMC Microbiol 9:123. https://doi.org/10.1186/1471-2180-9-123

Mattison JA, Colman RJ, Beasley TM, Allison DB, Kemnitz JW, Roth GS, Ingram DK, Weindruch R, de Cabo R, Anderson RM (2017) Caloric restriction improves health and survival of rhesus monkeys. Nat Commun 8:14063. https://doi.org/10.1038/ncomms14063

Maeyer RPH, Chambers ES (2021) The impact of ageing on monocytes and macrophages. Immunol Lett 230:1–10. https://doi.org/10.1016/j.imlet.2020.12.003

Mohajeri MH, Brummer RJM, Rastall RA, Weersma RK, Harmsen HJM, Faas M, Eggersdorfer M (2018) The role of the microbiome for human health: from basic science to clinical applications. Eur J Nutr 57:1–14. https://doi.org/10.1007%2Fs00394-018-1703-4

Molloy MJ, Bouladoux N, Belkaid Y (2012) Intestinal microbiota: shaping local and systemic immune responses. Seminal Imunol 24:58–66. https://doi.org/10.1016/j.smim.2011.11.008

Monaghan TM, Duggal NA, Rosati E, Griffin R, Hughes J, Roach B, Yang YD, Wang C, Wong K, Saxinger L, Pucic-Bakovic M, Vuckovic F, Klicek F, Lauc G, Tighe P, Mullish BH, Blanco JM, McDonald JAK, Marchesi JR, Xue N et al (2021) A multi-factorial observational study on sequential fecal microbiota transplant in patients with medically refractory *Clostridiodes difficile* infection. Cells 10:3234. https://doi.org/10.3390%2Fcells10113234

Mu Q, Kirby J, Reilly CM, Luo XM (2017) Leaky gut as a danger signal for autoimmune diseases. Front Immunol 8:598. https://doi.org/10.3389/fimmu.2017.00598

Nastasi C, Fredholm S, Willerslev-Olsen A, Hansen M, Menné Bonefeld C, Geisler C, Hald Andersen M, Ødum N, Woetmann A (2017) Butyrate and propionate inhibit antigen-specific $CD8^+$ T cell activation by suppressing IL-12 production by antigen-presenting cells. Sci Rep 7:14516. https://doi.org/10.1038/s41598-017-15099-w

Ni YF, Wang J, Yan XL, Tian F, Zhao JB, Wang YJ, Jiang T (2010) Histone deacetylase inhibitor, butyrate, attenuates lipopolysaccharide-induced acute lung injury in mice. Respir Res 11:33. https://doi.org/10.1186/1465-9921-11-33

Ong S, Hadadi E, Dang T, Yeap W, Tan CT, Ng T, Larbi A, Wong S (2018) The pro-inflammatory phenotype of the human non-classical monocyte subset is attributed to senescence. Cell Death Dis 9:266. https://doi.org/10.1038/s41419-018-0327-1

Ott B, Skurk T, Hastreiter L, Lagkouvardos I, Fischer S, Büttner J, Kellerer T, Clavel T, Rychlik M, Haller D, Hauner H (2017) Effect of caloric restriction on gut permeability, inflammation markers, and fecal microbiota in obese women. Sci Rep 7:11955. https://doi.org/10.1038/s41598-017-12109-9

Ouyang X, Yang Z, Zhang R, Arnaboldi P, Lu G, Li Q, Wang W, Zhang B, Cui M, Zhang H, Liang-Chen J, Qin L, Zheng F, Huang B, Xiong H (2011) Potentiation of Th17 cytokines in aging process contributes to the development of colitis. Cell Immunol 266:208–217. https://doi.org/10.1016/j.cellimm.2010.10.007

Pallikuth S, Russell KB, Sirupangi T, Mendez R, Kvistad DK, Pahwa R, Pahwa S (2021) Age associated microbiome modulation and its association with systemic inflammation in a rhesus macaque model. Front Immunol 12. https://doi.org/10.1101/813667

Pan Q, Li YQ, Guo K, Xue M, Gan Y, Wang K, Xu DB, Tu QY (2021) Elderly patients with mild cognitive impairment exhibit altered gut microbiota profiles. J Immunol Res 5578958 https://doi.org/10.1155/2021/5578958

Panda A, Arjona A, Sapey E, Bai F, Fikrig E, Montgomery RR, Lord JM, Shaw AC (2010) Human innate immunesenescence: causes and consequences for immunity in old age. https://doi.org/10.1016/j.it.2009.05.004

Peterson MJ, Thompson DK, Pieper CF, Morey MC, Kraus VB, Kraus WE, Sullivan P, Fillenbaum G, Cohen HJ (2016) A novel analytic technique to measure associations between circulating biomarkers and physical performance across the adult life span. J Gerontol Ser A Biol Sci Med Sci 71:196–202. https://doi.org/10.1093/gerona/glv007

Picca A, Ponziani FR, Calvani R, Marini F, Biancolillo A, Coelho-Júnior HJ, Gervasoni J, Primiano A, Putignani L, Del Chierico F, Reddel S, Gasbarrini A, Landi F, Bernabei R, Marzetti E (2020) Gut microbial, inflammatory and metabolic signatures in older people with

physical frailty and sarcopenia: Results from the BIOSPHERE study. Nutrients 12:65. https://doi.org/10.3390%2Fnu12010065
Prelog M (2006) Aging of the immune system: a risk factor for autoimmunity? Autoimmun Rev 5:136–139. https://doi.org/10.1016/j.autrev.2005.09.008
Rampelli S, Candela M, Turroni S, Biagi E, Collino S, Franceschi C, O'Toole PW, Brigidi P (2013) Functional metagenomic profiling of intestinal microbiom e in extreme ageing. Aging 5:902–12. https://doi.org/10.18632/aging.100623
Riley RL, Khomtchouk K, Blomberg BB (2017) Age-associated B cells (ABC) Inhibit B Lymphopoiesis and alter antibody repertoires in old age. Cell Immunol 321:61–67. https://doi.org/10.1016/j.cellimm.2017.04.008
Roager HM, Licht TR (2018) Microbial tryptophan catabolites in health and disease. Nat Commun 9:3294. https://doi.org/10.1038%2Fs41467-018-05470-4
Rosser EC, Piper CJM, Matei DE, Blair PA, Rendeiro AF, Orford M, Alber DG, Krausgruber T, Catalan D, Klein N, Manson JJ, Drozdov I, Bock C, Wedderburn LR, Eaton S, Mauri C (2020) Microbiota-derived metabolites suppress arthritis by amplifying aryl-hydrocarbon receptor activation in regulatory B cells. Cell Metab 31:837–851. https://doi.org/10.1016/j.cmet.2020.03.003
Sapey E, Greenwood H, Walton G, Mann E, Love A, Aaronson N, Insall RH, Stockley RA, Lord JM (2014) Phosphoinositide 3-kinase inhibition restores neutrophil accuracy in the elderly: toward targeted treatments for immunosenescence. Blood 123:239–248. https://doi.org/10.1182/blood-2013-08-519520
Salazar N, Arboleya S, Fernández-Navarro T, de los Reyes-Gavilán CG, Gonzalez S, Gueimonde M (2019) Age-associated changes in gut microbiota and dietary components related with the immune system in adulthood and old age: a cross-sectional study. Nutrients 11:1765. https://doi.org/10.3390/nu11081765
Sayed N, Huang Y, Nguyen K, Krejciova-Rajaniemi Z, Grawe A P, Gao T, Tibshirani R, Hastie T, Alpert A, Cui L, Kuznetsova T, Rosenberg-Hasson Y, Oston R, Monti D, Lehallier B, Shen-Orr S, Maecker HT, Dekker CL, Wyss-Coray T, Franceschi C, Jojic V, Haddad F, Montoya JG, Wu CJ , Davis MM, Furman D (2021) An inflammatory aging clock (iAge) based on deep learning tracks multimorbidity, immunosenescence , frailty and cardiovascular aging. Nat Aging 1:598–615. https://doi.org/10.1038%2Fs43587-021-00082-yorg/
Shen Y, Giardino Torchia ML, Lawson GW, Karp CL, Ashwell JD, Mazmanian SK (2012) Outer membrane vesicles of a human commensal mediate immune regulation and disease protection. Cell Host Microbe 12:509–520. https://doi.org/10.1016/j.chom.2012.08.004
Scheppach W, Sommer H, Kirchner T, Paganelli GM, Batram P, Christl S, Richter F, Dusel G, Kasper H (1992) Effect of butyrate enema on the colonic mucosa in distal ulcerative colitis. Gastroenterology 103:51–56. https://doi.org/10.1016/0016-5085(92)91094-k
Schetters STT, Jong WSP, Horrevorts SK, Kruijssen LJW, Engels S, Stolk D, Daleke-Schermerhorn MH, Garcia-Vallejo J, Houben D, Unger WWJ, den Haan JMM, Luirink J, van Kooyk Y (2109) Outer membrane vesicles engineered to express membrane-bound antigen program dendritic cells for cross-presentation to CD8(+) T cells. Acta Biomater 91:248–57. https://doi.org/10.1016/j.actbio.2019.04.033
Schulthess J, Pandey S, Capitani M, Rue-Albrecht KC, Arnold I, Franchini F, Chomka A, Ilott NE, Johnston DGW, Pires E, McCullagh J, Sansom SN, Arancibia-Cárcamo CV, Uhlig HH, Powrie F (2019) The short chain fatty acid butyrate imprints an antimicrobial program in macrophages. Immunity 50:432–445. https://doi.org/10.1016/j.immuni.2018.12.018
Shokryazdan P, Jahromi MF, Navidshad B, Liang JB (2017) Effects of prebiotics on immune system and cytokine expression. Med Microb Immuno 206:1–9. https://doi.org/10.1007/s00430-016-0481-y
Sina C, Gavrilova O, Förster M, Till A, Derer S, Hildebrand F, Raabe B, Chalaris A, Scheller J, Rehmann A, Franke A, Ott S, Häsler R, Nikolaus S, Fölsch UR, Rose-John S, Jiang H, Li J, Schreiber S, Rosenstiel P (2009) G protein-coupled receptor 43 is essential for neutrophil recruitment during intestinal inflammation. J Immunol 183:7514–7522. https://doi.org/10.4049/jimmunol.0900063

Sinha SR, Haileselassie Y, Nguyen LP, Tropini C, Wang M, Becker LS, Sim D, Jarr K, Spear ET, Singh G, Namkoong H, Bittinger K, Fischbach MA, Sonnenburg JL, Habtezion A (2020) Dysbiosis-induced secondary bile acid deficiency promotes intestinal inflammation. Cell Host Microbe 27:659–670. https://doi.org/10.1016/j.chom.2020.01.021

Stebegg M, Silva-Cayetano A, Innocentin S, Jenkins TP, Cantacessi C, Gilbert C, Linterman MA (2019) Heterochronic faecal transplantation boosts gut germinal centres in aged mice. Nat Commun 10:2443. https://doi.org/10.1038/s41467-019-10430-7

Tanca A, Abbondio M, Palomba A, Fraumene C, Marongiu F, Serra M, Pagnozzi D, Laconi E, Uzzau S (2018) Caloric restriction promotes functional changes involving short-chain fatty acid biosynthesis in the rat gut microbiota. Sci Rep 8:14778. https://doi.org/10.1038/s41598-018-33100-y

Thevaranjan N, Puchta A, Schulz C, Naidoo A, Szamosi JC, Verschnoor CP, Loukov D, Schenck LP, Jury J, Foley KP, Schertzer JD, Larche MJ, Davidson DJ, Verdu EF, Surette MG, Bowdish DME (2017) Age-associated microbial dysbiosis promotes intestinal permeability systemic inflammation, and macrophage dysfunction. Cell Host Microbe 21:455–466. https://doi.org/10.1016/j.chom.2017.03.002

Tiihonen K, Ouwehand AC, Rautonen N (2010) Human intestinal microbiota and healthy ageing. Ageing Res Rev 9:107–116. https://doi.org/10.1016/j.arr.2009.10.004

Thomas-Crusells J, McElhaney JE, Aguado MT (2012) Report of the ad-hoc consultation on aging and immunization for a future WHO research agenda on life-course immunization. Vaccine 30:6007–6012. https://doi.org/10.1016/j.vaccine.2012.07.025

Tran L, Greenwood-van Meerland B (2013) Age-associated remodeling of intestinal epithelial barrier. J Gerontol Biol Sci Med 68:1045–1056. https://doi.org/10.1093/gerona/glt106

Usami M, Kishimoto K, Ohata A, Miyoshi M, Aoyama M, Fueda Y (2008) Kotani J (2008) Butyrate and trichostatin A attenuate nuclear factor kappaB activation and tumor necrosis factor alpha secretion and increase prostaglandin E2 secretion in human peripheral blood mononuclear cells. Nutr Res 28:321–328. https://doi.org/10.1016/j.nutres.2008.02.012

Valls-Pedret C, Sala-Vila A, Serra-Mir M, Corella D, de la Torre R, Martínez-González MÁ, Martínez-Lapiscina EH, Fitó M, Pérez-Heras A, Salas-Salvadó J, Estruch R, Ros E (2015) Mediterranean diet and age-related cognitive decline: a randomized clinical trial. JAMA Intern Med 175:1094–1103. https://doi.org/10.1001/jamainternmed.2015.1668

Van der Beek CM, Bloemen JG, Van den Broek MA, Lenaerts K, Venema K, Buurman WA, Dejong CH (2015) Hepatic uptake of rectally administered butyrate prevents an increase in systemic butyrate concentrations in humans. J Nutr 145:2019–2024. https://doi.org/10.3945/jn.115.211193

Van Epps P, Oswald D, Higgins PA, Hornick TR, Aung H, Banks RE, Wilson BM, Burant C, Gravenstein S, Canaday DH (2016) Frailty has a strong association with inflammation than age in older veterans. Immun Ageing 13:27. https://doi.org/10.1186/s12979-016-0082-z

Visser M, Pahor M, Taaffe DR, Goodpaster BH, Simonsick EM, Newmann AB, Nevitt M, Harris TB (2002) Relationship of interleukin-6 and tumor necrosis factor-alpha with muscle mass and muscle strength in elderly men and women: the Health ABC Study. J Gerontol A Biol Sci Med Sci 57:326–332. https://doi.org/10.1093/gerona/57.5.M326

Wang Z, Klipfell E, Bennett BJ, Koeth R, Levison BS, DuGar B, Feldstein AE, Britt EB, Fu X, Chung Y, Wu Y, Schauer P, Smith JD, Allayee H, Tang WHW, DiDonato JA, Lusis AJ, Hazen SL (2011) Gut flora metabolism of phosphatidylcholine promotes cardiovascular disease. Nature 472:57–63. https://doi.org/10.1038/nature09922

Wang L, Zhu Q, Lu A, Liu X, Zhang L, Xu C, Liu X, Li H, Yang T (2017) Sodium butyrate suppresses angiotensin II-induced hypertension by inhibition of renal (pro)renin receptor and intrarenal renin-angiotensin system. J Hypertens 35:1899–1908. https://doi.org/10.1097/hjh.0000000000001378

Wilmanski T, Diener C, Rappaport N, Patwardhan S, Wiedrick J, Lapidus J, Earls JC, Zimmer A, Glusman G, Robinson M, Yurkovich JT, Kado DM, Cauley JA, Zmuda J, Lane NE, Magis AT, Lovejoy JC, Hood L, Gibbons SM, Orwoll ES, Price ND (2021) Gut microbiome pattern reflects healthy ageing and predicts survival in humans. Nat Metab 3:274–286. https://doi.org/10.1038/s42255-021-00348-0

Wammers M, Schupp A, Bode JG, Ehlting C, Wolf S, Deenen R, Köhrer K, Häussinger D (2018) Graf D (2018) Reprogramming of pro-inflammatory human macrophages to an anti-inflammatory phenotype by bile acids. Sci Rep 8:255. https://doi.org/10.1038/s41598-017-18305-x

Wang L, Gong Z, Zhang X, Zhu F, Liu Y, Jin C, Du X, Xu C, Chen Y, Cai W, Tian C, Wu J (2020) Gut microbial bile acid metabolite skews macrophage polarization and contributes to high-fat diet-induced colonic inflammation. Gut Microbes 12:e1819155. https://doi.org/10.1080/19490976.2020.1819155

Wilms E, Jonkers DMAE, Savelkoul HFJ, Elizalde M, Tischmann L, de Vos P, Masclee AAM, Troost FJ (2019) The impact of pectin supplementation on intestinal barrier function in healthy young adults and healthy elderly. Nutrients 11:1554. https://doi.org/10.3390/nu11071554

Wilson QN, Wells M, Davis AT, Sherrill C, Tsilimigras M CB, Jones RB, Fodor AA, Kavanagh K (2018) Greater microbial translocation and vulnerability to metabolic disease in healthy aged female monkeys. 8:11373.https://doi.org/10.1038/s41598-018-29473-9

Wlodarska M, Thaiss CA, Nowarski R, Henao-Mejia J, Zhang J, Brown EM, Frankel G, Levy M, Katz MN, Philbrick WM, Elinav E, Finlay BB, Flavell RA (2014) NLRP6 inflammasome orchestrates the colonic host-microbial interface by regulating goblet cell mucus secretion. Cell 156:1045–1059. https://doi.org/10.1016/j.cell.2014.01.026

Wu K, Yuan Y, Yu H, Dai X, Wang S, Sun Z, Wang F, Fei H, Lin Q, Jiang H, Chen T (2020) The gut microbial metabolite trimethylamine N-oxide aggravates GVHD by inducing M1 macrophage polarization in mice. Blood 136:501–515. https://doi.org/10.1182/blood.2019003990

Yatsunenko T, Rey FE, Manary MJ, Trehan I, Dominguez-Bello MG, Contreras M, Magris M, Hidalgo G, Baldassano RN, Anokhin AP, Heath AC, Warner B, Reeder J, Kuczynski J, Gregory JC, Lozupone CA, Clemente JC, Knights D, Knight R, Gordon JI (2012) Human gut microbiome viewed across age and geography. Nature 486:222–227. https://doi.org/10.1038/nature11053

Ye J, Qiu J, Bostick JW, Ueda A, Schjerven H, Li S, Jobin C, Chen ZE, Zhou L (2017) Aryl hydrocarbon receptor preferentially marks and promotes gut regulatory T cells. Cell Rep 21:2277–2290. https://doi.org/10.1016/j.celrep.2017.10.114

Zhan X, Stamova B, Jin LW, DeCarli C, Phinney B, Sharp FR (2016) Gram-negative bacterial molecules associate with Alzheimer disease pathology. Neurology 87:2324–2332. https://doi.org/10.1212/WNL.0000000000003391

Zheng D, Liwinski T, Elinav E (2020a) Interaction between microbiota and immunity in health

Zheng F, Huang Y, Guo Y, Chen X, Xiao L, Deng G (2020b) Association of inflammatory markers with the severity of COVID-19: a meta-analysis. Int J Infec Dis 96:467–474. https://doi.org/10.1016/j.ijid.2020.05.055

Zhang H, Alsaleh G, Feltham J, Sun Y, Napolitano G, Riffelmacher T, Charles P, Frau L, Hubiltz P, Yu Z, Mohammed S, Ballabio A, Balabanov S, Mellor J, Simon AK (2019) Polyamines control elF5A Hypusination, TFEB translation and autophagy to reverse B cell senescence. Mol Cell 76:110–125. https://doi.org/10.1016/j.molcel.2019.08.005

Zhang LT, Yao YM, Lu JQ, Yan XJ, Yu Y, Sheng ZY (2007) Sodium butyrate prevents lethality of severe sepsis in rats. Shock 27:672–677. https://doi.org/10.1097/SHK.0b013e31802e3f4c

Zhang M, Zhou Q, Dorfman RG, Huang X, Fan T, Zhang H, Zhang J, Yu C (2016) Butyrate inhibits interleukin-17 and generates Tregs to ameliorate colorectal colitis in rats. BMC Gastroenterol 16:84. https://doi.org/10.1186/s12876-016-0500-x

Chapter 4
Gut Microbiome and Virome: Anti-Aging Interventions

Birbal Singh, Gorakh Mal, Dixit Sharma, Ahmad Al Nakshabandi, and Francesco Marotta

Abstract Aging in biological systems is an inevitable, multifactorial and irreversible process impelled by anomalous cellular and molecular biological processes. In addition to gradual loss of functioning of multiple organs, immunosenescence, and low-grade chronic inflammation, the gut microbiota also change in aging persons. Ensuing microbial dysbiosis leads to inflammatory responses, bowel diseases, and conjointly hasten the age-related health problems. Human gut virome affects host health directly or via autoimmunity-associated gut bacteria. Resurrection of gut microbiome through microbial, virome and dietary interventions may improve immune homeostasis, prevent oxidative stress and slow the progression of aging.

Keywords Gut microbiome · Virome · Aging · Detoxification · Oxidative stress · Novel therapies

B. Singh (✉) · G. Mal
ICAR-Indian Veterinary Research Institute, Regional Station, Palampur 176 061, India
e-mail: birbal.singh@icar.gov.in; bsbpalampur@yahoo.co.in

D. Sharma
Centre for Computational Biology and Bioinformatics, School of Life Sciences, Central University of Himachal Pradesh, Shahpur Campus 176026, India

A. Al Nakshabandi
Gastroenterology & Hepatology Department, MedStar Georgetown University Hospital, Washington, DC 20007, USA

F. Marotta
ReGenera R&D International for Aging Intervention, Milano, Italy

F. Marotta (ed.), *Gut Microbiota in Aging and Chronic Diseases*, Healthy Ageing and Longevity 17, https://doi.org/10.1007/978-3-031-14023-5_4

4.1 Introduction

Aging involves a complex network of interacting signalling pathway, processes and events (Campisi et al. 2019). During past three decades, the research relating to aging research has transitioned from identification of aging phenotypes to understand the genetic determinants and molecular biological pathways that underlie these phenotypes.

Understanding the process of aging has remained a challenge to gerontologists as well as developmental biologists. Initial understanding of the phenomenon of 'aging' began with observations in 1939 which showed that restricting the calorie intake could increase lifespan of experimental mice and rats (McCay et al. 1975). Afterward, the concept was corroborated in other species such as non-human primates (Dal-Pan et al. 2011; Mattison et al. 2017; Pifferi et al. 2018). Dietary restraint was found to enhance life span and suppressed age-associated diseases (Omodei and Fontana 2011). These observations led to evolution of the dogma that life-style extension associates with slowed aging and increased health span (Campisi et al. 2019).

Aging can be considered as a strong risk factor for multiple health-deterring conditions such as myocardial infarction, stroke, few age-related cancers, macular degeneration, sarcopenia, osteoarthritis, neurodegeneration, besides other health-associated morbidities (Campisi et al. 2019).

There are several factors which accelerate the process of aging. Environmental stress, imbalanced nutrition, malnutrition, exposure to radiations, pollutants, ingestion of xenobiotics through foods and air, and burden of diseases accelerate the process of aging (Fig. 4.1). In addition, dysbiosis of human microbiome and virome caused by above factors has serious repercussions.

Fig. 4.1 Factors responsible for fast-paced aging. Internal microbial dysbiosis due to antibiotics and human viruses has adverse effects on health

4.2 Gut Microbiome, Virome and Aging

Humans and many other animals, non-vertebrates and vertebrates, have co-evolved with microorganisms to live in symbiosis. Particularly, the gut microbiota is a highly complex, yet partially explored ecosystem comprising of bacteria, yeasts, archaea and viruses. A healthy gut ecosystem comprises of trillions of commensal bacteria, archaea, yeasts and fungi, few protozoa, and a complex virome consisting of viruses and virus gene (Liang and Bushman 2021; Bai et al. 2022).

The alteration in body functioning is a steady process (O'Toole and Jeffery 2015, 2018). Studies have revealed that gut microbiota change in response to aging, dietary habits, medication, ethnicity and active lifestyle (Cella et al. 2021; Sepp et al. 2022). However, these variables are complex, may be unique to individuals.

Changes in gut microbiota of younger and older people are due to immunosenescence i.e., age-related decline in immune functioning and simultaneously existing inflammaging, the low-grade chronic inflammation, and aging-related pathologies (Vaiserman et al. 2017).

Gut microbiota interact with host, transform ingested ingredients including botanicals and drugs and affect physiology and overall functioning.

As majority of gut microorganisms are unculturable, culture-independent next generation sequencing (NGS), microarray technology and molecular biological procedures are preferred to determine the composition and role of gut microorganism and bacteriophages. Multi-omics approached including genomics and metagenomics are progressively applied to molecular research in nutrients-gene interactions, host-microbe interaction, and mechanisms of aging (Solovev et al. 2020). Therapeutic and restorative strategies have been developed to control age-associated illnesses such as metabolic chronic diseases and cardiovascular health complications (Yu et al. 2010).

Composition of the human microbiome is strongly intertwined with health, immunity, and susceptibility to diseases (Thaiss et al. 2016). The gut viruses are the prospective effectors of immune variations, both locally as well as out of gut ecosystem. Among various endogenous viruses, the human endogenous retroviruses (HERVs) are of paramount genetic and clinical significance. Most elements of HERVs genome sequences are highly degenerated, epigenetically regulated, and constitute around 8% of the total human genome (Attermann et al. 2018).

4.3 Gut Microbial Metabolism and Aging

Diet and nutrition affect composition and functioning of intestinal microbiota. Vegetarian diets enriched with plant fiber, cell wall polysaccharides and phytochemicals such as tannin-polyphenols, saponins, alkaloids and phytoestrogens increase metabolic competence of gut microbiota (Singh et al. 2003; Wilson et al.

2020). Metabolites such as short chain fatty acids (SCFAs) (acetate, propionate and butyrate) generated from enteric fermentation of dietary fibre, and lactic acid produced by resident lactic acid bacteria (LAB) maintain a low intestinal luminal pH, promote mucus production and strengthen the gut barrier integrity.

The gut microbes regulate production of several signaling molecules such as serotonin (Stasi et al. 2019), hormones and neurotransmitters (Strandwitz 2018; Chen et al. 2021).

Decline in production of enteric SCFAs is associated with neuro-immune activation. Reduction in butyrate-producing bacteria leads to proliferation of microbial consortia which promote mucin degradation, thereby increase probability of entry of pathobionts into gut mucosa, and develop aging-promoting inflammatory diseases such as inflammatory bowel disease (IBD), Crohn's diseases (CD), and ulcerative colitis (UC) in old age. Reduced level of propionate and propionate-producing bacteria affects intestinal gluconeogenesis and gut-brain neural circuit via GPR-41 (de Vadder et al. 2014).

4.4 Host Aging-Microbial Connection

Certain gut dominant microbial species maintain almost a persistent profile throughout human life. Nonetheless, the homeostasis is perturbed under certain conditions and produces aberrant metabolites, signals and host-microbe interaction (Fig. 4.2). The situation is characterized by various health problems (Fig. 4.3), and overexpression of pro-inflammatory responses shared by the so-called senescence-associated secretory phenotype (SASP) (Gurău et al. 2018).

Aging is characterized by enhanced incidences of dysbiosis or imbalance in gut ecosystem, skin and genitourinary microbiota, diminution in gut-barrier integrity, the condition known as leaky gut (Camilleri 2021), which allows passage of microorganisms, endotoxin and other microbial metabolites (e.g., indoxyl sulfate) into blood circulatory system (Grosicki et al. 2018). Compared to healthy younger people aging between 20 to 35 years, the bacteria in elder humans (60–75 years), have a different clustered β-diversity and α-diversity as well as phylogenetic diversity (Bufford et al. 2018).

4.5 Anti-Aging Effects of Gut Microbiome

The gut microbiota has a strong impact on energy homeostasis in the colon besides being an inexhaustible source of bioactive metabolites some of which use used to suppress the invading pathogens (Wilson et al. 2017; Shine and Crawford 2021).

In addition, gut microbes regulate production of signaling molecules such as serotonin (5-HT), hormones and neurotransmitters. SCFAs have varying effects in different tissues. They are utilized by enterocytes as energy sources and contribute

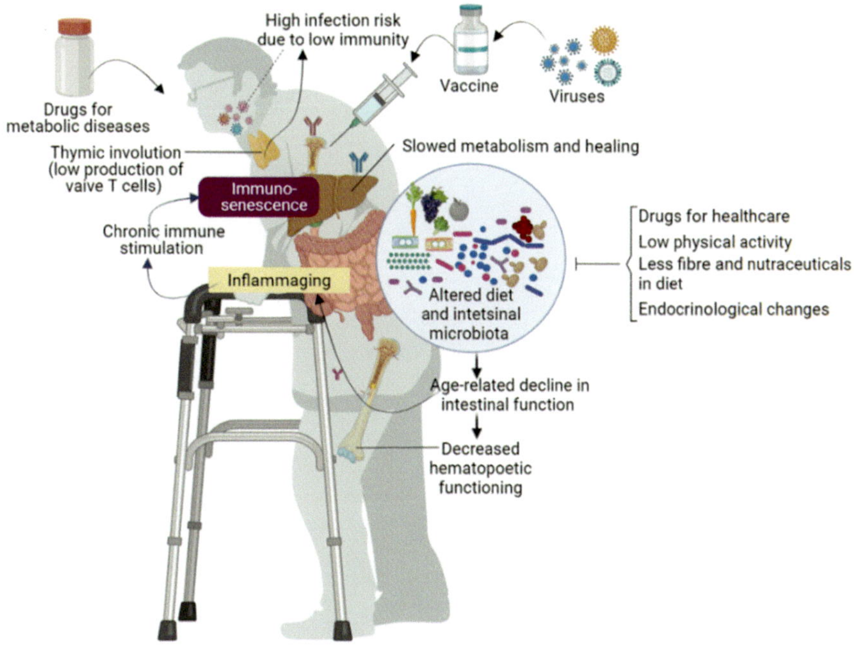

Fig. 4.2 Effects of aging on human health and commensal microorganisms. Microorganisms affect health through several mechanisms. On contrary, altered life style, low physical activities, and prolonged medication to control diseases diminish intestinal microbial diversity and metabolic activities which are vital to health of the host. Image drawn with commercial BioRender

to 5–10% of total energy requirement of human body (Donohoe et al. 2011; Makki et al. 2018). Acetate serves as substrate for hepatic lipogenesis, cholesterol synthesis and prevents liver carcinogenesis (Makki et al. 2018). Propionate is important component of hepatic gluconeogenesis, regulates cholesterol synthesis, and prevents development of cancer. Butyrate acts as epigenetic regulator of cellular genome (Kumar et al. 2013), and activates intestinal gluconeogenesis gene expression through cAMP-dependent mechanism.

The aging has adverse effects on bacterial consortia that produce beneficial metabolites such as butyrate (Rampelli et al. 2013), vitamins, antimicrobial peptides (AMPs) and bacteriocins, and therefore, creates a favorable milieu for pathogenic proteobacteria such as *Enterobacteriaceae*.

Predominating elderly-type gut bacteria in aged persons produce metabolites such as choline, trimethylamine, and N8-acetylspermidine which act as risk factors for age-associated diseases such as arteriosclerosis and colorectal cancer, and suppress the genes related to tight junctions in colonic epithelial cells (Yoshimoto et al. 2021).

Taking together, the aging process leads to increased intestinal permeability, and decrease in integrity of epithelial cell tight junctions (Thevaranjan et al. 2017, 2018).

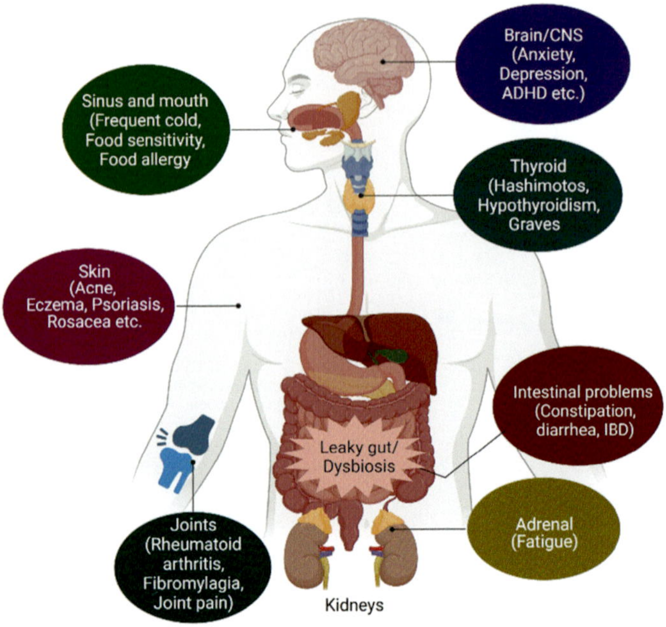

Fig. 4.3 Effects of dysbiosis on human ailments and aging. Intestinal bacteriophages play dual role- maintain intestinal bacterial population by killing bacteria which grow very fast, and induce dysbiosis by abolishing or killing bacteria which are beneficial. Image drawn with commercial BioRender

This pauci-symptomatic phenomenon leads to translocation of pro-inflammatory microorganisms and microbial metabolites including lipopolysaccharides (LPS) synthesized by Gram-negative bacteria (Zhao et al. 2017a, b; Kouzu et al. 2022).

Beneficial gut bacteria such as bifidobacteria undergo alterations in aged persons. For instance, *Bifidobacterium longum*, *Bifidobacterium bifidum* and *Bifidobacterium breve*, dominate in infants. On contrary, *Bifidobacterium adolescents* and *Bifidobacterium catenulatum* and *Bifidobacterium longum* predominate in adults and elderly people (Matsuki et al. 2004).

Probiotics and their metabolites (postbiotics), functional foods, and gut microbial enrichment through fecal microbial transplantation (FMT) (Fujimoto et al. 2021; Lam et al. 2022), effectively reinstate healthy gut microbiota, hence provide opportunities to prevent the diseases where conventional therapies are less effective (Kumar et al. 2016, 2021; Singh et al. 2017; Fujimoto et al. 2021). Given that normal microbiota plays a fundamental role in most aspects of human health and disease, it can be envisioned that novel alternative interventions such as probiotics, postbiotics, FMT would be the futuristic microbial interventions to manage health of elderly persons (Fujimoto et al. 2021).

Concerning the role of gut microbiota in health and diseases, and aging, various essential queries are still unresolved. For instance, it is not clear whether aging is a disease mediated by gut dysbiosis, or can it be delayed by timely nutritional and therapeutic intercessions (Nagpal et al. 2018).

4.6 Microbial Alleviation of Oxidative Stress

Reactive oxygen species (ROS) such as superoxide radicals ($O_2^{\cdot-}$), H_2O_2, hydroxyl radicals (·OH), singlet oxygen($^1O_2$), lipid peroxides, hypochlorite are generated as metabolic by-products in biological systems (Sato et al. 2013; Rajendran et al. 2014; Pizzino et al. 2017).

At low levels, ROS have a requisite role in intracellular signaling, defense against pathogens, molecular biological processes such as protein phosphorylation, activation of some transcriptional factors, apoptosis, immunity, and cellular differentiation (Rajendran et al. 2014). However, elevated levels of ROS cause oxidative stress that damages vital cellular components such as proteins, lipids, nucleic acids, cellular processes, and causes serious illnesses (Wu et al. 2013).

There are evidence that oxidative stress is responsible, albeit with different degrees of importance in onset or progression of diseases such as cancer, diabetes, metabolic disorders, atherosclerosis, ischemia, disruption of endocrine and immune functions, and physiological and pathophysiological signal transduction in cardiovascular system (Taniyama and Griendling 2003; Papaharalambus and Griendling 2007; Müller et al. 2022).

Environmental stressors like UV radiations, ionizing radiations, pollutants, and xenobiotics such as pesticides, mycotoxins, and heavy metals increased the production of ROS, hence causes oxidative stress in living system (Hofer 2021; McCann et al. 2021).

Intestinal LAB and Bifidobacteria protect from damage caused by oxidative stress. Bacterial strains with anti-oxidative properties have been used in dietary formulations (Lin et al. 2022). Lin and colleagues have postulated that doses of probiotics, namely *Bifidobacterium animalis* subsp. *infantis* BL1-02, *Bifidobacterium bereve* Bv889, *Bifidobacterium bifidum* VDD088, *Bifidobacterium animalis* subsp. *lactis* CP-9, and *Lactobacillus plantarum* PL-02 and their metabolites may be beneficial to slow down the process of aging, or may confer additional health benefits in aged persons (Lin et al. 2022).

4.7 Gut Detoxification of Genotoxicants

Xenobiotic substances entering human food chain through contaminated foods, milk and meat cause high oxidative stress, mutation, genotoxicity and cancer (Chirinos-Peinado et al. 2020, 2021). In addition, xenobiotics interrupt gut microflora, and may cause intestinal dysbiosis. Mice exposed to a mixture of polychlorinated biphenyls (PCBs) had a significantly reduced gut microbial composition (Choi et al. 2013), whereas arsenic (As) not only reduced the abundance of the gut Firmicutes, but also distressed the gut metabolic profile and functioning (Lu et al. 2014).

Gut microbial dysbiosis is linked to metabolic diseases such as diabetes (Wang et al. 2018), obesity (Everard et al. 2013; Liu et al. 2017; Amabebe et al. 2020), CVDs (Ley et al. 2006), allergies (Zhao et al. 2019), inflammatory bowel disease (IBD) (Nishida et al. 2018; Schirmer et al. 2018), pulmonary fibrotic changes or pulmonary wall thickening, hepatocellular inflammation and carcinoma (Chu et al. 2018; Giraud and Saleh 2021).

Despite prominent role of human microbiome to transform the xenobiotics, drugs, and phytochemicals, the microbial genes and enzymes responsible the versatile metabolic activities are not aptly understood (Koppel et al. 2017, 2018; Singh et al. 2020). Notably, intestinal bacteria play a key role detoxification and biotransformation of phytochemicals into metabolites which are more bioactive than their precursors (Singh et al. 2020). Enrichment of metabolically active gut microorganisms will detoxify toxic phytochemicals, and prevent the individuals from anti-nutritional effects of dietary components (Singh et al. 2020).

4.8 Age-Related Cognitive Diseases

Increase in endemic age-related diseases such as autism, strokes, dementia, Alzheimer's diseases (AD), Parkinson's disease (PD), and associated neurodegenerative disorders are of substantial biomedical concern. There are roughly 600 known diseases including brain tumors, epilepsy, AD, PD, stroke, and less known diseases such as frontotemporal and other dementias. In addition, cerebrovascular diseases such as migraine, stroke, multiple sclerosis, multiple neuroinfections, brain trauma are also classified as neurological diseases of substantial concern (Araujo 2015; Garkowski et al. 2015; Ishwarya and Narendhirakannan 2016).

The LAB, namely *Lactobacillus acidophilus*, *L. casei*, *L. casei* Shirota, *L. rhamnosus*, *L. helveticus*, *L. hilgardii*, *L. plantarum*, *L. pentosus*, *Bifidobacterium breve*, *B. longum*, *B. bifidum*, *Lactococcus lactis* and their metabolites are among the important bacterial cell biotherapeutics to alleviate or ease stress, anxiety and neurological problems. The results of trials involving probiotics in human volunteers have shown that *Bifidobacterium logum* 1714 could reduce stress and improve memory, albeit the basic mechanisms need to be resolved (Allen et al. 2016). Information is scarce on direct involvement of gut virome or phageome on progression of cognitive disorders.

4.9 Viruses as Indispensable Components of Human Microbiome and Health

A range of prokaryotic (bacterial and archaeal) and eukaryotic (human) viruses (such as influenza virus, Ebola and HIV) are found in humans. Thus, human virome represents the total crew of viruses and their genes found in multiples organs (Wylie et al. 2014; Cadwell 2015). Majority of the viruses include single (ssDNA) or double-stranded DNA (dsDNA) bacteriophages, RNA viruses, and endogenous retroviral elements.

Certain viruses do integrate into the human genome as proviruses or exist as endogenous viral elements (Wylie et al. 2012). Metagenomics, high-throughput sequencing of viruses assisted by bioinformatic analysis of massive sequence data have made it possible to examine the types and roles of dsDNA, ssDNA, and RNA virus-like particles (Carding et al. 2017).

Metagenomic screening using shotgun of uncultured human intestinal viral community from feces was first reported in 2003 (Breitbart et al. 2003). Notably, human virome is unique for each individual and influence the health either by affecting the microbiome or directly infect the humans which results in acute, persistent, or latent infections. Mechanisms of mutualism and symbiosis of viruses with co-existing bacteria, archaea, fungi and yeasts is not fully understood. The viruses promote human resistance to biotic and abiotic stress, and modulate the changes which are necessary for adaptation, development and survival (Jagdale and Joshi 2018; Stockdale and Hill 2021).

4.10 Gut Virome *Vis-À-Vis* Health and Aging

Despite being highly diverse and predominant, human virome is not yet fully understood. The possible reasons for it the fact that gut virome is interconnected with various cells, hence it is impractical to study the gut viruses in isolation (Mukhopadhya et al. 2019).

Intestinal virome, especially the phages directly influence human immunity. Phages modulate both innate and adaptive immunity (Van Belleghem et al. 2018). The bacteriophage adherence to mucus (BAM) models, reduce bacterial colonization to mucus surface, thus protect the host from pathogenic bacteria (Barr et al. 2013a, b). Bacteriophages predominate in body sites which are prone to bacterial infections, but role of bacteriophage enrichment is not clear (Sweere et al. 2019).

Intestinal viruses play beneficial as well as detrimental role. Indeed, host-gut virus relation is more akin to host-microbiome interactions. Approximately 10^{15} bacteriophages in the intestine (Carding et al. 2017), and 10^{8}–10^{9} viruses and virus-like particles are detected in per gram of human stool (Kim et al. 2011). dsDNA Caudovirales, and ssDNA Microviridale are the predominant bacteriophages in the intestine (Nayfach et al. 2021).

Human gut viruses are the prospective sources of immune variations at internal as well as external levels (Neil and Cadwell 2018). Human phageome is of high importance to host, but it is not fully understood (Dalmasso et al. 2014; Fitzgerald et al. 2021). Natural viral communities of humans are the focus of contemporary virology and vaccinology.

Different mechanisms have been suggested to explain the role of phages to affect intestinal bacteria. According to 'Kill the Winner' model, the phages target and kill intestinal bacteria which proliferate faster than others and tend to predominate, thus maintain a balance among different microbial genera and species. According to 'Biological Weapon' model, symbiotic commensal bacteria use phages associated with them kill the competing bacteria. However, it can decline the bacterial population, and may lead to intestinal dysbiosis (Mukhopadhya et al. 2019). In addition, viruses present in intestine are essential for homeostasis of the intestinal intraepithelial lymphocytes (IELs) (Liu et al. 2019).

Enteric eukaryotic viruses and bacteriophages kill host eukaryotic cells, and gut bacteria, respectively, and cause release of pro-inflammatory activators (Mukhopadhya et al. 2019). Dysbiosis of gut viruses has adverse consequences, generally associated with inflammatory diseases. It has been realized that cytosolic viral RNA-sensing receptor RIG-I in antigen-presenting cells can distinguish commensal viruses, and maintains IELs via a type I interferon-independent, but MAVS-IRF1-IL-15 axis-dependent manner (Liu et al. 2019). Recovery of IELs by interleukin-15 administration was found to reverse the susceptibility of the commensal virus-depleted mice to the dextran sulfate sodium-induced colitis model. It is proof of the concept that commensal viruses maintain the intraepithelial lymphocytes (IELs), and consequently sustain the intestinal homeostasis via non-canonical RIG-I signalling (Liu et al. 2019).

Studies carried out in model animals show that intestinal virome is important to prevent against colitis (Mukhopadhya et al. 2019). Mounting evidence suggest the safety and efficacy of novel therapies such as probiotic interventions, fecal virome transplantation (FVT) (Rasmussen et al. 2020b), fecal microbiota transplantation (FMT) to cure recurrent *Clostridial difficile* infection (rCDI), treat diseases related to gut dysbiosis (Lam et al. 2022), and prevent age-related pathological conditions such as atherosclerosis, T2DM, and Parkinson's disease (Vaiserman et al. 2017).

Stability of phage properties after FVT can be used as indicator of success of FMT (Broecker et al. 2017). As FVT is devoid of fecal bacterial components of donor feces, it circumvents the risk of invasive bacterial infections. Gut virome, mycobiome and bacteriome collectively have promising therapeutic prospects for FMT and FVT based on microorganisms as live biotherapeutics (Lam et al. 2022). However, threats like alterations in host gut microbiota, dysbiosis, host metabolome, posed by transplanted eukaryotic viruses, and prophage-encoded virulence should be dealt cautiously (Rasmussen et al. 2020a).

4.11 Outlook and Challenges

While we plan to move to introduce the inferences from in vitro or animal models, in humans, some questions arise. Most observations are obtained from in vitro experiments or from model animals. Concept of dietary restriction has been studied in insects, experimental mice or non-human primates in a specified environment and genetic strains of experimental animals. Same results cannot be expected from trials in humans.

Health in older age depends on the appropriate function of the homeostatic systems including nervous, endocrine and immune system, and a precise interaction between host and microbiota of gut ecosystem. Most organs and their functioning are altered during growing age, hence advance towards "inflammaging," characterized by low-grade inflammatory state which in involved in aetiology of age-related chronic pathological conditions.

Bifidobacterium and LAB especially, the lactobacilli are the most studied microorganisms in clinical trials in elderly populations. Intestinal bacteria enriched with *Bifidobacterium* and LAB may restore healthy ecosystem with anti-oxidative and antimicrobial properties, hence may prevent immunosenescence, control oxidative stress and inflammatory processes.

Future of medicine is strongly related to efficacy and quality of microbiota existing inside humans. Probiotics, probiotic metabolites and bacteriophages are already being used as alternative therapies to combat infectious and non-infectious diseases (Singh et al. 2017, 2020, Kumar et al. 2021), and age-related diseases. Explicit role of gut viruses in intestinal diseases, aging, and the strategies to alter human aging through normal microbiota deserves a prioritized attention of geriatric biological scientists. Role of intestinal phages in human health is a poorly understood, but it is a prospective field.

Undoubtedly, targeting gut microbes is a huge challenge. Double-blind placebo-controlled design should continue to develop complete blueprint and understanding of microbiota-host interaction, causes and mechanisms phage-mediated dysbiosis in elderly persons. The inferences obtained will provide a basis to develop anti-aging interventions.

References

Allen AP, Hutch W, Borre YE, Kennedy PJ, Temko A, Boylan G, Murphy E, Cryan JF, Dinan TG, Clarke G (2016) *Bifidobacterium longum* 1714 as a translational psychobiotic: modulation of stress, electrophysiology and neurocognition in healthy volunteers. Transl Psychiatry 6(11):e939. https://doi.org/10.1038/tp.2016.191

Amabebe E, Robert FO, Agbalalah T, Orubu ESF (2020) Microbial dysbiosis-induced obesity: role of gut microbiota in homoeostasis of energy metabolism. Br J Nutr 123(10):1127–1137. https://doi.org/10.1017/S0007114520000380 Epub 2020 Feb 3PMID: 32008579

Araujo AQ (2015) Update on neurological manifestations of HTLV-1 infection. Curr Infect Dis Rep 17(2):459. https://doi.org/10.1007/s11908-014-0459-0

Attermann AS, Bjerregaard AM, Saini SK, Grønbæk K, Hadrup SR (2018) Human endogenous retroviruses and their implication for immunotherapeutics of cancer. Ann Oncol 29(11):2183–2191. https://doi.org/10.1093/annonc/mdy413 PMID: 30239576

Bai GH, Lin SC, Hsu YH, Chen SY (2022) The human virome: viral metagenomics, relations with human diseases, and therapeutic applications. Viruses 14(2):278. https://doi.org/10.3390/v14020278 PMID: 35215871

Barr JJ, Auro R, Furlan M, Whiteson KL, Erb ML, Pogliano J, Stotland A, Wolkowicz R, Cutting AS, Doran KS, Salamon P, Youle M, Rohwer F (2013a) Bacteriophage adhering to mucus provide a non-host-derived immunity. Proc Natl Acad Sci U S A 110(26):10771–10776. https://doi.org/10.1073/pnas.1305923110 Epub 2013 May 20PMID: 23690590

Barr JJ, Youle M, Rohwer F (2013b) Innate and acquired bacteriophage-mediated immunity. Bacteriophage 3(3):e25857. https://doi.org/10.4161/bact.25857. Epub 2013 Jul 29.PMID: 24228227.

Breitbart M, Hewson I, Felts B, Mahaffy JM, Nulton J, Salamon P, Rohwer F (2003) Metagenomic analyses of an uncultured viral community from human feces. J Bacteriol 185(20):6220–6223. https://doi.org/10.1128/JB.185.20.6220-6223.2003 PMID: 14526037

Broecker F, Russo G, Klumpp J, Moelling K (2017) Stable core virome despite variable microbiome after fecal transfer. Gut Microbes 8(3):214–220. https://doi.org/10.1080/19490976.2016.1265196 Epub 2016 Dec 9PMID: 27935413

Bufford TW, Carter CS, VanDerPol WJ, Chen D, Lefkowitz EJ, Eipers P, Morrow CD, Bamman MM (2018) Composition and richness of the serum microbiome differ by age and link to systemic inflammation. GeroScience 40:257–268. https://doi.org/10.1007/s11357-018-0026-y

Cadwell K (2015) Expanding the role of the virome: commensalism in the gut. J Virol 89(4):1951–1953. https://doi.org/10.1128/JVI.02966-14 Epub 2014 Dec 10

Camilleri M (2021) What is the leaky gut? Clinical considerations in humans. Curr Opin Clin Nutr Metab Care 24(5):473–482. https://doi.org/10.1097/MCO.0000000000000778 PMID: 34138767

Campisi J, Kapahi P, Lithgow GJ, Melov S, Newman JC, Verdin E (2019) From discoveries in ageing research to therapeutics for healthy ageing. Nature 571(7764):183–192. https://doi.org/10.1038/s41586-019-1365-2 Epub 2019 Jul 10PMID: 31292558

Carding SR, Davis N, Hoyles L (2017) Review article: the human intestinal virome in health and disease. Aliment Pharmacol Ther 46(9):800–815. https://doi.org/10.1111/apt.14280 Epub 2017 Sep 4PMID: 28869283

Cella V, Bimonte VM, Sabato C, Paoli A, Baldari C, Campanella M, Lenzi A, Ferretti E, Migliaccio S (2021) Nutrition and physical activity-induced changes in gut microbiota: possible implications for human health and athletic performance. Foods 10(12):3075. https://doi.org/10.3390/foods10123075 PMID: 34945630

Chen Y, Xu J, Chen Y (2021) Regulation of neurotransmitters by the gut microbiota and effects on cognition in neurological disorders. Nutrients 13(6):2099. https://doi.org/10.3390/nu13062099 PMID: 34205336

Chirinos-Peinado DM, Castro-Bedriñana JI (2020) Lead and cadmium blood levels and transfer to milk in cattle reared in a mining area. Heliyon 6(3):e03579. https://doi.org/10.1016/j.heliyon.2020.e03579. eCollection 2020 Mar.PMID: 32195399

Chirinos-Peinado D, Castro-Bedriñana J, García-Olarte E, Quispe-Ramos R, Gordillo-Espinal S (2021) Transfer of lead from soil to pasture grass and milk near a metallurgical complex in the Peruvian Andes. Transl Anim Sci 5(1):txab003. https://doi.org/10.1093/tas/txab003. eCollection 2021 Jan.PMID: 34859197

Choi JJ, Eum SY, Rampersaud E, Daunert S, Abreu MT, Toborek M (2013) Exercise attenuates PCB-induced changes in the mouse gut microbiome. Environ Health Perspect 121(6):725–730. https://doi.org/10.1289/ehp.1306534 Epub 2013 Apr 26 PMID: 23632211

Chu H, Williams B, Schnabl B (2018) Gut microbiota, fatty liver disease, and hepatocellular carcinoma. Liver Res 2(1):43–51. https://doi.org/10.1016/j.livres.2017.11.005 Epub 2018 Feb 21PMID: 30416839

Dalmasso M, Hill C, Ross RP (2014) Exploiting gut bacteriophages for human health. Trends Microbiol 22(7):399–405. https://doi.org/10.1016/j.tim.2014.02.010 Epub 2014 Mar 20PMID: 24656964

Dal-Pan A, Terrien J, Pifferi F, Botalla R, Hardy I, Marchal J, Zahariev A, Chery I, Zizzari P, Perret M, Picq JL, Epelbaum J, Blanc S, Aujard F (2011) Caloric restriction or resveratrol supplementation and ageing in a non-human primate: first-year outcome of the RESTRIKAL study in Microcebus murinus. Age (dordr) 33(1):15–31. https://doi.org/10.1007/s11357-010-9156-6 Epub 2010 Jun 9PMID: 20532988

De Vadder F, Kovatcheva-Datchary P, Goncalves D, Vinera J, Zitoun C, Duchampt A, Bäckhed F, Mithieux G (2014) Microbiota-generated metabolites promote metabolic benefits via gut-brain neural circuits. Cell 156(1–2):84–96. https://doi.org/10.1016/j.cell.2013.12.016 Epub 2014 Jan 9PMID: 24412651

Donohoe DR, Garge N, Zhang X, Sun W, O'Connell TM, Bunger MK, Bultman SJ (2011) The microbiome and butyrate regulate energy metabolism and autophagy in the mammalian colon. Cell Metab 13(5):517–526. https://doi.org/10.1016/j.cmet.2011.02.018 PMID: 21531334

Everard A, Belzer C, Geurts L, Ouwerkerk JP, Druart C, Bindels LB, Guiot Y, Derrien M, Muccioli GG, Delzenne NM, de Vos WM, Cani PD (2013) Cross-talk between Akkermansia muciniphila and intestinal epithelium controls diet-induced obesity. Proc Natl Acad Sci U S A 110(22):9066–9071. https://doi.org/10.1073/pnas.1219451110 Epub 2013 May 13PMID: 23671105

Fitzgerald CB, Shkoporov AN, Upadrasta A, Khokhlova EV, Ross RP, Hill C (2021) Probing the "dark matter" of the human gut phageome: culture assisted metagenomics enables rapid discovery and host-linking for novel bacteriophages. Front Cell Infect Microbiol 11:616918. https://doi.org/10.3389/fcimb.2021.616918. eCollection 2021.PMID: 33791236

Fujimoto K, Kimura Y, Allegretti JR, Yamamoto M, Zhang YZ, Katayama K, Tremmel G, Kawaguchi Y, Shimohigoshi M, Hayashi T, Uematsu M, Yamaguchi K, Furukawa Y, Akiyama Y, Yamaguchi R, Crowe SE, Ernst PB, Miyano S, Kiyono H, Imoto S, Uematsu S (2021) Functional restoration of bacteriomes and viromes by fecal microbiota transplantation. Gastroenterology 160(6):2089-2102.e12. https://doi.org/10.1053/j.gastro.2021.02.013 Epub 2021 Feb 9PMID: 33577875

Garkowski A, Zajkowska J, Moniuszko A, Czupryna P, Pancewicz S (2015) Infectious causes of stroke. Lancet Infect Dis 15(6):632. https://doi.org/10.1016/S1473-3099(15)00020-1. Epub 2015 May 17. No abstract available

Giraud J, Saleh M (2021) Host-microbiota interactions in liver inflammation and cancer. Cancers (basel) 13(17):4342. https://doi.org/10.3390/cancers13174342 PMID: 34503151

Grosicki GJ, Fielding RA, Lustgarten MS (2018) Gut microbiota contribute to age-related changes in skeletal muscle size, composition, and function: biological basis for a gut-muscle axis. Calcif Tissue Int 102(4):433–442. https://doi.org/10.1007/s00223-017-0345-5 Epub 2017 Oct 20PMID: 29058056

Gurău F, Baldoni S, Prattichizzo F, Espinosa E, Amenta F, Procopio AD, Albertini MC, Bonafè M, Olivieri F (2018) Anti-senescence compounds: a potential nutraceutical approach to healthy aging. Ageing Res Rev 46:14–31. https://doi.org/10.1016/j.arr.2018.05.001 Epub 2018 May 6PMID: 29742452

Hofer T (2021) Oxidative stress in human toxicology. Antioxidants (basel) 10(8):1159. https://doi.org/10.3390/antiox10081159 PMID: 34439406

Ishwarya M, Narendhirakannan RT (2016) The advances in neurobiology. Adv Neurobiol 12:293–306. https://doi.org/10.1007/978-3-319-28383-8_15 PMID: 27651259

Jagdale SS, Joshi RS (2018) Enemies with benefits: mutualistic interactions of viruses with lower eukaryotes. Arch Virol 163(4):821–830. https://doi.org/10.1007/s00705-017-3686-5 Epub 2018 Jan 6

Kim MS, Park EJ, Roh SW, Bae JW (2011) Diversity and abundance of single-stranded DNA viruses in human feces. Appl Environ Microbiol 77(22):8062–70. https://doi.org/10.1128/AEM.06331-11

Koppel N, Maini Rekdal V, Balskus EP (2017) Chemical transformation of xenobiotics by the human gut microbiota. Science 356(6344):eaag2770. https://doi.org/10.1126/science.aag2770 PMID: 28642381

Koppel N, Bisanz JE, Pandelia ME, Turnbaugh PJ, Balskus EP (2018) Discovery and characterization of a prevalent human gut bacterial enzyme sufficient for the inactivation of a family of plant toxins. Elife 15(7):e33953. https://doi.org/10.7554/eLife.33953 PMID: 29761785

Kouzu K, Tsujimoto H, Kishi Y, Ueno H, Shinomiya N (2022) Bacterial translocation in gastrointestinal cancers and cancer treatment. Biomedicines 10(2):380. https://doi.org/10.3390/biomedicines10020380 PMID: 35203589

Kumar M, Sarma DK, Shubham S, Kumawat M, Verma V, Nina PB, Jp D, Kumar S, Singh B, Tiwari RR (2021) Futuristic non-antibiotic therapies to combat antibiotic resistance: a review. Front Microbiol 12:609459. https://doi.org/10.3389/fmicb.2021.609459. eCollection 2021. PMID: 33574807

Kumar M, Nagpal R, Verma V, Kumar A, Kaur N, Hemalatha R, Gautam SK, Singh B (2013) Probiotic metabolites as epigenetic targets in the prevention of colon cancer. Nutr Rev 71 (1):23–34. https://doi.org/10.1111/j.1753-4887.2012.00542.x Epub 2012 Nov 9PMID: 23282249

Kumar M, Yadav AK, Verma V, Singh B, Mal G, Nagpal R, Hemalatha R (2016) Bioengineered probiotics as a new hope for health and diseases: an overview of potential and prospects. Future Microbiol 11(4):585–600. https://doi.org/10.2217/fmb.16.4 Epub 2016 Apr 12PMID: 27070955

Lam S, Bai X, Shkoporov AN, Park H, Wu X, Lan P, Zuo T (2022) Roles of the gut virome and mycobiome in faecal microbiota transplantation. Lancet Gastroenterol Hepatol 7(5):472–484. https://doi.org/10.1016/S2468-1253(21)00303-4 Epub 2022 Mar 8PMID: 35276080

Ley RE, Turnbaugh PJ, Klein S, Gordon JI (2006) Microbial ecology: human gut microbes associated with obesity. Nature 444(7122):1022–1023. https://doi.org/10.1038/4441022a PMID: 17183309

Liang G, Bushman FD (2021) The human virome: assembly, composition and host interactions. Nat Rev Microbiol 19(8):514–527. https://doi.org/10.1038/s41579-021-00536-5 Epub 2021 Mar 30PMID: 33785903

Lin WY, Lin JH, Kuo YW, Chiang PR, Ho HH (2022) Probiotics and their metabolites reduce oxidative stress in middle-aged mice. Curr Microbiol 79(4):104. https://doi.org/10.1007/s00284-022-02783-y PMID: 35157139

Liu L, Gong T, Tao W, Lin B, Li C, Zheng X, Zhu S, Jiang W, Zhou R (2019) Commensal viruses maintain intestinal intraepithelial lymphocytes via noncanonical RIG-I signaling. Nat Immunol 20(12):1681–1691. https://doi.org/10.1038/s41590-019-0513-z Epub 2019 Oct 21PMID: 31636462

Liu R, Hong J, Xu X, Feng Q, Zhang D, Gu Y, Shi J, Zhao S, Liu W, Wang X, Xia H, Liu Z, Cui B, Liang P, Xi L, Jin J, Ying X, Wang X, Zhao X, Li W, Jia H, Lan Z, Li F, Wang R, Sun Y, Yang M, Shen Y, Jie Z, Li J, Chen X, Zhong H, Xie H, Zhang Y, Gu W, Deng X, Shen B, Xu X, Yang H, Xu G, Bi Y, Lai S, Wang J, Qi L, Madsen L, Wang J, Ning G, Kristiansen K, Wang W (2017) Gut microbiome and serum metabolome alterations in obesity and after weight-loss intervention. Nat Med 23(7):859–868. https://doi.org/10.1038/nm.4358 Epub 2017 Jun 19PMID: 28628112

Lu K, Abo RP, Schlieper KA, Graffam ME, Levine S, Wishnok JS, Swenberg JA, Tannenbaum SR, Fox JG (2014) Arsenic exposure perturbs the gut microbiome and its metabolic profile in mice: an integrated metagenomics and metabolomics analysis. Environ Health Perspect 122(3):284–291. https://doi.org/10.1289/ehp.1307429 Epub 2014 Jan 10PMID: 24413286

Makki K, Deehan EC, Walter J, Bäckhed F (2018) The impact of dietary fiber on gut microbiota in host health and disease. Cell Host Microbe 23(6):705–715. https://doi.org/10.1016/j.chom.2018.05.012 PMID: 29902436

Matsuki T, Watanabe K, Fujimoto J, Kado Y, Takada T, Matsumoto K, Tanaka R (2004) Quantitative PCR with 16S rRNA-gene-targeted species-specific primers for analysis of human intestinal bifidobacteria. Appl Environ Microbiol 70(1):167–173. https://doi.org/10.1128/AEM.70.1.167-173.2004 PMID: 14711639

Mattison JA, Colman RJ, Beasley TM, Allison DB, Kemnitz JW, Roth GS, Ingram DK, Weindruch R, de Cabo R, Anderson RM (2017) Caloric restriction improves health and survival of rhesus monkeys. Nat Commun 17(8):14063. https://doi.org/10.1038/ncomms14063 PMID: 28094793

McCann MS, Fernandez HR, Flowers SA, Maguire-Zeiss KA (2021) Polychlorinated biphenyls induce oxidative stress and metabolic responses in astrocytes. Neurotoxicology 86:59–68. https://doi.org/10.1016/j.neuro.2021.07.001 Epub 2021 Jul 12PMID: 34265337

McCay CM, Maynard LA, Sperling G, Barnes LL (1939) J Nutr. 18 July—December:1–13. Retarded growth, life span, ultimate body size and age changes in the albino rat after feeding diets restricted in calories. Nutr Rev 1975 Aug;33(8):241–3. https://doi.org/10.1111/j.1753-4887.1975.tb05227.x.PMID: 1095975

Mukhopadhya I, Segal JP, Carding SR, Hart AL, Hold GL (2019) The gut virome: the 'missing link' between gut bacteria and host immunity? Therap Adv Gastroenterol 12:1756284819836620. https://doi.org/10.1177/1756284819836620. eCollection 2019

Müller N, Warwick T, Noack K, Malacarne PF, Cooper AJL, Weissmann N, Schröder K, Brandes RP, Rezende F (2022) Reactive oxygen species differentially modulate the metabolic and transcriptomic response of endothelial cells. Antioxidants (basel) 11(2):434. https://doi.org/10.3390/antiox11020434 PMID: 35204316

Nagpal R, Mainali R, Ahmadi S, Wang S, Singh R, Kavanagh K, Kitzman DW, Kushugulova A, Marotta F, Yadav H (2018) Gut microbiome and aging: physiological and mechanistic insights. Nutr Healthy Aging 4(4):267–285. https://doi.org/10.3233/NHA-170030 PMID: 29951588

Nayfach S, Páez-Espino D, Call L, Low SJ, Sberro H, Ivanova NN, Proal AD, Fischbach MA, Bhatt AS, Hugenholtz P, Kyrpides NC (2021) Metagenomic compendium of 189,680 DNA viruses from the human gut microbiome. Nat Microbiol 6(7):960–970. https://doi.org/10.1038/s41564-021-00928-6 Epub 2021 Jun 24PMID: 34168315

Neil JA, Cadwell K (2018) The intestinal virome and immunity. J Immunol 201(6):1615–1624. https://doi.org/10.4049/jimmunol.1800631

Nishida A, Inoue R, Inatomi O, Bamba S, Naito Y, Andoh A (2018) Gut microbiota in the pathogenesis of inflammatory bowel disease. Clin J Gastroenterol 11(1):1–10. https://doi.org/10.1007/s12328-017-0813-5 Epub 2017 Dec 29PMID: 29285689

Omodei D, Fontana L (2011) Calorie restriction and prevention of age-associated chronic disease. FEBS Lett 585(11):1537–1542. https://doi.org/10.1016/j.febslet.2011.03.015 Epub 2011 Mar 12PMID: 21402069

O'Toole PW, Jeffery IB (2015) Gut microbiota and aging. Science 350(6265):1214–1215. https://doi.org/10.1126/science.aac8469 PMID: 26785481

O'Toole PW, Jeffery IB (2018) Microbiome-health interactions in older people. Cell Mol Life Sci 75(1):119–128. https://doi.org/10.1007/s00018-017-2673-z Epub 2017 Oct 6PMID: 28986601

Papaharalambus CA, Griendling KK (2007) Basic mechanisms of oxidative stress and reactive oxygen species in cardiovascular injury. Trends Cardiovasc Med 17(2):48–54. https://doi.org/10.1016/j.tcm.2006.11.005 PMID: 17292046

Pifferi F, Terrien J, Marchal J, Dal-Pan A, Djelti F, Hardy I, Chahory S, Cordonnier N, Desquilbet L, Hurion M, Zahariev A, Chery I, Zizzari P, Perret M, Epelbaum J, Blanc S, Picq JL, Dhenain M, Aujard F (2018) Caloric restriction increases lifespan but affects brain integrity in grey mouse lemur primates. Commun Biol 1:30. https://doi.org/10.1038/s42003-018-0024-8. eCollection 2018.PMID: 30271916

Pizzino G, Irrera N, Cucinotta M, Pallio G, Mannino F, Arcoraci V, Squadrito F, Altavilla D, Bitto A (2017) Oxidative stress: harms and benefits for human health. Oxid Med Cell Longev 2017:8416763. https://doi.org/10.1155/2017/8416763 Epub 2017 Jul 27PMID: 28819546

Rajendran P, Nandakumar N, Rengarajan T, Palaniswami R, Gnanadhas EN, Lakshminarasaiah U, Gopas J, Nishigaki I (2014) Antioxidants and human diseases. Clin Chim Acta 25(436):332–347. https://doi.org/10.1016/j.cca.2014.06.004 Epub 2014 Jun 13PMID: 24933428

Rampelli S, Candela M, Turroni S, Biagi E, Collino S, Franceschi C, O'Toole PW, Brigidi P (2013) Functional metagenomic profiling of intestinal microbiome in extreme ageing. Aging (albany NY) 5(12):902–912. https://doi.org/10.18632/aging.100623 PMID: 24334635

Rasmussen TS, Koefoed AK, Jakobsen RR, Deng L, Castro-Mejía JL, Brunse A, Neve H, Vogensen FK, Nielsen DS (2020a) Bacteriophage-mediated manipulation of the gut microbiome-promises and presents limitations. FEMS Microbiol Rev 44(4):507–521. https://doi.org/10.1093/femsre/fuaa020 PMID: 32495834

Rasmussen TS, Mentzel CMJ, Kot W, Castro-Mejía JL, Zuffa S, Swann JR, Hansen LH, Vogensen FK, Hansen AK, Nielsen DS (2020b) Faecal virome transplantation decreases symptoms of type 2 diabetes and obesity in a murine model. Gut 69(12):2122–2130. https://doi.org/10.1136/gutjnl-2019-320005 Epub 2020 Mar 12PMID: 32165408

Sato H, Shibata M, Shimizu T, Shibata S, Toriumi H, Ebine T, Kuroi T, Iwashita T, Funakubo M, Kayama Y, Akazawa C, Wajima K, Nakagawa T, Okano H, Suzuki N (2013) Differential cellular localization of antioxidant enzymes in the trigeminal ganglion. Neuroscience 17 (248):345–358. https://doi.org/10.1016/j.neuroscience.2013.06.010 Epub 2013 Jun 15PMID: 23774632

Schirmer M, Franzosa EA, Lloyd-Price J, McIver LJ, Schwager R, Poon TW, Ananthakrishnan AN, Andrews E, Barron G, Lake K, Prasad M, Sauk J, Stevens B, Wilson RG, Braun J, Denson LA, Kugathasan S, McGovern DPB, Vlamakis H, Xavier RJ, Huttenhower C (2018) Dynamics of metatranscription in the inflammatory bowel disease gut microbiome. Nat Microbiol 3(3):337–346. https://doi.org/10.1038/s41564-017-0089-z Epub 2018 Jan 8PMID: 29311644

Sepp E, Smidt I, Rööp T, Štšepetova J, Kõljalg S, Mikelsaar M, Soidla I, Ainsaar M, Kolk H, Vallas M, Jaagura M, Mändar R (2022) Comparative analysis of gut microbiota in centenarians and young people: impact of eating habits and childhood living environment. Front Cell Infect Microbiol 12:851404. https://doi.org/10.3389/fcimb.2022.851404. eCollection 2022.PMID: 35372105

Shine EE, Crawford JM (2021) Molecules from the microbiome. Annu Rev Biochem 20(90):789–815. https://doi.org/10.1146/annurev-biochem-080320-115307 Epub 2021 Mar 26

Singh B, Bhat TK, Singh B (2003) Potential therapeutic applications of some antinutritional plant secondary metabolites. J Agric Food Chem 51(19):5579–5597. https://doi.org/10.1021/jf021150r PMID: 12952405

Singh B, Mal G, Sharma D, Sharma R, Antony CP, Kalra RS (2020) Gastrointestinal biotransformation of phytochemicals: towards futuristic dietary therapeutics and functional foods. Trends Food Sci Technol 106:64–77. https://doi.org/10.1016/j.tifs.2020.09.022

Singh B, Mal G, Marotta F (2017) Designer probiotics: paving the way to living therapeutics. Trends Biotechnol 35(8):679–682. https://doi.org/10.1016/j.tibtech.2017.04.001 Epub 2017 May 5PMID: 28483159

Singh B, Mal G, Sharma D, Sharma R, Antony CP, Kalra RS (2020) Gastrointestinal biotransformation of phytochemicals: towards futuristic dietary therapeutics and functional foods. Trends Food Sci Technol 106:64–77. https://doi.org/10.1016/j.tifs.2020.09.022

Solovev I, Shaposhnikov M, Moskalev A (2020) Multi-omics approaches to human biological age estimation. Mech Ageing Dev 185:111192. https://doi.org/10.1016/j.mad.2019.111192 Epub 2019 Nov 28PMID: 31786174

Stasi C, Sadalla S, Milani S (2019) The relationship between the serotonin metabolism, gut-microbiota and the gut-brain axis. Curr Drug Metab 20(8):646–655. https://doi.org/10.2174/1389200220666190725115503 PMID: 31345143

Stockdale SR, Hill C (2021) Progress and prospects of the healthy human gut virome. Curr Opin Virol 51:164–171. https://doi.org/10.1016/j.coviro.2021.10.001 Epub 2021 Oct 29PMID: 34742036

Strandwitz P (2018) Neurotransmitter modulation by the gut microbiota. Brain Res 1693(Pt B):128–133. https://doi.org/10.1016/j.brainres.2018.03.015 PMID: 29903615

Sweere JM, Van Belleghem JD, Ishak H, Bach MS, Popescu M, Sunkari V, Kaber G, Manasherob R, Suh GA, Cao X, de Vries CR, Lam DN, Marshall PL, Birukova M, Katznelson E, Lazzareschi DV, Balaji S, Keswani SG, Hawn TR, Secor PR, Bollyky PL (2019) Bacteriophage trigger antiviral immunity and prevent clearance of bacterial infection. Science 363(6434):eaat9691. https://doi.org/10.1126/science.aat9691.PMID: 30923196

Taniyama Y, Griendling KK (2003) Reactive oxygen species in the vasculature: molecular and cellular mechanisms. Hypertension 42(6):1075–1081. https://doi.org/10.1161/01.HYP.0000100443.09293.4F Epub 2003 Oct 27PMID: 14581295

Thaiss CA, Levy M, Korem T, Dohnalová L, Shapiro H, Jaitin DA, David E, Winter DR, Gury-BenAri M, Tatirovsky E, Tuganbaev T, Federici S, Zmora N, Zeevi D, Dori-Bachash M, Pevsner-Fischer M, Kartvelishvily E, Brandis A, Harmelin A, Shibolet O, Halpern Z, Honda K, Amit I, Segal E, Elinav E (2016) Microbiota diurnal rhythmicity programs host transcriptome oscillations. Cell 167(6):1495-1510.e12. https://doi.org/10.1016/j.cell.2016.11.003 PMID: 27912059

Thevaranjan N, Puchta A, Schulz C, Naidoo A, Szamosi JC, Verschoor CP, Loukov D, Schenck LP, Jury J, Foley KP, Schertzer JD, Larché MJ, Davidson DJ, Verdú EF, Surette MG, Bowdish DME (2017) Age-associated microbial dysbiosis promotes intestinal permeability, systemic inflammation, and macrophage dysfunction. Cell Host Microbe 21(4):455-466.e4. https://doi.org/10.1016/j.chom.2017.03.002 PMID: 28407483

Thevaranjan N, Puchta A, Schulz C, Naidoo A, Szamosi JC, Verschoor CP, Loukov D, Schenck LP, Jury J, Foley KP, Schertzer JD, Larché MJ, Davidson DJ, Verdú EF, Surette MG, Bowdish DME (2018) Age-associated microbial dysbiosis promotes intestinal permeability, systemic inflammation, and macrophage dysfunction. Cell Host Microbe 23(4):570. https://doi.org/10.1016/j.chom.2018.03.006 PMID: 29649447

Vaiserman AM, Koliada AK, Marotta F (2017) Gut microbiota: a player in aging and a target for anti-aging intervention. Ageing Res Rev 35:36–45. https://doi.org/10.1016/j.arr.2017.01.001 Epub 2017 Jan 18PMID: 28109835

Van Belleghem JD, Dąbrowska K, Vaneechoutte M, Barr JJ, Bollyky PL (2018) Interactions between bacteriophage, bacteria, and the mammalian immune system. Viruses 11(1):10. https://doi.org/10.3390/v11010010 PMID: 30585199

Wang J, Zheng J, Shi W, Du N, Xu X, Zhang Y, Ji P, Zhang F, Jia Z, Wang Y, Zheng Z, Zhang H, Zhao F (2018) Dysbiosis of maternal and neonatal microbiota associated with gestational diabetes mellitus. Gut 67(9):1614–1625. https://doi.org/10.1136/gutjnl-2018-315988 Epub 2018 May 14PMID: 29760169

Wilson AS, Koller KR, Ramaboli MC, Nesengani LT, Ocvirk S, Chen C, Flanagan CA, Sapp FR, Merritt ZT, Bhatti F, Thomas TK, O'Keefe SJD (2020) Diet and the human gut microbiome: an international review. Dig Dis Sci 65(3):723–740. https://doi.org/10.1007/s10620-020-06112-w PMID: 32060812

Wilson MR, Zha L, Balskus EP (2017) Natural product discovery from the human microbiome. J Biol Chem 292(21):8546–8552. https://doi.org/10.1074/jbc.R116.762906 Epub 2017 Apr 7PMID: 28389564

Wu JQ, Kosten TR, Zhang XY (2013) Free radicals, antioxidant defense systems, and schizophrenia. Prog Neuropsychopharmacol Biol Psychiatry 1(46):200–206. https://doi.org/10.1016/j.pnpbp.2013.02.015 Epub 2013 Mar 5PMID: 23470289

Wylie KM, Weinstock GM, Storch GA (2012) Emerging view of the human virome, version 2. Transl Res 160(4):283–90. https://doi.org/10.1016/j.trsl.2012.03.006. Epub 2012 Apr 24

Wylie KM, Mihindukulasuriya KA, Zhou Y, Sodergren E, Storch GA, Weinstock GM (2014) BMC metagenomic analysis of double-stranded DNA viruses in healthy adults. Biol 10(12):71. https://doi.org/10.1186/s12915-014-0071-7 PMID: 25212266

Yoshimoto S, Mitsuyama E, Yoshida K, Odamaki T, Xiao JZ (2021) Enriched metabolites that potentially promote age-associated diseases in subjects with an elderly-type gut microbiota. Gut Microbes 13(1):1–11. https://doi.org/10.1080/19490976.2020.1865705.PMID: 33430687

Yu LX, Yan HX, Liu Q, Yang W, Wu HP, Dong W, Tang L, Lin Y, He YQ, Zou SS, Wang C, Zhang HL, Cao GW, Wu MC, Wang HY (2010) Endotoxin accumulation prevents carcinogen-induced apoptosis and promotes liver tumorigenesis in rodents. Hepatology 52 (4):1322–1333. https://doi.org/10.1002/hep.23845 PMID: 20803560

Zhao S, Liu W, Wang J, Shi J, Sun Y, Wang W, Ning G, Liu R, Hong J (2017a) Akkermansia muciniphila improves metabolic profiles by reducing inflammation in chow diet-fed mice. J Mol Endocrinol 58(1):1–14. https://doi.org/10.1530/JME-16-0054 Epub 2016 Nov 7PMID: 27821438

Zhao Y, Jaber V, Lukiw WJ (2017b) Secretory products of the human GI tract microbiome and their potential impact on Alzheimer's disease (AD): detection of Lipopolysaccharide (LPS) in AD hippocampus. Front Cell Infect Microbiol 7:318. https://doi.org/10.3389/fcimb.2017.00318. eCollection 2017.PMID: 28744452

Zhao W, Ho HE, Bunyavanich S (2019) The gut microbiome in food allergy. Ann Allergy Asthma Immunol 122(3):276–282. https://doi.org/10.1016/j.anai.2018.12.012 Epub 2018 Dec 20PMID: 30578857

Chapter 5
Bile Microbiota Profile in Aging and Age-Related Disease

Consolato M. Sergi

Abstract Bile has been considered sterile for a long time, probably derived from the concept of the pH and chemical compounds that may have some aseptic features. On the other side, recent knowledge has indicated that bile may harbor bacteria, and some of them may be related to age-related diseases. This chapter attempts to tackle the most recent data on bile microbiota profile in aging and age-related disease.

Keywords Bile · Microbiome · Aging · Taxonomy · Classification · Cancer

5.1 Introduction

The gallbladder is a vesicle located on the inferior site of the liver with the function of draining the bile from the hepatocytes (Keplinger and Bloomston 2014; Sergi 2020). The bile, which is also called gall, is composed of, mainly, cholesterol ($C_{27}H_{46}O$), bile salts, or also often called, in the gastrointestinal literature, bile acids, bilirubin ($C_{33}H_{36}N_4O_6$), which is the principal mammalian bile pigment, water, potassium, sodium (*natrium*), copper (*cuprum*), and, among other components, other metals. Bile is crucial for the digestion of mammals. In fact, it degrades or breaks down lipidic molecules (fats) or lipids into fatty acids. It can be taken into the body by the digestive tract. The gallbladder exhibits several disorders with a range from

C. M. Sergi (✉)
Division of Anatomic Pathology, Children's Hospital of Eastern Ontario, Ottawa, ON, Canada
e-mail: csergi@cheo.on.ca

C. M. Sergi
Department of Laboratory Medicine and Pathology, University of Ottawa, Ottawa, ON, Canada

C. M. Sergi
Department of Laboratory Medicine and Pathology, University of Alberta, Edmonton, AB, Canada

F. Marotta (ed.), *Gut Microbiota in Aging and Chronic Diseases*, Healthy Ageing and Longevity 17, https://doi.org/10.1007/978-3-031-14023-5_5

infectious diseases to chronic degenerative conditions and even cancer at signaling level and microRNA level (Goldman and Pranikoff 2011; Al-Bahrani et al. 2013; Bahitham et al. 2014; Ebata et al. 2016; Di Ciaula et al. 2017, 2018; Sauerbruch et al. 2021; Sergi 2021; Zailaie and Sergi 2022). The improvement of laboratory information systems has been vital in tracking the amount of data related to microbiology and correlating this data with clinical and surgical pathology data (Sergi 2022b). There is an intriguing consideration involving the gut and brain, and this can go through the bile and the hormonal functions of the liver. Bile acids are signaling hormonal molecules involved in regulating different metabolic signaling pathways (Monteiro-Cardoso et al. 2021). Bile acids can bind and signal through their receptors. The gut microbiome finely modulates this capability. Much is known about the gut-brain axis, but this axis may be expanded in the future involving the bile microbiome. Bacterial microorganisms have been demonstrated to dispatch signals to the brain (central nervous system) via their metabolite molecules (Monteiro-Cardoso et al. 2021). There is a strong possibility that bile acids may be a direct link between the bile and gut microbiome. In this consideration and current thoughts, proper investigation of the bile microbiome in aging and diseases is crucial (Di Carlo et al. 2018, 2019, 2021; Serra et al. 2018, 2021). This chapter examines the bile microbiome and scrutinizes the relationship between aging and diseases.

5.2 Gallbladder and Disease

The gallbladder is a vesicle located on the inferior site of the liver with the function of draining the bile from the hepatocytes (Keplinger and Bloomston 2014; Sergi 2020). The bile, which is also called gall, is composed of cholesterol ($C_{27}H_{46}O$), bile salts, also often called, in the gastrointestinal literature, bile acids, bilirubin ($C_{33}H_{36}N_4O_6$), which is the principal mammalian bile pigment, water, potassium, sodium (*natrium*), copper (*cuprum*), and, among other components, other metals. Bile is crucial for digestion. It separates fats into fatty acids. It can be taken slowly into the human body by the gastrointestinal tract. The gallbladder exhibits several disorders ranging from infectious diseases to chronic degenerative conditions and even cancer. This small pouch collects bile produced by the liver. Still, a change in bile mixture, particularly in cholesterol, is considered accountable for one of the most common gallbladder diseases, which is labeled as cholelithiasis or gallstones (Lamont and Carey 1992; Apstein and Carey 1996).

5.3 Microbiome Venues

In the most recent years, and probably approximately 1–2 decades, microorganisms in the body's reservoirs, mostly in the gut, have acquired tremendous importance for the health of individuals of any age. The change in the gut microbiota has been associated with ulcerative colitis or Crohn's disease (inflammatory bowel disease) and neoplastic conditions ranging from epithelial to mesenchymal tissue of origin and located in the colon and the rectum (Di Carlo et al. 2018, 2019; Serra et al. 2018, 2021; Vaishnavi et al. 2018; Granado-Serrano et al. 2019; Gutierrez-Repiso et al. 2019; Wirbel et al. 2019; Ortega et al. 2022). The dysbiosis is critical in determining a change in the milieu that may undermine the correct proliferation of the cells. It triggers a change in the genome or epigenome that, in conditions of p53 tumor suppressor gene or other tumor suppressor genes altered, can have dreadful consequences. It may elicit a chain reaction to develop neoplastic cells and immortalize cancer stem cells (Breton et al. 2022; Song and Zhang 2022). Most recent studies have been carried out to verify the link between the community of microorganisms detected in an organ called the microbiome and a particular disease. The advances in molecular biology tools to study the genome, transcriptome, and epigenome have been galvanizing the scientific community (Mars et al. 2020; Ohno 2020; Vasaikar et al. 2021). The number of gene controllers and dysbiosis effectors may be the target of the most imminent therapeutic frontline. In Fig. 5.1, the role of the microbiome in protecting the human body and the potential alterations are depicted.

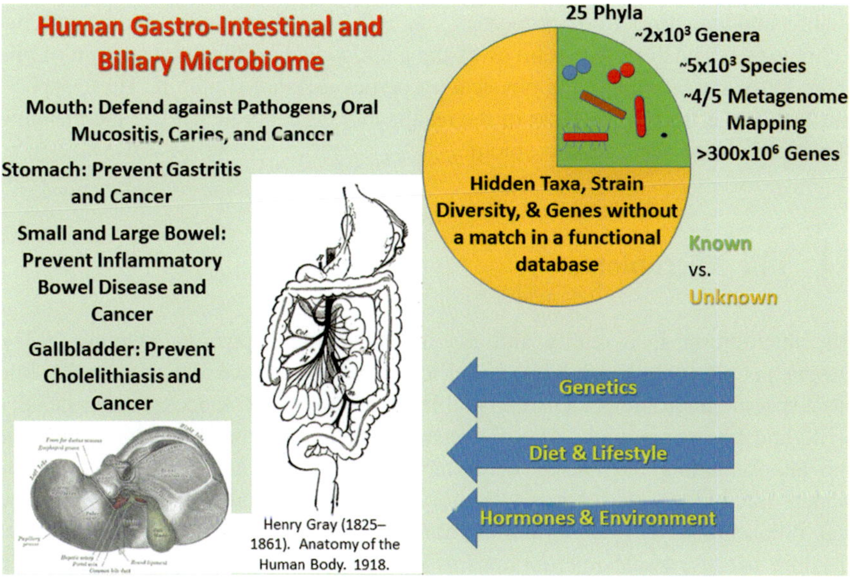

Fig. 5.1 Human Gastro-Intestinal and Biliary Microbiome. They are critical for the prevention of diseases at different anatomic levels. Still, genetics, diet components, lifestyle, hormonal status, and environmental factors may alter the microbiota and their preventive action against diseases. A complicated picture is demonstrated by our relative knowledge about the microorganisms, with only one-fourth known. About 75% of the phyla, genera, species, and genes remain unknown and need further investigation.

All this big data and single-cell sequencing data have been vital to enchanneled in several platforms. Some of them are on the clouds (Members 2017; Rea et al. 2020). These data, either generated by microbiological investigation followed by sequencing and transcriptomic analysis or directly derived by the next-generation sequencing (NGS) analysis, are paramount in science. These data have been crucial to use by computational analytics experts. Deep neural network applications have exponentially grown in the last couple of years. We can wisely and thoroughly explore the microbiome and gene expression profiles, labeled as metatranscriptomics, proteins, metaproteomics, and metabolites, which are tackled in metabolomics (Sergi 2022a; Shiroma et al. 2022). Metagenomics includes not recent technologies to investigate the detailed composition of the microbial community. However, metagenomics also targets the microbiome in an organ system. It occurs by identifying genetic information of microorganisms by NGS. Specifically, it takes place after the DNA is extracted from the samples preserved at the bedside (Anto and Blesso 2022; Qu et al. 2022; Sergi 2022a). As far as we know, the technologies that we have in our hands allow us to explore the gallbladder microbiome using sequencing technologies and computational methods to serve in the diagnosis and therapy of both benign and malignant conditions. Metagenomic methodologies are accurate in gathering data on several different communities of microorganisms in both healthy and sick patients. However, despite the number of technologies that show a refined and meticulous target, there is still the challenge of the sample quality and quantity. Both properties can influence the results of the applying technologies. The disease that involves the gallbladder can affect the amount of bile taken for examination and the physical properties associated with it. These aspects can undermine the effectiveness of the results and probably jeopardize the meaning of research using the bile microbiome.

5.4 Bile Microbiome

Bile microbiome is a reality and not an ephemeral happening (Manos 2022). Despite people assuming that bile is sterile for many years, we now assume that bile has a specific microbiome. There is real interaction between mucosa, either healthy or ill, and bile compounds. The discrepancy in bile composition, mainly bile salt and fats, including fatty acids, cholesterol, and lecithin, may influence the gallbladder microbiome. In addition, the story may be more complicated if we assume that bile salts have precise antimicrobial properties. These properties are not absolute because some microorganisms can survive even rigid and inhospitable milieus. In the old literature, it has been described the survival of a few microorganisms in the biliary environment. They include Salmonella spp and other enteric commensal and pathogenic microorganisms. Notably, *Vibrio cholera*, *Listeria monocytogenes*, *Campylobacter jejuni*, and of course, *Escherichia coli* have been characterized. Some studies have pointed out that their role as triggering or co-triggering factors influencing diseases involving the gall bladder (van

Velkinburgh and Gunn 1999; Prouty and Gunn 2000; Bina and Mekalanos 2001; Sleator et al. 2005; Torres et al. 2007; Crawford et al. 2008). However, the story does not seem to end here because other studies on bacterial bile composition have identified microorganisms that imply a quite likely carcinogenetic role of these or other bacteria and their waste products in human bile through the chronic inflammation pathway (Sharma et al. 2007; Nath et al. 2010; Koshiol et al. 2016). There is, however, still debate regarding the contribution of initiating factors in continuing the pathways that promote carcinogenesis. In other words, it is not clear if cancer stem cells may be elicited by either the non-bacterial initiating factors or the microorganisms. The gastrointestinal tract only conceals over 500 distinctive microbial taxa containing an estimated 1014 microorganisms in humans. This number significantly exceeds the number of cells in the human body. Since most of these microbes cannot be cultivated in vitro, the operating bases of these interactions remain challenging to investigate. These interactions would be very helpful in examining the human gastrointestinal microbiome-related carcinogenesis but have mainly persisted elusive. Finally, there is no safe haven in thinking about which role the fungal elements or mycetoma. This aspect is quite important because *Aspergillus flavus* and the derived aflatoxins as well as other fungi, have been linked to liver cancer (McCullough and Lloyd 2019; Benkerroum 2020; Mamo et al. 2021; Dhakal and Sbar 2022).

5.5 Bile Microbiome in Aging and Cancer

A number of studies have identified specific pathogens linked to cancer in the biliary and gastrointestinal tract. Studies involving the culture of bile samples collected by endoscopic retrograde cholangiopancreatography (ERCP) have clearly detected a pattern of bacteria isolated in patients with biliary and pancreas disease linked with the survival rate of individuals suffering from neoplastic processes (Di Carlo et al. 2018, 2019; Serra et al. 2018, 2021; Sergi et al. 2019). In addition, other bile elements, such as cholesterol and its derivates, have clearly indicated a confused determination of cancer and microbiome. Biochemical mechanisms also involving microbial activities on sterols, such as cholesterol in the genesis of benign and malignancy gallbladder disorders, are elusive (Crawford et al. 2008; Nath et al. 2010; Phelan et al. 2017; Sergi et al. 2019; Kiss et al. 2020; Gruner and Mattner 2021). The immunoregulation role of bile acids has been examined and is under the lens for hepatic and colonic diseases (Li and Lu 2018; Sipe et al. 2020). Recent proof implies that both biological sex and age may affect variations in the microbial arrangement of the gastrointestinal microbiota. The relationship between these vital parameters on bile microbiota can be studied both pro- and retrospectively. However, prospective investigations may last decades to conclude without the option of an animal model. Thus, retrospective investigations may elucidate better associations betweesn gastrointestinal microbiota and cancer because of the nature of these long-term studies. Despite the potential bias of confounder factors, there is probably a compromise if these studies may involve patients that are followed up for decades, and bile samples are either immediately analyzed or snap-frozen for future studies. It may also be critical that such studies should involve a thorough nutritional analysis of some patients with data on daily activities. Such a tremendous amount of information may be vital in studying confounder factors using computational analysis.

In a quality assurance (QA)-certified academic unit of a single institution, 449 positive biliary cultures from 172 subjects with several conditions of the biliopancreatic system (in-patients vs. out-patients) were thoroughly explored for aerobic, anaerobic, and fungal microorganisms in Serra et al. (Serra et al. 2021). The subjects were graded into four age intervals. The age intervals were 22−66, 67 −74, then, 75−81, and, ultimately, 82−93 years. The follow-up lasted five years. In this study, Gram-positive organisms (bacteria) were negatively linked with age only in the multivariate analysis with an astounding α-level of significance of 0.017. In this study, younger patients were more prone to harbor Gram-positive bacteria. In the elderly, patients were more likely to present with Gram-negative bacteria. In this investigation, there was a well-definite significant association with the male sex analyzing the data with both univariate and multivariate analysis. The most common strain found in subjects with Gram-positive bacteria was *Enterococcus spp* except for individuals aged 67 − 74 years for both males (95.2%) and females (80.9%). In each group taken under the lens, *Escherichia coli* and *Klebsiella spp* were most often seen. There is an analogy of the results discovered for non-fermenting and Gram-negative bacterial bacilli, including *Stenotrophomonas* and *Pseudomonas spp*. This remarkable investigation boosts the bond of chronological age and sex with the composition of bile microbiota. It suggests that additional inquiries may be needed in aiming properly at the aging microbiome.

Substantially, the human gastrointestinal tract and bile microbiota can be modified by the daily diet, age, exposure to drugs (e.g., antibiotics), and diseases. With regard to the mucosal whole barrier depletion and microorganismic "diversity loss" are essential conditions in some forms of ulcerative colitis and Crohn's disease (inflammatory bowel disease)s (Cai et al. 2022; Fobofou and Savidge 2022; Manos 2022; Thomas et al. 2022)). There is some crucial evidence pointing out that age and sex may impact fluctuations in the mixture of microorganisms of gut and bile microbiota (Di Carlo et al. 2018, 2019; Serra et al. 2018, 2021). However, we do not know yet how much correlation may be present between gut microbial diversity and bile microbial diversity. Thus, it may be important to check and target these fields in the future. The increased aging of the population worldwide, but mainly in the Western countries, China, and Japan, would indicate that the elderly are in good health.

In the last two decades of this age group, it may be key to investigate the alteration of the biliary flow determined by cancer or inflammation, including anastomotic stenosis that can occur following orthotopic liver transplantation. Not to forget the mentioned concept that biliary disease with gallstones and chronic pancreatitis are two very frequent causes in the elderly for an elective or emergency based endoscopic ERCP. Infection of the biliary tract is a widespread cause of bacteremia. It is associated with high morbidity and mortality. Biliary bacteremia occurs predominantly in patients affected with underlying co-morbidities (Agrusa et al. 2014, 2016). At this time, bile should not be considered a sterile fluid anymore. It relies on reports that a microbial ecosystem happens in individuals with and without hepatobiliary disorders [(Sergi et al. 2017; Molinero et al. 2019a, b). It is important to emphasize that the biliary system continuously communicates with the gut microbiota. Microbial products have been suggested as probable triggers for biliary diseases. However, the role of initiation and promotion and the separation of these roles for the microorganisms may acquire challenges currently and in the future (Agrusa et al. 2016). The influences of sex and aging on gut microbiota composition have been substantially described, but no comprehensive data are present for the bile (Boesjes and Brufau, 2014). Practically inexistent is the contemporaneous evaluation of the gut and bile microbiota in either animal models or humans. Senescence can indeed affect gut bacterial species in composition and metabolic functions (Kelly 2019). If this alteration can be linked to several gut-related diseases in the host, it remains to be correlated with further studies. These studies may be the principle to study microbiologic colonization in perinatology, interpret epigenetic data in inflammatory bowel disease, and shape new drugs in the nearest future with or without pandemic (Khan and Sergi 2020; Khan et al. 2021). In line with these studies, Di Carlo et al. carried out a cross-sectional survey of human bile samples to assess microbiota typing (Di Carlo et al. 2019). These authors found several strains that were the most frequent bacteria identified by post-hoc Z-test. They were *Pseudomonas spp* (27.6%), *Escherichia coli* (12.9%), *Stenotrophomonas spp* (11.4%), *Klebsiella spp* (9.8%), and *Enterococcus spp* (8.2%). On the other hand, the lowest isolates were *Brevundimonas spp* (1.1%), *Streptococcus spp* (0.9%), *Proteus spp* (0.7%), *Serratia spp* (0.7%), *Sphingomonas spp* (0.7%), *Delftia spp* (0.7%), *Elizabethkingia spp* (0.7%), *Sphingobacterium spp* (0.2%), *Staphylococcus spp* (0.2%), *Morganella spp* (0.2%), *Pantoea spp* (0.2%), *Aeromonas spp* (0.2%), and the fungal strain represented by *Candida spp* (1.3%). In addition, these authors considered bacilli classes including Gram-positive, *Enterobacteriaceae*, and no *Enterobacteriaceae*. The most frequently recognized microorganisms were *Pseudomonas spp* (47.5%) and *Stenotrophomonas spp* (19.5%) among non-fermenting Gram-negative bacteria, *Escherichia coli* (41.7%) and *Klebsiella spp* (31.6%), among *Enterobacteriaceae*. Moreover, *Enterococcus* spp (88.1%) was the most frequent strain among Gram-positive bacteria. The lowest isolated microorganisms identified were *Achromobacter spp* (5.4%), *Brevundimonas spp* (1.9%), *Delftia spp* (1.1%), and *Sphingomonas spp* (1.1%), as well as *Elizabethkingia spp.* (1.1%), and *Sphingobacterium spp* (0.4%) among non-fermenting Gram-negative microorganisms. In the group of *Enterobacteriaceae*, *Proteus spp* (2.2%), *Serratia spp* (2.2%), *Morganella spp.* (0.7%), and *Pantoea spp* (0.7%) have been identified. Finally, *Streptococcus spp* (9.5%) and *Staphylococcus spp* (2.4%) were the culprit bacteria among Gram-positive bacteria.

The analysis into groups showed that among the Gram-positive bacteria, the most common strain was *Enterococcus spp* (Group M: 95.2%; Group F: 80.9%). On the other side, *Staphylococcus sp.* was the less represented strain (Group M: < 0.1%;

Group F: 4.8%). Conversely, among *Enterobacteriaceae*, Gram-negative bacteria, the most frequently identified strains were *Escherichia coli* (Group M: 36.9%; Group F: 49.1%) and Klebsiella spp. (Group M: 33.3%; Group F: 29.1%). On the other side, *Proteus spp.* (Group M: 2.4%; Group F: 1.8%), *Morganella spp.* (Group M: < 0.1%; Group F: 1.8%), *Serratia spp.* (Group M: 3.6%; Group F: < 0.1%), and *Pantoea spp* (Group M:1.2%; Group F: < 0.1%) were the less represented bacteria. Concerning the non-fermenting Gram-negative bacteria, the most frequent strains were Pseudomonas spp. (Group M: 45.7%; Group F: 50.0%) and *Stenotrophomonas sp.* (Group M: 20.3%; Group F:18.5%), while the less frequent bacterial microorganisms were *Sphingobacterium spp* (Group M: < 0.1%; Group F: 0.9%), *Sphingomonas spp* (Group M: 0.6%; Group F: 1.8%), *Brevundimonas spp* (Group M: 2.0%; Group F:1.8%), in addition to *Delphia spp.* (Group M: < 0.1%; Group F:0.9%) and *Elizabethkingia spp* (Group M: 1.3%; Group F: 0.9%).

About the relationship between Groups of males and females, no noteworthy variations among different varieties were identified. Analyses into age cohorts demonstrated that the Enterococcus spp was the variety most meaningfully present in all groups apart from Group B among Gram-positive bacteria. Among Gram-negative bacteria, the highest proportion was for gut commensal E. coli and the bacterium *Klebsiella spp.* Among non-fermenting Gram-negative bacteria, the highest proportion was *Pseudomonas spp.*, and *Stenotrophomonas spp*; while *Sphingobacterium*, *Brevundimonas*, *Delftia*, *Elizabethkingia*, and *Sphingomonas*, were the less epitomized bacteria. *Sphingobacterium spp.*, *Delftia*, and *Elizabethkingia* existed with the lowest frequency in all categories.

Regarding the Gram-negative bacteria, the bacteriological species analysis demonstrated *Klebsiella pneumoniae* sequence. The inter-assay statistical test between the different cohorts by age did not disclose any significant difference in rates of bacteriological strains. The mortality rate was 81.98%, although co-morbidities were not tackled in the paper. Co-morbidities are frequent a complicating factor in the elderly. The follow-up analysis of patients by groups was also particularly triggering. It disclosed that one-third of individuals died in the ED group, 21.52% in the non-ED-1, and 23.26% in the non-ED-2 group. No significant variation was identified between groups of survival (ED groups).

These results harmonize with preceding investigations that consider older individuals at risk of altering the bowel microbiome with a tremendous and disease-prone loss of permeability. This loss of permeability will induce the possibility of influence dysbiosis (Hao et al. 2022; Liu et al. 2022). This data seems established by the findings of the low frequency of *Sphingobacterium spp*, *Delftia spp*, and *Elizabethkingia spp* by aging. It confirms that aging and its progress open the doors to the potential consequence for health and age-related disease [32,33]. There is a bunch of literature promoting yogurt and probiotics in the elderly. Both yogurt and probiotics have been associated with health benefits (Breton et al. 2022). In the Serra et al. investigation (Serra et al. 2021), biliary obstruction caused by either bile duct stones or malignancy accounts for more than 90% of all surgical procedures performed in their series and increases specifically with age [6,9]. *Enterococcus spp.* was the prevalent strain in the bacterial Gram-positive (genera).

The percentage of *Enterococcus spp.* was substantial in all patients with the exception of patients with ages in the range of 67 − 74 years.

In our understanding of gut physiology and microbiology, enterococci are comparatively minor elements of the human gut microbiota (probably less than 1% of all gut commensals). We barely begin to understand the impact of the host's characteristics (age, nutrition, food additive/preservatives, healthiness status, and antibiotic treatment) explains for the miscellany and people incidence of various bacteriological groups. Other investigations recognized Enterococcus spp. as the most common bile isolated bacterium in individuals with biliary tract infection poisoned with bacteremia. Despite the statistical and epidemiological analysis on predominant species, the Enterococcus genus disclosed a low pervasiveness of vancomycin-related antibiotic resistance in regions of Southern Europe (Ayobami et al. 2020). Additionally, other classes are evolving as pathogens in both hospital and long-term healthcare facilities-acquired infections [36]. There have been individual reports of dominant *E. gallinarum* in the geographical area of Southern Europe (Tsuchiya et al. 2018). Therefore, the incidence of Enterococcus spp. Any of these studies should be interpreted as a red flag, mainly if the bile sample originates from the elderly, and prompt the physicians to initiate a specific antibiotic regimen. Moreover, in this age group, co-morbidities are quite often present. It has been indicated that the insertion of a surgical stent using ERCP for palliative management of an obstructed bile duct is accountable for the highest percentage of *Enterococcus spp.* in cancer patients suffering from cancer. *Enterococcus spp.* represents a large genus of specific bacteria with lactic acid production of the phylum *Bacillota*. *Enterococci* are gram-positive cocci that frequently arise in pairs, which are also called "diplococci", or short chains. *Enterococci* are challenging to differentiate from *Streptococci* on physical features alone. In the human gut, two strains are common commensal organisms: *E. faecalis* (90–98%) and *E. faecium* (2–10%). There are rare constellations of infections identifying other species, including *E. gallinarum*, *E. raffinosus*, and *E. casseliflavus*. Of note, *Enterococci* are facultative anaerobic microorganisms. These bacteria are skilled in cellular respiration in both oxygen-poor and oxygen-rich environments. *Enterococci* are not capable of developing spores and demonstrate tolerance to a wide range of environmental conditions, including cold and hot temperatures (10–45 °C), variable tolerance to a range of pH (4.6–9.9), and high concentrations of sodium chloride, and display γ-hemolysis on blood agar.

In the past two decades, the literature emphasized the severe role of *Streptococcus spp.* Choosing antibiotics wisely for therapy and preventing biliary septicemia is critical mainly because this strain can involve the cardiovascular system (Serra et al. 2020). In their study, *Streptococcus spp.* has been sequestered at the lowest rate. This was the case, especially in male patients. *Streptococcus spp.* is common in gastrointestinal studies that include patients with biliary gallstones (Alves et al. 2016). Among Gram-negative bacteria, the authors found a prevalence of *Pseudomonas spp.* and *Stenotrophomonas spp* in addition to *E. coli*, *Klebsiella spp*. These bacteria are characteristically reported in investigations led in individuals with biliary tract infections. They are often having the same pathogens isolated from blood. In their study, the data of bacteria, including multidrug-resistant (MDR) pathogens, underlines how close we are to a hospital pathogenic flora. In recent years, the epidemiological analysis of bacteria isolated from bile samples with positive culture has changed. Although some bacteria, such as *Escherichia coli* and *Klebsiella spp.*, have been reported in the last decade of this century, their specific resistance to drugs (antibiotics resistance) has recently shifted, especially in the Mediterranean region (Serra et al. 2021).

It has been indicated that the epidemic of *K. pneumoniae* and *Acinetobacter baumannii* in a surgical setting and intensive care units in Sicily, Italy, was life-threatening. It is important to emphasize that it is not a remote possibility only limited to the Southern regions of Europe, but it can involve other areas of the Mediterranean Sea (Falagas et al. 2008; Tran et al. 2017; Zhao et al. 2017; Di Carlo et al. 2021). There may be no surprise that Gram-negative bacteria are more often isolated in female patients than in male patients. This is because Gram-negative bacteria play a significant role in several diseases of the women's reproductive system, and the vicinity of the anus to the vagina and clitoris is a predisposing factor for such conditions of infections. In fact, the urinary tract is the most common main cause of early gram-negative disease. The most recent investigations point out that the intestinal barrier is made of several integrated physical, biochemical, immunological, and microbiological constituents. Such components evolve across the entire life (Agrusa et al. 2014, 2016; Rodolico et al. 2017). Apart from the indiscriminate use of antibiotics in childhood and adulthood, antibiotic prophylaxis in surgery has led to a change in the gut vegetation of surgical patients. Immunocompromised individuals have a scant gut microbiome repertory. It may promote colonization by multi-frug-resistant (MDR) bacteria quite quickly. It remains a huge challenge in modern healthcare [45]. There are also uncommon bacteria, and the authors describe *Sphingobacterium spp* and *Sphingomonas spp*. *Sphingobacterium* is a rare genus, which is located in the family *Sphingobacteriaceae*. This genus is characterized by the high rate of sphingophospholipids as lipid components. On the other hand, *Sphingomonas* was defined in 1990 as a group of Gram-negative bacteria. They are rod-shaped, chemotrophic, heterotrophic, and strictly aerobic bacteria. They retain ubiquinone 10 as their primary respirational quinone. *Sphingomonas* contain glycosphingolipids, specifically ceramide, as a replacement for lipopolysaccharide in their cell envelopes. Typically, they produce yellow-pigmented colonies. The glycosphingolipids help to protect the microorganisms from antibacterial substances. Different most Gram-negative bacteria, *Sphingomonas* harbors endotoxins. *Sphingomonas* have a hydrophobic surface characterized by the glycosphingolipids' carbohydrate part's short nature. *Sphingomonas* genus includes more than 20 strains that are quite distinct in terms of their phylogenesis, ecology, and physiology. As an end result, *Sphingomonas* are subdivided into different genera: *Sphingomonas*, *Sphingobium*, *Sphingopyxis*, *Sphingosinicella*, and *Novosphingobium*. These genera are often referred to together as *Sphingomonads*. Both *Sphingobacterium spp* and *Sphingomonas spp* have been reported as constituents of the biliary microbiome in a subgroup of patients with biliary lithiasis and a characteristic Mediterranean diet composed of caciocavallo cheese (De Pasquale et al. 2014). This cheese is a type of stretched-curd cheese made from sheep's or cow's milk. It is produced throughout Southern Italy.

In conclusion, the bile microbiome is a recent field of intense investigation with bactibilia patterns described that may signify the emergence and assortment of some bacteria such as *K. pneumoniae* and *E. coli*. These bacteria have been blamed for hospital outbreaks in Southern Europe (Di Carlo et al. 2018, 2019; Serra et al. 2018, 2021; Sergi et al. 2019). In the future, metagenomics and metatranscriptomics will be critical. They will play a significant role in accumulating big data, hopefully delivering the promise of a profile characteristic for health status, generational nutrition, and diseases (Aguiar-Pulido et al. 2016; Just et al. 2018; Pratt et al. 2021). Bayesian approaches and mathematical investigations have been increasingly applied in our current experiments. However, the challenges of predicting routes of contamination remain ahead. It will be essential to associate liquid biopsy procedures with the existing approaches. In this sense, liquid chromatography-mass spectrometry is an analytical chemistry method that conglomerates the physical separation critical capabilities of liquid chromatography associated with the mass analysis capabilities of mass spectrometry. The genetic landscape and hospital-based environments may bring several vital attitudes, which will have profound consequences and practical repercussions for patients, healthcare suppliers, and drug companies.

Acknowledgements Dr. C. Sergi's research has been funded by the generosity of the Children's Hospital of Eastern Ontario, Ottawa, Ontario, and the Stollery Children's Hospital Foundation and supporters of the Lois Hole Hospital for Women through the Women and Children's Health Research Institute (WCHRI, Grant ID #: 2096), Natural Science Foundation of Hubei Province for Hubei University of Technology (100-Talent Grant for Recruitment Program of Foreign Experts Total Funding: Digital PCR and NGS-based diagnosis for infection and oncology, 2017-2022), Österreichische Krebshilfe Tyrol (Krebsgesellschaft Tirol, Austrian Tyrolean Cancer Research Institute, 2007 and 2009 - "DMBTI and cholangiocellular carcinomas" and "Hsp70 and HSPBP1 in carcinomas of the pancreas"), Austrian Research Fund (Fonds zur Förderung der wissenschaftlichen Forschung, FWF, Grant ID L313-B13), Canadian Foundation for Women's Health ("Early Fetal Heart-RES0000928"), Cancer Research Society (von Willebrand factor gene expression in cancer cells), Canadian Institutes of Health Research (Omega-3 Fatty Acids for Treatment of Intestinal Failure Associated Liver Disease: A Translational Research Study, 2011-2014, CIHR 232514), and the Saudi Cultural Bureau, Ottawa, Canada. The funders had no role in study design, data collection, and analysis, decision to publish, or preparation of the current manuscript.

References

Agrusa A, Frazzetta G, Chianetta D, Di Giovanni S, Gulotta L, Di Buno G et al (2016) "Relaparoscopic" management of surgical complications: The experience of an Emergency Center. Surg Endosc 30(7):2804–2810. https://doi.org/10.1007/s00464-015-4558-2

Agrusa A, Romano G, Frazzetta G, Chianetta D, Sorce V, Di Buono G et al (2014) Role and outcomes of laparoscopic cholecystectomy in the elderly. Int J Surg 12(Suppl 2):S37-39. https://doi.org/10.1016/j.ijsu.2014.08.385

Aguiar-Pulido V, Huang W, Suarez-Ulloa V, Cickovski T, Mathee K, Narasimhan G (2016) Metagenomics, metatranscriptomics, and metabolomics approaches for microbiome analysis. Evol Bioinform Online 12(Suppl):5–16. https://doi.org/10.4137/EBO.S36436

Al-Bahrani R, Abuetabh Y, Zeitouni N, Sergi C (2013) Cholangiocarcinoma: risk factors, environmental influences and oncogenesis. Ann Clin Lab Sci 43(2):195–210

Alves JR, Silva Rdo C, Guerra SC, Freitas TT, Souza DL, Amico EC (2016) Microbiological analysis of bile in patients with benign and malignant biliopancreatic diseases and its consequences. Arq Gastroenterol 53(3):156-162. https://doi.org/10.1590/S0004-28032016000300007

Anto L, Blesso CN (2022) Interplay between diet, the gut microbiome, and atherosclerosis: role of dysbiosis and microbial metabolites on inflammation and disordered lipid metabolism. J Nutr Biochem 105:108991. https://doi.org/10.1016/j.jnutbio.2022.108991

Apstein MD, Carey MC (1996) Pathogenesis of cholesterol gallstones: a parsimonious hypothesis. Eur J Clin Invest 26(5):343–352. https://doi.org/10.1046/j.1365-2362.1996.148287.x

Ayobami O, Willrich N, Reuss A, Eckmanns T, Markwart R (2020) The ongoing challenge of vancomycin-resistant Enterococcus faecium and Enterococcus faecalis in Europe: an epidemiological analysis of bloodstream infections. Emerg Microbes Infect 9(1):1180–1193. https://doi.org/10.1080/22221751.2020.1769500

Bahitham W, Liao X, Peng F, Bamforth F, Chan A, Mason A et al (2014) Mitochondriome and cholangiocellular carcinoma. PLoS ONE 9(8):e104694. https://doi.org/10.1371/journal.pone.0104694

Benkerroum N (2020) Aflatoxins: producing-molds, structure, health issues and incidence in Southeast Asian and Sub-Saharan African Countries. Int J Environ Res Public Health 17(4). https://doi.org/10.3390/ijerph17041215

Bina JE, Mekalanos JJ (2001) Vibrio cholerae tolC is required for bile resistance and colonization. Infect Immun 69(7):4681–4685. https://doi.org/10.1128/IAI.69.7.4681-4685.2001

Boesjes M, Brufau G (2014) Metabolic effects of bile acids in the gut in health and disease. Curr Med Chem 21(24):2822–2829

Breton J, Galmiche M, Dechelotte P (2022) Dysbiotic gut bacteria in obesity: an overview of the metabolic mechanisms and therapeutic perspectives of next-generation probiotics. Microorganisms 10(2). https://doi.org/10.3390/microorganisms10020452

Cai J, Sun L, Gonzalez FJ (2022) Gut microbiota-derived bile acids in intestinal immunity, inflammation, and tumorigenesis. Cell Host Microbe 30(3):289–300. https://doi.org/10.1016/j.chom.2022.02.004

Crawford RW, Gibson DL, Kay WW, Gunn JS (2008) Identification of a bile-induced exopolysaccharide required for Salmonella biofilm formation on gallstone surfaces. Infect Immun 76(11):5341–5349. https://doi.org/10.1128/IAI.00786-08

De Pasquale I, Di Cagno R, Buchin S, De Angelis M, Gobbetti M (2014) Microbial ecology dynamics reveal a succession in the core microbiota involved in the ripening of pasta filata caciocavallo pugliese cheese. Appl Environ Microbiol 80(19):6243–6255. https://doi.org/10.1128/AEM.02097-14

Dhakal A, Sbar E (2022) Aflatoxin toxicity. In: StatPearls (Treasure Island (FL))

Di Carlo P, Serra N, D'Arpa F, Agrusa A, Gulotta G, Fasciana T et al (2019) The microbiota of the bilio-pancreatic system: a cohort, STROBE-compliant study. Infect Drug Resist 12:1513–1527. https://doi.org/10.2147/IDR.S200378

Di Carlo P, Serra N, Gulotta G, Giammanco A, Colomba C, Melfa G et al (2018) Bactibilia in diseases of the biliary tract and pancreatic gland in patients older than 80 years: a STROBE-retrospective cohort study in a teaching hospital in Italy. Eur J Clin Microbiol Infect Dis 37(5):953–958. https://doi.org/10.1007/s10096-018-3213-y

Di Carlo P, Serra N, Lo Sauro S, Carelli VM, Giarratana M, Signorello JC et al (2021) Epidemiology and pattern of resistance of gram-negative bacteria isolated from blood samples in hospitalized patients: a single center retrospective analysis from Southern Italy. Antibiotics (Basel) 10(11). https://doi.org/10.3390/antibiotics10111402

Di Ciaula A, Garruti G, Lunardi Baccetto R, Molina-Molina E, Bonfrate L, Wang DQ et al (2017) Bile acid physiology. Ann Hepatol 16(Suppl. 1: s3–105):s4–s14. https://doi.org/10.5604/01.3001.0010.5493

Di Ciaula A, Wang DQ, Portincasa P (2018) An update on the pathogenesis of cholesterol gallstone disease. Curr Opin Gastroenterol 34(2):71–80. https://doi.org/10.1097/MOG.0000000000000423

Ebata T, Ercolani G, Alvaro D, Ribero D, Di Tommaso L, Valle JW (2016) Current status on cholangiocarcinoma and gallbladder cancer. Liver Cancer 6(1):59–65. https://doi.org/10.1159/000449493

Falagas ME, Rafailidis PI, Matthaiou DK, Virtzili S, Nikita D, Michalopoulos A (2008) Pandrug-resistant klebsiella pneumoniae, pseudomonas aeruginosa and acinetobacter baumannii infections: characteristics and outcome in a series of 28 patients. Int J Antimicrob Agents 32 (5):450–454. https://doi.org/10.1016/j.ijantimicag.2008.05.016

Fobofou SA, Savidge T (2022) Microbial metabolites: cause or consequence in gastrointestinal disease? Am J Physiol Gastrointest Liver Physiol 322(6):G535–G552. https://doi.org/10.1152/ajpgi.00008.2022

Goldman M, Pranikoff T (2011) Biliary disease in children. Curr Gastroenterol Rep 13(2):193–201. https://doi.org/10.1007/s11894-010-0169-1

Granado-Serrano AB, Martin-Gari M, Sanchez V, Riart Solans M, Berdun R, Ludwig IA , et al (2019) Faecal bacterial and short-chain fatty acids signature in hypercholesterolemia. Sci Rep 9 (1):1772. https://doi.org/10.1038/s41598-019-38874-3

Gruner N, Mattner J (2021) Bile acids and microbiota: multifaceted and versatile regulators of the liver-gut axis. Int J Mol Sci 22(3). https://doi.org/10.3390/ijms22031397

Gutierrez-Repiso C, Moreno-Indias, de Hollanda A, Martin-Nunez GM, Vidal J, Tinahones FJ (2019) Gut microbiota specific signatures are related to the successful rate of bariatric surgery. Am J Transl Res 11(2):942–952

Hao Y, Zhou P, Zhu YJ, Zou S, Zhao Q, Yu J et al (2022) Gut microbiota dysbiosis and altered bile acid catabolism lead to metabolic disorder in psoriasis mice. Front Microbiol 13:853566. https://doi.org/10.3389/fmicb.2022.853566

Just S, Mondot S, Ecker J, Wegner K, Rath E, Gau L et al (2018) The gut microbiota drives the impact of bile acids and fat source in diet on mouse metabolism. Microbiome 6(1):134. https://doi.org/10.1186/s40168-018-0510-8

Kelly PN (2019) Microbiota and age-related disease. Science 365(6457):996–996. https://doi.org/10.1126/science.365.6457.996-a

Keplinger KM, Bloomston M (2014) Anatomy and embryology of the biliary tract. Surg Clin North Am 94(2):203–217. https://doi.org/10.1016/j.suc.2014.01.001

Khan A, Das S, Sergi C (2021) Therapeutic potential of neu1 in Alzheimer's disease via the immune system. Am J Alzheimers Dis Other Demen 36:1533317521996147. https://doi.org/10.1177/1533317521996147

Khan A, Sergi C (2020) SAMHD1 as the Potential Link Between SARS-CoV-2 Infection and Neurological Complications. Front Neurol 11:562913. https://doi.org/10.3389/fneur.2020.562913

Kiss B, Miko E, Sebo E, Toth J, Ujlaki G, Szabo J et al (2020) Oncobiosis and microbial metabolite signaling in pancreatic adenocarcinoma. Cancers (Basel) 12(5). https://doi.org/10.3390/cancers12051068

Koshiol J, Wozniak A, Cook P, Adaniel C, Acevedo J, Azocar L et al (2016) Salmonella enterica serovar Typhi and gallbladder cancer: a case-control study and meta-analysis. Cancer Med 5 (11):3310–3235. https://doi.org/10.1002/cam4.915

Lamont JT, Carey MC (1992) Cholesterol gallstone formation. 2 Pathobiology and Pathomechanics. Prog Liver Dis 10:165–191

Li Y, Lu LG (2018) Therapeutic roles of bile acid signaling in chronic liver diseases. J Clin Transl Hepatol 6(4):425–430. https://doi.org/10.14218/JCTH.2018.00025

Liu J, Yang D, Wang X, Asare PT, Zhang Q, Na L et al (2022) Gut microbiota targeted approach in the management of chronic liver diseases. Front Cell Infect Microbiol 12:774335. https://doi.org/10.3389/fcimb.2022.774335

Mamo FT, Abate BA, Zheng Y, Nie C, He M, Liu Y (2021) Distribution of aspergillus fungi and recent aflatoxin reports, health risks, and advances in developments of biological mitigation strategies in China. Toxins (Basel) 13(10). https://doi.org/10.3390/toxins13100678.

Manos J (2022) The human microbiome in disease and pathology. APMIS. https://doi.org/10.1111/apm.13225

Mars RAT, Yang Y, Ward T, Houtti M, Priya S, Lekatz HR et al (2020) Longitudinal multi-omics reveals subset-specific mechanisms underlying irritable bowel syndrome. Cell 182(6):1460–1473 e1417. https://doi.org/10.1016/j.cell.2020.08.007

McCullough AK, Lloyd RS (2019) Mechanisms underlying aflatoxin-associated mutagenesis-Implications in carcinogenesis. DNA Repair (amst) 77:76–86. https://doi.org/10.1016/j.dnarep.2019.03.004

Members BIGDC (2017) The BIG Data Center: from deposition to integration to translation. Nucleic Acids Res 45(D1):D18–D24. https://doi.org/10.1093/nar/gkw1060

Molinero N, Ruiz L, Milani C, Gutierrez-Diaz I, Sanchez B, Mangifesta M et al (2019a) The human gallbladder microbiome is related to the physiological state and the biliary metabolic profile. Microbiome 7(1):100. https://doi.org/10.1186/s40168-019-0712-8

Molinero N, Ruiz L, Sanchez B, Margolles A, Delgado S (2019b) Intestinal bacteria interplay with bile and cholesterol metabolism: implications on host physiology. Front Physiol 10:185. https://doi.org/10.3389/fphys.2019.00185

Monteiro-Cardoso VF, Corliano M, Singaraja RR (2021) Bile acids: a communication channel in the gut-brain axis. Neuromolecular Med 23(1):99–117. https://doi.org/10.1007/s12017-020-08625-z

Nath G, Gulati AK, Shukla VK (2010) Role of bacteria in carcinogenesis, with special reference to carcinoma of the gallbladder. World J Gastroenterol 16(43):5395–5404. https://doi.org/10.3748/wjg.v16.i43.5395

Ohno H (2020) The impact of metabolites derived from the gut microbiota on immune regulation and diseases. Int Immunol 32(10):629–636. https://doi.org/10.1093/intimm/dxaa041

Ortega MA, Alvarez-Mon MA, Garcia-Montero C, Fraile-Martinez O, Guijarro LG, Lahera G et al (2022) Gut microbiota metabolites in major depressive disorder-deep insights into their pathophysiological role and potential translational applications. Metabolites 12(1). https://doi.org/10.3390/metabo12010050

Phelan JP, Reen FJ, Caparros-Martin JA, O'Connor R, O'Gara F (2017) Rethinking the bile acid/gut microbiome axis in cancer. Oncotarget 8(70):115736–115747. https://doi.org/10.18632/oncotargets.22803

Pratt M, Forbes JD, Knox NC, Bernstein CN, Van Domselaar G (2021) Microbiome-mediated immune signaling in inflammatory bowel disease and colorectal cancer: support from meta-omics data. Front Cell Dev Biol 9:716604. https://doi.org/10.3389/fcell.2021.716604

Prouty AM, Gunn JS (2000) Salmonella enterica serovar typhimurium invasion is repressed in the presence of bile. Infect Immun 68(12):6763–6769. https://doi.org/10.1128/IAI.68.12.6763-6769.2000

Qu Y, Su C, Zhao Q, Shi A, Zhao F, Tang L et al (2022) Gut microbiota-mediated elevated production of secondary bile acids in chronic unpredictable mild stress. Front Pharmacol 13:837543. https://doi.org/10.3389/fphar.2022.837543

Rea T, Guillari A, Sergi C, Serra N (2020) Mathematical models in nursing research. J Public Health Res 9(4):1952. https://doi.org/10.4081/jphr.2020.1952

Rodolico V, Di Carlo P, Gulotta G, D'Arpa F, Salamone G, Cocorullo G et al (2017) Intra-abdominal candida spp infection in acute abdomen in a quality assurance (QA)-certified academic setting. J Clin Pathol 70(7):579–583. https://doi.org/10.1136/jclinpath-2016-203936

Sauerbruch T, Hennenberg M, Trebicka J, Beuers U (2021) Bile acids, liver cirrhosis, and extrahepatic vascular dysfunction. Front Physiol 12:718783. https://doi.org/10.3389/fphys.2021.718783

Sergi C, Di Carlo P, Gulotta G, D'Arpa F (2019) Biliary microbiota in pancreatic cancer. HPB (oxford) 21(12):1790. https://doi.org/10.1016/j.hpb.2019.06.001
Sergi C, Shen F, Lim DW, Liu W, Zhang M, Chiu B et al (2017) Cardiovascular dysfunction in sepsis at the dawn of emerging mediators. Biomed Pharmacother 95:153–160. https://doi.org/10.1016/j.biopha.2017.08.066
Sergi CM (2020) Pathology of childhood and adolescence. Springer, An Illustrated Guide
Sergi CM (2021) Carcinoma of the liver in children and adolescents. In: Sergi CM (ed) Liver cancer (Brisbane (AU))
Sergi CM (2022a) Biorepository-a key component of research studies. Contemp Clin Trials 112:106655. https://doi.org/10.1016/j.cct.2021.106655
Sergi CM (2022b) Implementing epic beaker laboratory information system for diagnostics in anatomic pathology. Risk Manag Healthc Policy 15:323–330. https://doi.org/10.2147/RMHP.S332109
Serra N, Colomba C, Di Carlo P, Palermo G, Fasciana T, Giammanco A et al (2020) Infective endocarditis: preliminary results of a cohort study in the Southern Italian population. Cureus 12 (5):e8338. https://doi.org/10.7759/cureus.8338
Serra N, Di Carlo P, D'Arpa F, Battaglia E, Fasciana T, Gulotta G et al (2021) Human bile microbiota: a retrospective study focusing on age and gender. J Infect Public Health 14(2):206–213. https://doi.org/10.1016/j.jiph.2020.11.005
Serra N, Di Carlo P, Gulotta G, d' Arpa F, Giammanco A, Colomba C et al (2018) Bactibilia in women affected with diseases of the biliary tract and pancreas: a strobe guidelines-adherent cross-sectional study in Southern Italy. J Med Microbiol 67(8):1090–1095. https://doi.org/10.1099/jmm.0.000787
Sharma V, Chauhan VS, Nath G, Kumar A, Shukla VK (2007) Role of bile bacteria in gallbladder carcinoma. Hepatogastroenterology 54(78):1622–1625
Shiroma H, Shiba S, Erawijantari PP, Takamaru H, Yamada M, Sakamoto T et al (2022) Surgical treatment for colorectal cancer partially restores gut microbiome and metabolome traits. mSystems 7(2): e0001822. https://doi.org/10.1128/msystems.00018-22
Sipe LM, Chaib M, Pingili AK, Pierre JF, Makowski L (2020) Microbiome, bile acids, and obesity: how microbially modified metabolites shape anti-tumor immunity. Immunol Rev 295 (1):220–239. https://doi.org/10.1111/imr.12856
Sleator RD, Wemekamp-Kamphuis HH, Gahan CG, Abee T, Hill C (2005) A PrfA-regulated bile exclusion system (BilE) is a novel virulence factor in Listeria monocytogenes. Mol Microbiol 55(4):1183–1195. https://doi.org/10.1111/j.1365-2958.2004.04454.x
Song Q, Zhang X (2022) The role of gut-liver axis in gut microbiome dysbiosis associated NAFLD and NAFLD-HCC. Biomedicines 10(3). https://doi.org/10.3390/biomedicines10030524
Thomas JP, Modos D, Rushbrook SM, Powell N, Korcsmaros T (2022) The emerging role of bile acids in the pathogenesis of inflammatory bowel disease. Front Immunol 13:829525. https://doi.org/10.3389/fimmu.2022.829525
Torres AG, Tutt CB, Duval L, Popov V, Nasr AB, Michalski J et al (2007) Bile salts induce expression of the afimbrial LDA adhesin of atypical enteropathogenic Escherichia coli. Cell Microbiol 9(4):1039–1049. https://doi.org/10.1111/j.1462-5822.2006.00850.x
Tran GM, Ho-Le TP, Ha DT, Tran-Nguyen CH, Nguyen TSM, Pham TTN et al (2017) Patterns of antimicrobial resistance in intensive care unit patients: a study in Vietnam. BMC Infect Dis 17 (1):429. https://doi.org/10.1186/s12879-017-2529-z
Tsuchiya Y, Loza E, Villa-Gomez G, Trujillo CC, Baez S, Asai T et al (2018) Metagenomics of microbial communities in gallbladder bile from patients with gallbladder cancer or cholelithiasis. Asian Pac J Cancer Prev 19(4):961–967. https://doi.org/10.22034/APJCP.2018.19.4.961
Vaishnavi C, Samanta J, Kochhar R (2018) Characterization of biofilms in biliary stents and potential factors involved in occlusion. World J Gastroenterol 24(1):112–123. https://doi.org/10.3748/wjg.v24.i1.112

van Velkinburgh JC, Gunn JS (1999) PhoP-PhoQ-regulated loci are required for enhanced bile resistance in Salmonella spp. Infect Immun 67(4):1614–1622. https://doi.org/10.1128/IAI.67.4.1614-1622.1999

Vasaikar SV, Deshmukh AP, den Hollander P, Addanki S, Kuburich NA, Kudaravalli S et al (2021) EMTome: a resource for pan-cancer analysis of epithelial-mesenchymal transition genes and signatures. Br J Cancer 124(1):259–269. https://doi.org/10.1038/s41416-020-01178-9

Wirbel J, Pyl PT, Kartal E, Zych K, Kashani A, Milanese A et al (2019) Meta-analysis of fecal metagenomes reveals global microbial signatures that are specific for colorectal cancer. Nat Med 25(4):679–689. https://doi.org/10.1038/s41591-019-0406-6

Zailaie SA, Sergi CM (2022) MiR-126 in hepatocellular carcinoma and cholangiocellular carcinoma: a reappraisal with an in situ detection of miR-126. Ann Clin Lab Sci 52(1):73–85

Zhao J, Nian L, Kwok LY, Sun T, Zhao J (2017) Reduction in fecal microbiota diversity and short-chain fatty acid producers in Methicillin-resistant Staphylococcus aureus infected individuals as revealed by PacBio single molecule, real-time sequencing technology. Eur J Clin Microbiol Infect Dis 36(8):1463–1472. https://doi.org/10.1007/s10096-017-2955-2

Chapter 6
Gut Microbiota and Eating Disorders on the Extremes of Aging

Brandi Miller, Sidharth P. Mishra, Shalini Jain, Francesco Marotta, and Hariom Yadav

Abstract Eating disorders (EDs) are defined by persistent disturbances in eating behaviors and affect at least 9% of the population worldwide. Over time, they place physical, psychological, and social burdens on affected individuals and increase the risk for several chronic diseases. EDs (or "disordered eating") may present at any age, but they are most common in young adolescents and the aged population. The literature suggests that the bacteria living in the gut (gut microbiota) are linked with EDs. In particular, perturbations in the gut microbiota composition (gut dysbiosis) and changes in metabolite production influence brain health (and vice versa) via gut-brain communication, which may lead to severe disturbances in eating patterns. Gut-brain communication is facilitated by the vagus nerve (physical connection) and humoral molecules, including neurons/neurotransmitters, metabolites, gut hormones, and immune cells. This chapter aims to describe evidence of the link between the gut microbiota and ED prevalence at the extremes of aging. Understanding this link will aid in developing microbiome modulators as potential treatments for EDs in various populations.

Keywords Gut microbiota · Aging · Eating disorders · Gut · Brain · Gut-brain axis

Abbreviations

A	Akkermansia
AgRP	Agouti-related peptide
AN	Anorexia nervosa

B. Miller · S. P. Mishra · S. Jain · H. Yadav (✉)
USF Center for Microbiome Research, Microbiomes Institute, Center of Excellence for Aging and Brain Repair, Department of Neurosurgery and Brain Repair, University of South Florida, Tampa, FL, USA
e-mail: hyadav@usf.edu

F. Marotta
ReGenera R&D International for Aging Intervention, Milano, Lombardia, Italy

F. Marotta (ed.), *Gut Microbiota in Aging and Chronic Diseases*, Healthy Ageing and Longevity 17, https://doi.org/10.1007/978-3-031-14023-5_6

B	Bifidobacterium
BDNF	Brain-derived neurotrophic factor
BED	Binge eating disorder
BN	Bulimia nervosa
BSH	Bile salt hydrolase
C	Clostridium
CA	Cholinic acid
CCK	Cholecystokinin
ClpB	Caseinolytic protease B
CNS	Central nervous system
DCA	Deoxycholinic acid
E	Escherichia
ED	Eating disorder
F	Faecalibacterium
FFAR	Free fatty acid receptor
FXR	Farnesoid X receptor
GABA	Gamma aminobutyric acid
GLP-1	Glucagon-like peptide-1
GPCR	G-protein coupled receptor
IL	Interleukin
L	Lactobacillus
LCA	Lithocholic acid
LPS	Lipopolysaccharide
M	Methanobrevibacter
NPY	Neuropeptide Y
OSFED	Other specified feeding or eating disorder
POMC	Pro-opiomelanocortin
PVN	Paraventricular hypothalamic nucleus
PYY	Peptide YY
SCFA	Short-chain fatty acid
TGR5	Takeda G-protein coupled receptor 5
TNF-α	Tumor necrosis factor-α
α-MSH	α-Melanocyte stimulating hormone

6.1 Introduction

Eating disorders (EDs) manifest as severe, persistent disruptions in normal eating behaviors, which lead to physical, psychological, and social impairments over time and may culminate in many chronic diseases, or even death (Galmiche et al. 2019; Santomauro et al. 2021). The pathophysiology of EDs is complex, with dysregulated eating patterns being diagnosed on a spectrum. Therefore, "disordered eating" can define a broad range of conditions with aberrant eating patterns and psychiatric impairment (Wu et al. 2019b). EDs and disordered eating affect at least 9% of the global population and can impact both males and females of any age group (Ward et al. 2019). EDs place significant psychological and economic burdens on individuals and constitute a paramount public health concern due to their growing prevalence. According to the 2019 Global Burden of Diseases, Injuries, and Risk Factors Study, anorexia nervosa (AN) and bulimia nervosa (BN) were the only recognized burdensome EDs (Santomauro et al. 2021; Vos et al. 2020). However, additional research has suggested that binge-eating disorder (BED) and other specified feeding or eating disorders (OSFED) are common, with up to half of the population receiving ED treatment being diagnosed with BED or OSFED (Lindvall Dahlgren et al. 2017). Although EDs can affect any age group, they more commonly persist in young (adolescent) and aging individuals (Rohde et al. 2017; Fostinelli et al. 2020; Donini et al. 2013). It is hypothesized that changes in the gut microbiota during the aging process are linked to the occurrence of EDs. Because EDs affect so many people worldwide and are continuing to emerge as a global health concern, it is essential to develop treatments for these debilitating disorders. Understanding the link between the gut microbiota, its metabolites, and ED prevalence will aid in developing such therapies.

The gut microbiota is defined as the collection of bacteria, viruses, fungi, and archaea residing in the gut. The microbiota is comprised of both aerobic and anaerobic species from 12 bacterial phyla, with Firmicutes and Bacteroidetes being the most abundant. Gut microbes are known to ferment indigestible fibers into beneficial metabolites, like short-chain fatty acids (SCFAs; i.e., acetate, butyrate, and propionate), which are essential for improving gut health by promoting mucus production, maintaining the integrity of the intestinal epithelium, and protecting from chronic inflammation (Tolhurst et al. 2012; Lewis et al. 2010; Peng et al. 2009). Due to these prominent roles and others, many studies in animals and humans have implicated the role of the gut microbiota composition in health and disease. For example, it is well-known that detrimental changes in the gut microbiota composition (gut dysbiosis) and their metabolites are associated with metabolic disorders (like obesity and diabetes) and neurological diseases (Turnbaugh et al. 2009, 2006). Age is one of the most significant factors impacting the gut microbiota composition, thus leading to the onset of such diseases, especially later in life. The young microbiota is characterized by an increased abundance of Actinobacteria (such as bifidobacteria). In contrast, the aged microbiota is characterized by reduced microbial diversity and increased abundances of pathogenic or disease-causing species, such as Proteobacteria and *Enterobacteriaceae* (Odamaki et al. 2016). These microbiota signatures can also affect eating behaviors via their metabolites and crosstalk between the gut and the brain ("gut-brain axis").

Communication between the gut and the brain is bidirectional, and changes in the gut can affect brain behaviors and vice versa (Carabotti et al. 2015; Cryan et al. 2019). Figure 6.1 shows the communication networks between the gut and the brain ("gut-brain axis") and how food intake behaviors are regulated by the gut microbiome composition and vice versa. The hypothalamus is the brain region responsible for energy balance (i.e., energy intake and expenditure). It contains specific classes of orexigenic and anorexigenic neurons that fire in response to certain cues from energy regulating hormones and metabolites, which are sensed in the enteric nervous system. These sensory signals are transmitted from the gut to the brain (and vice versa) via physical wirings, such as nerves, or humoral routes, including immune cells, neurons/neurotransmitters, and gut hormones. The vagus nerve physically connects the gut and brain and has known roles in energy balance (Bravo et al. 2011; Cryan et al. 2019). Changes in the microbiota and their metabolites modulate eating behaviors due to the secretion of gut hormones, such as glucagon-like peptide-1 (GLP-1) or ghrelin, and the stimulation of orexigenic/anorexigenic neurons. The gut-brain axis has been suggested as a central component contributing to ED pathology because changes in the microbiota affect eating behaviors and different eating behaviors modulate the microbiota composition (Fig. 6.1) [15]. However, the link between gut microbiota and the prevalence of EDs is not well known and warrants further study. Further understanding of this bidirectionality between the gut microbiota and EDs will allow for the development of therapies or regimens to treat EDs or disordered eating.

This chapter describes the differences in EDs and their prevalence in the young and elderly, focusing on the link between the gut microbiota composition, function, and the prevalence of EDs at both extremes of aging.

6.2 Eating Disorders as a Growing Public Health Concern

EDs are chronic, severe, and persistent disturbances in eating behaviors. These chronic perturbations in eating behaviors lead to psychiatric, nutritional, and social impairments, which require medical interventions or management. The precise definition and pathophysiology of EDs are complex, and aberrant eating patterns are classified using a spectrum—the term "disordered eating" is often used to define a condition where eating behaviors or cognition negatively impact health or are psychologically distressing (Wu et al. 2019b). There is a complex intertwining of genetic, metabolic, and environmental factors contributing to the onset of dysregulated eating behaviors or the diagnosis of EDs. The most prevalent EDs include AN, BN, and BED (Peters et al. 2021). AN is characterized by severe caloric/dietary restriction due to distortion in body image; significant weight loss is also a notable feature of the AN (Ghenciulescu et al. 2021).

On the other hand, BN is classified by recurrent binging (or compulsive eating), loss of control over eating, and compensatory behaviors, including self-induced vomiting, aberrant fasting, and excessive exercise (Hail and Le Grange 2018). Similarly, BED is characterized by excessive, compulsive, and rapid food intake behaviors, but compensatory behaviors are absent (Williams and First 2013). Although EDs can impact anyone of any age group, they occur more frequently in younger individuals, females, and those who experience severe gastrointestinal

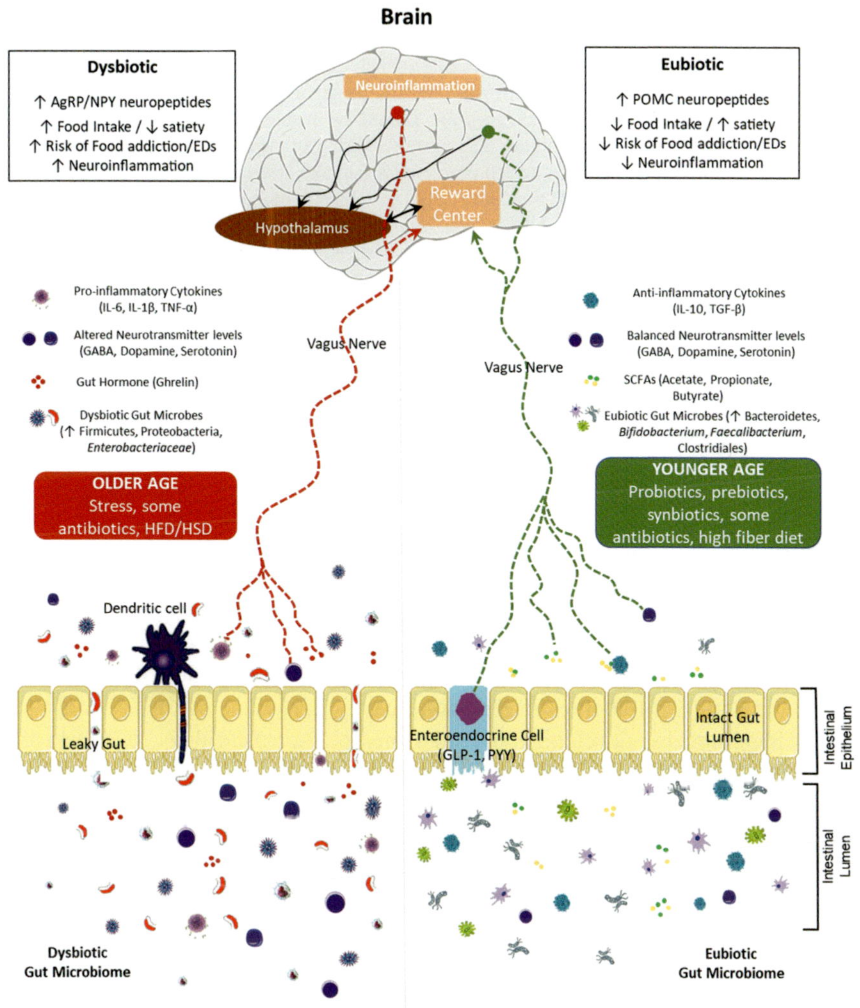

symptoms, suffer from anxiety and depression, or have a lower quality of life, according to a recent systematic review (Peters et al. 2021).

From a psychological perspective, EDs place profound burdens on the quality of life regarding an individual's mental health and social life. Apart from this, EDs also set financial obligations on the healthcare system, especially in the United States and Western countries, where EDs are more prevalent. In the United States alone, the total economic costs of EDs surpassed $64 billion, according to a recent analysis, and these costs continue to rise (Streatfeild et al. 2021). Based on these data, EDs constitute a growing public health concern globally. However, there are still large gaps in the literature regarding the pathophysiology of EDs. The gut microbiota has emerged as a modulator of health and disease. The microbiota

◂**Fig. 6.1** Bidirectional communications between the gut and the brain ("gut-brain axis") regulate food intake behaviors. Perturbations in the gut microbiota composition, such as an enrichment of Firmicutes, Proteobacteria, and *Enterobacteriacae* (which can be associated with older age), lead to a pro-inflammatory state, alterations in neurotransmitter levels, and an increase in ghrelin, the orexigenic hunger hormone. These changes are sensed by mechanoreceptors in the enteroendocrine cells in the intestinal epithelium, leading to the firing of AgRP and NPY neurons in the arcuate nucleus of the hypothalamus, which promote food intake, decrease satiety, and increase the risk of food addiction or other EDs. Conversely, enrichment of beneficial microbes, such as Bacteroidetes, *Bifidobacterium*, and *Faecalibacterium*, leads to increased concentrations of SCFAs like acetate and butyrate, balances in neurotransmitter levels, secretion of anorexigenic hormones like GLP-1 and PYY, and an anti-inflammatory state. This microbiota signature is associated with increased firing of POMC neurons in the arcuate nucleus, leading to decreased food intake (increased satiety) and a lower risk of food addictions or EDs. *Abbreviations in figure*: AgRP/NPY: Agouti-related peptide/Neuropeptide Y; ED: Eating disorder; GABA: Gamma-amino butyric amino acid; GLP-1: Glucagon-like peptide-1; HFD: High-fat diet; HSD: High-sugar diet; IL-1β: Interleukin-1β; IL-6: Interleukin-6; IL-10: Interleukin-10; POMC: Pro-opiomelanocortin; PYY: Peptide YY; SCFAs: Short-chain fatty acids; TGF-β: Transforming growth factor-β; TNF-α: Tumor necrosis factor-α

signature and microbiota-derived metabolites are thought to have indispensable roles in food intake behaviors, which are at the core of ED pathology. This chapter will review the current literature regarding the gut microbiota composition, its metabolites, and EDs, with an effort to show the link between the gut microbiota and ED pathology at the extremes of aging. Understanding the role of the gut microbiota and its metabolites in regulating eating behaviors and deciphering this link is important for developing potential therapies or screenings to reduce the occurrence and burdens of these debilitating conditions.

6.3 Microbiota and Its Metabolites

The gut microbiota is comprised of over 10^{14} bacteria, viruses, archaea, and fungi that colonize the gut. Metagenomic sequencing and 16S rRNA sequencing, two of the most common methods to study the gut microbiota, have identified over 2000 species in the human gut microbiota comprising 12 bacterial phyla (Li et al. 2014). Firmicutes, Bacteroidetes, Actinobacteria, and Proteobacteria are the four most represented bacterial phyla, constituting approximately 94% of the microbes. Although higher levels of species diversity in the gut microbiota are generally associated with a healthy phenotype, and the microbiota varies significantly from person to person, it is important to note that the microbiota also demonstrates a functional redundancy (Tian et al. 2020). This means that the genetic composition and functions of the microbiota are highly conserved, despite drastic differences in composition among people. The microbiota has important roles in maintaining intestinal epithelial integrity, protecting against pathogens, and providing critical vitamins to the host (Thursby and Juge 2017). It also has an important role in developing the intestinal and systemic immune systems, neuronal systems, brain, and metabolism. This has been studied extensively using germ-free mice; mice lacking microbiota have striking differences in their development and health span compared to mice with microbiota colonization (such as specific-pathogen-free mice) (Kawase et al. 2017; Pan et al. 2018; Sudo et al. 2004; Nishino et al. 2013). The microbiota has a profound impact on disease pathology, and perturbations in the gut microbiota (called "gut dysbiosis") are associated with disease. Lowered microbiota diversity and gut dysbiosis are linked to many diseases in mouse models and humans (Schwimmer et al. 2019; Kelley et al. 2016; Lindheim et al. 2017). The ratio of Firmicutes to Bacteroidetes, the two most predominant phyla, is another valuable index for assessing the health of the microbiota, and imbalances in this ratio may be associated with metabolic conditions, such as obesity or diabetes, and impaired energy harvest (Ley et al. 2005; Turnbaugh et al. 2006). Obesity, which results from increased energy intake and decreased energy expenditure, is characterized by increased Firmicutes (decreased Bacteroidetes) in mice and humans and an increase in the Firmicutes: Bacteroidetes ratio (Turnbaugh et al. 2008; Ley et al. 2006, 2005; Bervoets et al. 2013). In obese humans, decreases in mucin-degrading *Akkermansia (A.)*, *Faecalibacterium* (*F.*), *Oscillibacter*, and *Alistipes* were seen; similar taxonomic changes were seen in type-2 diabetes, with increases also observed in *Escherichia* and *Shigella*, which are pathobionts (Thingholm et al. 2019). These bacterial taxa changes were correlated with serum metabolites, indicating that gut-derived metabolites are also critical in health and disease.

The gut microbiota imparts significant effects on host health and disease via the production of bioactive compounds, such as SCFAs and bile acids. Microbiota-derived metabolites are indispensable for maintaining host physiology because they allow communication (via immune cells, neurons/neurotransmitters, and gut hormones) between the gut and peripheral organs/tissues, including the brain, liver, heart, adipose tissue, and lungs (Schroeder and Bäckhed 2016). Furthermore, perturbations in the gut microbiota composition may lead to chronic, systemic inflammation, which contributes to several disease etiologies, including diabetes, cancer, cardiovascular disease, neurodegenerative diseases and dementia, psychiatric disorders, and accelerated aging (Furman et al. 2019). Therefore, understanding how the gut microbiota regulates energy harvest through the production of metabolites is important for uncovering their essential role in ED pathologies.

6.3.1 Short-Chain Fatty Acids

SCFAs are classified as major end products of bacterial fermentation of indigestible dietary fibers (prebiotics), such as inulin, fructooligosaccharides, and galactooligosaccharides, in the large intestine (colon) and are important signaling molecules. SCFAs (particularly butyrate) are also critical for providing energy to colonic cells, constituting approximately 10% of the daily energy requirements for these cells in animals and humans (Bergman 1990; Roediger 1982). The synthesis pathways of SCFAs have been described in detail elsewhere (Nogal et al. 2021; Den Besten et al. 2013). Butyrate, propionate, and acetate are three of the most common SCFAs abundant in the gut. *F. prausnitzii*, *Eubacterium* species, and *Coprococcus* species are among the most well-known butyrate producers, while *A. muciniphila* is a potent propionate and acetate producer (Louis et al. 2010; Belzer et al. 2017). *Bifidobacterium* is also a critical acetate producer, especially in the infant gut (Alcon-Giner et al. 2020). SCFAs can regulate energy balance by binding to transmembrane G-protein coupled receptors (GPCRs), namely free-fatty acid receptors (FFAR)-2 and 3, which are expressed in intestinal epithelial cells (such as L cells) (Tolhurst et al. 2012; Karaki et al. 2006). This binding of SCFAs leads to the secretion of gut hormones, such as GLP-1 and peptide YY (PYY), which regulate energy harvest and satiety. Butyrate and propionate have been shown to strongly stimulate the secretion of GLP-1 and PYY in human enteroendocrine cell lines, as well as in mice and Wistar rats via GPCR activation (Larraufie et al. 2018; Psichas et al. 2015; Forbes et al. 2015; Brown et al. 2003). Butyrate, propionate, and acetate (via injection or oral gavage) all suppressed food intake in fasted mice (Goswami et al. 2018; Li et al. 2018), and butyrate upregulated energy expenditure (Gao et al. 2009). Furthermore, $FFAR2^{-/-}$ (Gpr43) or $FFAR3^{-/-}$ (Gpr41) mice showed reductions in SCFA-mediated secretion of GLP-1 and PYY, as well as an obese phenotype (Tolhurst et al. 2012; Kimura et al. 2013; Samuel et al. 2008). In a human pilot study, the role of acetate on gut hormone production was implicated—acetate significantly upregulated plasma levels of PYY and GLP-1, thus demonstrating their essential role in energy balance (Freeland and Wolever 2010). These studies support the critical role of microbiota-derived SCFAs in regulating energy harvest via receptor binding and gut hormone secretion.

6.3.2 Bile Acids

In the liver, bile acids are synthesized from cholesterol before secretion into the gut (small intestine) before being modified by the gut microbiota in the colon. More specifically, cholesterol is converted to primary bile acids, such as cholic acid (CA), which undergo conjugation before secretion into the bile (Russell 2003). After eating, bile and bile acids are secreted into the gut, absorbed (in the ileum), or reach the colon (Xie et al. 2021). Colonic gut microbiota plays a critical role in bile acid metabolism in multiple ways, including deconjugation via bile salt hydrolase (BSH) activity and conversion of primary deconjugated bile acids into secondary bile acids via 7a-dehydroxylase activity (Molinero et al. 2019). These pathways have been described in more detail elsewhere (Wu et al. 2021b). BSH-related genes have been identified in *Bacteroides*, *Clostridium*, *Lactobacillus* (*L.*), and *Bifidobacterium* (*B.*); in vivo, increased BSH activity improved weight, cholesterol, and host physiology (Joyce et al. 2014).

On the other hand, *Eubacterium* and *Clostridium* species are the only genera with identified 7a-hydroxylase activity (Molinero et al. 2019). Deoxycholic acid

(DCA) and lithocholic acid (LCA) are the two most abundant secondary bile acids in the human gut. They can modulate energy metabolism via Takeda G-protein coupled receptor 5 (TGR5), a transmembrane bile acid-specific GPCR, as well as the nuclear farnesoid X receptor (FXR) (Duboc et al. 2014; Kuhre et al. 2018). Like SCFAs, bile acids regulate food intake behaviors via receptor binding and the secretion of satiety-related gut hormones. LCA and DCA were potent stimulators of GLP-1 and PYY in mice; conversely, this stimulatory effect was absent in $TGR5^{-/-}$ mice, recapitulating the importance of this receptor binding (Kuhre et al. 2018). In humans, bile acid mixtures stimulated the secretion of cholecystokinin (CCK; a satiety-related hormone) and GLP-1 (Koop et al. 1996; Wu et al. 2013). Recent in vitro, in vivo, and human studies have revealed the TGR5 and FXR receptors as therapeutic targets for energy balance via modulation of GLP-1 secretion by enteroendocrine cells (Trabelsi et al. 2015; Zheng et al. 2021; Li et al. 2019; Prawitt et al. 2011). Overall, these findings demonstrate the critical role of microbiota-derived bile acids in energy homeostasis via their involvement with GPCRs.

6.3.3 Other Gut-Derived Metabolites

The gut microbiota is also responsible for producing other functional metabolites, such as endotoxins, which have known roles in stimulating the enteric immune system and affecting health and disease. Lipopolysaccharide (LPS) is an endotoxin derived from the cell wall of Gram-negative bacteria. It has known roles in inflammation and is a potent trigger for insulin resistance and the upregulation of pro-inflammatory cytokines, such as tumor necrosis factor-alpha (TNF-α), interleukin (IL)-6, and IL-1, in mice (Cani et al. 2007). Enrichment of Gram-negative species, such as Proteobacteria and *Enterobacteriaceae*, is linked to increased circulation of LPS in the blood, which subsequently increases inflammation via the innate immune system and toll-like receptor signaling (Stevens et al. 2018; Palazzo et al. 2007).

Taken together, it can be seen that the gut microbiota have a very intricate and vital role in health and disease via the production of functional metabolites, such as SCFAs and bile acids. These metabolites, which change depending on the microbiome signature, have essential roles in energy balance (i.e., energy intake and expenditure) and are severely dysregulated in EDs. Thus, understanding how the microbiota changes over life can elucidate the trends in EDs among different age groups and attempt to establish a link between the microbiota and ED diagnosis at the extremes of aging.

6.4 Microbiota at the Extremes of Aging

The high plasticity of the gut microbiota allows it to be modulated by many different environmental and lifestyle factors, such as diet, exercise, age, and medications such as antibiotics (Human Microbiome Project Consortium 2012). Age is known to have a significant impact on the gut microbiota composition, and the abundances of different microbes change throughout life. The microbes colonizing the gut are different at all stages of life, including birth, infancy, childhood, adolescence, early adulthood, late adulthood, and old age. They are also shaped by genetic and environmental factors (Bosco and Noti 2021; Nagpal et al. 2018a). Although factors such as delivery mode (vaginal versus Cesarean section), term length (pre-term versus full-term), feeding, and exposure to antibiotics can affect the microbiome composition and diversity at birth, the infant microbiome is generally characterized by large amounts of Actinobacteria and Firmicutes at the phylum level, and *Bifidobacterium*, *Bacteroides*, *Enterococcus*, *Streptococcus*, *Staphylococcus*, and *Clostridia* at the genus level (Hill et al. 2017; Nagpal et al. 2016). Using quantitative real-time PCR and 16S rRNA sequencing, previous studies have reported bifidobacteria, enterobacteria, and *Bacteroides*/*Prevotella* as the three main bacteria present in neonates, with the detection of *Lactobacillus* colonization significantly higher in vaginally born babies (Hopkins et al. 2001; Nagpal et al. 2016). Depending on the mode of delivery, the Firmicutes:Bacteroidetes ratio begins to stabilize within the first 24 weeks of life and until 18 months of age (Hill et al. 2017; Dogra et al. 2015). During the first months and years of life, the microbiome undergoes significant changes, with levels of many bacteria fluctuating (Odamaki et al. 2016). As such, the typical "healthy" microbiota composition during adulthood is comprised of high abundances of Bacteroidetes and Firmicutes and decreased abundances of Proteobacteria, Actinobacteria, Fusobacteria, and Verrucomicrobia, compared to younger "healthy" counterparts (Human Microbiome Project Consortium 2012).

As we age, the abundances of core bacterial taxa (including *Bacteroides* and *Prevotella*) decline, with significant changes in α-diversity and decreases in the abundances of *Bifidobacterium, Faecalibacterium,* and *Clostridiales* species; these changes are associated with increased frailty and physical decline in subjects over 65 years old (Castro-Mejía et al. 2020; Jackson et al. 2016; Salazar et al. 2013). The abundances of core phyla are also unusual, with high variability from subject to subject in people over 65 years old (Claesson et al. 2011). A low relative abundance of *Clostridiales* (i.e., *Clostridium* cluster IV, *Ruminococcus,* and *Saccharofermentans*) and *A. muciniphila* is linked to a decline in SCFA production, which alters energy regulation, promotes gut leakiness, and activates inflammatory pathways (Bodogai et al. 2018; Biagi et al. 2010; Kong et al. 2016). In comparison to the young microbiome, the aging microbiome is associated with an enrichment of pathobionts such as Proteobacteria and *Enterobacteriaceae*, and a decrease in beneficial microbes such as Clostridiales and *Bifidobacterium* (Jeffery et al. 2016). In centenarians (individuals > 104 years old) and super-centenarians (individuals 104 years of age or older), *Bacteroides*, *Roseburia*, and *Faecalibacterium* are severely downregulated (Biagi et al. 2010). Age-related differences in Proteobacteria, Firmicutes, and the Firmicutes:Bacteroidetes ratio have also been observed in monkeys (rhesus macaques), with significant increases in geriatric monkeys (Adriansjach et al. 2020; Pallikkuth et al. 2021). Studies in rodents have shown no marked differences in microbiota diversity, but *Lachnospiraceae* and *Ruminococcaceae* abundances declined in 74-week to 2-year-old rats, and Firmicutes were significantly downregulated in mice (Choi et al. 2021; Wu et al. 2021a). *Lachnospiraceae* and *Ruminococcaceae* are known SCFA producers, while decreases in phylum Firmicutes may indicate impaired energy harvest mechanisms. Microbiome sequencing of multiple cohorts has revealed marked differences between the microbiome structure of the elderly and that of younger adults (Claesson et al. 2012; Park et al. 2015). Furthermore, shotgun sequencing has shown significant differences in the microbiota between elderly and centenarian cohorts. Compared to the elderly subjects, centenarians demonstrated decreased levels of SCFA-producers such as *Eubacterium*, *F. prauznitzii*, and *Bifidobacterium*, while there were significant increases in *Ruminococcus* and *Escherichia* (Rampelli et al. 2013; Biagi et al. 2010). *Methanobrevibacter* (*M.*) *smithii* was also upregulated in centenarians, with marked decreases in microbes responsible for carbohydrate degradation and fermentation (i.e., SCFA producers) (Wu et al. 2019a). Decreases in these beneficial microbes and increases in pathobionts were associated with elevated levels of pro-inflammatory markers (such as IL-6, C-reactive protein, and TNF-α) in centenarians through a phenomenon commonly called "inflammaging" (Biagi et al. 2010; Dogra et al. 2015). However, not all of the studies corroborate these findings; that is, some studies have elucidated marked microbiota remodeling in elderly individuals more significant than 80 years old or centenarians (compared to younger elderly) that protects them from major age-related changes (i.e., enrichment of *Akkermansia*, *Lactobacillus*, and *Bifidobacterium*) (Biagi et al. 2016; Kong et al. 2016; Salazar et al. 2019). These changes may be prevalent for multiple reasons, including individual dietary/health habits or genetics. Because of these discrepancies in the literature, more cross-sectional or longitudinal studies are necessary to fully delineate the key differences in the core gut microbiome between elderly individuals and centenarians.

6.5 Microbiota Metabolites at the Extremes of Aging

Microbiota perturbations observed during aging and at different times in life are also associated with changes in the production of essential metabolites, including SCFAs, bile acids, and endotoxins, which can impair energy harvest and activate inflammatory pathways. The infant gut lacks significant SCFA concentrations (especially propionate and butyrate); however, during the first 2 years of life, the levels of acetate stabilize, with concentrations of butyrate and propionate increasing and abundances of lactate decreasing (largely due to changes in lactic acid-producing bacteria, such as *Lactobacillus* and *Bifidobacterium*) (Tsukuda et al.

2021). The shifts in the microbiota composition and their correlated changes in metabolites have been studied using in vitro and in vivo models, and recent studies have elucidated them in humans. Using an in vitro batch fermentation model, it has been shown that taxonomic changes in the microbiota may occur as early as in midlife (between 40 and 60 years old), with a significant reduction in *Bifidobacterium* colonization but no notable changes in fecal SCFAs (Chen et al. 2021). In aging models, rodent studies have shown that cecal butyrate concentrations are significantly decreased at 74 weeks and even more so at 2 years of age, accompanied by an increase in TNF-α (Choi et al. 2021). In monkeys, the aging microbiome is associated with the upregulation of systemic pro-inflammatory biomarkers, such as C-reactive protein, TNF-α, IL-6, IL-8, and IFN-γ. It decreases metabolites critical for lipid and amino acid metabolism and increases susceptibility to infection or disease (Pallikkuth et al. 2021).

Similarly, elderly subjects exhibited decreases in fecal SCFAs (compared to their young adult counterparts) as well as higher circulation of TNF-α, IL-12, and IL-8 (a chemokine for neutrophils); these changes were linked with decreases in *Faecalibacterium*, *Bacteroides*, and *Blautia* and increases in *Lactobacillus* (Salazar et al. 2013). Decreases in known butyrate producers, including *Clostridium* (*C.*) cluster XIVa (i.e., *Roseburia* and *Ruminococcus* species), have also been observed in elderly subjects (Mäkivuokko et al. 2010). Furthermore, in a cohort including males and females between 30 and 90 years old, increases in *Akkermansia* and *Lactobacillus* (which increased linearly with age) were associated with a significant decrease in the total fecal SCFAs as well as individual decreases in acetate, propionate, and butyrate as the subjects became older (Salazar et al. 2019). Furthermore, gut dysbiosis in aging is associated with an increase in the intestinal barrier permeability ("leaky gut"), and this has been shown in multiple models, including *Caenorhabditis elegans* and rodents (Thevaranjan et al. 2017; Mullin et al. 2002; Gelino et al. 2016). This phenotype is linked with higher abundances in Proteobacteria (specifically Gammaproteobacteria) as well as an increase in the systemic circulation of pro-inflammatory molecules (such as cell wall components of both Gram-negative and Gram-positive bacteria), higher antigenic load, more translocation of bacterial cells from the lumen to the intestinal epithelia, and immune system activation (Thevaranjan et al. 2017; Sato et al. 2014; Rera et al. 2012; Man et al. 2014). These studies highlight the changes in microbiota-derived metabolites associated with aging. Although the findings suggest a link between gut microbiota composition, the production of metabolites, and aging, more studies are necessary to fully elucidate this link.

6.6 Microbiota and Metabolite Changes in Eating Disorders

Gut microbiota and their metabolites are significantly different between young and aged subjects and different in ED pathology. Conversely, ED pathologies can affect and re-shape the microbiota signature, especially in the first years of life. Gut dysbiosis at this critical developmental time due to severe malnutrition leads to "microbiota immaturity," severely altered development of vital organs/tissues, and a higher predisposition for developing metabolic, neurological, and psychiatric disorders later in life (Subramanian et al. 2014; Goyal et al. 2015). Clinical trials with young to middle-aged (age 18–40 years) female AN subjects have revealed gut dysbiosis, a significant decline in the abundance of *Clostridium* and *Bacteroides*, and reductions in potential probiotics *Streptococcus* and *Lactobacillus*. Furthermore, reductions in SCFA-producing bacteria, such as butyrate-producing *Roseburia* (Mondot et al. 2022) and carbohydrate-fermenter *Ruminococcus,* were decreased, which correlated with reductions in fecal acetate and propionate and butyrate (Morita et al. 2015; Borgo et al. 2017; Armougom et al. 2009; Million et al. 2013). Reductions in *Roseburia* were also associated with increases in *Clostridium* clusters I, XI, and XVIII, Actinobacteria (such as *Bifidobacterium*), *Enterobacteriaceae*, *Klebsiella*, *Salmonella*, *M. smithii*, which is known to digest the end products of bacterial fermentation, and mucin-degrading bacteria. The increase in pathobionts and mucin-degraders significantly perturbs the gut barrier, which leads to chronic inflammation, as indicated by an upregulation in IL-6 (Mörkl et al. 2017). Reductions in microbiota α-diversity have also been observed in multiple cohorts of AN patients. This is considered one of the hallmarks of gut dysbiosis; reduced diversity is also prevalent in malnutrition (particularly undereating) (Kleiman et al. 2015; Mörkl et al. 2017; Hanachi et al. 2019; Genton et al. 2015). At the metabolite level, AN patients, in general, exhibited decreases in SCFAs acetate, propionate, and butyrate; however, some studies reported reductions in only one SCFA or no change in total SCFA concentrations (Morita et al. 2015; Mack et al. 2016; Borgo et al. 2017; Speranza et al. 2018). Transplantation of human feces (from healthy controls and AN subjects) into gnotobiotic mice revealed that the AN microbiota significantly slowed weight gain and reduced food intake; these findings were also linked with increases in anxious behaviors, which were measured using open-field testing and the marble burial test (Hata et al. 2019). Decreases in *Bacteroides* in the AN gut were strongly correlated with increased anxiousness during the marble burial test (Hata et al. 2019).

Bacteria-derived compounds, endotoxins, and bacterial proteins are prevalent in EDs because of their impact on appetite-related pathways. In particular, *Escherichia* (*E.*) *coli* infusion (to mimic its growth during regular nutrition) in rodents increased levels of PYY, which was associated with a decline in food intake (Breton et al. 2016b; Tennoune et al. 2014). Caseinolytic protease B (ClpB), a bacterial-derived protein mimetic of α-melanocyte-stimulating hormone (α-MSH), was detected in the feces and upregulated the firing rate of pro-opiomelanocortin (POMC) neurons, which induced satiety (Breton et al. 2016b). Therefore, high exposure to *E. coli* proteins may impair eating patterns and signal chronic meal termination; thus, ClpB may be central to AN pathology. *Enterobacteriaceae* commonly produces ClpB, so enrichment of these species in the gut may contribute to the onset of EDs; in particular, ClpB was elevated in the plasma of ED patients (Breton et al. 2016a). Anti-ClpB autoantibodies were significantly upregulated in the plasma of patients with AN, BN, and BED (Tennoune et al. 2014). Similarly, food restriction in male and female rats increased ClpB fecal levels (Breton et al. 2020). Based on the results of these studies, ClpB may be at the core of ED pathology. However, this link warrants further investigation, especially in BN, BED, and other disordered eating behaviors.

The gut microbiota also affects the release of cytokines, which are important signaling molecules observed in ED pathology. Classic studies in animals indicated that injection of IL-1, IL-6, and TNF-α suppresses food intake in AN via the central nervous system (CNS); these cytokines have been implicated as natural appetite suppressors and facilitate weight loss in rodents (Plata-Salamán et al. 1988, 1996; Langstein et al. 1991; Schiller et al. 1991). A recent meta-analysis, which included 23 studies on AN and four studies on BN, revealed that levels of TNF-α and IL-6 were significantly increased in patients with AN compared to healthy controls (Dalton et al. 2018). On the other hand, there was not a significant difference in levels of these cytokines in BN patients (Dalton et al. 2018). Increased pro-inflammatory markers could have been partly due to increases in gut barrier permeability and decreased mucus, which was also prevalent in the AN mouse model (Jésus et al. 2014).

These findings highlight how the gut microbiota, and its metabolites, change in EDs, but the results are minimal. One limitation is that most of the available literature studies EDs or disordered eating in females only; this is likely because EDs are much more common in females (i.e., 90% of AN diagnoses are female) (Hoek 2006). However, it would be helpful to establish the link between aberrant eating and gut microbiota in males. Furthermore, the available literature is limited in that most of the studies discuss the microbiota composition in AN. The literature surrounding other EDs, such as BN and BED, is sparse, and the link between microbiota perturbations and their prevalence remains elusive. Thus, it is necessary to conduct more extensive scale studies in (male and female) animals and humans to understand how different ED pathologies are prevalent based on the gut microbiota composition and its metabolites and determine how age impacts these pathologies.

6.7 Gut-Brain Axis: How Changes in the Gut Microbiota and Its Metabolites Affect the Brain

Gut microbiota perturbations in aging and their link with ED occurrence only tell half of the story. Age changes the gut microbiota composition and its metabolites, and these signatures are directly linked to ED prevalence. However, the question remains: how do these microbiota changes affect eating behaviors, which are controlled by the brain (such as the hypothalamic control of food intake). Changes in the gut microbiota and its metabolites are pivotal in regulating neuroendocrine functions. They can even affect blood–brain barrier (BBB) permeability, which is central to controlling food intake signals via the onset of neuroinflammation (Silva et al. 2020; Braniste et al. 2014). The detailed mechanisms central to gut-brain communication have been reviewed extensively elsewhere and are largely outside the scope of this chapter (Cryan et al. 2019). In brief, the gut microbiota can influence the CNS and brain via two main routes, including (1) physical wiring, such as the vagus nerve, and (2) humoral routes, which involve gut hormones neurotransmitters, and immune cells [16]. There are two classes of appetite-related neurons present in the arcuate nucleus of the hypothalamus, which vagus nerve and humoral signals can stimulate. These classes are orexigenic (appetite-stimulating) neuropeptide Y (NPY)/Agouti-related peptide (AgRP)-expressing neurons and anorexigenic (appetite-suppressing) POMC-expressing neurons (Gropp et al. 2005; Balthasar et al. 2005). They project to second-order neurons in the paraventricular hypothalamic nucleus (PVN) and multiple neurocircuits to stimulate energy intake and expenditure via the secretion of neuropeptides (Roh et al. 2016). Impairment of the PVN and other hypothalamic regions contributes to overeating/hyperphagia or undereating/hypophagia, thus suggesting the indispensable role of AgRP/POMC neurons and second-order neurons in energy regulation (Leibowitz et al. 1981; Shimizu et al. 1987; Milam et al. 1980).

The vagus nerve physically connects the gut and brain and transmits nutrient-related sensory signals via immune system modulation, neurotransmitters (such as dopamine and serotonin), and microbial metabolites. Afferent neurons in the intestine and enteroendocrine cells expressing chemical and mechanoreceptors sense the contents of the gut and release gut hormones, such as CCK, ghrelin, and GLP-1, in response to energy needs, via the vagus nerve (Forsythe et al. 2014; Luyer et al. 2005; Reimann et al. 2012). Classic studies in mice and rats have demonstrated the role of the vagus nerve via subdiaphragmatic vagotomy; vagotomy reduced behavioral deficits instigated by LPS, IL-1, or IL-1β, which are all pro-inflammatory markers (Luheshi et al. 2000; Konsman et al. 2000; Hansen et al. 2001; Bluthé et al. 1994). The role of the vagus nerve has also been implicated in the context of pro-inflammatory molecules and food intake behaviors in mice (Bretdibat et al. 1995).

SCFAs are a critical component of this interplay, and many studies have elucidated their roles in neuroendocrine regulation, gut motility, and even immunomodulation. After synthesis in the colon, SCFAs can be found in the portal or hepatic circulation or in the blood, where they serve as signaling molecules in various tissues (Murase et al. 1995). For example, acetate has demonstrated BBB permeability, thus demonstrating a plausible mechanism of SCFAs in neuroendocrine functions (Frost et al. 2014). SCFAs can activate the vagal receptors (or afferents) in the portal circulation, hormones such as GLP-1, PYY, and CCK, or neurotransmitters, such as serotonin (and its receptor 5-HT) or dopamine, which are all products of gut microbiota (Forsythe et al. 2014). Direct evidence in mice reveals that gut microbiota, including pathogens (i.e., *Salmonella typhimurium* and *Campylobacter jejuni*) (Wang et al. 2002; Goehler et al. 2005) and beneficial bacteria (probiotics such as *L. johnsonii*), can modulate the vagus nerve via gut-brain communications (Tanida et al. 2005). Food intake behaviors and appetite are regulated by secretion of anorectic gut hormones in enteroendocrine cells in rodents; in particular, PYY and GLP-1 suppress food intake (Batterham et al. 2002; Cani et al. 2005, 2004; Drucker 2018). Gut microbes can regulate CNS metabolites, such as serotonin and dopamine (neurotransmitters involved in food reward and addiction), and neurotrophic factors (i.e., brain-derived neurotrophic factor [BDNF], which is vital in neuronal survival and differentiation) (Matsumoto et al. 2013; Desbonnet et al. 2008; Murray et al. 2014; Bercik et al. 2011a). Additionally, the inhibitory neurotransmitter gamma-aminobutyric acid (GABA) has been regulated by *L. rhamnosus*, a notable beneficial bacteria); magnetic resonance microscopy revealed that this bacteria significantly increased GABA concentrations (Janik et al. 2016).

These studies highlight the mechanisms of gut-brain communication underpinning food intake behaviors and the reported role of the microbiome in these behaviors. However, more comprehensive studies are necessary to further elucidate the interplay between the gut microbiota, its metabolites, and energy intake/ expenditure regulation via gut-brain communications.

6.8 How the Gut-Brain Axis Communication Affects Eating Disorders

The role of the gut-brain axis in EDs is still under investigation, and there are limited studies available implicating this role. Hypothalamic neuron firing and projections and peripheral gut hormones (ghrelin, PYY, GLP-1, and insulin) are very much involved in EDs, similar to their roles in addiction. Therefore, due to the parts of these hormones in food reward, addiction, and impulsivity, EDs are classified as psychiatric/psychological disorders (Cardi et al. 2015; Murray et al. 2014). Females with constitutional thinness and AN had significantly higher PYY and GLP-1 but lower amounts of leptin (Germain et al. 2007). Ghrelin was also increased in AN patients compared to their controls and thin counterparts (Germain et al. 2007); this finding has been confirmed in additional studies by the same researchers (Germain et al. 2009, 2010). Ghrelin resistance (or reduced sensitivity to ghrelin) has been detected in subjects with AN—this could strongly contribute to the severe decline in eating patterns in AN. In particular, AN patients may become resistant to ghrelin levels. They may experience chronic anorexia due to the modulation of the dopaminergic reward system in the ventral tegmental area (involved in reward) and the hypothalamus (Abizaid et al. 2006; Kern et al. 2012; Sharpe et al. 2016). Thus, these gut hormonal imbalances directly affect eating behaviors, leading to disordered eating, extreme thinness, and a severely underweight phenotype in AN.

Furthermore, it has been shown that BDNF is decreased in patients with AN, and autoantibodies that attack the neuropeptides involved in food intake and appetite (such as ghrelin) have also been observed (Monteleone et al. 2005; Terashi et al. 2011). Decreased food intake in AN is correlated with more rapid firing of POMC neurons (promoting satiety), especially in the presence of *Enterobacteriaceae* or *E. coli* proteins (Breton et al. 2016a). Evidence-based studies demonstrating the link between the gut microbiota/its metabolites and ED pathology via modulation of the gut-brain axis are minimal, warranting further investigation. It is also essential to conduct comprehensive studies on other EDs, such as BN and BED, to establish the role of gut-brain communication in these pathologies.

6.9 Translational Perspectives

The current chapter has provided evidence of the microbiota in the extremes of aging and how these changes in the microbiota and its metabolites affect eating behaviors, which are severely dysregulated in EDs. Gut dysbiosis has a very causative role in many human diseases, including metabolic, neurological, and psychiatric disorders, and its metabolites are central to energy regulation and harvest. Thus, the gut microbiota may be a feasible target for treating EDs. Probiotics, prebiotics, and synbiotics are three emerging microbiome modulators that may benefit ED pathologies. Probiotics are live microorganisms that confer beneficial host effects when administered adequately (i.e., lactobacilli and bifidobacteria species), while prebiotics are the indigestible fibers that are fermented by the gut microbiota to promote the secretion of beneficial metabolites (like SCFAs) and the colonization of beneficial microbes in the gut (Schrezenmeir and de Vrese 2001). Synbiotics are the combination of probiotics and prebiotics and may exert synergistic effects on the host (Schrezenmeir and de Vrese 2001; Swanson et al. 2020).

These dietary interventions have demonstrated promising benefits in mouse and human diabetic and aging models by suppressing the gut microbiota and its metabolites (like SCFAs). Still, their precise roles in ED pathology remain elusive (Zou et al. 2020; Nagpal et al. 2018b; Ahmadi et al. 2019, 2020; Miller et al. 2021; Yadav et al. 2013). Probiotics *B. longum*, *C. symbosium*, *F. prausnitzii*, and *L. fermentum* have reversed neurological deficits and neuroinflammation following a stroke in aged mice and have restored the gut microbiota and SCFA levels (Lee et al. 2020). Furthermore, there are studies indicating how probiotic bacteria, such as *L. rhamnosus*, *L. reuteri*, and *B. longum*, may affect brain behavior (called "psychobiotics") via vagal pathways, GABA receptor expression, gut motility changes, neurotrophic factors, and neurons (Bercik et al. 2011b; Bravo et al. 2011; Kunze et al. 2009; Wang et al. 2010; Savignac et al. 2015, 2013; Liang et al. 2015). The difficulty lies in developing a therapeutic agent that can penetrate the BBB or exert psychobiotic effects from the periphery. Emerging studies have been designed to elucidate the role of gut microbiota modulation by probiotics in adolescents with AN (Gröbner et al. 2022). Prebiotics have been implicated for their role in increasing satiety and reducing food intake, but the findings remain inconsistent (Kellow et al. 2014). Thus, more studies are necessary to uncover the psychobiotic effects of probiotics, prebiotics, and synbiotics in ED pathology. Doing so will allow for feasible regimens to counter EDs, such as AN, BN, and BED, which are becoming increasingly prevalent globally and place significant social and economic burdens on affected individuals.

6.10 Conclusions

This chapter has elucidated gut microbiota changes at the extremes of aging as critical components of ED pathology. EDs are classified by chronic and severe perturbations in eating patterns, which place significant physical, psychological, and social constraints on individuals and financial burdens on healthcare systems. Therefore, it is critical to understand the mechanisms by which EDs occur in different age groups to develop possible therapeutics and regimens to reduce their burden. The gut microbiota is highly plastic, meaning that many genetic and environmental factors shape its composition, affecting the production of essential metabolites (SCFAs and bile acids) and subsequently impacting energy balance mediated by the hypothalamus. Age shapes the microbiota composition, with individuals of different ages demonstrating apparent differences in their microbiota composition. For example, younger individuals have higher colonization of *Bifidobacterium* and other Actinobacteria, while aged individuals show a higher relative abundance of pathobionts, such as Proteobacteria and *Enterobacteriaceae*. Via the gut-brain axis, which includes physical and humoral components, the gut microbiota and its metabolites can send sensory signals to the hypothalamus (and vice versa) to regulate food intake behaviors. Modulating the gut-brain axis via changes in the gut microbiota composition, its metabolites, and other biomarkers such as cytokines and gut hormones has indispensable roles in ED pathology. Emerging studies uncover these links, but additional studies are warranted to understand the precise mechanisms in EDs or disordered eating. Modulating the gut microbiota via therapeutic agents, such as probiotics, prebiotics, and synbiotics, may be a viable approach for treating EDs and reducing their burden.

References

Abizaid A, Liu Z-W, Andrews ZB, Shanabrough M, Borok E, Elsworth JD, Roth RH, Sleeman MW, Picciotto MR, Tschöp MH (2006) Ghrelin modulates the activity and synaptic input organization of midbrain dopamine neurons while promoting appetite. J Clin Investig 116:3229–3239

Adriansjach J, Baum ST, Lefkowitz EJ, Van Der Pol WJ, Buford TW, Colman RJ (2020) Age-related differences in the gut microbiome of rhesus macaques. J Geront Series A 75:1293–1298

Ahmadi S, Nagpal R, Wang S, Gagliano J, Kitzman DW, Soleimanian-Zad S, Sheikh-Zeinoddin M, Read R, Yadav H (2019) Prebiotics from acorn and sago prevent high-fat-diet-induced insulin resistance via microbiome–gut–brain axis modulation. J Nutr Biochem 67:1–13

Ahmadi S, Wang S, Nagpal R, Wang B, Jain S, Razazan A, Mishra SP, Zhu X, Wang Z, Kavanagh K (2020). A human-origin probiotic cocktail ameliorates aging-related leaky gut and inflammation via modulating the microbiota/taurine/tight junction axis. JCI Insight, 5

Alcon-Giner C, Dalby MJ, Caim S, Ketskemety J, Shaw A, Sim K, Lawson MA, Kiu R, Leclaire C, Chalklen L (2020) Microbiota supplementation with Bifidobacterium and Lactobacillus modifies the preterm infant gut microbiota and metabolome: an observational study. Cell Reports Med 1:100077

Armougom F, Henry M, Vialettes B, Raccah D, Raoult D (2009) Monitoring bacterial community of human gut microbiota reveals an increase in Lactobacillus in obese patients and Methanogens in anorexic patients. PLoS ONE 4:e7125

Balthasar N, Dalgaard LT, Lee CE, Yu J, Funahashi H, Williams T, Ferreira M, Tang V, Mcgovern RA, Kenny CD (2005) Divergence of melanocortin pathways in the control of food intake and energy expenditure. Cell 123:493–505

Batterham RL, Cowley MA, Small CJ, Herzog H, Cohen MA, Dakin CL, Wren AM, Brynes AE, Low MJ, Ghatei MA (2002) Gut hormone PYY3-36 physiologically inhibits food intake. Nature 418:650–654

Belzer C, Chia LW, Aalvink S, Chamlagain B, Piironen V, Knol J, De Vos WM (2017) Microbial metabolic networks at the mucus layer lead to diet-independent butyrate and vitamin B12 production by intestinal symbionts. Mbio 8:e00770-e817

Bercik P, Denou E, Collins J, Jackson W, Lu J, Jury J, Deng Y, Blennerhassett P, Macri J, Mccoy KD (2011a) The intestinal microbiota affect central levels of brain-derived neurotropic factor and behavior in mice. Gastroenterology 141(599–609):e3

Bercik P, Park A, Sinclair D, Khoshdel A, Lu J, Huang X, Deng Y, Blennerhassett P, Fahnestock M, Moine D (2011b) The anxiolytic effect of Bifidobacterium longum NCC3001 involves vagal pathways for gut–brain communication. Neurogastroenterol Motil 23:1132–1139

Bergman E (1990) Energy contributions of volatile fatty acids from the gastrointestinal tract in various species. Physiol Rev 70:567–590

Bervoets L, Van Hoorenbeeck K, Kortleven I, Van Noten C, Hens N, Vael C, Goossens H, Desager KN, Vankerckhoven V (2013) Differences in gut microbiota composition between obese and lean children: a cross-sectional study. Gut Pathogens 5:1–10

Biagi E, Nylund L, Candela M, Ostan R, Bucci L, Pini E, Nikkïla J, Monti D, Satokari R, Franceschi C (2010) Through ageing, and beyond: gut microbiota and inflammatory status in seniors and centenarians. PLoS ONE 5:e10667

Biagi E, Franceschi C, Rampelli S, Severgnini M, Ostan R, Turroni S, Consolandi C, Quercia S, Scurti M, Monti D (2016) Gut microbiota and extreme longevity. Curr Biol 26:1480–1485

Bluthé R-M, Walter V, Parnet P, Laye S, Lestage J, Verrier D, Poole S, Stenning BE, Kelley KW, Dantzer R (1994) Lipopolysaccharide induces sickness behaviour in rats by a vagal mediated mechanism. Comptes Rendus De l'acad Des Sci Serie III, Sci De La Vie 317:499–503

Bodogai M, O'connell J, Kim K, Kim Y, Moritoh K, Chen C, Gusev F, Vaughan K, Shulzhenko N, Mattison JA (2018) Commensal bacteria contribute to insulin resistance in aging by activating innate B1a cells. Sci Transl Med 10:eaat4271

Borgo F, Riva A, Benetti A, Casiraghi MC, Bertelli S, Garbossa S, Anselmetti S, Scarone S, Pontiroli AE, Morace G (2017) Microbiota in anorexia nervosa: the triangle between bacterial species, metabolites and psychological tests. PLoS ONE 12:e0179739

Bosco N, Noti M (2021) The aging gut microbiome and its impact on host immunity. Genes Immun 22:289–303

Braniste V, Al-Asmakh M, Kowal C, Anuar F, Abbaspour A, Tóth M, Korecka A, Bakocevic N, Ng LG, Kundu P (2014) The gut microbiota influences blood-brain barrier permeability in mice. Sci Transl Med 6:263ra158–263ra158

Bravo JA, Forsythe P, Chew MV, Escaravage E, Savignac HM, Dinan TG, Bienenstock J, Cryan JF (2011) Ingestion of lactobacillus strain regulates emotional behavior and central GABA receptor expression in a mouse via the vagus nerve. Proc Natl Acad Sci 108:16050–16055

Bretdibat J-L, Bluthe R-M, Kent S, Kelley KW, Dantzer R (1995) Lipopolysaccharide and interleukin-1 depress food-motivated behavior in mice by a vagal-mediated mechanism. Brain Behav Immun 9:242–246

Breton J, Legrand R, Akkermann K, Järv A, Harro J, Déchelotte P, Fetissov SO (2016a) Elevated plasma concentrations of bacterial ClpB protein in patients with eating disorders. Int J Eat Disord 49:805–808

Breton J, Tennoune N, Lucas N, Francois M, Legrand R, Jacquemot J, Goichon A, Guérin C, Peltier J, Pestel-Caron M (2016b) Gut commensal E. coli proteins activate host satiety pathways following nutrient-induced bacterial growth. Cell Metab 23:324–334

Breton J, Jacquemot J, Yaker L, Leclerc C, Connil N, Feuilloley M, Déchelotte P, Fetissov SO (2020) Host starvation and female sex influence enterobacterial clpb production: a possible link to the etiology of eating disorders. Microorganisms 8:530

Brown AJ, Goldsworthy SM, Barnes AA, Eilert MM, Tcheang L, Daniels D, Muir AI, Wigglesworth MJ, Kinghorn I, Fraser NJ (2003) The orphan G protein-coupled receptors GPR41 and GPR43 are activated by propionate and other short chain carboxylic acids. J Biol Chem 278:11312–11319

Cani PD, Dewever C, Delzenne NM (2004) Inulin-type fructans modulate gastrointestinal peptides involved in appetite regulation (glucagon-like peptide-1 and ghrelin) in rats. Br J Nutr 92:521–526

Cani PD, Neyrinck AM, Maton N, Delzenne NM (2005) Oligofructose promotes satiety in rats fed a high-fat diet: involvement of glucagon-like peptide-1. Obes Res 13:1000–1007

Cani PD, Amar J, Iglesias MA, Poggi M, Knauf C, Bastelica D, Neyrinck AM, Fava F, Tuohy KM, Chabo C (2007) Metabolic endotoxemia initiates obesity and insulin resistance. Diabetes 56:1761–1772

Carabotti M, Scirocco A, Maselli MA, Severi C (2015) The gut-brain axis: interactions between enteric microbiota, central and enteric nervous systems. Annals Gastroent Quart Public Hellenic Soc Gastroent 28:203

Cardi V, Leppanen J, Treasure J (2015) The effects of negative and positive mood induction on eating behaviour: a meta-analysis of laboratory studies in the healthy population and eating and weight disorders. Neurosci Biobehav Rev 57:299–309

Castro-Mejía JL, Khakimov B, Krych Ł, Bülow J, Bechshøft RL, Højfeldt G, Mertz KH, Garne ES, Schacht SR, Ahmad HF (2020) Physical fitness in community-dwelling older adults is linked to dietary intake, gut microbiota, and metabolomic signatures. Aging Cell 19:e13105

Chen J, Pi X, Liu W, Ding Q, Wang X, Jia W, Zhu L (2021) Age-related changes of microbiota in midlife associated with reduced saccharolytic potential: an in vitro study. BMC Microbiol 21:1–14

Choi SI, Son JH, Kim N, Kim YS, Nam RH, Park JH, Song C-H, Yu JE, Lee DH, Yoon K (2021) Changes in cecal microbiota and short-chain fatty acid during lifespan of the rat. J Neurogastroent Motility 27:134

Claesson MJ, Cusack S, O'sullivan O, Greene-Diniz R, De Weerd H, Flannery E, Marchesi JR, Falush D, Dinan T, Fitzgerald G (2011) Composition, variability, and temporal stability of the intestinal microbiota of the elderly. Proc Natl Acad Sci 108:4586–4591

Claesson MJ, Jeffery IB, Conde S, Power SE, O'connor EM, Cusack S, Harris HM, Coakley M, Lakshminarayanan B, O'sullivan O (2012) Gut microbiota composition correlates with diet and health in the elderly. Nature 488:178–184

Cryan JF, O'riordan KJ, Cowan CS, Sandhu K, Bastiaanssen TF, Boehme M, Codagnone MG, Cussotto S, Fulling C, Golubeva AV (2019) The microbiota-gut-brain axis. Physiological Reviews

Dalton B, Bartholdy S, Robinson L, Solmi M, Ibrahim MA, Breen G, Schmidt U, Himmerich H (2018) A meta-analysis of cytokine concentrations in eating disorders. J Psychiatr Res 103:252–264

Den Besten G, Van Eunen K, Groen AK, Venema K, Reijngoud D-J, Bakker BM (2013) The role of short-chain fatty acids in the interplay between diet, gut microbiota, and host energy metabolism. J Lipid Res 54:2325–2340

Desbonnet L, Garrett L, Clarke G, Bienenstock J, Dinan TG (2008) The probiotic Bifidobacteria infantis: an assessment of potential antidepressant properties in the rat. J Psychiatr Res 43:164–174

Dogra S, Sakwinska O, Soh S-E, Ngom-Bru C, Brück WM, Berger B, Brüssow H, Lee YS, Yap F, Chong Y-S (2015) Dynamics of infant gut microbiota are influenced by delivery mode and gestational duration and are associated with subsequent adiposity. Mbio 6:e02419-e2514

Donini LM, Poggiogalle E, Piredda M, Pinto A, Barbagallo M, Cucinotta D, Sergi G (2013) Anorexia and eating patterns in the elderly. PLoS ONE 8:e63539

Drucker DJ (2018) Mechanisms of action and therapeutic application of glucagon-like peptide-1. Cell Metab 27:740–756

Duboc H, Taché Y, Hofmann AF (2014) The bile acid TGR5 membrane receptor: from basic research to clinical application. Dig Liver Dis 46:302–312

Forbes S, Stafford S, Coope G, Heffron H, Real K, Newman R, Davenport R, Barnes M, Grosse J, Cox H (2015) Selective FFA2 agonism appears to act via intestinal PYY to reduce transit and food intake but does not improve glucose tolerance in mouse models. Diabetes 64:3763–3771

Forsythe P, Bienenstock J, Kunze WA (2014) Vagal pathways for microbiome-brain-gut axis communication. Microbial endocrinology: the microbiota-gut-brain axis in health and disease, pp 115–133

Fostinelli S, De Amicis R, Leone A, Giustizieri V, Binetti G, Bertoli S, Battezzati A, Cappa SF (2020) Eating behavior in aging and dementia: the need for a comprehensive assessment. Frontiers in Nutrition 307

Freeland KR, Wolever TM (2010) Acute effects of intravenous and rectal acetate on glucagon-like peptide-1, peptide YY, ghrelin, adiponectin and tumour necrosis factor-α. Br J Nutr 103:460–466

Frost G, Sleeth ML, Sahuri-Arisoylu M, Lizarbe B, Cerdan S, Brody L, Anastasovska J, Ghourab S, Hankir M, Zhang S (2014) The short-chain fatty acid acetate reduces appetite via a central homeostatic mechanism. Nat Commun 5:1–11

Furman D, Campisi J, Verdin E, Carrera-Bastos P, Targ S, Franceschi C, Ferrucci L, Gilroy DW, Fasano A, Miller GW (2019) Chronic inflammation in the etiology of disease across the life span. Nat Med 25:1822–1832

Galmiche M, Déchelotte P, Lambert G, Tavolacci MP (2019) Prevalence of eating disorders over the 2000–2018 period: a systematic literature review. Am J Clin Nutr 109:1402–1413

Gao Z, Yin J, Zhang J, Ward RE, Martin RJ, Lefevre M, Cefalu WT, Ye J (2009) Butyrate improves insulin sensitivity and increases energy expenditure in mice. Diabetes 58:1509–1517

Gelino S, Chang JT, Kumsta C, She X, Davis A, Nguyen C, Panowski S, Hansen M (2016) Intestinal autophagy improves healthspan and longevity in C. elegans during dietary restriction. PLoS Genetics, 12:e1006135

Genton L, Cani PD, Schrenzel J (2015) Alterations of gut barrier and gut microbiota in food restriction, food deprivation and protein-energy wasting. Clin Nutr 34:341–349

Germain N, Galusca B, Le Roux CW, Bossu C, Ghatei MA, Lang F, Bloom SR, Estour B (2007) Constitutional thinness and lean anorexia nervosa display opposite concentrations of peptide YY, glucagon-like peptide 1, ghrelin, and leptin. Am J Clin Nutr 85:967–971

Germain N, Galusca B, Grouselle D, Frere D, Tolle V, Zizzari P, Lang F, Epelbaum J, Estour B (2009) Ghrelin/obestatin ratio in two populations with low bodyweight: constitutional thinness and anorexia nervosa. Psychoneuroendocrinology 34:413–419

Germain N, Galusca B, Grouselle D, Frere D, Billard S, Epelbaum J, Estour B (2010) Ghrelin and obestatin circadian levels differentiate bingeing-purging from restrictive anorexia nervosa. J Clin Endocrinol Metab 95:3057–3062

Ghenciulescu A, Park RJ, Burnet PW (2021) The Gut microbiome in Anorexia nervosa: friend or foe? Front Psych 11:1463

Goehler LE, Gaykema RP, Opitz N, Reddaway R, Badr N, Lyte M (2005) Activation in vagal afferents and central autonomic pathways: early responses to intestinal infection with Campylobacter jejuni. Brain Behav Immun 19:334–344

Goswami C, Iwasaki Y, Yada T (2018) Short-chain fatty acids suppress food intake by activating vagal afferent neurons. J Nutr Biochem 57:130–135

Goyal MS, Venkatesh S, Milbrandt J, Gordon JI, Raichle ME (2015) Feeding the brain and nurturing the mind: linking nutrition and the gut microbiota to brain development. Proc Natl Acad Sci 112:14105–14112

Gröbner EM, Zeiler M, Fischmeister FPS, Kollndorfer K, Schmelz S, Schneider A, Haid-Stecher N, Sevecke K, Wagner G, Keller L (2022) The effects of probiotics administration on the gut microbiome in adolescents with anorexia nervosa—a study protocol for a longitudinal, double-blind, randomized, placebo-controlled trial. Eur Eat Disord Rev 30:61–74

Gropp E, Shanabrough M, Borok E, Xu AW, Janoschek R, Buch T, Plum L, Balthasar N, Hampel B, Waisman A (2005) Agouti-related peptide–expressing neurons are mandatory for feeding. Nat Neurosci 8:1289–1291

Human Microbiome Project Consortium (2012) Structure, function and diversity of the healthy human microbiome. Nature 486:207. https://www.ncbi.nlm.nih.gov/pmc/articles/PMC3564958/. PMID: 22699609

Hail L, Le Grange D (2018) Bulimia nervosa in adolescents: prevalence and treatment challenges. Adolesc Health Med Ther 9:11

Hanachi M, Manichanh C, Schoenenberger A, Pascal V, Levenez F, Cournède N, Doré J, Melchior J-C (2019) Altered host-gut microbes symbiosis in severely malnourished anorexia nervosa (AN) patients undergoing enteral nutrition: an explicative factor of functional intestinal disorders? Clin Nutr 38:2304–2310

Hansen MK, O'connor KA, Goehler LE, Watkins LR, Maier SF (2001) The contribution of the vagus nerve in interleukin-1β-induced fever is dependent on dose. Am J Physiol –regul Integ Comp Physiol 280:R929–R934

Hata T, Miyata N, Takakura S, Yoshihara K, Asano Y, Kimura-Todani T, Yamashita M, Zhang X-T, Watanabe N, Mikami K (2019) The gut microbiome derived from anorexia nervosa patients impairs weight gain and behavioral performance in female mice. Endocrinology 160:2441–2452

Hill CJ, Lynch DB, Murphy K, Ulaszewska M, Jeffery IB, O'shea CA, Watkins C, Dempsey E, Mattivi F, Tuohy K (2017) Evolution of gut microbiota composition from birth to 24 weeks in the INFANTMET Cohort. Microbiome 5:1–18

Hoek HW (2006) Incidence, prevalence and mortality of anorexia nervosa and other eating disorders. Curr Opin Psychiatry 19:389–394

Hopkins M, Sharp R, Macfarlane G (2001) Age and disease related changes in intestinal bacterial populations assessed by cell culture, 16S rRNA abundance, and community cellular fatty acid profiles. Gut 48:198–205

Jackson MA, Jeffery IB, Beaumont M, Bell JT, Clark AG, Ley RE, O'toole PW, Spector TD, Steves CJ (2016) Signatures of early frailty in the gut microbiota. Genome Med 8:1-11

Janik R, Thomason LA, Stanisz AM, Forsythe P, Bienenstock J, Stanisz GJ (2016) Magnetic resonance spectroscopy reveals oral Lactobacillus promotion of increases in brain GABA, N-acetyl aspartate and glutamate. Neuroimage 125:988–995

Jeffery IB, Lynch DB, O'toole PW (2016) Composition and temporal stability of the gut microbiota in older persons. ISME J 10:170-182

Jésus P, Ouelaa W, François M, Riachy L, Guérin C, Aziz M, Do Rego J-C, Déchelotte P, Fetissov SO, Coëffier M (2014) Alteration of intestinal barrier function during activity-based anorexia in mice. Clin Nutr 33:1046–1053

Joyce SA, Macsharry J, Casey PG, Kinsella M, Murphy EF, Shanahan F, Hill C, Gahan CG (2014) Regulation of host weight gain and lipid metabolism by bacterial bile acid modification in the gut. Proc Natl Acad Sci 111:7421–7426

Karaki S-I, Mitsui R, Hayashi H, Kato I, Sugiya H, Iwanaga T, Furness JB, Kuwahara A (2006) Short-chain fatty acid receptor, GPR43, is expressed by enteroendocrine cells and mucosal mast cells in rat intestine. Cell Tissue Res 324:353–360

Kawase T, Nagasawa M, Ikeda H, Yasuo S, Koga Y, Furuse M (2017) Gut microbiota of mice putatively modifies amino acid metabolism in the host brain. Br J Nutr 117:775–783

Kelley ST, Skarra DV, Rivera AJ, Thackray VG (2016) The gut microbiome is altered in a letrozole-induced mouse model of polycystic ovary syndrome. PLoS ONE 11:e0146509

Kellow NJ, Coughlan MT, Reid CM (2014) Metabolic benefits of dietary prebiotics in human subjects: a systematic review of randomised controlled trials. Br J Nutr 111:1147–1161

Kern A, Albarran-Zeckler R, Walsh HE, Smith RG (2012) Apo-ghrelin receptor forms heteromers with DRD2 in hypothalamic neurons and is essential for anorexigenic effects of DRD2 agonism. Neuron 73:317–332

Kimura I, Ozawa K, Inoue D, Imamura T, Kimura K, Maeda T, Terasawa K, Kashihara D, Hirano K, Tani T (2013) The gut microbiota suppresses insulin-mediated fat accumulation via the short-chain fatty acid receptor GPR43. Nat Commun 4:1–12

Kleiman SC, Watson HJ, Bulik-Sullivan EC, Huh EY, Tarantino LM, Bulik CM, Carroll IM (2015) The intestinal microbiota in acute anorexia nervosa and during renourishment: relationship to depression, anxiety, and eating disorder psychopathology. Psychosom Med 77:969

Kong F, Hua Y, Zeng B, Ning R, Li Y, Zhao J (2016) Gut microbiota signatures of longevity. Curr Biol 26:R832–R833

Konsman JP, Luheshi GN, Bluthé RM, Dantzer R (2000) The vagus nerve mediates behavioural depression, but not fever, in response to peripheral immune signals; a functional anatomical analysis. Eur J Neurosci 12:4434–4446

Koop I, Schindler M, Bosshammer A, Scheibner J, Stange E, Koop H (1996) Physiological control of cholecystokinin release and pancreatic enzyme secretion by intraduodenal bile acids. Gut 39:661–667

Kuhre RE, Albrechtsen NJW, Larsen O, Jepsen SL, Balk-Møller E, Andersen DB, Deacon CF, Schoonjans K, Reimann F, Gribble FM (2018) Bile acids are important direct and indirect regulators of the secretion of appetite-and metabolism-regulating hormones from the gut and pancreas. Molecular Metabolism 11:84–95

Kunze WA, Mao YK, Wang B, Huizinga JD, Ma X, Forsythe P, Bienenstock J (2009) Lactobacillus reuteri enhances excitability of colonic AH neurons by inhibiting calcium-dependent potassium channel opening. J Cell Mol Med 13:2261–2270

Langstein HN, Doherty GM, Fraker DL, Buresh CM, Norton JA (1991) The roles of γ-interferon and tumor necrosis factor α in an experimental rat model of cancer cachexia. Can Res 51:2302–2306

Larraufie P, Martin-Gallausiaux C, Lapaque N, Dore J, Gribble F, Reimann F, Blottiere H (2018) SCFAs strongly stimulate PYY production in human enteroendocrine cells. Sci Rep 8:1–9

Lee J, D'aigle J, Atadja L, Quaicoe V, Honarpisheh P, Ganesh BP, Hassan A, Graf J, Petrosino J, Putluri N (2020) Gut microbiota–derived short-chain fatty acids promote poststroke recovery in aged mice. Circ Res 127:453–465

Leibowitz SF, Hammer NJ, Chang K (1981) Hypothalamic paraventricular nucleus lesions produce overeating and obesity in the rat. Physiol Behav 27:1031–1040

Lewis K, Lutgendorff F, Phan V, Söderholm JD, Sherman PM, Mckay DM (2010) Enhanced translocation of bacteria across metabolically stressed epithelia is reduced by butyrate. Inflamm Bowel Dis 16:1138–1148

Ley RE, Bäckhed F, Turnbaugh P, Lozupone CA, Knight RD, Gordon JI (2005) Obesity alters gut microbial ecology. Proc Natl Acad Sci 102:11070–11075

Ley RE, Turnbaugh PJ, Klein S, Gordon JI (2006) Human gut microbes associated with obesity. Nature 444:1022–1023

Li J, Jia H, Cai X, Zhong H, Feng Q, Sunagawa S, Arumugam M, Kultima JR, Prifti E, Nielsen T (2014) An integrated catalog of reference genes in the human gut microbiome. Nat Biotechnol 32:834–841

Li Z, Yi C-X, Katiraei S, Kooijman S, Zhou E, Chung CK, Gao Y, Van Den Heuvel JK, Meijer OC, Berbée JF (2018) Butyrate reduces appetite and activates brown adipose tissue via the gut-brain neural circuit. Gut 67:1269–1279

Li P, Zhu L, Yang X, Li W, Sun X, Yi B, Zhu S (2019) Farnesoid X receptor interacts with cAMP response element binding protein to modulate glucagon-like peptide-1 (7–36) amide secretion by intestinal L cell. J Cell Physiol 234:12839–12846

Liang S, Wang T, Hu X, Luo J, Li W, Wu X, Duan Y, Jin F (2015) Administration of Lactobacillus helveticus NS8 improves behavioral, cognitive, and biochemical aberrations caused by chronic restraint stress. Neuroscience 310:561–577

Lindheim L, Bashir M, Münzker J, Trummer C, Zachhuber V, Leber B, Horvath A, Pieber TR, Gorkiewicz G, Stadlbauer V (2017) Alterations in gut microbiome composition and barrier function are associated with reproductive and metabolic defects in women with polycystic ovary syndrome (PCOS): a pilot study. PLoS ONE 12:e0168390

Lindvall Dahlgren C, Wisting L, Rø Ø (2017) Feeding and eating disorders in the DSM-5 era: a systematic review of prevalence rates in non-clinical male and female samples. J Eat Disord 5:1–10

Louis P, Young P, Holtrop G, Flint HJ (2010) Diversity of human colonic butyrate-producing bacteria revealed by analysis of the butyryl-CoA: acetate CoA-transferase gene. Environ Microbiol 12:304–314

Luheshi GN, Bluthé R-M, Rushforth D, Mulcahy N, Konsman J-P, Goldbach M, Dantzer R (2000) Vagotomy attenuates the behavioural but not the pyrogenic effects of interleukin-1 in rats. Auton Neurosci 85:127–132

Luyer MD, Greve JWM, Hadfoune MH, Jacobs JA, Dejong CH, Buurman WA (2005) Nutritional stimulation of cholecystokinin receptors inhibits inflammation via the vagus nerve. J Exp Med 202:1023–1029

Mack I, Cuntz U, Grämer C, Niedermaier S, Pohl C, Schwiertz A, Zimmermann K, Zipfel S, Enck P, Penders J (2016) Weight gain in anorexia nervosa does not ameliorate the faecal microbiota, branched chain fatty acid profiles and gastrointestinal complaints. Sci Rep 6:1–16

Mäkivuokko H, Tiihonen K, Tynkkynen S, Paulin L, Rautonen N (2010) The effect of age and non-steroidal anti-inflammatory drugs on human intestinal microbiota composition. Br J Nutr 103:227–234

Man AL, Gicheva N, Nicoletti C (2014) The impact of ageing on the intestinal epithelial barrier and immune system. Cell Immunol 289:112–118

Matsumoto M, Kibe R, Ooga T, Aiba Y, Sawaki E, Koga Y, Benno Y (2013) Cerebral low-molecular metabolites influenced by intestinal microbiota: a pilot study. Front Syst Neurosci 7:9

Milam KM, Stern JS, Storlien LH, Keesey RE (1980) Effect of lateral hypothalamic lesions on regulation of body weight and adiposity in rats. Am J Physiol Regul Integ Comp Physiol 239: R337–R343

Miller B, Mainali R, Nagpal R, Yadav H (2021) A newly developed synbiotic yogurt prevents diabetes by improving the microbiome—intestine—pancreas axis. Int J Mol Sci 22:1647

Million Á, Angelakis E, Maraninchi M, Henry M, Giorgi R, Valero R, Vialettes B, Raoult D (2013) Correlation between body mass index and gut concentrations of lactobacillus reuteri, bifidobacterium animalis, methanobrevibacter smithii and escherichia coli. Int J Obes 37:1460–1466

Molinero N, Ruiz L, Sánchez B, Margolles A, Delgado S (2019) Intestinal bacteria interplay with bile and cholesterol metabolism: implications on host physiology. Frontiers in Physiology 185

Mondot S, Lachkar L, Doré J, Blottière HM, Hanachi M (2022) Roseburia, a decreased bacterial taxon in the gut microbiota of patients suffering from anorexia nervosa. European J Clin Nutr 1–4

Monteleone P, Fabrazzo M, Martiadis V, Serritella C, Pannuto M, Maj M (2005) Circulating brain-derived neurotrophic factor is decreased in women with anorexia and bulimia nervosa but not in women with binge-eating disorder: relationships to co-morbid depression, psychopathology and hormonal variables. Psychol Med 35:897–905

Morita C, Tsuji H, Hata T, Gondo M, Takakura S, Kawai K, Yoshihara K, Ogata K, Nomoto K, Miyazaki K (2015) Gut dysbiosis in patients with anorexia nervosa. PLoS ONE 10:e0145274

Mörkl S, Lackner S, Müller W, Gorkiewicz G, Kashofer K, Oberascher A, Painold A, Holl A, Holzer P, Meinitzer A (2017) Gut microbiota and body composition in anorexia nervosa inpatients in comparison to athletes, overweight, obese, and normal weight controls. Int J Eat Disord 50:1421–1431

Mullin J, Valenzano M, Verrecchio J, Kothari R (2002) Age-and diet-related increase in transepithelial colon permeability of Fischer 344 rats. Dig Dis Sci 47:2262–2270

Murase M, Kimura Y, Nagata Y (1995) Determination of portal short-chain fatty acids in rats fed various dietary fibers by capillary gas chromatography. J Chromatogr B Biomed Sci Appl 664:415–420

Murray S, Tulloch A, Gold MS, Avena NM (2014) Hormonal and neural mechanisms of food reward, eating behaviour and obesity. Nat Rev Endocrinol 10:540–552

Nagpal R, Tsuji H, Takahashi T, Kawashima K, Nagata S, Nomoto K, Yamashiro Y (2016) Sensitive quantitative analysis of the meconium bacterial microbiota in healthy term infants born vaginally or by cesarean section. Front Microbiol 7:1997

Nagpal R, Mainali R, Ahmadi S, Wang S, Singh R, Kavanagh K, Kitzman DW, Kushugulova A, Marotta F, Yadav H (2018a) Gut microbiome and aging: Physiological and mechanistic insights. Nutrition and Healthy Aging 4:267–285

Nagpal R, Wang S, Ahmadi S, Hayes J, Gagliano J, Subashchandrabose S, Kitzman DW, Becton T, Read R, Yadav H (2018b) Human-origin probiotic cocktail increases short-chain fatty acid production via modulation of mice and human gut microbiome. Sci Rep 8:1–15

Nishino R, Mikami K, Takahashi H, Tomonaga S, Furuse M, Hiramoto T, Aiba Y, Koga Y, Sudo N (2013) Commensal microbiota modulate murine behaviors in a strictly contamination-free environment confirmed by culture-based methods. Neurogastroenterol Motil 25:521-e371

Nogal A, Valdes AM, Menni C (2021) The role of short-chain fatty acids in the interplay between gut microbiota and diet in cardio-metabolic health. Gut Microbes 13:1897212

Odamaki T, Kato K, Sugahara H, Hashikura N, Takahashi S, Xiao J-Z, Abe F, Osawa R (2016) Age-related changes in gut microbiota composition from newborn to centenarian: a cross-sectional study. BMC Microbiol 16:1–12

Palazzo M, Balsari A, Rossini A, Selleri S, Calcaterra C, Gariboldi S, Zanobbio L, Arnaboldi F, Shirai YF, Serrao G (2007) Activation of enteroendocrine cells via TLRs induces hormone, chemokine, and defensin secretion. J Immunol 178:4296–4303

Pallikkuth S, Mendez R, Russell K, Sirupangi T, Kvistad D, Pahwa R, Villinger F, Banerjee S, Pahwa S (2021) Age associated microbiome and microbial metabolites modulation and its association with systemic inflammation in a rhesus macaque model. Front Immunol, 4300

Pan W-H, Sommer F, Falk-Paulsen M, Ulas T, Best P, Fazio A, Kachroo P, Luzius A, Jentzsch M, Rehman A (2018) Exposure to the gut microbiota drives distinct methylome and transcriptome changes in intestinal epithelial cells during postnatal development. Genome Med 10:1–15

Park S-H, Kim K-A, Ahn Y-T, Jeong J-J, Huh C-S, Kim D-H (2015) Comparative analysis of gut microbiota in elderly people of urbanized towns and longevity villages. BMC Microbiol 15:1–9

Peng L, Li Z-R, Green RS, Holzman IR, Lin J (2009) Butyrate enhances the intestinal barrier by facilitating tight junction assembly via activation of AMP-activated protein kinase in Caco-2 cell monolayers. J Nutr 139:1619–1625

Peters JE, Basnayake C, Hebbard GS, Salzberg MR, Kamm MA (2021) Prevalence of disordered eating in adults with gastrointestinal disorders: a systematic review. Neurogastroent Motility, e14278

Plata-Salamán CR, Oomura Y, Kai Y (1988) Tumor necrosis factor and interleukin-1β: suppression of food intake by direct action in the central nervous system. Brain Res 448:106–114

Plata-Salamán CR, Sonti G, Borkoski JP, Wilson CD, Ffrench-Mullen JM (1996) Anorexia induced by chronic central administration of cytokines at estimated pathophysiological concentrations. Physiol Behav 60:867–875

Prawitt J, Abdelkarim M, Stroeve JH, Popescu I, Duez H, Velagapudi VR, Dumont J, Bouchaert E, Van Dijk TH, Lucas A (2011) Farnesoid X receptor deficiency improves glucose homeostasis in mouse models of obesity. Diabetes 60:1861–1871

Psichas A, Sleeth M, Murphy K, Brooks L, Bewick G, Hanyaloglu A, Ghatei M, Bloom S, Frost G (2015) The short chain fatty acid propionate stimulates GLP-1 and PYY secretion via free fatty acid receptor 2 in rodents. Int J Obes 39:424–429

Rampelli S, Candela M, Turroni S, Biagi E, Collino S, Franceschi C, O'toole PW, Brigidi P (2013) Functional metagenomic profiling of intestinal microbiome in extreme ageing. Aging (Albany NY) 5:902

Reimann F, Tolhurst G, Gribble FM (2012) G-protein-coupled receptors in intestinal chemosensation. Cell Metab 15:421–431

Rera M, Clark RI, Walker DW (2012) Intestinal barrier dysfunction links metabolic and inflammatory markers of aging to death in Drosophila. Proc Natl Acad Sci 109:21528–21533

Roediger W (1982) Utilization of nutrients by isolated epithelial cells of the rat colon. Gastroenterology 83:424–429

Roh E, Song DK, Kim M-S (2016) Emerging role of the brain in the homeostatic regulation of energy and glucose metabolism. Exp Mol Med 48:e216–e216

Rohde P, Stice E, Shaw H, Gau JM, Ohls OC (2017) Age effects in eating disorder baseline risk factors and prevention intervention effects. Int J Eat Disord 50:1273–1280

Russell DW (2003) The enzymes, regulation, and genetics of bile acid synthesis. Annu Rev Biochem 72:137–174

Salazar N, López P, Valdés L, Margolles A, Suárez A, Patterson AM, Cuervo A, Reyes-Gavilán CGDL, Ruas-Madiedo P, Gonzalez S (2013) Microbial targets for the development of functional foods accordingly with nutritional and immune parameters altered in the elderly. J Am Coll Nutr 32:399–406

Salazar N, Arboleya S, Fernández-Navarro T, De Los Reyes-Gavilán CG, Gonzalez S, Gueimonde M (2019) Age-associated changes in gut microbiota and dietary components related with the immune system in adulthood and old age: a cross-sectional study. Nutrients 11:1765

Samuel BS, Shaito A, Motoike T, Rey FE, Backhed F, Manchester JK, Hammer RE, Williams SC, Crowley J, Yanagisawa M (2008) Effects of the gut microbiota on host adiposity are modulated by the short-chain fatty-acid binding G protein-coupled receptor, Gpr41. Proc Natl Acad Sci 105:16767–16772

Santomauro DF, Melen S, Mitchison D, Vos T, Whiteford H, Ferrari AJ (2021) The hidden burden of eating disorders: an extension of estimates from the Global Burden of Disease Study 2019. Lancet Psyc 8:320–328

Sato J, Kanazawa A, Ikeda F, Yoshihara T, Goto H, Abe H, Komiya K, Kawaguchi M, Shimizu T, Ogihara T (2014) Gut dysbiosis and detection of "live gut bacteria" in blood of Japanese patients with type 2 diabetes. Diabetes Care 37:2343–2350

Savignac HM, Corona G, Mills H, Chen L, Spencer JP, Tzortzis G, Burnet PW (2013) Prebiotic feeding elevates central brain derived neurotrophic factor, N-methyl-D-aspartate receptor subunits and D-serine. Neurochem Int 63:756–764

Savignac H, Tramullas M, Kiely B, Dinan T, Cryan J (2015) Bifidobacteria modulate cognitive processes in an anxious mouse strain. Behav Brain Res 287:59–72

Schiller JH, Storer BE, Witt PL, Alberti D, Tombes MB, Arzoomanian R, Proctor RA, Mccarthy D, Brown RR, Voss SD (1991) Biological and clinical effects of intravenous tumor necrosis factor-α administered three times weekly. Can Res 51:1651–1658

Schrezenmeir J, De Vrese M (2001) Probiotics, prebiotics, and synbiotics—approaching a definition. Am J Clin Nutr 73:361s–364s

Schroeder BO, Bäckhed F (2016) Signals from the gut microbiota to distant organs in physiology and disease. Nat Med 22:1079–1089

Schwimmer JB, Johnson JS, Angeles JE, Behling C, Belt PH, Borecki I, Bross C, Durelle J, Goyal NP, Hamilton G (2019) Microbiome signatures associated with steatohepatitis and moderate to severe fibrosis in children with nonalcoholic fatty liver disease. Gastroenterology 157:1109–1122

Sharpe M, Clemens K, Morris M, Westbrook R (2016) Daily exposure to sucrose impairs subsequent learning about food cues: a role for alterations in ghrelin signaling and dopamine D2 receptors. Neuropsychopharmacology 41:1357–1365

Shimizu N, Oomura Y, Plata-Salamán CR, Morimoto M (1987) Hyperphagia and obesity in rats with bilateral ibotenic acid-induced lesions of the ventromedial hypothalamic nucleus. Brain Res 416:153–156

Silva YP, Bernardi A, Frozza RL (2020) The role of short-chain fatty acids from gut microbiota in gut-brain communication. Front Endocrinol 11:25

Speranza E, Cioffi I, Santarpia L, Del Piano C, De Caprio C, Naccarato M, Marra M, De Filippo E, Contaldo F, Pasanisi F (2018) Fecal short chain fatty acids and dietary intake in Italian women with restrictive anorexia nervosa: a pilot study. Front Nutr 119

Stevens BR, Goel R, Seungbum K, Richards EM, Holbert RC, Pepine CJ, Raizada MK (2018) Increased human intestinal barrier permeability plasma biomarkers zonulin and FABP2 correlated with plasma LPS and altered gut microbiome in anxiety or depression. Gut 67:1555–1557

Streatfeild J, Hickson J, Austin SB, Hutcheson R, Kandel JS, Lampert JG, Myers EM, Richmond TK, Samnaliev M, Velasquez K (2021) Social and economic cost of eating disorders in the United States: evidence to inform policy action. Int J Eat Disord 54:851–868

Subramanian S, Huq S, Yatsunenko T, Haque R, Mahfuz M, Alam MA, Benezra A, Destefano J, Meier MF, Muegge BD (2014) Persistent gut microbiota immaturity in malnourished Bangladeshi children. Nature 510:417–421

Sudo N, Chida Y, Aiba Y, Sonoda J, Oyama N, Yu XN, Kubo C, Koga Y (2004) Postnatal microbial colonization programs the hypothalamic–pituitary–adrenal system for stress response in mice. J Physiol 558:263–275

Swanson KS, Gibson GR, Hutkins R, Reimer RA, Reid G, Verbeke K, Scott KP, Holscher HD, Azad MB, Delzenne NM (2020) The international scientific association for probiotics and prebiotics (ISAPP) consensus statement on the definition and scope of synbiotics. Nat Rev Gastroenterol Hepatol 17:687–701

Tanida M, Yamano T, Maeda K, Okumura N, Fukushima Y, Nagai K (2005) Effects of intraduodenal injection of Lactobacillus johnsonii La1 on renal sympathetic nerve activity and blood pressure in urethane-anesthetized rats. Neurosci Lett 389:109–114

Tennoune N, Chan P, Breton J, Legrand R, Chabane Y, Akkermann K, Järv A, Ouelaa W, Takagi K, Ghouzali I (2014) Bacterial ClpB heat-shock protein, an antigen-mimetic of the anorexigenic peptide α-MSH, at the origin of eating disorders. Transl Psychiatry 4:e458–e458

Terashi M, Asakawa A, Harada T, Ushikai M, Coquerel Q, Sinno MH, Déchelotte P, Inui A, Fetissov SO (2011) Ghrelin reactive autoantibodies in restrictive anorexia nervosa. Nutrition 27:407–413

Thevaranjan N, Puchta A, Schulz C, Naidoo A, Szamosi J, Verschoor CP, Loukov D, Schenck LP, Jury J, Foley KP (2017) Age-associated microbial dysbiosis promotes intestinal permeability, systemic inflammation, and macrophage dysfunction. Cell Host Microbe 21(455–466):e4

Thingholm LB, Rühlemann MC, Koch M, Fuqua B, Laucke G, Boehm R, Bang C, Franzosa EA, Hübenthal M, Rahnavard A (2019) Obese individuals with and without type 2 diabetes show different gut microbial functional capacity and composition. Cell Host Microbe 26(252–264): e10

Thursby E, Juge N (2017) Introduction to the human gut microbiota. Biochem J 474:1823–1836

Tian L, Wang X-W, Wu A-K, Fan Y, Friedman J, Dahlin A, Waldor MK, Weinstock GM, Weiss ST, Liu Y-Y (2020) Deciphering functional redundancy in the human microbiome. Nat Commun 11:1–11

Tolhurst G, Heffron H, Lam YS, Parker HE, Habib AM, Diakogiannaki E, Cameron J, Grosse J, Reimann F, Gribble FM (2012) Short-chain fatty acids stimulate glucagon-like peptide-1 secretion via the G-protein–coupled receptor FFAR2. Diabetes 61:364–371

Trabelsi M-S, Daoudi M, Prawitt J, Ducastel S, Touche V, Sayin SI, Perino A, Brighton CA, Sebti Y, Kluza J (2015) Farnesoid X receptor inhibits glucagon-like peptide-1 production by enteroendocrine L cells. Nat Commun 6:1–13

Tsukuda N, Yahagi K, Hara T, Watanabe Y, Matsumoto H, Mori H, Higashi K, Tsuji H, Matsumoto S, Kurokawa K (2021) Key bacterial taxa and metabolic pathways affecting gut short-chain fatty acid profiles in early life. ISME J 15:2574–2590

Turnbaugh PJ, Ley RE, Mahowald MA, Magrini V, Mardis ER, Gordon JI (2006) An obesity-associated gut microbiome with increased capacity for energy harvest. Nature 444:1027–1031

Turnbaugh PJ, Bäckhed F, Fulton L, Gordon JI (2008) Diet-induced obesity is linked to marked but reversible alterations in the mouse distal gut microbiome. Cell Host Microbe 3:213–223

Turnbaugh PJ, Hamady M, Yatsunenko T, Cantarel BL, Duncan A, Ley RE, Sogin ML, Jones WJ, Roe BA, Affourtit JP (2009) A core gut microbiome in obese and lean twins. Nature 457:480–484

Vos T, Lim SS, Abbafati C, Abbas KM, Abbasi M, Abbasifard M, Abbasi-Kangevari M, Abbastabar H, Abd-Allah F, Abdelalim A (2020) Global burden of 369 diseases and injuries in 204 countries and territories, 1990–2019: a systematic analysis for the Global Burden of Disease Study 2019. The Lancet 396:1204–1222

Wang X, Wang B-R, Zhang X-J, Xu Z, Ding Y-Q, Ju G (2002) Evidences for vagus nerve in maintenance of immune balance and transmission of immune information from gut to brain in STM-infected rats. World J Gastroenterol 8:540

Wang B, Mao YK, Diorio C, Pasyk M, Wu RY, Bienenstock J, Kunze WA (2010) Luminal administration ex vivo of a live Lactobacillus species moderates mouse jejunal motility within minutes. FASEB J 24:4078–4088

Ward ZJ, Rodriguez P, Wright DR, Austin SB, Long MW (2019) Estimation of eating disorders prevalence by age and associations with mortality in a simulated nationally representative US cohort. JAMA Netw Open 2:e1912925–e1912925

Williams JB, First M (2013) Diagnostic and statistical manual of mental disorders. Encyclopedia of Social Work

Wu T, Bound MJ, Standfield SD, Jones KL, Horowitz M, Rayner CK (2013) Effects of taurocholic acid on glycemic, glucagon-like peptide-1, and insulin responses to small intestinal glucose infusion in healthy humans. J Clin Endocrinol Metab 98:E718–E722

Wu XY, Yin WQ, Sun HW, Yang SX, Li XY, Liu HQ (2019b) The association between disordered eating and health-related quality of life among children and adolescents: A systematic review of population-based studies. PLoS ONE 14:e0222777

Wu C-S, Muthyala SDV, Klemashevich C, Ufondu AU, Menon R, Chen Z, Devaraj S, Jayaraman A, Sun Y (2021a) Age-dependent remodeling of gut microbiome and host serum metabolome in mice. Aging (albany NY) 13:6330

Wu J, Wang K, Wang X, Pang Y, Jiang C (2021b) The role of the gut microbiome and its metabolites in metabolic diseases. Protein Cell 12:360–373

Wu L, Zeng T, Zinellu A, Rubino S, Kelvin DJ, Carru C (2019a) A cross-sectional study of compositional and functional profiles of gut microbiota in Sardinian centenarians. mSystems 4: e00325–19

Xie C, Huang W, Young RL, Jones KL, Horowitz M, Rayner CK, Wu T (2021) Role of bile acids in the regulation of food intake, and their dysregulation in metabolic disease. Nutrients 13:1104

Yadav H, Lee J-H, Lloyd J, Walter P, Rane SG (2013) Beneficial metabolic effects of a probiotic via butyrate-induced GLP-1 hormone secretion. J Biol Chem 288:25088–25097

Zheng X, Chen T, Jiang R, Zhao A, Wu Q, Kuang J, Sun D, Ren Z, Li M, Zhao M (2021) Hyocholic acid species improve glucose homeostasis through a distinct TGR5 and FXR signaling mechanism. Cell Metab 33(791–803):e7

Zou Y, Ju X, Chen W, Yuan J, Wang Z, Aluko RE, He R (2020) Rice bran attenuated obesity via alleviating dyslipidemia, browning of white adipocytes and modulating gut microbiota in high-fat diet-induced obese mice. Food Funct 11:2406–2417

Part II
Chronic Diseases and Gut Microbiota

Chapter 7
Diabetes Mellitus and Microbiota: Knowledge and Perspectives

Gaetano Emanuele Rizzo, Alessandro Rizzi, Linda Tartaglione, Maria Laura Leo, Luca Viti, Mauro Di Leo, and Dario Pitocco

Abstract Diabetes mellitus is a global epidemic disease constantly increasing, due to genetic predisposition and behavioral risk factors, with etiological mechanisms not completely defined. Moreover, diabetes can induce critical complications that damage the heart, vessels, kidney, eye and nerves, with important alterations in the quality and quantity of life. Gut microbiota represents a topic of growing interest in clinical research and several studies have focused their attention on the possible interplay between metabolic diseases and alteration in gut microbiota. A different composition in bacterial phyla and species among patients with type 1 and type 2 diabetes mellitus and healthy people has been observed, with some differences in patients with complications. The purpose of this chapter is to show current knowledge on the composition of gut microbiota in diabetic patients, the relationship between diabetic treatments and gut microbiota and the possible advantages in the prevention and therapy of diabetes through the action on the microbiota by the use of prebiotics, probiotics and fecal transplant.

Keywords Gut microbiota · Type 1 diabetes · Type 2 diabetes · Dysbiosis · Diabetes complications · Prebiotics · Probiotics · Fecal transplant

7.1 Introduction

Diabetes mellitus (DM) is a global epidemic disease with a constantly increasing incidence and prevalence that affects about 537 million people worldwide (ADA 2022).

DM is a syndrome characterized by altered glucose metabolism (hyperglycemia, insulin resistance and/or impairment in insulin secretion).

G. E. Rizzo · A. Rizzi · L. Tartaglione · M. L. Leo · L. Viti · M. Di Leo · D. Pitocco (✉)
Diabetes Care Unit Fondazione Policlinico Universitario, Agostino Gemelli IRCCs-Catholic University, Rome, Italy
e-mail: dario.pitocco@unicatt.it

F. Marotta (ed.), *Gut Microbiota in Aging and Chronic Diseases*, Healthy Ageing and Longevity 17, https://doi.org/10.1007/978-3-031-14023-5_7

According to the American Diabetes Association (ADA), DM is classified into these categories (Diabetes Care 2022):

- Type 1 diabetes (T1DM), due to autoimmune destruction of pancreatic β-cell with a progressive absolute insulin deficiency (5–10% of the cases of DM);
- Type 2 diabetes (T2DM), characterized by insulin resistance with a progressive loss of adequate secretion of insulin (about 90%);
- Specific types of DM due to other causes (monogenic diabetes syndrome, drugs, exocrine pancreas disease);
- Gestational diabetes mellitus (GDM).

The mechanisms leading to the dysfunction of insulin secretion by pancreatic β-cell and the altered response of tissues to insulin are not all well known. Several authors have focused their investigation on the link between DM and the "gut microbiota", an "organ" composed of all the microorganisms that live in the human intestine.

The purpose of this chapter is to show the possible etiological and pathophysiological role of alteration in gut microbiota ("dysbiosis") in the developing of DM and its complications, interaction among the treatment of diabetes and gut microbiota and potential therapeutic strategies directed to GUT microbiota.

7.2 Metabolic Role of Gut Microbiota

The human intestinal tract is colonized by more than 100 trillion microorganisms, subdivided into nearly 1000 species, most of them being bacteria, but there are also fungi, parasites and viruses.

Microbiota lives in harmony with the host and colonizes not only the gut, but also skin oral cavity and genital mucosa.

Gut microbiota counts more than 3 million genes, a number 150 times higher than human genome with a total weight of about 1.5 kg.

Gut microbiota is composed of more than 50 phyla, with two dominant strains that cover about 90% of intestinal bacteria: *Firmicutes* and *Bacteroidetes.*

Firmicutes are gram-positive bacteria that encode for about 60–80% of gut microbiota; in this phylum are included more than 200 genera, among which the most represented are *Lactobacillus, Clostridium, Bacillus, Ruminococcus* and *Enterococcus*. *Bacteroidetes* are gram-negative bacteria that constitute about 20–30% of gut microbiota and include within it *Bacteroides, Xylanibacter* and *Prevotella.* Sub-dominant strains are *Actinobacteria* (including *Bifidobacterium*), *Proteobacteria* (including *Enterobacteriaceae* and *Escherichia*), *Fusobacteria, Cyanobacteria* and *Verrucomicrobia* (Pitocco et al. 2020) (Fig. 7.1).

Based on the prevailing composition of bacterial genera, a classification in three enterotypes that does not depend on age, ethnicity, body mass index (BMI) and sex has been proposed. Enterotypes have been identified by the amount of three genera:

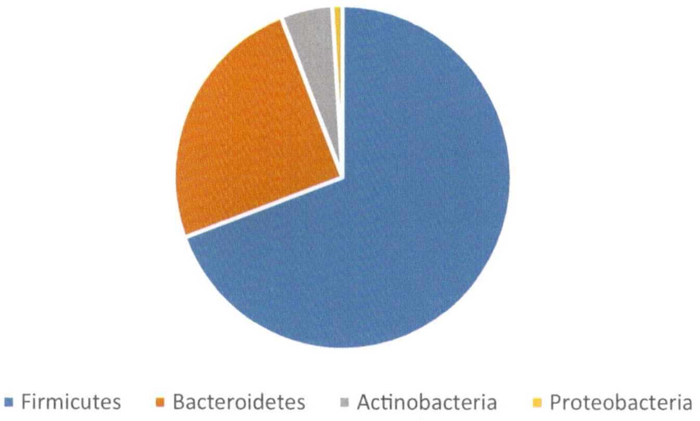

Fig. 7.1 Dominant bacteria in human gut microbiota

Bacteroides for enterotype 1, *Prevotella* for enterotype 2 and *Ruminococcus* for enterotype 3. Bacterial genera in enterotype 1 obtain energy through fermentation of carbohydrates and protein, with a broad gene expression of enzymes like galactosidases, hexosaminidases, proteases and enzymes involved in glycolysis and pentose phosphate pathways; the main vitamin metabolism pathway in this enterotype concerns biotin, riboflavin, pantothenate and ascorbate. In enterotype 2, the prevailing genus *Prevotella* is composed of bacteria able to degrade mucin glycoproteins of the intestinal mucosa and, unlike enterotype 1, prevails vitamin metabolism of thiamine and folate. Genera in enterotype 3 (*Ruminococcus* and *Akkermansia*) are also able to degrade mucins and express membrane transporters involved in the uptake of simple sugars (Arumugam et al. 2011).

The composition of microbiota begins at birth by the colonization of microorganisms from the mother and continues during the first year of life; it can be influenced by maternal or neonatal exposition to antibiotics, type of delivery and milk feeding. Changes in the microbiota can also occur in childhood and can be due to various factors such as diseases, drug exposure, lifestyle changes and exposure to smoke (Craciun et al. 2022).

Gut microbiota lives in symbiosis with the host, participating in processes of energy production and modulation of the immune system. Polysaccharides taken with the diet are hydrolyzed and fermented by bacteria with resulting production of monosaccharides and short-chain unsaturated fats (SCFAs), like butyrate, acetate and propionate, that represent the 5–10% of total human energy intake (McNeil 1984).

Butyrate and propionate are produced through glycolysis and also by the metabolism of amino acids; acetate is produced by acetyl-CoA; it can also be transformed into butyrate and it is the most present SCFA in gut with a molar ratio in the colon of nearly 60:20:20 for acetate:propionate:butyrate (Parada Venegas et al. 2019). The main bacterial sources of butyrate are *Faecalibacterium prausnitzii* and *Clostridium leptum* (belonging to the family *Ruminococcaceae*), *Eubacterium rectale* and *Roseburia* spp. (family *Lachnospiraceae*), all belonging to the phylum *Firmicutes* (Louis et al. 2009).

The main producers of acetate are *Akkermansia muciniphila, Bifidobacterium* spp., *Bacteroides* spp., *Lactobacillus* spp., *Prevotella* spp., *Streptococcus* spp. and *Ruminococcus* spp.; propionate is produced by *Bacteroides* spp., *Salmonella* spp. and *Ruminococcus obeum*, all these bacteria producers of acetate and propionate belonging to the *Bacteroidetes* phylum (Feng et al. 2018).

Butyrate is mainly used in the gut, propionate and acetate flows in blood circulation reaching peripheral tissues like the brain, adipose tissue and liver where it can have metabolic effects (Koh et al. 2016).

Several in vivo and in vitro studies have shown that SCFAs play different roles in lipid metabolism. Acetate acts as a substrate in the synthesis of cholesterol and has an antilipolytic effect that can improve insulin sensitivity due to a reduction of lipid overflow to peripheral tissues sensitive to insulin; it also plays a regulatory role in adipogenesis and browning of white adipose tissue increasing mRna levels of peroxisome proliferator-activated receptor gamma (PPARγ), peroxisome proliferator-activated receptor-gamma coactivator-1 alpha (PGC-1α) and uncoupling protein 1 (UCP1), two markers of browning mechanism (Hernández et al. 2019).

SCFAs are also signaling molecules with an effect on glucose metabolism, in fact, SCFAs act like endogenous ligands of G protein-coupled receptors 41 and 43 (GPR41, GPR43), expressed in adipose tissue with an important role on lipid metabolism and in enteroendocrine L cell of the small intestine where they stimulate the secretion of glucagon-like peptide hormone (GLP-1) and peptide YY (Manoff et al. 2015). The first is an incretin hormone involved in glucose homeostasis by stimulation of insulin secretion induced by nutrients and by reduction of glucagon secretion; it also reduces gastric motility by slowing down its emptying and reduces appetite (Sandoval et al. 2015). Peptide YY has an inhibitory action on gut function; it inhibits gastric acid secretion, gastrointestinal motility and electrolyte and water secretion; it also has a promoting role on mucosal cell differentiation and has an incretin-like effect (Holzer et al. 2012).

Gut microbiota is involved in the metabolism of bile acids, which in turn control intestinal bacterial overgrowth and act as regulators of metabolism. Liver produces primary bile acids (like colic acid and chenodeoxycholic acid) that undergo a conjugation process in which hydroxylation is required with the involvement of bacteria like *Bacteroides, Lactobacillus* and *Listeria.* By a process of dihydroxylation, intestinal bacteria, like *Clostridium,* convert primary bile acids into secondary bile acid (like lithocholic acid and deoxycholic acid). Bile acids influence glucose metabolism binding to the nuclear farnesoid X receptor (FXR) and Takeda G protein-coupled receptor 5 (TGR5). FXR regulates hepatic glucose production suppressing some enzymes involved in gluconeogenesis like phosphoenolpyruvate carboxykinase (PEPCK), glucose 6-phosphatase (G6Pase) and fructose 1,6-bisphosphatase 1 (FBPase); it also induces intestinal postprandial expression of fibroblast growth factor 19 (FGF-19) that increases glycogen synthesis, reduces gluconeogenesis and promotes energetic expenditure. TGR5 promotes the secretion of GLP-1 increasing levels of cAMP that induce an intracellular calcium influx with a resulting release of GLP-1 (Alberto González-Regueiro et al. 2017).

7.3 Gut Microbiota and Type 2 Diabetes Mellitus: Possible Pathophysiological Link

The role of gut microbiota on developing diabetes mellitus and their interaction is a topic of great interest for a lot of authors.

One of the questions in several studies has been the possible existence of differences in the composition of the microbiota between diabetics and healthy individuals and what the possible pathophysiological mechanisms underlying this are.

In diabetic patients, it has been shown a higher ratio *Firmicutes/Bacteroidetes* compared to healthy people (Table 7.1).

Karlsson et al. have characterized fecal metagenome of 145 European women looking for differences among healthy people, patients with impaired glucose control and T2DM. They have found an abundance of *Lactobacillus gasseri.* and *Streptococcus mutans*, with a reduction of *Roseburia*, one *Bacteroides intestinalis* and two unknown *Clostridium* species (Karlsson et al. 2013).

In an analysis of gut microbiota in obese people with type 2 diabetes, it has been found a predominance of *Firmicutes, Clostridia* and *Negativicutes, Verrucomicrobia, Bacteroidetes, Proteobacteria,* and *Elusimicrobia,* instead, have been resulted less abundant (Ahmad et al. 2019).

Egshatyan et al. have analyzed the composition of gut microbiota in 92 patients based on the type of diet and glucose control; they have found a higher prevalence of *Blautia* genus in subjects with glucose intolerance while consuming less fat and carbohydrates compared to healthy people, *Serratia* has also resulted higher in

Table 7.1 Occurrence of bacterial phyla and species in T2DM

	Decreased in T2DM	Increased in T2DM
Phylum		
Firmicutes		X
Bacteroidetes	X	
Species		
Escherichia coli		X
Faecalibacterium prauznitzii	X	
Roseburia	X	
Streptococccus mutans		X
Lactobacillus gasseri		X

people with glucose intolerance compared to healthy donors with the same consumption of carbohydrates and calories; *Verrucomicrobia* phylum representation has resulted in a decrease in patients with glucose intolerance instead (Egshatyan et al. 2016). In other two studies, an increase in *Lactobacillus* in patients with type 2 diabetes compared to heathy donors has been shown; *Faecalibacterium prausnitzii* and *Bacteroides* have been less represented, although the latter not significantly, in diabetic patients, without significant differences in quantity for *Prevotella* and *Fusobacterium* (Sedighi et al. 2017; Navab-Moghadam et al. 2017).

T2DM is characterized by increased blood levels of glucose (mainly produced by liver) with a variable grade of impaired insulin secretion and action in peripheral tissue; its pathogenesis is influenced both by hereditary and environmental factors (Rhodes et al. 2005).

Moreover, T2DM is characterized by a pro-inflammatory state that can alter the insulin signaling pathway by an overproduction of cytokines like interleukyne-1 (IL-1), interleukyne-6 (IL-6) and tumor necrosis factor alpha (TNFα).

It has been observed as a damage to the intestinal mucosa in diabetic subjects that may be at the basis of a passage of molecules of bacterial origin, lipopolysaccharide (LPS), into the systemic circulation, a phenomenon called "endotoxemia" (Cani et al. 2008).

The increased intestinal permeability could be the result of a process of dysbiosis that impairs the integrity of the tight-junction of intestinal mucosa. In fact, butyrate has been shown to be the most important SCFA involved in avoiding an increased intestinal permeability and species of bacteria producing butyrate (like *Roseburia intestinalis* and *Faecalibacterium prausnitzii*), as mentioned above, have resulted in a decrease in diabetic patients compared to healthy people (Couto et al. 2020). Moreover, diabetic people have a minor quantity of bacteria with anti-inflammatory properties, like *Bifidobacterium* and *F. prausnitzii*, and a higher content of bacterial species associated with oxidative stress and inflammation (Palau-Rodriguez et al. 2015).

Dysbiosis and bacterial translocation in T2DM could be also related to diabetic microangiopathy and neuropathy that might induce physical damage to intestinal cell permeability and alter intestinal motility (Wellen et al. 2005). Studies in obese mice and type 2 diabetic subjects have shown increased levels of LPS that can get worse inflammation and metabolic diseases (Creely et al. 2007). LPS is recognized in the blood by lipopolysaccharide-binding protein (LBP), an acute-phase protein produced by the liver, that can be a marker of immune response to LPS and it is increased in T2DM, obesity and metabolic syndrome (Lepper et al. 2007). LPS binds the CD14/TLR4 receptor expressed by macrophages and can play a causative role in hepatic insulin resistance and obesity inducing a pro-inflammatory state. This has been demonstrated in a study where the infusion of LPS in mice with the hereditary nonappearance of the CD14/TLR4 receptor has not caused metabolic changes unlike what has been observed in normal mice, where there has been an increase in insulin resistance in liver and adipose tissue after LPR administration (Pussinen et al. 2011).

Even if the exact mechanisms are not yet clearly defined, several experimental data lead to focus attention on the relationship between type 2 diabetes and immune response. Cytokines like IL-1β, produced after local or generalized immune system response through activation of tool-like receptors (TLRs), can interfere with pancreatic β cell function (Donath et al. 2011). For example, TLR5 seems to be involved in metabolic syndrome, in fact, TLR5-deficient mice develop hyperphagia, insulin resistance, hypertension, hyperlipidemia and obesity, with the development of similar features in wild-type mice after transplantation of gut microbiota from TLR5-deficient mice (Vijay-Kumar et al. 2010).

Another pathogenetic interaction between type 2 diabetes and gut microbiota concerns the microbial amino acid metabolism. It has been shown that gut microbiota in diabetic patients, compared to healthy people, produces more imidazole propionate (ImP), a metabolite derived from histidine, with higher circulating levels of Imp in patients with type 2 diabetes. ImP impairs the function of insulin receptors by activation of the mechanistic target of rapamycin complex 1 (mTORC1), with a reduction of glucose tolerance when administered to mice (Koh et al. 2018).

7.4 Gut Microbiota and Type 1 Diabetes Mellitus

Type 1 diabetes mellitus is characterized by hyperglycemia caused by a deficit in insulin secretion due to chronic autoimmune destruction of insulin-producing pancreatic beta cells, with cytotoxic T lymphocytes that cause necrotic cell death. Environmental factors seem to have a triggering role in developing autoimmune diseases, and several studies have investigated the potential role of microbiota in the pathogenesis of autoimmune diseases like type 1 diabetes mellitus, rheumatoid arthritis, multiple sclerosis, coeliac disease and inflammatory bowel diseases (Hotamisligil 2006; Vojdani 2014). These clinical conditions tend to have, in principle, a decreasing incidence from north to south, occurring more in rich countries, probably due to climatic differences and hygiene and food habits.

Gut microbiota modules immune system by an interaction with innate and adaptive immune cells, especially it has been shown that segmented filamentous bacteria (SFG) induce expression of Th17 (CD4 + T lymphocytes that produce IL-17) and T regulatory lymphocytes (T cells), pro-inflammatory cells involved in guaranteeing integrity of intestinal mucosal barrier with a documented protective role in non-obese diabetic (NOD) mice for developing of T1DM (Kriegel et al. 2011).

Alteration in microbiota could influence autoimmunity through a mechanism of loss of immune tolerance. It has been shown a significant difference in gut microbiota diversity among healthy people and patients with T1DM (De Groot et al. 2017).

Longitudinal study of large cohorts at T1DM risk (like DIABIMMUNE, BABYDIAB, TEDDY, ABIS, FINDIA and TRIGR) have given some data about the composition of microbiota in diabetic patients.

In DIABIMMUNE study, about 1000 newborns with HLA haplotype at high risk for diabetes have been observed, with characterization of their GUT microbiota. An increase in *Blautia, Streptococcus* and *Ruminococcus* has been found in children that developed type 1 diabetes, with fewer *Veillonellaceae*, *Lachnospiraceae* and *Verrucomicrobiaceae (Akkermansia)*; moreover, in infants with antibody positivity of T1DM, compared to seronegative subjects, a higher prevalence of bacterial genes involved in transport of sugar and less expression of gene involved in biosynthesis of amino acids has been observed (Kostic et al. 2015). The authors have observed a different prevalence of *Bacteroidetes* species among Russian, Finnish and Estonian children, with the last two having a greater incidence of type 1 diabetes compared to Russians, and it has been speculated that this was due to increased early-life exposure to LPS from *Escherichia* in Russian children and from *Bacteroides* species for Finns and Estonians (Vatanen et al. 2016).

In the longitudinal study The Environmental Determinants of Diabetes (TEDDY) T1DM, patients have had a higher prevalence of *Parabacteroides*, *Alistipes shahii, Roseburia hominis* and *Bifidobacterium pseudocatenulatum,* with a decreased prevalence of *Akkermansia, Lactococcus, Ruminococcaceae* and *Streptococcus;* also subjects with positive autoimmunity have had a higher abundance of *Streptococcus* (Stewart et al. 2018).

Several studies have investigated how environmental factors like birth method, breastfeeding and nutrition in infants could influence gut microbiota and the consequent role on risk of type 1 diabetes.

Birth method influences newborn's microbiota, but it has been postulated that already during pregnancy there is a fetal exposure to maternal microbiota that concern above all butyrate-producing bacteria that increase T regulatory cells' levels (Blaser et al. 2016). It has been observed that newborns with vaginal delivery have a microbiota similar to the vaginal microbiota of the mother, dominated by *Bifidobacterium* and *Lactobacillus*, unlike infants born by cesarean section that have a predominance of species that typically live on maternal skin, such as *Clostridium* and *Staphylococcus* and lack of *Bifidobacterium, Bacteroides* and *Lactobacillus* (Dominguez-Bello et al. 2010). Data on birth method and onset of type 1 diabetes are a bit discordant. However, in a meta-analysis based on results from 20 studies, a 20% increased risk of type 1 diabetes for children born by cesarean section compared to vaginally delivery has been shown (Cardwell et al. 2008).

Breastfeeding is another important modulatory factor of microbiota's composition. A transfer of bacteria able to metabolize human milk oligosaccharides (HMOs), like *Lactobacillus* and *Bifidobacterium*, through breast milk has been shown. These bacteria have an important role in maintaining the integrity of the intestinal mucosa barrier, in producing SCFAs and in modulating the immune system (Mueller et al. 2015). Breastfeeding seems to play a protective role in the development of type 1 diabetes, as seen in the MIDIA study, that have investigated association between T1DM, breastfeeding and age of introduction of solid foods (Lund-Blix et al. 2015). In several longitudinal studies, a decreased ratio *Bifidobacterium:Bacteroides* in people at increased risk of T1DM has been observed, and at the same time infants breastfed have a greater abundance of *Bifidobacterium* compared to formula-fed infants, richer in *Bacteroides* (Sadauskaite-Kuehne et al. 2004; Marcobal et al. 2011).

Several authors have investigated the role of dietary factors in T1DM risk, like gluten and cow-milk proteins. Gluten exposition can induce an inflammation that can interest intestinal mucosa permeability through gut microbiota modulation. An increase in bacterial species involved in the regulation of epithelial tight junction of intestinal cells, like *Akkermansia*, with a possible protective role in T1DM development has been seen in gluten-free diets (Karczewski et al. 2010). An important factor is the timing of gluten exposition, with an early exposition, before the third month, associated with an increased risk of islet autoimmunity and with a decreased risk when gluten exposition occurs during breastfeeding (Norris et al. 2003).

A positive correlation between the incidence rate of T1DM and cow milk consumption has been found. In particular, in the Finnish Diabetes Prediction and Prevention Project (DIPP) study, consumption of cow milk has resulted to be a risk factor for islet autoimmunity (Virtanen et al. 2012).

Like dietary factors, also antibiotic exposition influences gut microbiota, with potential implication on T1DM onset. In an English study based on a medical record database, it has been found that taking five or more antibiotics, mostly for penicillin, cephalosporins, macrolides and quinolones, has been associated with an increased risk of T1DM (Boursi et al. 2015).

In conclusion, even if data are not often concordant, environmental factors seem to have a great role in gut microbiota modulation with a potential role in T1DM risk.

7.5 Gestational Diabetes Mellitus and Microbiota

Gestational diabetes mellitus (GDM) is a condition of reduced glucose tolerance with onset or first recognition during pregnancy.

It's a complication of about 5–18% of all pregnancies due to insulin resistance caused by secretion of diabetogenic hormones like cortisol, progesterone, placental lactogen and growth hormone and a various degree of deficit in insulin secretion.

GDM is a risk factor both for the mother (hypertension, pre-eclampsia, cesarean delivery, preterm birth) and the newborn (macrosomia, polycythemia, shoulder dystocia, hypoglycemia, neonatal asphyxia), and it also increases maternal T2DM risk.

A possible link between gut dysbiosis and GDM, looking for differences in microbiota composition among GDM and normal pregnancies, has been investigated.

In GDM women, a reduction in alpha-diversity compared to healthy pregnant women, with an increase in *Firmicutes/Bacteroides* ratio, especially in late pregnancy has been shown. In particular, a greater composition in gram-negative bacteria (*Prevotella, Parabacteroides, Haemophilus*) in GDM, with a positive correlation with blood glucose levels, has been observed (Cortez et al. 2019). SCFAs-producing bacteria, instead, have resulted to be reduced in GDM women, like *Akkermansia, Ruminococcus, Faecalibacterium* and *Roseburia* (Crussel et al. 2018).

At the same time, the alterations in the microbiota also affect newborns from GDM mothers. A reduction in alpha-diversity with an increase in *Escherichia, Parabacteroides* and *Actinobacteria* and a reduction in *Bacteroidetes* has been observed in these newborns (Wang et al. 2018; Su et al. 2018).

Several randomized clinical trials have been conducted by use of probiotics and synbiotics (a combination of probiotics and prebiotics) compared to placebo, with some benefits in metabolic features like insulin resistance and lipid metabolism in GDM women (Karamali et al. 2016, 2018).

7.6 Complications of Diabetes and Gut Microbiota

Diabetes, both type 1 and type 2, is characterized by the development of serious complications that impair several organs like the eye, kidney and heart.

These complications are generally classified as microvascular and macrovascular complications: the first are diabetic neuropathy, diabetic nephropathy and diabetic retinopathy, the latter are cardiovascular diseases like stroke, coronary heart disease and peripheral artery disease (PAD).

Diabetic retinopathy (DR) is one of the first causes of impaired vision in the world and is strictly related to the duration of diabetes and poor glycaemic control. It's divided into two major forms, non-proliferative and proliferative, according to the presence or not of new blood vessels in the retina. Patients affected by DR show a decreased proportion of *Akkermansia, Bacteroidetes* and *Mucoromycota* compared to healthy people, with an increase in *Escherichia, Acidaminococcus* and *Enterobacter* (Das et al. 2021). Study in mice with administration of probiotics (like *Lactobacillus paracasei* and *rhamnosus*) has shown a reduction in ocular inflammation and an improvement of eye disease (Verma et al. 2019; Petit Homme et al. 2020).

Diabetes is also the first cause of chronic and end-stage kidney disease worldwide. The classical term "diabetic nephropathy" has been replaced by "diabetic kidney disease" that includes a clinical condition characterized by albuminuria and/or decreased estimated glomerular filtration rate in diabetic patients. There is a dual relationship between dysbiosis and renal failure, on the one hand, it has been seen that some products of bacterial metabolism can worsen renal function, on the other hand, chronic kidney disease (CKD) causes a worsening of dysbiosis (Mahmoodpoor et al. 2017; Vaziri et al. 2013). In patients with CKD, there is a reduction of bacterial species with anti-inflammatory properties, like *Bifidobacterium, Lactobacillus* and *Prevotella,* with the proliferation of bacteria associated with systemic inflammation, such as *Clostridiaceae* and *Bacteroidaceae* (Kanbay et al. 2018). Administration of probiotic in patients with CKD have reduced levels of urea nitrogen and uric acid (Ranganathan et al. 2009).

Diabetic neuropathy is the most common complication of diabetes, concerning nearly 50% of patients during their life. It's due to damage to the peripheral nervous system, in particular to sensory, motor or autonomic fibers. In diabetic patients with neuropathy, a difference in gut microbiota compared to diabetics without this complication has been shown, with an increase in *Actinobacteria* and *Firmicutes* and a decrease in *Bacteroidetes* (Wang et al. 2020).

Diabetes is also one of the most important risk factors for cardiovascular diseases. Dysbiosis seems to be related to vascular diseases in animal and human studies and some benefits have been found by administration of probiotics.

For example in a study in patients with diabetic foot ulcer, due to PAD, administration of probiotics (*Bifidobacterium bifidum, Lactobacillus casei, L. acidophilus* and *L. fermentum*) has led to an improvement of ulcer and metabolic parameters like blood glucose and serum insulin (Mohseni et al. 2018).

7.7 Interaction Between Treatment of Diabetes and Gut Microbiota

Nowadays, there are various options for the treatment of diabetes. If it is true that for T1DM insulin remains the only therapy available, administered by multiple daily injections or by continuous subcutaneous infusion, in T2DM there are various classes of drugs that allow treatment strategies aimed not only at lowering blood sugar, but also at personalizing the therapy based on the patient's metabolic phenotype and at preventing complications of diabetes.

The first line of therapy in T2DM is metformin, a molecule belonging to the biguanide class. It inhibits hepatic gluconeogenesis, decreasing so hepatic glucose output, and increases insulin sensitivity in peripheral tissues (DeFronzo et al. 1991). Metformin reduces gluconeogenesis by inhibition of a specific mitochondrial isoform of glycerophosphate dehydrogenase (mGDP) and lowers serum lipid levels activating in hepatocytes AMP-activated protein kinase (AMPK) (Madiraju et al. 2014; Hawley et al. 2002).

In several studies, metformin seems to modulate gut microbiota composition. It decreases the abundance of *Intestinibacter* and increases levels of *Escherichia species* (Forslund et al. 2015). Metformin seems to improve the production of butyrate and propionate, increasing the abundance of bacteria like *Akkermansia, Bifidobacterium, Lactobacillus, Prevotella* and *Blautia*, with beneficial metabolic effects (Wu et al. 2017).

In studies on rodents fed with high fat diet, metformin has increased levels of *A. muciniphila* that is negatively related to blood glucose levels, and has decreased levels of *Clostridium orbiscindens*, negatively correlated with the expression of GLUT2 and PPARα (Lee et al. 2014).

Metformin is supposed to increase secretion of GLP-1 and decreases DPP-4 enzyme activity by its modulatory effect on the gut microbiota and to improve intestinal sensing of glucose by regulation of intestinal mucosa permeability (Mannucci et al. 2001; Forslund et al. 2015).

GLP-1 receptor agonists (GLP-1 RAs) are a class of drugs including molecules administered subcutaneously (liraglutide, dulaglutide, semaglutide, lixisenatide, exenatide) or orally (semaglutide) with a beneficial effect in cardiovascular prevention and body weight reduction.

In animal studies, liraglutide has increased the abundance of SCFAs producing bacteria, like *Bacteroides* and *Bifidobacterium*, with a decrease in *Firmicutes* (Madsen et al. 2019; Zhang et al. 2017a). In a recent human study on 52 patients receiving GLP1-RA, liraglutide or dulaglutide, the difference in gut microbiota between responder and not responder patients to the treatment has been evaluated. A positive correlation has been shown, in responders, between the decrease of blood glucose levels and *Bacteroides dorei, Roseburia inulinivorans, Lachnoclostridium sp.* and *Butyricicoccus sp.,* and a negative correlation with *Prevotella copri, Ruminococcaceae sp, Eubacterium coprostanoligenes sp., Bacteroidales sp., Alistipes obesi, Lactobacillus mucosae* and *Butyricimonas virosa* (Tsai et al. 2022).

Dipeptidyl-peptidase IV inhibitors (DPP-4i) are drugs that indirectly improve the metabolism of incretins inhibiting the enzyme DPP-4 involved in their degradation. Vildagliptin and sitagliptin have resulted in an increase in rats' abundance of bacteria producing SCFAs (in particular, the first has increased quantity of *Bacteroides* and *Erysipelotrichaceae*, the latter *Clostridium, Roseburia* and *Blautia*) (Zhang et al. 2017b; Yan et al. 2016). In mice, vildagliptin has increased *Lactobacillus spp.* and decreased *Oscillibacter spp.* with the resulting effect of increased propionate and reduced expression of TLR-2 and TLR-4 (Olivares et al. 2018). Saxagliptin, instead, in rats has induced an increase in *Firmicutes* (Wang et al. 2016).

Another class of diabetic drugs with proven cardiovascular and renal benefits are sodium-glucose co-transporter 2 inhibitors (SGLT-2i) that lower blood glucose levels by inhibiting SGLT-2 expressed in the proximal tubule. One of these drugs, dapagliflozin, has reduced *Firmicutes/Bacteroidetes* ratio in diabetic mice compared to controls and increased *Proteobacteria* and *Ruminococcaceae* in diabetic rats (Lee et al. 2018; Yang et al. 2020). Potential greater effects on gut microbiota modulation could be seen in the future with the use of sotagliflozin that inhibits not only SGLT-2 but also SGLT-1, a transporter expressed in the gut (Sims et al. 2018).

Acarbose is an anti-diabetic drug with interesting effects on gut microbiota modulation due to its mechanism of action that consists in inhibition of alpha-glucosidase, reducing degradation of oligosaccharides in monosaccharides and disaccharides and so reducing intestinal absorption of glucose, giving nutrients to intestinal bacteria. Acarbose seems to increase butyrate-producing bacteria and improve bile acids metabolism, increasing the proportion of *Bifidobacterium* and *Lactobacillus* and reducing *Ruminococcus* and *Butyricicoccus* (Su et al. 2015; Gu et al. 2017).

There is little data about the relation between insulin treatment and gut microbiota. Maybe its subcutaneous administration could be a reason of little influence on the intestinal bacterial population, however, it has been shown that dysbiosis could influence response to insulin therapy in patients with type 2 diabetes (Li et al. 2019).

7.8 Therapeutic Role of Gut Microbiota in Diabetes Mellitus

Several studies have focused their attention on possible therapeutic strategies for diabetes by modulation of gut microbiota. Nowadays, gut microbiota modulation takes place through lifestyle changes or directly through the administration of prebiotics (food components, mainly dietary fibers that promote growth of bacteria with beneficial effects for humans), probiotics (i.e. the direct administration of microorganisms with beneficial effects) or fecal microbiota transplant (FMT). Diet is an important factor that influences microbiota composition. It has been seen that a diet rich in plants and fibers in African children, compared to the diet of European children, gives important differences in microbiota composition, with a greater representation of SCFAs producing bacteria (*Bacteroidetes*) and reduction of *Firmicutes* and *Enterobacteriaceae* (De Filippo et al. 2010). Also, proteins can have a modulatory effect on microbiota, in particular diets rich in protein derived from plants increase *Lactobacillus* and *Bifidobacterium* and decrease *Clostridium perfrigens* and *Bacteroides fragilis* (Singh et al. 2017).

Mediterranean diet is associated with metabolic beneficial effects that could in part be explained by a change in microbiota. It has been seen that this alimentary regimen promotes an abundance of SCFAs bacteria and could also retard the onset of T1DM and improve glycaemic control in patients with T1DM by modulating immune system response preserving intestinal permeability (Calabrese et al. 2021).

Prebiotics are substances that modify gut microbiota with advantages for the host. Examples of prebiotics are inulin, lactulose, fructo-oligosaccharides and galacto-oligosaccharides (Pineiro et al. 2008). In a study involving some T2DM obese patients, a strict vegetarian diet for a month induced an improvement in glucose control, triglyceride levels and body weight, with a reduction of *Firmicutes/Bacteroides* ratio (Kum et al. 2013).

Probiotics are microorganisms ingested as food or as supplements that offer advantages to the host, especially in terms of the immune response, glucose and lipid metabolism, and protection against intestinal and respiratory diseases. *Lactobacillus* and *Bifidobacterium* are the most used probiotics with documented metabolic benefits (Prasad et al. 1998). Other microorganisms with beneficial effects are *A. muciniphila* and *Faecalibacterium prausnitzii*, both have anti-inflammatory properties and the first has been associated with improvement of glucose control and features of metabolic syndrome (Plovier et al. 2017).

FMT has been evaluated for a therapeutic role in diabetes. In mice, the possibility to transfer the metabolic microbiome profile to germ-free animals has been shown (Turnbaugh et al. 2006).

In a clinical trial, it has been seen that transfer of intestinal microbiota from lean donors to males with metabolic syndrome has given an improvement in insulin sensitivity and an increase in butyrate-producing bacteria (Vrieze et al. 2012). However, to date, results obtained with FMT have been transient and are related to baseline microbiome features of recipient subjects (Kootte et al. 2017).

7.9 Aging: The Interplay with Diabetes and Microbiota

DM, particularly type 2, mainly affects adults and the elderly, with most of the diabetic patients over 65 years old (Wild et al. 2004). Diabetes seems to be a pro-aging factor, in fact, it increases risk of age-related conditions like cardiovascular diseases, mild cognitive impairment, urinary incontinence, periodontal disease and Alzheimer's disease; moreover, cellular senescence is a possible etiological factor for diabetes, due to an abundance in senescent cells in adipose tissue and beta pancreatic cells (Palmer et al. 2019). For example, tumor suppressor p53 is related to cellular senescence and is a regulator of adipogenesis, in fact, p53 has to be inhibited to allow differentiation of adipogenic precursor cells into adipocytes responsive to insulin and it also impairs glucose transport and increases lipolysis in fat cells (Vergoni et al. 2016).

With age, a change of gene expression in beta cells has been shown, with increased expression of genes like CDKN2a and CDKN2b, related with cellular senescence (Helman et al. 2016); moreover, in a T1DM murine model, an improvement of insulin secretion by elimination of senescent beta cells has been obtained (Thompson et al. 2019).

There is a mutual connection between gut microbiota and aging with a shown change in microbiota composition until old age, even if it is not still clear if microbiota alteration is cause or result of aging (Nagpal et al. 2018). There are a lot of causes of microbiota alteration associated with aging, such as diet (reduction in fruit and vegetable assumption, reduction of appetite, masticatory dysfunction), drugs, hospitalizations, hygiene, infections, reduced intestinal activity and chronic diseases (Odamaki et al. 2016). It has been supposed that such species of bacteria could be related to a healthy aging. For example, in semi-supercentenarians (aged 105–109 years), a prevalence of symbiotic bacterial taxa from *Bacteroidaceae*, *Ruminococcaceae* and *Lachnospiraceae*, probably related to healthy aging has been reported (Biagi et al. 2016).

Interventions on the microbiota in old age can also have benefits in terms of healthy aging, as demonstrated by the NU-AGE project. In this study, the authors have profiled gut microbiota of more than 600 non-frail or pre-frail people before and after administration of a Mediterranean diet for one year. The dietary intervention has improved health status, with a reduction of markers of lower frailty and inflammatory markers and an improvement in cognitive function; microbiota analysis has shown an enrichment in bacterial species associated with short-chained fatty acid production and lower carbon dioxide and secondary bile acids production (Ghosh et al. 2020).

7.10 Conclusions

A growing number of scientific evidence suggests that the microbiota constitutes a real metabolically active organ within the human organism and dysbiosis can be related to the onset or worsening of several chronic diseases like diabetes mellitus.

Modulation of microbiota is a field of great interest in medical research.

Nowadays, there are emerging promising data about the effectiveness of techniques of microbiota modulation in the improvement of diabetes mellitus and healthy aging, but numerous large-scale studies are needed to obtain more robust evidence.

References

Ahmad A, Yang W, Chen G et al. (2019) Analysis of gut microbiota of obese individuals with type 2 diabetes and healthy individuals. PLoS One 14(12):e0226372. Published 2019 Dec 31. doi: https://doi.org/10.1371/journal.pone.0226372

Alberto González-Regueiro J, Moreno-Castañeda L, Uribe M, Carlos C-T (2017) The role of bile acids in glucose metabolism and their relation with diabetes. Ann Hepatol 16(Suppl 1):S15–S20. https://doi.org/10.5604/01.3001.0010.5494 PMID: 31196630

Arumugam M, Raes J, Pelletier E, Le Paslier D, Yamada T, Mende DR, Fernandes GR, Tap J, Bruls T, Batto JM, Bertalan M, Borruel N, Casellas F, Fernandez L, Gautier L, Hansen T, Hattori M, Hayashi T, Kleerebezem M, Kurokawa K, Leclerc M, Levenez F, Manichanh C, Nielsen HB, Nielsen T, Pons N, Poulain J, Qin J, Sicheritz-Ponten T, Tims S, Torrents D, Ugarte E, Zoetendal EG, Wang J, Guarner F, Pedersen O, de Vos WM, Brunak S, Doré J; MetaHIT Consortium, Antolín M, Artiguenave F, Blottiere HM, Almeida M, Brechot C, Cara C, Chervaux C, Cultrone A, Delorme C, Denariaz G, Dervyn R, Foerstner KU, Friss C, van de Guchte M, Guedon E, Haimet F, Huber W, van Hylckama-Vlieg J, Jamet A, Juste C, Kaci G, Knol J, Lakhdari O, Layec S, Le Roux K, Maguin E, Mérieux A, Melo Minardi R, M'rini C, Muller J, Oozeer R, Parkhill J, Renault P, Rescigno M, Sanchez N, Sunagawa S, Torrejon A, Turner K, Vandemeulebrouck G, Varela E, Winogradsky Y, Zeller G, Weissenbach J, Ehrlich SD, Bork P (2011) Enterotypes of the human gut microbiome. Nature 473(7346):174–180

Biagi E, Franceschi C, Rampelli S, Severgnini M, Ostan R, Turroni S, Consolandi C, Quercia S, Scurti M, Monti D, Capri M, Brigidi P, Candela M (2016) Gut microbiota and extreme longevity. Curr Biol 26(11):1480–1485. https://doi.org/10.1016/j.cub.2016.04.016 Epub 2016 May 12 PMID: 27185560

Blaser MJ, Dominguez-Bello MG (2016) The human microbiome before birth. Cell Host Microbe 20:558–560. https://doi.org/10.1016/j.chom.2016.10.014

Boursi B, Mamtani R, Haynes K, Yang YX (2015) The effect of past antibiotic exposure on diabetes risk. Eur J Endocrinol 172:639–648. https://doi.org/10.1530/EJE-14-1163

Calabrese CM, Valentini A, Calabrese G (2021) Gut Microbiota and type 1 diabetes mellitus: the effect of mediterranean diet. Front Nutr 7:329

Cani PD, Bibiloni R, Knauf C, Waget A, Neyrinck AM, Delzenne NM, Burcelin R (2008) Changes in gut microbiota control metabolic endotoxemia-induced inflammation in high-fat diet-induced obesity and diabetes in mice. Diabetes 57:1470–1481

Cardwell CR, Stene LC, Joner G, Cinek O, Svensson J, Goldacre MJ et al (2008) Caesarean section is associated with an increased risk of childhood onset type 1 diabetes mellitus: a meta-analysis of observational studies. Diabetologia 51:726–735. https://doi.org/10.1007/s00125-008-0941-z

Classification and diagnosis of diabetes: Standards of Medical Care in Diabetes (2022). Diabetes Care 45(Suppl. 1):S17–S38

Cortez RV, Taddei CR, Sparvoli LG, Ângelo AG, Padilha M, Mattar R, Daher S (2019) Microbiome and its relation to gestational diabetes. Endocrine 64(2):254–264

Couto MR, Gonçalves P, Magro F, Martel F (2020) Microbiota-derived butyrate regulates intestinal inflammation: focus on inflammatory bowel disease. Pharmacol Res 159:104947

Craciun CI, Neag MA, Catinean A, Mitre AO, Rusu A, Bala C, Roman G, Buzoianu AD, Muntean DM, Craciun AE (2022) The relationships between gut microbiota and diabetes mellitus, and treatments for diabetes mellitus. Biomedicines 10(2):308. https://doi.org/10.3390/biomedicines10020308.PMID:35203519;PMCID:PMC8869176

Creely SJ, McTernan PG, Kusminski CM et al. (2007) Lipopolysaccharide activates an innate immune system response in human adipose tissue in obesity and type 2 diabetes. Am J Physiol Endocrinol Metab 292:e740e7

Crusell MKW, Hansen TH, Nielsen T, Allin KH, Rühlemann MC, Damm P, Vestergaard H, Rørbye C, Jørgensen NR, Christiansen OB, Heinsen F-A, Franke A, Hansen T, Lauenborg J, Pedersen O (2018) Gestational diabetes is associated with change in the gut microbiota composition in third trimester of pregnancy and post-partum. Microbiome

Das T, Jayasudha R, Chakravarthy S, Prashanthi GS, Bhargava A, Tyagi M, Rani PK, Pappuru RR, Sharma S, Shivaji S (2021) Alterations in the gut bacterial microbiome in people with type 2 diabetes mellitus and diabetic retinopathy. Sci Rep 11:2738

De Filippo C, Cavalieri D, Di Paola M, Ramazzotti M, Poullet JB, Massart S, Collini S, Pieraccini G, Lionetti P (2010) Impact of diet in shaping gut microbiota revealed by a comparative study in children from Europe and rural Africa. Proc Natl Acad Sci USA 107:14691–14696

De Groot PF, Belzer C, Aydin O, Levin E, Levels JH, Aalvink S et al (2017) Distinct fecal and oral microbiota composition in human type 1 diabetes, an observational study. PLoS ONE 12: e0188475. https://doi.org/10.1371/journal.pone.0188475

DeFronzo RA, Barzilai N, Simonson DC (1991) Mechanism of metformin action in obese and lean noninsulin-dependent diabetic subjects. J Clin Endocrinol Metab 73:1294

Dominguez-Bello MG, Costello EK, Contreras M, Magris M, Hidalgo G, Fierer N et al (2010) Delivery mode shapes the acquisition and structure of the initial microbiota across multiple body habitats in newborns. Proc Natl Acad Sci USA 107:11971–11975. https://doi.org/10.1073/pnas.1002601107

Donath MY, Shoelson SE (2011) Type 2 diabetes as an inflammatory disease. Nat Rev Immunol 11:98–107

Egshatyan L, Kashtanova D, Popenko A, Tkacheva O, Tyakht A, Alexeev D, Karamnova N, Kostryukova E, Babenko V, Vakhitova M, Boytsov S (2016) Gut microbiota and diet in patients with different glucose tolerance. Endocr Connect 5(1):1–9. https://doi.org/10.1530/EC-15-0094. Epub 2015 Nov 10. PMID: 26555712; PMCID: PMC4674628

Feng W, Ao H, Peng C (2018) Gut Microbiota, short-chain fatty acids, and herbal medicines. Front Pharmacol 9:1354. Published 2018 Nov 23. doi:https://doi.org/10.3389/fphar.2018.01354
Forslund K, Hildebrand F, Nielsen T, Falony G, Le Chatelier E, Sunagawa S, Prifti E, Vieira-Silva S, Gudmundsdottir V, Krogh Pedersen H et al (2015) Disentangling type 2 diabetes and metformin treatment signatures in the human gut microbiota. Nature 528:262–266
Ghosh TS, Rampelli S, Jeffery IB et al (2020) Mediterranean diet intervention alters the gut microbiome in older people reducing frailty and improving health status: the NU-AGE 1-year dietary intervention across five European countries. Gut 69(7):1218–1228. https://doi.org/10.1136/gutjnl-2019-319654
Gu Y, Wang X, Li J, Zhang Y, Zhong H, Liu R, Zhang D, Feng Q, Xie X, Hongmei Z et al (2017) Analyses of gut microbiota and plasma bile acids enable stratification of patients for antidiabetic treatment. Nat Commun 8:1785
Hawley SA, Gadalla AE, Olsen GS, Hardie DG (2002) The antidiabetic drug metformin activates the AMP-activated protein kinase cascade via an adenine nucleotide-independent mechanism. Diabetes 51:2420
Hernández MAG, Canfora EE, Jocken JWE, Blaak EE (2019) The short-chain fatty acid acetate in body weight control and insulin sensitivity. Nutrients 11(8):1943. Published 2019 Aug 18. doi: https://doi.org/10.3390/nu11081943
Holzer P, Reichmann F, Farzi A (2012) Neuropeptide Y, peptide YY and pancreatic polypeptide in the gut-brain axis. Neuropeptides 46(6):261-274. doi:https://doi.org/10.1016/j.npep.2012.08.005
Hotamisligil GS (2006) Inflammation and metabolic disorders. Nature 444:860–867
IDF Diabetes Atlas 2021, 10th edition https://diabetesatlas.org/atlas/tenth-edition/. Accessed on January 17, 2022. American Diabetes Association Professional Practice Committee
Kanbay M, Onal EM, Afsar B, Dagel T, Yerlikaya A, Covic A, Vaziri ND (2018) The crosstalk of gut microbiota and chronic kidney disease: role of inflammation, proteinuria, hypertension, and diabetes mellitus. Int Urol Nephrol 50:1453–1466
Karamali M, Dadkhah F, Sadrkhanlou M, Jamilian M, Ahmadi S, Tajabadi-Ebrahimi M, Jafari P, Asemi Z (2016) Efects of probiotic supplementation on glycaemic control and lipid profles in gestational diabetes: a randomized, double-blind, placebocontrolled trial. Diabetes Metab 42 (4):234–324
Karamali M, Nasiri N, Taghavi Shavazi N, Jamilian M, Bahmani F, Tajabadi-Ebrahimi M, Asemi Z (2018) The efects of synbiotic supplementation on pregnancy outcomes in gestational diabetes. Probiotics Antimicrob Proteins 10(3):4
Karczewski J, Troost FJ, Konings I, Dekker J, Kleerebezem M, Brummer RJ et al (2010) Regulation of human epithelial tight junction proteins by Lactobacillus plantarum in vivo and protective effects on the epithelial barrier. Am J Physiol Gastrointest Liver Physiol 298:G851–G859. https://doi.org/10.1152/ajpgi.00327.2009
Karlsson FH, Tremaroli V, Nookaew I, Bergström G, Behre CJ, Fagerberg B, Nielsen J, Bäckhed F (2013) Gut metagenome in European women with normal, impaired and diabetic glucose control. Nature 498(7452):99–103. https://doi.org/10.1038/nature12198 Epub 2013 May 29 PMID: 23719380
Kim MS, Hwang SS, Park EJ, Bae JW (2013) Strict vegetarian diet improves the risk factors associated with metabolic diseases by modulating gut microbiota and reducing intestinal inflammation. Environ Microbiol Rep 5:765–775
Koh A, Molinaro A, Ståhlman M, Khan MT, Schmidt C, Mannerås-Holm L, Wu H, Carreras A, Jeong H, Olofsson LE, Bergh PO, Gerdes V, Hartstra A, de Brauw M, Perkins R, Nieuwdorp M, Bergström G, Bäckhed F (2018) Microbially produced imidazole propionate impairs insulin signaling through mTORC1. Cell 175(4):947-961.e17. https://doi.org/10.1016/j.cell.2018.09.055 Epub 2018 Oct 25 PMID: 30401435
Koh A, De Vadder F, Kovatcheva-Datchary P, Backhed F (2016) From dietary fiber to host physiology: Short-chain fatty acids as key bacterial metabolites. Cell 165:1332e45

Kootte RS, Levin E, Salojarvi J, et al. (2017) Improvement of insulin sensitivity after lean donor feces in metabolic syndrome is driven by baseline intestinal microbiota composition. Cell Metab 26:611e619.e6

Kostic AD, Gevers D, Siljander H, Vatanen T, Hyotylainen T, Hamalainen AM et al (2015) The dynamics of the human infant gut microbiome in development and in progression toward type 1 diabetes. Cell Host Microbe 17:260–273. https://doi.org/10.1016/j.chom.2015.01.001

Kriegel MA, Sefik E, Hill JA, Wu HJ, Benoist C, Mathis D (2011) Naturally transmitted segmented filamentous bacteria segregate with diabetes protection in nonobese diabetic mice. Proc Natl Acad Sci USA 108:11548–11553. https://doi.org/10.1073/pnas.1108924108

Lee H, Ko G (2014) Effect of metformin on metabolic improvement and gut microbiota. Appl Environ Microbiol 80:5935–5943

Lee DM, Battson ML, Jarrell DK, Hou S, Ecton KE, Weir TL, Gentile CL (2018) SGLT2 inhibition via dapagliflozin improves generalized vascular dysfunction and alters the gut microbiota in type 2 diabetic mice. Cardiovasc Diabetol 17:1–14

Lepper P, Schumann C, Triantafilou K, Rasche F, Schuster T, Frank H, Schneider E, Triantafilou M, von Eynatten M (2007) Association of lipopolysaccharide-binding protein and coronary artery disease in men. J Am Coll Cardiol 50:25–31

Li T, Li H, Li W, Chen S, Feng T, Jiao W, Wu C, Dong J, Li Y, Li S et al (2019) Interleukin-37 sensitize the elderly type 2 diabetic patients to insulin therapy through suppressing the gut microbiota dysbiosis. Mol Immunol 112:322–329

Louis P, Flint HJ (2009) Diversity, metabolism and microbial ecology of butyrate-producing bacteria from the human large intestine. FEMS Microbiol Lett 294:1–8. https://doi.org/10.1111/j.1574-6968.2009.01514.x

Lund-Blix NA, Stene LC, Rasmussen T, Torjesen PA, Andersen LF, Ronningen KS (2015) Infant feeding in relation to islet autoimmunity and type 1 diabetes in genetically susceptible children: the MIDIA study. Diabetes Care 38:257–263. https://doi.org/10.2337/dc14-1130

Madiraju AK, Erion DM, Rahimi Y et al (2014) Metformin suppresses gluconeogenesis by inhibiting mitochondrial glycerophosphate dehydrogenase. Nature 510:542

Madsen MSA, Holm JB, Pallejà A, Wismann P, Fabricius K, Rigbolt K, Mikkelsen M, Sommer M, Jelsing J, Nielsen HB et al (2019) Metabolic and gut microbiome changes following GLP-1 or dual GLP-1/GLP-2 receptor agonist treatment in diet-induced obese mice. Sci Rep 9:1–12

Mahmoodpoor F, Rahbar Saadat Y, Barzegari A, Ardalan M, Zununi Vahed S (2017) The impact of gut microbiota on kidney function and pathogenesis. Biomed Pharm 93:412–419

Mannucci E, Ognibene A, Cremasco F, Bardini G, Mencucci A, Pierazzuoli E, Ciani S, Messeri G, Rotella CM (2001) Effect of metformin on glucagon-like peptide 1 (GLP-1) and Leptin levels in obese nondiabetic subjects. Diabetes Care 24:489–494

Manoff M (2015) Gut microbiome: the peacekeepers. Nature 518:S3–S11

Marcobal A, Barboza M, Sonnenburg ED, Pudlo N, Martens EC, Desai P et al (2011) Bacteroides in the infant gut consume milk oligosaccharides via mucus-utilization pathways. Cell Host Microbe 10:507–514. https://doi.org/10.1016/j.chom.2011.10.007

McNeil NI (1984) The contribution of the large intestine to energy supplies in man. Am J Clin Nutr 39:338–342

Mohseni S, Bayani M, Bahmani F, Tajabadi-Ebrahimi M, Bayani MA, Jafari P, Asemi Z (2018) The beneficial effects of probiotic administration on wound healing and metabolic status in patients with diabetic foot ulcer: a randomized, double-blind, placebo-controlled trial. Diabetes Metab Res 34(3). https://doi.org/10.1002/dmrr.2970. Epub 2017 Dec 21. PMID: 29193662

Mueller NT, Bakacs E, Combellick J, Grigoryan Z, Dominguez-Bello MG (2015) The infant microbiome development: mom matters. Trends Mol Med 21:109–117. https://doi.org/10.1016/j.molmed.2014.12.002

Nagpal R, Mainali R, Ahmadi S, Wang S, Singh R, Kavanagh K, Kitzman DW, Kushugulova A, Marotta F, Yadav H (2018) Gut microbiome and aging: physiological and mechanistic insights. Nutr Healthy Aging 4(4):267–285. https://doi.org/10.3233/NHA-170030.PMID:29951588; PMCID:PMC6004897

Navab-Moghadam F, Sedighi M, Khamseh ME, Alaei-Shahmiri F, Talebi M, Razavi S, Amirmozafari N (2017) The association of type II diabetes with gut microbiota composition. Microb Pathog 110:630–636. https://doi.org/10.1016/j.micpath.2017.07.034 Epub 2017 Jul 21 PMID: 28739439

Norris JM, Barriga K, Klingensmith G, Hoffman M, Eisenbarth GS, Erlich HA et al (2003) Timing of initial cereal exposure in infancy and risk of islet autoimmunity. JAMA 290:1713–1720. https://doi.org/10.1001/jama.290.13.1713

Odamaki T, Kato K, Sugahara H, Hashikura N, Takahashi S, Xiao JZ, Abe F, Osawa R (2016) Age-related changes in gut microbiota composition from newborn to centenarian: a cross-sectional study. BMC Microbiol 16(1):90

Olivares M, Neyrinck A, Pötgens SA, Beaumont M, Salazar N, Cani PD, Bindels LB, Delzenne NM (2018) The DPP-4 inhibitor vildagliptin impacts the gut microbiota and prevents disruption of intestinal homeostasis induced by a Western diet in mice. Diabetologia 61:1838–1848

Palau-Rodriguez M, Tulipan S, Queipo-Ortuño MI, Urpi-Sarda M, Tinahones FJ, Andres-Lacueva C (2015) Metabolomic insight into the intricate gut microbial host interaction in the development of obesity and type 2 diabetes. Front Microbiol 6:1151

Palmer AK, Gustafson B, Kirkland JL, Smith U (2019) Cellular senescence: at the nexus between ageing and diabetes. Diabetologia 62(10):1835–1841. https://doi.org/10.1007/s00125-019-4934-x

Parada Venegas D, De la Fuente MK, Landskron G, González MJ, Quera R, Dijkstra G, Harmsen HJM, Faber KN, Hermoso MA (2019) Short chain fatty acids (SCFAs)-mediated gut epithelial and immune regulation and its relevance for inflammatory bowel diseases. Front Immunol 11(10):277. https://doi.org/10.3389/fimmu.2019.00277.Erratum.In:FrontImmunol. 2019Jun28;10:1486.PMID:30915065;PMCID:PMC6421268

Petit Homme R, George AK, Stanisic DN, Malonee C, Molnar J, Smolenkova I, Sandhu HAS, Tyagi SC, Singh M (2020) Effects of probiotic on the development of diabetic retinopathy. Investig Ophthalmol Vis Sci 61:4961

Pineiro M, Asp NG, Reid G, Macfarlane S, Morelli L, Brunser O, Tuohy K (2008) FAO technical meeting on prebiotics. J Clin Gastroenterol 42:S156–S159

Pitocco D, Di Leo M, Tartaglione L, De Leva F, Petruzziello C, Saviano A, Pontecorvi A, Ojetti V (2020) The role of gut microbiota in mediating obesity and diabetes mellitus. Eur Rev Med Pharmacol Sci 24(3):1548–1562. https://doi.org/10.26355/eurrev_202002_20213 PMID: 32096204

Plovier H, Everard A, Druart C, et al. (2017) A purified membrane protein from Akkermansia muciniphila or the pasteurized bacterium improves metabolism in obese and diabetic mice. Nat Med 23:107e13

Prasad J, Gill HS, Smart JB, Gopal PK (1998) Selection and characterization of Lactobacillus and Bifidobacterium strains for use as probiotics. Int Dairy J 8:993–1002

Pussinen PJ, Havulinn AS, Lehto M, Sundvall J, Salomma V (2011) Endotoxemia is associated with an increased risk of incident diabetes. Diabetes Care 34:392–397

Ranganathan N, Friedman EA, Tam P, Rao V, Ranganathan P, Dheer R (2009) Probiotic dietary supplementation in patients with stage 3 and 4 chronic kidney disease: a 6-month pilot scale trial in Canada. Curr Med Res Opin 25:1919–1930

Rhodes CJ (2005) Type 2 diabetes a matter of beta-cell life and death? Science 307:380–384

Sadauskaite-Kuehne V, Ludvigsson J, Padaiga Z, Jasinskiene E, Samuelsson U (2004) Longer breastfeeding is an independent protective factor against development of type 1 diabetes mellitus in childhood. Diabetes Metab Res Rev 20:150–157. https://doi.org/10.1002/dmrr.425

Sandoval DA, D'Alessio DA (2015) Physiology of proglucagon peptides: role of glucagon and GLP-1 in health and disease. Physiol Rev 95(2):513–548. https://doi.org/10.1152/physrev.00013.2014 PMID: 25834231

Sedighi M, Razavi S, Navab-Moghadam F, Khamseh ME, Alaei-Shahmiri F, Mehrtash A, Amirmozafari N (2017) Comparison of gut microbiota in adult patients with type 2 diabetes and healthy individuals. Microb Pathog 111:362–369. https://doi.org/10.1016/j.micpath.2017.08.038 Epub 2017 Sep 11 PMID: 28912092

Sims H, Smith KH, Bramlage P, Minguet J (2018) Sotagliflozin: a dual sodium-glucose co-transporter-1 and -2 inhibitor for the management of Type 1 and Type 2 diabetes mellitus. Diabet Med 35(8):1037–1048. https://doi.org/10.1111/dme.13645 Epub 2018 Jul 3 PMID: 29637608

Singh RK, Chang H-W, Yan D, Lee KM, Ucmak D, Wong K, Abrouk M, Farahnik B, Nakamura M, Zhu TH et al (2017) Influence of diet on the gut microbiome and implications for human health. J Transl Med 15:73

Stewart CJ, Ajami NJ, O'brien JL, Hutchinson DS, Smith DP, Wong MC, et al (2018) Temporal development of the gut microbiome in early childhood from the TEDDY study. Nature 562:583–8. https://doi.org/10.1038/s41586-018-0617-x

Su B, Liu H, Li J, Sunli Y, Liu B, Liu D, Zhang P, Meng X (2015) Acarbose treatment affects the serum levels of inflammatory cytokines and the gut content of bifidobacteria in Chinese patients with type 2 diabetes mellitus. J Diabetes 7:729–739

Su M, Nie Y, Shao R, Duan S, Jiang Y, Wang M, Xing Z, Sun Q, Liu X, Xu W (2018) Diversified gut microbiota in newborns of mothers with gestational diabetes mellitus. PLoS ONE 13(10): e0205695

Thompson PJ, Shah A, Ntranos V, Van Gool F, Atkinson M, Bhushan A (2019) Targeted elimination of senescent beta cells prevents type 1 diabetes. Cell Metab 29(5):1045–1060

Tsai CY, Lu HC, Chou YH, et al. (2022) Gut microbial signatures for glycemic responses of GLP-1 receptor agonists in type 2 diabetic patients: a pilot study. Front Endocrinol (Lausanne). 12:814770. Published 2022 Jan 10. doi:https://doi.org/10.3389/fendo.2021.814770.

Turnbaugh PJ, Ley RE, Mahowald MA, Magrini V, Mardis ER, Gordon JI (2006) An obesity-associated gut microbiome with increased capacity for energy harvest. Nature 444:1027e31

Vatanen T, Kostic AD, D'hennezel E, Siljander H, Franzosa EA, Yassour M, et al. (2016) Variation in microbiome LPS immunogenicity contributes to autoimmunity in humans. Cell 165:842–53. https://doi.org/10.1016/j.cell.2016.04.007

Vaziri ND, Yuan J, Nazertehrani S, Ni Z, Liu S (2013) Chronic kidney disease causes disruption of gastric and small intestinal epithelial tight junction. Am J Nephrol 38:99–103

Vergoni B, Cornejo PJ, Gilleron J et al (2016) DNA damage and the activation of the p53 pathway mediate alterations in metabolic and secretory functions of adipocytes. Diabetes 65(10):3062–3074

Verma A, Xu K, Du T, Zhu P, Liang Z, Liao S, Zhang J, Raizada MK, Grant MB, Li Q (2019) Expression of human ACE2 in Lactobacillus and beneficial effects in diabetic retinopathy in mice. Mol Methods Clin Dev 14:161–170

Vijay-Kumar M, Aitken JD, Carvalho FA, Cullender TC, Mwangi S, Srinivasan S, Sitaraman SV, Knight R, Ley RE, Gewirtz AT (2010) Metabolic syndrome and altered gut microbiota in mice lacking Toll-like receptor 5. Science 328:228–231

Virtanen SM, Nevalainen J, Kronberg-Kippila C, Ahonen S, Tapanainen H, Uusitalo L et al (2012) Food consumption and advanced beta cell autoimmunity in young children with HLA-conferred susceptibility to type 1 diabetes: a nested case-control design. Am J Clin Nutr 95:471–478. https://doi.org/10.3945/ajcn.111.018879

Vojdani A (2014) A potential link between environmental triggers and autoimmunity. Autoimmune Dis 2014:437231. https://doi.org/10.1155/2014/437231

Vrieze A, Van Nood E, Holleman F, Salojärvi J, Kootte RS, Bartelsman JF, Dallinga-Thie GM, Ackermans MT, Serlie MJ, Oozeer R, Derrien M, Druesne A, Van Hylckama Vlieg JE, Bloks VW, Groen AK, Heilig HG, Zoetendal EG, Stroes ES, de Vos WM, Hoekstra JB, Nieuwdorp M (2012) Transfer of intestinal microbiota from lean donors increases insulin sensitivity in individuals with metabolic syndrome. Gastroenterology 143(4):913–6.e7. https://doi.org/10.1053/j.gastro.2012.06.031. Epub 2012 Jun 20. Erratum in: Gastroenterology. 2013 Jan;144(1):250. PMID: 22728514

Wang L, Li P, Tang Z, Yan X, Feng B (2016) Structural modulation of the gut microbiota and the relationship with body weight: compared evaluation of liraglutide and saxagliptin treatment. Sci Rep 6:33251

Wang J, Zheng J, Shi W, Du N, Xu X, Zhang Y, Ji P, Zhang F, Jia Z, Wang Y, Zheng Z, Zhang H, Zhao F (2018) Dysbiosis of maternal and neonatal microbiota associated with gestational diabetes mellitus. Gut 67(9):1614–1625

Wang Y, Ye X, Ding D, Lu Y (2020) Characteristics of the intestinal flora in patients with peripheral neuropathy associated with type 2 diabetes. J Int Med Res 48:300060520936806

Wellen KE, Hotamisligil GS (2005) Inflammation, stress, and diabetes. J Clin Invest 115:1111–1119

Wild S, Roglic G, Green A, Sicree R, King H (2004) Global prevalence of diabetes: estimates for the year 2000 and projections for 2030. Diabetes Care 27(5):1047–1053

Wu H, Esteve E, Tremaroli V, et al. (2017) Metformin alters the gut microbiome of individuals with treatment-naive type 2 diabetes, contributing to the therapeutic effects of the drug. Nat Med 23:850e8

Yan X, Feng B, Li P, Tang Z, Wang L (2016) Microflora disturbance during progression of glucose intolerance and effect of sitagliptin: an animal study. J Diabetes Res 2016:1–10

Yang M, Shi F-H, Liu W, Zhang M-C, Feng R-L, Qian C, Liu W, Ma J (2020) Dapagliflozin modulates the fecal microbiota in a type 2 diabetic rat model. Front Endocrinol 11:635

Zhang Q, Xiao X, Zheng J, Li M, Yu M, Ping F, Wang T, Wang X (2017a) Featured article: structure moderation of gut microbiota in liraglutide-treated diabetic male rats. Exp Biol Med 243:34–44

Zhang Q, Xiaojing W, Li M, Yu M, Ping F, Zheng J, Wang T, Wang X (2017b) Vildagliptin increases butyrate-producing bacteria in the gut of diabetic rats. PLoS ONE 12:e0184735

Chapter 8
Aging Gut Microbiota and Chronic Kidney Disease

Arunkumar Radhakrishnan, Abinaya Elango, and Divya Ilanchoorian

Abstract Chronic Kidney Disease (CKD) is one of the non-communicable diseases that affects millions of people worldwide with significant morbidity and mortality. The diagnosis, treatment and follow-up of CKD is challenging due to the lack of widespread availability of the resources including equipment, laboratories, and trained medical and paramedical staff. While this is a challenge on one side, the other side has the economic impact of CKD on the suffering patients, their families and the community, which is huge as the therapeutic interventions are costly and many patients especially in the low and middle-income countries could not afford to spend. The diagnostics, therapeutics and supportive measures for CKD are constantly evolving and coming to the aid of needy patients and at the same time, the basic biological, clinical, laboratory and genetic research conducted in the CKD domain is also increasing and improving. In the recent past, scientists have been focusing on the role of intestinal microbial flora in various disease conditions of the human body not only in the gastrointestinal system but also in several other systems including kidney function. This chapter gives an overview of CKD, epidemiology, etiopathology, clinical features, treatment modalities in a brief and detailed account of the gut microbial flora and how it influences the renal function in CKD and vice versa.

Keywords Aging · Chronic kidney disease · Dysbiosis · Microbiota · Uremia

A. Radhakrishnan (✉) · A. Elango
Department of Pharmacology, Chettinad Hospital and Research Institute, Chettinad Academy of Research and Education, Rajiv Gandhi Salai, Kelambakkam, Tamilnadu 603103, India
e-mail: arunrmbbs1978@gmail.com

D. Ilanchoorian
Department of Community Medicine, Chettinad Hospital and Research Institute, Chettinad Academy of Research and Education, Rajiv Gandhi Salai, Kelambakkam, Tamilnadu 603103, India

F. Marotta (ed.), *Gut Microbiota in Aging and Chronic Diseases*, Healthy Ageing and Longevity 17, https://doi.org/10.1007/978-3-031-14023-5_8

8.1 Overview of Chronic Kidney Disease (CKD)

Chronic Kidney Disease (CKD) manifests due to persistent and evolving abnormalities in the structure of the kidney or function of nephrons for more than 3 months. Sometimes, both the structural and functional aspects of kidneys are affected simultaneously while at other times, one will precede the other. A lot of factors can cause kidney dysfunction starting from epidemiological factors to genetic factors, environmental factors and co-morbid health conditions.

The global prevalence of CKD ranges from 7 to 12% (Hill et al. 2016). The prevalence is low in developed countries like the United States of America, Canada, Australia, England, Germany and China, whereas the prevalence is not well characterized in low- and middle-income countries (Romagnani et al. 2017).

CKD is common in geriatric people, especially after 65 years of age. However young patients with CKD progress to End Stage Renal Disease (ESRD) faster than elderly people. Women suffer more than men with CKD. People with co-morbid conditions like diabetes mellitus and hypertension are prone to developing CKD than people who do not have such co-morbid conditions. Certain ethnic populations 'Aboriginal Australians, African Americans, people of Spanish descent in central and South America, indigenous populations in Canada, South Asians, East Asians and Pacific Islanders' develop CKD more than others due to genetic predisposition. Regions in Latin America, Sri Lanka, India, Cameroon, Mexico and Australia that have environmental factors such as toxins and certain infections also record higher CKD prevalence (Romagnani et al. 2017).

The Kidney Disease Improving Global Outcomes (KDIGO) initiative classified CKD based on the functional and structural abnormalities of kidneys. Structural damage is indicated by albuminuria and functional damage by glomerular filtration rate (GFR). It is classified into G1–G5 based on a decrease in GFR and A1–A3 based on albuminuria as given in Fig. 8.1 (Levin et al. 2013).

The pathological factors that contribute to the development of CKD include loss of nephrons, hypertrophy of nephrons, impaired glomerular filtration and renal fibrosis. There are well-defined contributing factors that trigger these pathologies such as diabetes mellitus, systemic hypertension, aging, Acute Kidney Injury (AKI), obesity, pregnancy, genetic factors, prematurity and low birth weight that lead to CKD (Romagnani et al. 2017).

Appropriate treatment of CKD is important as deteriorating renal function is associated with a lot of complications such as fluid and electrolyte abnormalities, anemia, metabolic acidosis, hyperuricemia, arterial hypertension, dyslipidemia, cardiovascular dysfunction, endothelial dysfunction and uremia. The patients suffer due to these complications more often than CKD as such.

CKD is diagnosed with the measurements of Glomerular Filtration Rate (GFR) and albuminuria by collecting blood and urine samples. Other detailed lab investigations, imaging studies and histopathological studies are carried out on patients to identify the etiology and choose suitable treatment. The treatment not only includes the management of CKD but also the management of predisposing

Guide to Frequency of Monitoring (number of times per year) by GFR and Albuminuria Category				Persistent albuminuria categories Description and range		
				A1	A2	A3
				Normal to mildly increased	Moderately increased	Severely increased
				<30 mg/g <3 mg/mmol	30–300 mg/g 3–30 mg/mmol	>300 mg/g >30mg/mmol
GFR categories (ml/min/1.73 m^2) Description and range	G1	Normal or high	≥90	1 if CKD	1	2
	G2	Mildly decreased	60–89	1 if CKD	1	2
	G3a	Mildly to moderately decreased	45–59	1	2	3
	G3b	Moderately to severely decreased	30–44	2	3	3
	G4	Severely decreased	15–29	3	3	4+
	G5	Kidney failure	<15	4+	4+	4+

Fig. 8.1 Kidney disease improving global outcomes (KDIGO) classification of CKD. *Source* Kidney Disease: Improving Global Outcomes (KDIGO) CKD Work Group. KDIGO 2012 clinical practice guideline for the evaluation and management of chronic kidney disease. Kidney Int. Suppl. 3, 1–150 (2013)

factors and co-morbid conditions. Optimal control of blood sugar and blood pressure is vital in the management of CKD.

The goals of CKD management include controlling ongoing injury of nephrons, normalizing hyperfiltration of individual nephrons and controlling CKD complications. When CKD reaches end stage renal disease, patients are subjected to renal replacement therapies like peritoneal dialysis, hemodialysis and renal transplantation.

Restriction of fluid and salt intake and a healthy lifestyle are important in the management of CKD. Various pharmacological agents are used in CKD including diuretics, sodium bi-carbonate, calcium and vitamin D supplements, iron supplements, erythropoietin, antioxidants and drugs regulating sodium and potassium levels.

8.2 Gut Microbiota

The collection of microbes including bacteria and other prokaryotes as well as some eukaryotes in the human gastrointestinal (GI) tract is given the term 'gut microbiota'. In addition to the microbiota, the GI tract of humans is exposed to several other antigens and toxins, as more than sixty tons of food pass through it in an average lifetime. Thus, the GI tract is one of the largest interfaces in the human body that interacts with diverse antigens and other environmental factors. The estimated number of microbes that colonize the gut is greater than 10^{14}, which accounts for ten times more the number of bacterial cells compared to human cells. The genomic content (microbiome) of these microbes is nearly 100-fold that of the human genome (Thursby and Juge 2017). However, Sender et al. in 2016 estimated and reported that the ratio of the bacterial and human cells is almost equal (ratio closer to 1:1) (Sender et al. 2016).

The microbiota of the gut plays a vital role in maintaining the integrity of the gut, regulation of host immunity and protection against various pathogens. The disruption or alteration in the gut microbiota is termed as 'dysbiosis', which would predispose to several pathological processes including metabolic diseases in the host (Thursby and Juge 2017).

The microbiota of the gut is different from the microbial flora of other body parts, and it is not as diverse as seen in other body parts. They are also functionally redundant. Unlike in the past, where intensive culture methods were used to identify the gut microbiota, there are newer approaches available now like high throughput and low-cost screening methods to survey the composition of gut microbiota. One such method is targeting the 16S rRNA (ribosomal RNA) of the bacteria, as this gene is present in all the bacteria and has 9 highly variable regions, which would facilitate distinguishing the species. The other highly sensitive techniques that are more precise and give reliable information on the microbiota composition are provided by metagenomics like whole genome shotgun sequencing. Li Junhua et al. (2014), and Hugon et al. (2015), identified and reported the composition of the gut microbiota in humans. The data compiled from both these studies identified 2172 prokaryotic species categorized into 12 phyla. Among the species, 93.5% are *Proteobacteria, Actinobacteria, Bacteroidetes* and *Firmicutes*. 386 species are strict anaerobes and are usually seen in the oral mucosa and the GI tract (Hugon et al. 2015).

The development of gut microbiota usually begins from birth, though few studies identified the microbes in the placenta (Aagaard et al. 2014; Rodríguez et al. 2015). The mode of delivery has an impact on the composition of the microbiota. The infants born through normal vaginal delivery harbor abundant *lactobacilli* in their gut in the initial few days after birth, as the vaginal flora has *lactobacilli*. The composition is different in the infants born through cesarean section and they have facultative anaerobes like *Clostridium* species colonizing their gut. During the early stages of development in infants, two phyla—*Proteobacteria* and *Actinobacteria*—dominate the microbiota. Later, in the first year of life, the microbial diversity starts increasing and by the age of 2.5–3 years, the composition and diversity resemble the adult microbiota. The composition of microbiota is not stable and subject to variation due to several environmental factors including changes in the diet, diseases and treatment with antibiotics. In the elderly above 65 years, the microbiota changes with a predominance of *Bacteroides* and cluster IV *Clostridium* as against cluster XIVa in young adults. In people aged about 100 years, facultative anaerobes like *Escherichia coli* are abundant. In the elderly, the metabolic processes carried out by the microbiota such as the production of short chain fatty acid and amylolysis are reduced, whereas the proteolytic activity is increased. There is increasing evidence that short chain fatty acids play an important role in several metabolic and immune pathways (Thursby and Juge 2017). Hence, the change in microbiota in the elderly would affect the short chain fatty acid production that could predispose them to a few diseases.

8.3 Factors Altering the Composition of Gut Microbiota

There is a continuous threat to the gut microbiota from several host and environmental factors. The GI tract of humans offers protection against these factors by limiting their exposure to three types of barriers (Thursby and Juge 2017):

1. Physical barrier (intestinal wall including epithelial and mucous layers).
2. Biochemical barrier (including antimicrobial proteins and enzymes).
3. Immunological barrier (through immune cells of the epithelium and Immunoglobulin A-IgA).

The factors that could alter the gut microbiota are as follows:

(a) *Diet:*

The gut microbes derive their energy through biochemical processes such as sulfate reduction and fermentation of dietary carbohydrates. Depending on the nutrient availability in the gut, the microbiota alters the phenotypic trait. Studies report that the microbiota present in the ileum has the capacity to metabolize simple sugars, thus reflecting their adaptation to the dietary environment of the small intestine. If the diet is extremely 'plant-based' or 'animal-based', it would result in significant alteration in the gut microbiota. In infants, the type of milk they consume (human milk or formula feed) has an effect on the type of microbes, for example, *Bifidobacterium* and *Bacteroids* outnumber *E. coli, Clostridium* and Lactobacilli in breast-fed infants, as they can utilize the oligosaccharides present in human milk (Thursby and Juge 2017).

(b) *Intestinal mucosa:*

The intestinal mucosa of the host alters its microbiota through some molecules and immunoglobulins produced by the epithelial cells which include antimicrobial proteins (AMPs), mucus and IgA. These factors favor the growth of some bacteria while inhibiting others. The large intestine has 2 layers of mucus, namely inner and outer layers. The inner layer does not harbor any microbes while the outer layer does. It is because the outer layer is abundant in mucin made up of O-glycans which serve as a binding site and source of nutrition to the gut microbiota. The O-glycans of the mucins play a key role in selecting the microbes that are appropriate for the wellbeing of the host. In the small intestine, AMPs modulate the gut microbiota. The structural components like flagella and lipopolysaccharides of the microbes induce the production of AMPs (including defensins, ribonucleases, Reg family proteins and cathelicidins) by the Paneth cells present in the crypts of the small intestine. The commensal microbes of the gut are resistant to these AMPs, whereas the pathogenic microbes are not. In addition, the secretary IgA produced by the plasma cells of the small intestine helps in the biofilm formation and favors the commensals and the deficiency or absence of which leads to the colonization by pathogenic microbes (Hasan and Yang 2019).

(c) *Use of antibiotics:*

The use of antibiotics destroys not only the pathogenic bacteria but also the commensal bacteria present in the gut. The indiscriminate use of antibiotics would lead to dysbiosis and the growth of unwanted bacteria. Animal studies have shown that antibiotics affect the metabolism of serotonin and secondary bile acids in the gut leading to depletion of microbiota. It has been reported that Clindamycin, Clarithromycin, Metronidazole and Ciprofloxacin affect the microbiota for prolonged periods. Clindamycin decreases the number of *Bacteroides*, Clarithromycin —*Actinobacteria*, Ciprofloxacin–*Ruminococcus* and Vancomycin decreases *Bacteroides*, *Fuminococcus* and *Faecalibacterium* (Hasan and Yang 2019).

(d) *Probiotics and prebiotics:*

Probiotics are the living microorganisms that include bacteria like *Lactobacillus*, *Bifidobacterium* and yeast like *Saccharomyces boulardii*, which when taken promote the growth of favorable gut microbiota. Probiotics compete with the adhesion sites of the pathogenic bacteria and eliminate them. Some probiotics like *Lactobacillus reuteri* directly kill the pathogenic bacteria and enhance the host immunity. *Bifidobacterium* reduces the inflammation of the intestine, promotes the mucosal barrier function and increases IgA secretion. Probiotics are proven to be beneficial in many chronic diseases including urinary tract infections, dyslipidemia, necrotizing enterocolitis, type 2 diabetes and so on (Hasan and Yang 2019).

Prebiotics consist of oligosaccharides and short polysaccharides like oligofructose, galacto fructose and inulin. When taken, they are metabolized by the digestive enzymes in the intestine and promote the growth of microbiota that are already colonized in the colon. The prebiotics increase the production of short chain fatty acids and decrease the pH, favoring the growth of commensal microbiota. Prebiotic administration protects the intestine against diet-induced changes in the mucosal layer and metabolic changes (Hasan and Yang 2019).

8.4 Role of Gut Microbiota in Human Health

The gut microbiota, with its high genomic content and metabolic functions, provide a wide range of beneficial effects in the host like:

1. Maintaining the mucosal barrier;
2. Providing nutrients;
3. Protecting against pathogenic microbes;
4. Promoting the immune function.

The gut microbes metabolize complex carbohydrates and generate metabolites such as short chain fatty acids. Acetate, butyrate and propionate are the predominant short chain fatty acids of the GI tract present in a 3:1:1 ratio. These short chain fatty acids are involved in the regulation of chemotaxis, gene expression, cell differentiation, proliferation and apoptosis. In humans, *Bacteroides* produce propionate, *Firmicutes* are involved in the production of butyrate and acetate is produced by most of the gut anaerobes (Thursby and Juge 2017).

The short chain fatty acids (SCFAs) have the following functions (Corrêa-Oliveira et al. 2016):

- Anti-cancer and anti-inflammatory activity (Butyrate).
- Inhibition of translocation of bacteria and enhancement of the mucosal barrier function (butyrate).
- Cytokine production (IL-18—maintains epithelial integrity).
- Gluconeogenesis (Propionate) and lipogenesis (Acetate) in the liver.
- Regulation of gene expression (Butyrate and Propionate—histone deacetylase inhibitors).
- Modulation of appetite and energy intake.

The gut microbiota facilitates the synthesis of essential vitamins. Lactic acid bacteria present in the gut is involved in the synthesis of vitamin B12 and *Bifidobacterium* produces folate. The other vitamins synthesized by the gut microbiota include biotin, riboflavin, thiamine, pyridoxine, pantothenic acid, niacin and vitamin K (Hill 1997).

8.5 Chronic Kidney Disease Affecting Gut Microbiome

The relationship between the large intestine and kidney is bi-directional. Uremia has an influence on the metabolism of gut microbes, and some microbial metabolic products are involved in the worsening of renal disease. Changes in the number and the composition of gut microbiota are observed in patients with chronic kidney disease (CKD). When compared to healthy individuals, uremic patients have a higher number of aerobic and anaerobic microbes in the gut. The number of *Lactobacilli* and *Prevotella* is decreased in patients with CKD. In patients undergoing hemodialysis, there is an increase in the number of aerobic bacteria like *Enterobacteria,* while the number of anaerobic bacteria like *Bifidobacteria* is less. In CKD, the gut microbial alteration may be iatrogenic, or sometimes uremia per se can cause dysbiosis (Ramezani and Raj 2014).

The microbiota changes observed in the gut of CKD patients might be due to uremia as increased urea secreted into the intestine changes the gut carbohydrate–protein ratio thereby causing dysbiosis (Bryniarski et al. 2019). But uremia alone may not be the sole reason for gut dysbiosis though urease-producing bacteria are increased in the gut of CKD patients and animal CKD models (Vaziri et al. 2013; Chaves et al. 2018), as there could be other factors such as disease renal pathology (Tilg and Moschen 2014), treatments and nutritional supplements administered including antibiotics, immunotherapeutic agents, phosphate binders and iron supplements (Dostal et al. 2015) and dietary restrictions that may also play a role in the alteration of gut microbiota (Hooper et al. 2002).

Vaziri et al. (2013), conducted a study on humans and rats to assess and compare the composition of gut microbiota in CKD and normal population. For human study, 24 patients with chronic renal failure (CRF) due to various causes including glomerulonephritis, diabetic nephropathy, polycystic kidney disease and hypertensive nephrosclerosis were enrolled. 12 healthy adult subjects were included as the control. For the animal study, Sprague Dawley (SD) rats were used. CRF was induced in the test group by nephrectomy and the control group underwent sham surgery. Microarray analysis was done with the fecal samples collected from humans and rats, including both the CRF group and control group. Human stool sample analysis revealed that significantly different bacterial subclasses were seen between the CRF and control groups, especially the *Pseudomonadaceae* family, subphylum *Clostridia* and *Proteobacteria* phyla. Around 190 operational taxonomic units (OTU) of the bacteria were found to be abundant in the CRF group compared to healthy controls. Similarly, the animal study data showed that there were 175 OTUs that differed significantly between CRF rats and control rats. *Bacteroides* and *Firmicutes* like *Lactobacilli* and *Prevotella* were less in CRF rats (Vaziri et al. 2013).

Similarly, Ramezani et al. in their article, stated that increased urea results in the production of large amounts of ammonia, which in turn is harmful and could affect the growth of commensal microbes. Further, vitamin K deficiency seen in patients with CKD could be attributed to the reduced number of *Clostridia, Bifidobacterium* and *Streptococcus faecalis*, which are involved in vitamin K synthesis (Ramezani and Raj 2014).

The following biochemical mechanisms are attributed to the change in the gut flora in CKD:

1. Increased urea in CKD, undergoes hydrolysis by urease from gut bacteria leading to the formation of a large amount of ammonia, thus contributing to uremic enterocolitis (Vaziri et al. 1985; Kang 1993).

2. In CKD, the GI tract takes over the function of excreting uric acid and oxalate, which in normal situations are excreted via the kidney (Hatch et al. 1994).
3. Dietary restrictions in CKD lead to limited consumption of fruits, and vegetables rich in oxalate and potassium, which in turn could cause alteration in the composition of gut microbiota as these foods serve as a primary source of nutrition to the microbiota (Vaziri et al. 2013).
4. Chronic intake of phosphate binders could alter the luminal environment of the gut (Vaziri et al. 2013).
5. Frequent use of antibiotics in CKD to treat infections (Vaziri et al. 2013).

Disruption of the intestinal barrier is noted in patients with CKD. The changes include a reduction in villous height, crypt elongation and infiltration of inflammatory cells in the lamina propria. Uremia disrupts the epithelial tight junctions in the colon, which leads to bacterial translocation and movement of toxins across the gut wall. It is not just the CKD pathology causing dysbiosis or intestinal barrier disruption, even the systemic circulatory stress induced by hemodialysis may disrupt the intestinal barrier (Ramezani and Raj 2014).

The pathogenic bacteria such as *Klebsiella spp.*, *Proteus spp.*, *Escherichia spp.*, *Enterobacter spp.* and *Pseudomonas spp.* are seen more in the blood samples of patients with END stage Renal Disease (ESRD), and this could be due to the dysbiotic bacteria entering the systemic bloodstream via disrupted intestinal barrier (Ott et al. 2006). In fact, the atherosclerotic lesion of CKD patients has a few pathogenic organisms such as *Proteus vulgaris* and *Klebsiella pneumoniae* (Ott et al. 2006), and there is even a hypothesis that CKD-associated dysbiosis triggers leakage of bacteria and their toxins into the systemic circulation thereby leading to systemic inflammatory response that may be one of the reasons for coronary artery disease in CKD (Bryniarski et al. 2019).

Endotoxins are produced continuously by the gram-negative bacteria in the gut and transported via the intestinal capillaries, reach the liver and phagocyte cells and are finally cleared off. The increase in endotoxins due to the gut dysbiosis is one of the important causes for inflammation seen in CKD patients (Ramezani and Raj 2014).

The uremic toxins generated by the gut microbes are usually cleared by the kidneys. The most extensively studied uremic toxins are Indoxyl sulfate and p-Cresyl sulfate, which are the metabolic products produced from the fermentation of tryptophan and tyrosine, respectively, by the gut microbes. They are highly plasma protein bound and get accumulated when the renal function declines. Both these metabolites are associated with cardiovascular disease, progression of CKD and overall mortality (Corrêa-Oliveira et al. 2016; Hasan and Yang 2019). These uremic toxins cause damage to the tubular cells of the kidney. Indoxyl sulfate causes nephrotoxicity by entering the proximal tubular cells via an organic anion transporter. In addition to nephrotoxicity, Indoxyl sulfate induces oxidative damage in the vascular endothelial cells and predisposes them to cardiovascular disease (Ramezani and Raj 2014).

Moreover, dysbiosis activates the renin–angiotensin system (RAS), leading to hypertension which in turn can induce kidney damage. The other renal disorders associated with dysbiosis include IgA nephropathy, membranous nephropathy, diabetic kidney disease and SLE-associated renal dysfunction (Lu et al. 2018) (Fig. 8.2).

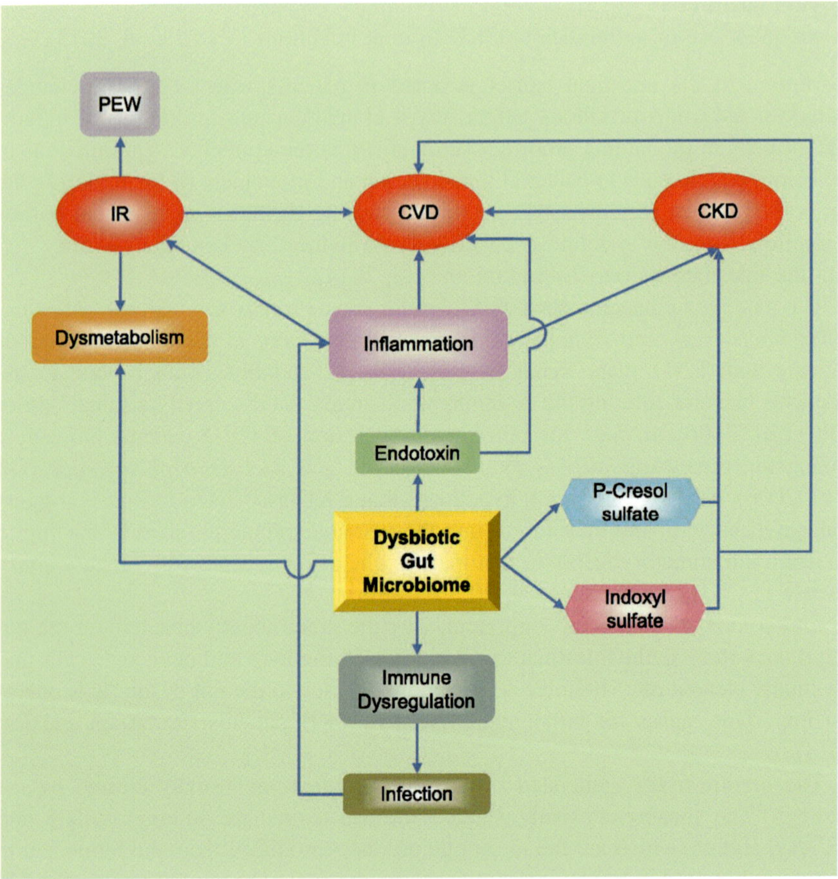

PEW: Protein energy wasting; IR: insulin resistance;
CVD: Cardio Vascular Disease; CDK: ChronicKidney Disease

Fig. 8.2 Gut Dysbiosis modulating CVD, CKD and IR. *Source* Ramezani A, Raj DS. The gut microbiome, kidney disease and targeted interventions. Journal of the American Society of Nephrology. 2014 Apr 1;25(4):657–70

8.6 Dysbiosis-Induced Inflammation Predisposing to CKD

Chronic and low-grade inflammation is seen in most of the patients with CKD, as reflected by the elevated levels of inflammatory markers in circulation (Silverstein 2009; Mihai Codrici et al. 2018). Kidneys are highly prone to the inflammatory injury due to the lack of well-developed anti-oxidant and anti-inflammatory mechanisms like the liver. The various factors predisposing to inflammation in CKD are acidosis and oxidative stress, frequent infections, altered adipose tissue metabolism and gut dysbiosis (Mihai et al. 2018).

Systemic and intrarenal inflammations lead to dysregulation of the microcirculatory regulatory mechanisms in the kidney, modulated by hormones and other vasoactive mediators like prostaglandins, kinins, endothelins and nitric oxide. This dysregulation favors the increased production of toxins and Reactive Oxygen Species (ROS) from the renal tubular cells causing tubular injury, nephron loss and onset of CKD (Qian 2017).

Gut dysbiosis is predisposing individuals to get into systemic inflammatory reactions that may lead to CKD (Khoury et al. 2017). Various renal pathologies are common in Inflammatory Bowel Disease (IBD) and incidentally IBD and CKD share similar changes in the gut microbiome. It is hypothesized that the gut microbial changes induce systemic inflammation in both IBD and CKD (Kanbay et al. 2018).

Gut dysbiosis triggers systemic inflammation by modulating Lipopolysaccharides, Short Chain Fatty Acids, bile acids, C reactive protein, cytokines, Tumor Necrosis factor alfa, Interleukin 6 and Trimethylamine N-Oxide (TMAO) (Chen et al. 2020). Gut dysbiosis is associated with several renal pathologies causing proteinuria including IgA nephropathy, lupus nephritis, idiopathic nephrotic syndrome and renal amyloidosis (Kanbay et al. 2018). Further, gut microbial changes may lead to hypertension (Yang et al. 2015) and diabetes mellitus (Roesch et al. 2009; Wen and Duffy 2017) that could also predispose to CKD by causing proteinuria and nephropathy.

Albuminuria is one of the hallmark features of leaking glomeruli and an important component of CKD diagnosis. The gut microbial flora undergoes changes wherein some bacteria are reduced in number and some increased in number. The anti-inflammatory bacteria *Lactobacillus* and *Bifidobacterium* genera are significantly reduced in proteinuric kidney disease. The genera from *Proteobacteria* phylum are enriched in proteinuric kidney disease (Kanbay et al. 2018). *Proteobacteria* can cause systemic inflammation and lead to proteinuria and ultimately CKD by disrupting the mucosal barrier and producing pro-inflammatory substances, serotonin and uremic toxins (Poveda et al. 2014).

The supplement with probiotics in animal models of CKD was found to be associated with reduced proteinuria and pro-inflammatory markers and restoration of renal histology (Yoshifuji et al. 2016) and thus can be evidence to indirectly commit that gut dysbiosis can lead to CKD.

8.7 Immunological Changes in Gut Dysbiosis and CKD

The gut microbiome and the mucosal barrier of the intestine together regulate innate and adaptive immunity (Craig et al. 2012). The Intestinal immune system is normally not active and is activated when the mucosal and microbial barriers are disrupted. The immune cells upon activation infiltrate kidneys and trigger pro-inflammatory and anti-inflammatory responses (Dhana et al. 2018).

The innate immunity serves as the first line of host defense and restores homeostasis through mechanisms like inflammation and apoptosis. These processes depend on the pattern recognition receptors (PRRs) that recognize danger-associated molecular patterns (DAMPs) and pathogen-associated molecular patterns (PAMPs) (Mihai et al. 2018). The innate immune cells like monocytes, neutrophils and macrophages induce an immune response by augmenting PAMPs and DAMPs (Eppensteiner et al. 2019).

The pattern recognition receptors (PRRs) like toll-like receptors (TLRs) present in the renal cells when activated by the pro-inflammatory cytokines like interleukin 1 and tumor necrosis factor- α induced by gut dysbiosis further activate the inflammatory pathways mediated by protein kinases, nuclear factor-κB (NF-κB) and activator protein-1 (Cowan and Storey 2003; De Heer and Peters 2008; Mihai et al. 2018). The involvement of PRR may be instrumental in the pathogenesis of infection-mediated renal disease (Shi et al. 1998). In addition, renal epithelial cells possess receptors that recognize complement proteins, resulting in adhesion and activation of inflammatory cells (De Heer and Peters 2008).

Figure 8.3 depicts the activation of renal epithelial cells in response to inflammation, toxic injury, metabolic and shear stress and the resultant innate immune responses.

The role of chronic inflammation in crystal-induced renal fibrosis and CKD has been reported by Dhana et al. (2018). Dendritic cells, macrophages, T- lymphocytes and several other inflammatory cells play an active role in the chronic inflammation of the kidney. Dendritic cells (DCs) are the most abundantly seen leucocytes in kidneys. DCs have both protective as well as adverse effects on the kidney. While they are important for activating the adaptive immune responses in the kidney, on the other side, the pro-inflammatory property contributes to chronic inflammation and fibrosis in the kidney (Dhana et al. 2018). The renal dendritic cells, when exposed to infective pathogens like *E. Coli*, fight against the infection by recruiting T cells at the site (Yatim et al. 2016).

Similar to DCs, macrophages contribute to inflammation in the kidney when exposed to mechanical stress, pathogens and other stimuli. The macrophages do so by secreting inflammatory mediators like inducible Nitric Oxide Synthase (iNOS), tumor necrosis factor (TNF)-alfa, Interleukins like IL-1beta, IL-6 and Reactive Oxygen Species (ROS) (Lee et al. 2011).

In addition to DCs and macrophages, regulatory T cells, myeloid-derived suppressor cells also contribute to renal inflammation through secretion of cytokines

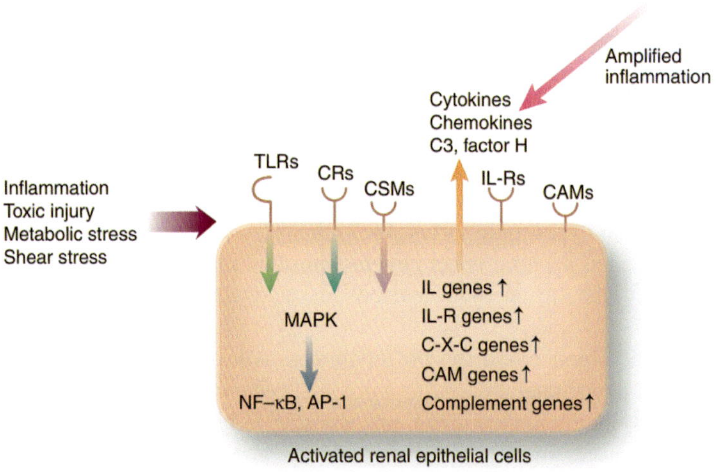

CAM-cellular adhesion molecule, IL- Interleukin, MAPK- Mitogen activated protein kinase, CR- complement receptors

Fig. 8.3 Activation of renal epithelial cells in response to inflammation, toxic injury, metabolic and shear stress and the resultant innate immune responses. *Source* De Heer E, Peters DJM. Innate immunity as a driving force in renal disease. Kidney Intern. (2008) 73:7–8. https://doi.org/10.1038/sj.ki.5002658

like DAMPs and other pro-inflammatory cytokines. These cytokines bind to PRRs, especially TLR, and activate the inflammatory pathway (Dhana et al. 2018).

The alteration in the gut microbiota exposes the kidney to various toxins and pathogens, thus activating any one or all of the inflammatory pathways discussed above, leading to inflammation and chronic kidney disease.

8.8 Aging and Gut Microbiome

It is a well-known fact that elderly people are fragile and more prone to infections compared to younger people. It is due to the dysfunction of immune function occurring in geriatric people termed as ‘immunosenescence’. The gut microbiota shows sequential changes as age advances. *Actinobacteria* decrease with age and *Firmicutes* are abundant in older children. In the elderly aged 70 years and above, the predominant microbial species found in the gut are *Bacteroides* and *Proteobacteria*. The alteration in the gut microbiota with age leads to loss of epithelial barrier function of the intestine causing ‘leaky gut’. It facilitates the bacterial toxins reaching the systemic circulation and various organs.

The immune cells that are activated due to endotoxins induce pro-inflammatory and anti-inflammatory reactions in various organs including the kidney. Neutrophils and macrophages are activated and protect the kidney against pathogens. If the macrophage function is impaired, it would result in chronic inflammation of the kidney. The renal tubular epithelial cells secrete pro-inflammatory cytokines, chemokines and antimicrobial substances to fight against microbial toxins. In this process, the tubular cells may get damaged and during dedifferentiation, inflammation and renal hypertrophy occur due to the recruitment of myofibroblasts at that site. The damaged nephron is not able to withstand the increased tubular transport, which in turn leads to anerobic metabolism, acidosis and hypoxia (Stavropoulou et al. 2021).

The gut microbiota modulates the inflammation through Short Chain Fatty Acids (SCFAs) and hence alteration in the microbial composition with age could lead to chronic inflammation in the kidney, predisposing to kidney disease. Advancing age leads to decreased immune function and activation of low-grade inflammation called 'inflammaging'. The gut microbiota plays an important role in the inflammaging process due to its ability to produce inflammatory processes and interaction with other organ systems (Stavropoulou et al. 2021).

Figure 8.4 depicts how gut microbiota and other factors contribute to immunosenescence and chronic kidney disease.

8.9 Modulation of Gut Microbiota as a Therapeutic Approach in CKD

Restoration of the balance of gut microbiota reverses dysbiosis that leads to infection, inflammation, immune dysfunction and chronic kidney disease. Different methods of restoring gut microbiome are discussed as follows.

(i) *Microorganisms:*

A lot of treatment modalities have been researched in the past couple of decades with the aim of getting an easy, convenient and affordable treatment for CKD as the presently existing renal replacement therapies are costlier and not comfortable for the patients. Many of these exploratory treatments attempt to regulate and balance the fluid and electrolytes and uremic toxins in the body. These options range from the combination of orally administered agents including adsorbants, osmotic agents, ion exchange agents and enzymes such as urease to microencapsulated multi-enzyme system—microbial organisms—and to stem cells (Jain et al. 2009).

The concept of using microbial organisms in managing uremia is not new, and it has been researched since 1990. Huge technological developments in molecular biology and genetics have made rapid inroads in the production of non-pathogenic bacteria which can be used for various disease conditions. The advantage of using microbial organisms in handling uremia compared to other agents is the ability of

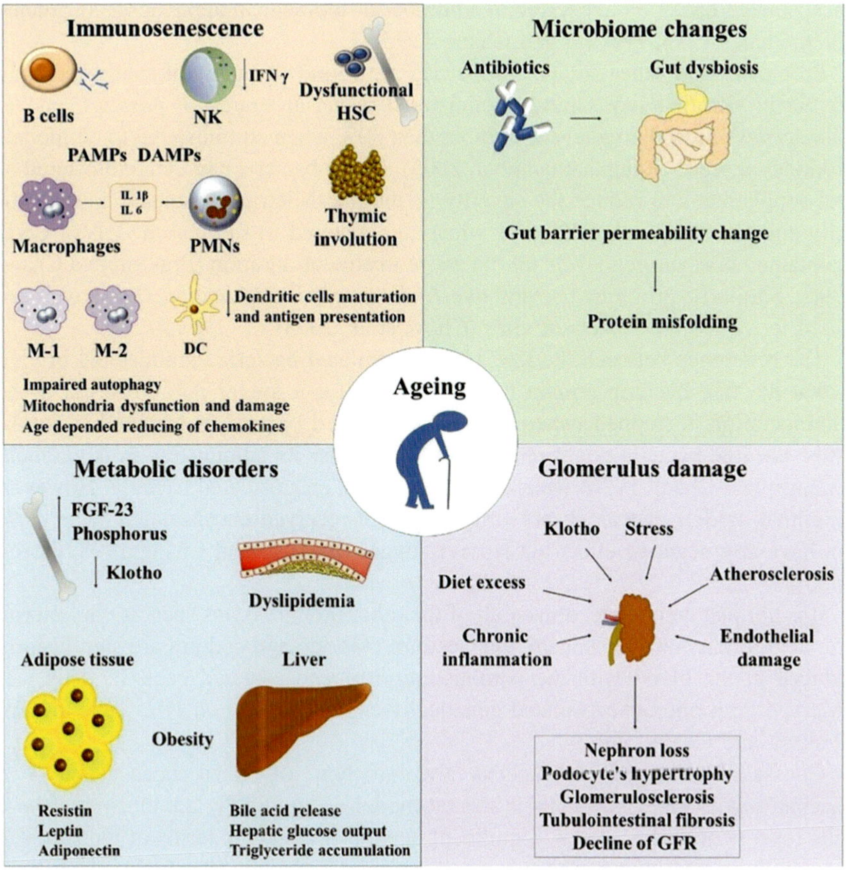

Fig. 8.4 Aging gut microbiota and Immunosenescence, glomerular changes and metabolic disorders. *Source* Stavropoulou E, Kantartzi K, Tsigalou C, Aftzoglou K, Voidarou C, Konstantinidis T, Chifiriuc MC, Thodis E, Bezirtzoglou E. Microbiome, Immunosenescence, and Chronic Kidney Disease. Frontiers in Medicine. 2021 Mar 19;8:287

the microbes to handle large amounts of uremic toxins. In CKD, uremic toxins including urea and creatinine keep accumulating and the efficiency of enzymes/adsorbants/osmotic agents is restricted to the quantity they can be clinically used (Jain et al. 2009).

Scientists have explored two options for delivering microbial organisms to gut with the aim of restoring gut microbiota and also balancing the uremia. One is giving the microorganisms as such and the other option is giving the organisms in the encapsulated form. In the 1990s, Chang has successfully designed models for delivering genetically modified bacteria in microencapsulated form and observed in his experiments in vivo and in vitro that such encapsulated organisms reduce the uremic burden. Under various experimental models, it was established that these encapsulated bacteria have a role in a number of biomedical applications (Cattaneo 1989; Chang 1994; Prakash and Chang 1995).

Probiotics and other strains of bacteria are found to be significantly reducing uremia in vitro/in vivo. Orally administered bacterial strains *B. pasteurii* and *L. Sporogenes* reduced urea levels by more than 30% when administered to rat models of kidney disease (Ranganathan et al. 2005). These bacteria can be administered as oral supplements to reduce the severity of uremia. Microencapsulated and genetically engineered *E. coli DH5* cells when administered to surgical models of CKD maintained normal urea levels till the entire treatment duration (Prakash and Chang 1996). Similarly, probiotic bacteria like *L. delbrueckii* and *Lactobacillus* have been found to reduce urea levels in vitro (Chow et al. 2003).

The difference between the free and encapsulated bacteria administered in vivo is that the free bacteria remain in the intestine for a longer duration even if the administration is stopped, whereas the encapsulated bacteria is excreted in stools. Thus, the free bacteria could reduce urea even after its administration is withheld (Prakash and Chang 1996) which is not seen with encapsulated bacteria. However, in animal models, it was observed in those that received encapsulated bacteria do not have any negative effect but survive longer (Prakash and Chang 1998; Chang 2000).

It is not just urea alone, almost all of the other uremic toxins such as cholesterol, uric acid, potassium, phosphate, magnesium, chloride and sodium are significantly reduced in the blood with the administration of encapsulated bacteria. This was observed with microencapsulated genetically engineered *E. coli DH5* cells in vitro (Prakash and Chang 1999).

Coussa R et al. investigated the efficacy of microencapsulated live Saccharomyces cerevisiae cells in the rat model of uremia, by administering these cells for a period of 8 weeks. Significant beneficial effects in terms of reduction in the uremic toxins over 8 weeks were observed. In addition, the body weight was also increased which is one of the surrogate markers indicating good clinical efficacy (Coussa et al. 2010).

Lactobacillus is one of the most commonly prescribed probiotic for several years. It is used in many gastrointestinal disorders and co-prescribed with high-end antibiotics to prevent diarrhea. Probiotics restore the intestinal flora balance by enhancing the gut barrier by increasing mucus integrity, enhancing the epithelial cells' survival and epithelial tight junction (ETJ). They control the pathobionts overgrowth by the production of antimicrobial peptides (AMPs), and defensins and reduction of local pH. Reduction in the gut pathobionts decreases the production of uremic toxins. They also stimulate the production of secretory IgA, provide anti-inflammatory effect and stimulate better immunity tolerance (Koppe et al. 2015).

(ii) *Diet*

High salt diet alters the gut microbial flora and may lead to worsening of CKD in two ways, first, by causing electrolyte imbalances like hypernatremia and second by affecting Gut–Kidney axis. Dietary restriction of salt is one of the main therapeutic

Table 8.1 Common drugs modulating gut microbiota

Drug	Effect on microbiota
Metformin	Increases SCFAs producing *Lactobacillus*, *Bacteroides* and *Bifidobacterium*
Sitagliptin	Increases *Bacteroides*, *Firmicutes* and *Bifidobacterium*
Dapagliflozin	Increases *Bacteroides*, *Firmicutes* and *Akkermansia*
Canagliflozin	Increases SCFAs producing bacteria
Acarbose	Increases *Lactobacillus* and *Bifidobacterium*
Captopril	Increases *Allobaculum*
Ferric citrate (Phosphate binder)	Increases *Verrucomicrobia*, *Clostridia* and *Enterobacteriaceae*

approaches in CKD as it helps restore gut microbial flora and also avoids hypernatremia. In addition, a high salt diet is associated with a reduced number of lactobacilli in the intestine. It was observed that Lactobacillus supplementation balances gut microbiome and modulates the immunological cascade triggered by dysbiosis by reducing Th17 lymphocyte levels (Stavropoulou et al. 2021). The decrease in Th17 lymphocytes would reduce the inflammation in the kidney as it is an important pro-inflammatory mediator.

(iii) *Fecal microbial transplantation (FMT)*

FMT is the transfer of the prescreened fecal material from the healthy donor to the patient. This will restore the normal gut microbiota. FMT has proven to be successful in the treatment of *Clostridium difficile*-associated diarrhea, where the efficacy is more than 80%. There are only limited data available on the use of FMT in other diseases including CKD. However, FMT might become a treatment option or adjuvant treatment in CKD in future, as it may restore intestinal flora (Caggiano et al. 2020).

(iv) *Drugs:*

Some of the commonly used anti-diabetic and anti-hypertensive drugs modulate the gut microbiota and the details are given in Table 8.1 (Caggiano et al. 2020).

These drugs can be preferred over the other drugs used for the same disease condition in case the treating physicians want to restore the gut microbiome.

8.10 Summary

Gut microbiome is attracting a lot of attention nowadays, and data available in the public domain interlinks the microbial flora with several organ systems and diseases. There is a broad agreement that the evolution of gut microbiome is complete in the early childhood, and it remains the same from early life till four or five decades, unless the environment, diet, drugs and other factors change it. In old age, multiple factors change the gut microbiota, and this is associated with many diseases. It is not clear whether the diseases primarily affect the gut microbiome or age-related dysbiosis triggers the disease pathology, first.

As CKD is associated with significant morbidity and mortality, newer treatment modalities are in need that could modulate the risk factors associated with CKD. One such approach currently under research is modulation of the gut microbiota in patients with CKD. The relationship between the gut microbiota and kidney is bidirectional. Uremia, due to CKD, has an influence on gut microbiota (dysbiosis) as reflected by a decrease in the commensal and increase in pathogenic bacteria. On the other way around, dysbiosis affects the SCFAs production that are very crucial for modulating the immune response. As a result, dysbiosis alters the immunological pathways leading to low-grade inflammation in the kidney.

A few animal and human studies support that the supplementation of prebiotics and probiotics benefits the patient with CKD. At present what is evident is that restoration of gut microbial flora may be an interesting therapeutic tool that can add benefit to the standard treatment of CKD along with drugs and renal replacement therapies.

Acknowledgements We would like to thank Chettinad Academy of Research and Education for their support.

Authors' Contributions Conceptualization: Arunkumar Radhakrishnan; Writing original draft: Arunkumar Radhakrishnan, Abinaya Elango, Divya Ilanchoorian; Revision and editing: Arunkumar Radhakrishnan. The content of the submitted manuscript was approved by all the authors.

Consent for Publication All the authors who participated in the study provided consent for publication

Competing Interests The authors declare that they have no known competing financial interests or personal relationships that could have appeared to influence the work reported in this article.

References

Aagaard K, Ma J, Antony KM, Ganu R, Petrosino J, Versalovic J (2014) The placenta harbors a unique microbiome. Sci Transl Med 6(237):237ra265–237ra265

Bryniarski MA, Hamarneh F, Yacoub R (2019) The role of chronic kidney disease-associated dysbiosis in cardiovascular disease. Exp Biol Med 244(6):514–525

Caggiano G, Cosola C, Di Leo V, Gesualdo M, Gesualdo L (2020) Microbiome modulation to correct uremic toxins and to preserve kidney functions. Curr Opin Nephrol Hypertens 29 (1):49–56

Cattaneo MV (1989) The potential of microencapsulated urease-zeolite oral sorbent for the removal of urea in uraemia

Chang TM (1994) Artificial cell including blood substitute and biomicroencapsulation: from ideas to applications, vol 22. Taylor and Francis, pp vii–xiv

Chang TMS (2000) Artificial cell biotechnology for medical applications. Blood Purif 18(2):91–96
Chaves LD, McSkimming DI, Bryniarski MA, Honan AM, Abyad S, Thomas SA, Wells S, Buck M, Sun Y, Genco RJ (2018) Chronic kidney disease, uremic milieu, and its effects on gut bacterial microbiota dysbiosis. Am J Physiol-Ren Physiol 315(3):F487–F502
Chen J, Yue Y, Wang L, Deng Z, Yuan Y, Zhao M, Yuan Z, Tan C, Cao Y (2020) Altered gut microbiota correlated with systemic inflammation in children with Kawasaki disease. Sci Rep 10(1):1–12
Chow KM, Liu ZC, Prakash S, Chang TM (2003) Free and microencapsulated Lactobacillus and effects of metabolic induction on urea removal. Artif Cells, Blood Substit, Biotechnol 31 (4):425–434
Corrêa-Oliveira R, Fachi JL, Vieira A, Sato FT, Vinolo MAR (2016) Regulation of immune cell function by short-chain fatty acids. Clin Transl Immunol 5(4):e73
Coussa R, Martoni C, Bhathena J, Urbanska AM, Prakash S (2010) Oral microencapsulated live Saccharomyces cerevisiae cells for use in renal failure uremia: preparation and in vivo analysis. J Biomed Biotechnol
Cowan KJ, Storey KB (2003) Mitogen-activated protein kinases: new signaling pathways functioning in cellular responses to environmental stress. J Exp Biol 206(7):1107–1115
Craig L, Charles O, Robin D, Casey T (2012) Reciprocal interactions of the intestinal microbiota and immune system. Nature 489(9):231–241
De Heer E, Peters D (2008) Innate immunity as a driving force in renal disease. Kidney Int 73 (1):7–8
Dhana E, Ludwig-Portugall I, Kurts, C (2018) Role of immune cells in crystal-induced kidney fibrosis. Matrix Biol 68:280–292
Dostal A, Lacroix C, Bircher L, Pham VT, Follador R, Zimmermann MB, Chassard C (2015) Iron modulates butyrate production by a child gut microbiota in vitro. Mbio 6(6):e01453-e1415
Eppensteiner J, Kwun J, Scheuermann U, Barbas A, Limkakeng AT, Kuchibhatla M, Elster EA, Kirk AD, Lee J (2019) Damage-and pathogen-associated molecular patterns play differential roles in late mortality after critical illness. JCI Insight 4(16)
Hasan N, Yang H (2019) Factors affecting the composition of the gut microbiota, and its modulation. Peer J 7:e7502
Hatch M, Freel RW, Vaziri N (1994) Intestinal excretion of oxalate in chronic renal failure. J Am Soc Nephrol 5(6):1339–1343
Hill M (1997) Intestinal flora and endogenous vitamin synthesis. Eur J Cancer Prev: Off J Eur Cancer Prev Organ (ECP) 6:S43-45
Hill NR, Fatoba ST, Oke JL, Hirst JA, O'Callaghan CA, Lasserson DS, Hobbs FR (2016) Global prevalence of chronic kidney disease–a systematic review and meta-analysis. PLoS ONE 11 (7):e0158765
Hooper LV, Midtvedt T, Gordon JI (2002) How host-microbial interactions shape the nutrient environment of the mammalian intestine. Annu Rev Nutr 22(1):283–307
Hugon P, Dufour J-C, Colson P, Fournier P-E, Sallah K, Raoult D (2015) A comprehensive repertoire of prokaryotic species identified in human beings. Lancet Infect Dis 15(10):1211–1219
Jain P, Shah S, Coussa R, Prakash S (2009) Potentials and limitations of microorganisms as renal failure biotherapeutics. Biol Targets and Ther 3:233
Kanbay M, Onal EM, Afsar B, Dagel T, Yerlikaya A, Covic A, Vaziri ND (2018) The crosstalk of gut microbiota and chronic kidney disease: role of inflammation, proteinuria, hypertension, and diabetes mellitus. Int Urol Nephrol 50(8):1453–1466
Kang J (1993) The gastrointestinal tract in uremia. Dig Dis Sci 38(2):257–268
Khoury T, Tzukert K, Abel R, Abu Rmeileh A, Levi R, Ilan Y (2017) The gut-kidney axis in chronic renal failure: A new potential target for therapy. Hemodial Int 21(3):323–334
Koppe L, Mafra D, Fouque D (2015) Probiotics and chronic kidney disease. Kidney Int 88 (5):958–966

Lee S, Huen S, Nishio H, Nishio S, Lee HK, Choi B-S, Ruhrberg C, Cantley LG (2011) Distinct macrophage phenotypes contribute to kidney injury and repair. J Am Soc Nephrol 22(2):317–326

Levin A, Stevens PE, Bilous RW, Coresh J, De Francisco AL, De Jong PE, Griffith KE, Hemmelgarn BR, Iseki K, Lamb EJ (2013) Kidney disease: improving global outcomes (KDIGO) CKD work group. KDIGO 2012 clinical practice guideline for the evaluation and management of chronic kidney disease. Kidney Int Suppl 3(1):1–150

Li J, Jia H, Cai X, Zhong H, Feng Q, Sunagawa S, Arumugam M, Kultima JR, Prifti E, Nielsen T (2014) An integrated catalog of reference genes in the human gut microbiome. Nat Biotechnol 32(8):834–841

Lu CC, Ma KL, Ruan XZ, Liu BC (2018) Intestinal dysbiosis activates renal renin-angiotensin system contributing to incipient diabetic nephropathy. Int J Med Sci 15(8):816

Mihai S, Codrici E, Popescu ID, Enciu A-M, Albulescu L, Necula LG, Mambet C, Anton G, Tanase, C (2018) Inflammation-related mechanisms in chronic kidney disease prediction, progression, and outcome. J Immunol Res

Ott SJ, El Mokhtari NE, Musfeldt M, Hellmig S, Freitag S, Rehman A, Kühbacher T, Nikolaus S, Namsolleck P, Blaut M (2006) Detection of diverse bacterial signatures in atherosclerotic lesions of patients with coronary heart disease. Circulation 113(7):929–937

Poveda J, Sanchez-Niño MD, Glorieux G, Sanz AB, Egido J, Vanholder R, Ortiz A (2014) p-Cresyl sulphate has pro-inflammatory and cytotoxic actions on human proximal tubular epithelial cells. Nephrol Dial Transplant 29(1):56–64

Prakash S, Chang T (1995) Preparation and in vitro analysis of microencapsulated genetically engineered *E. coli* DH5 cells for urea and ammonia removal. Biotechnol Bioeng 46(6):621–626

Prakash S, Chang T (1996) Microencapsulated genetically engineered *E. coli* DH5 cells for plasma urea and ammonia removal based on: 1. Column bioreactor and 2. Oral administration in uremic rats. Artif Cells, Blood SubstitS, Biotechnol 24(3):201–218

Prakash S, Chang T (1999) Artificial cell microcapsules containing genetically engineered *E. coli* dhs cells for in-vitro lowering of plasma potassium, phosphate, magnesium, sodium, chloride, uric acid, cholesterol, and creatinine: a preliminary report. Artif Cells Blood Substit Biotechnol 27(5–6):475–481

Prakash S, Chang TS (1998) Growth and survival of renal failure rats that received oral microencapsulated genetically engineered *E. coli* DH5 cells for urea removal. Artif Cells Blood Substit Biotechnol 26(1):35–51

Qian Q (2017) Inflammation: a key contributor to the genesis and progression of chronic kidney disease. Expand Hemodial 191:72–83

Ramezani A, Raj DS (2014) The gut microbiome, kidney disease, and targeted interventions. J Am Soc Nephrol 25(4):657–670

Ranganathan N, Patel B, Ranganathan P, Marczely J, Dheer R, Chordia T, Dunn SR, Friedman EA (2005) Probiotic amelioration of azotemia in 5/6th nephrectomized Sprague-Dawley rats. Sci World J 5:652–660

Rodríguez JM, Murphy K, Stanton C, Ross RP, Kober OI, Juge N, Avershina E, Rudi K, Narbad A, Jenmalm MC (2015) The composition of the gut microbiota throughout life, with an emphasis on early life. Microb Ecol Health Dis 26(1):26050

Roesch LF, Lorca GL, Casella G, Giongo A, Naranjo A, Pionzio AM, Li N, Mai V, Wasserfall CH, Schatz D (2009) Culture-independent identification of gut bacteria correlated with the onset of diabetes in a rat model. ISME J 3(5):536–548

Romagnani P, Remuzzi G, Glassock R, Levin A, Jager KJ, Tonelli M, Massy Z, Wanner C, Anders H-J (2017) Chronic kidney disease. Nat Rev Dis Prim 3(1):1–24

Sender R, Fuchs S, Milo R (2016) Revised estimates for the number of human and bacteria cells in the body. PLoS Biol 14(8):e1002533

Shi Y, Kaliyaperumal A, Lu L, Southwood S, Sette A, Michaels MA, Datta SK (1998) Promiscuous presentation and recognition of nucleosomal autoepitopes in lupus: role of autoimmune T cell receptor α chain. J Exp Med 187(3):367–378

Silverstein DM (2009) Inflammation in chronic kidney disease: role in the progression of renal and cardiovascular disease. Pediatr Nephrol 24(8):1445–1452

Stavropoulou E, Kantartzi K, Tsigalou C, Aftzoglou K, Voidarou C, Konstantinidis T, Chifiriuc MC, Thodis E, Bezirtzoglou E (2021) Microbiome, immunosenescence, and chronic kidney disease. Front Med 8:287

Thursby E, Juge N (2017) Introduction to the human gut microbiota. Biochemical Journal 474 (11):1823–1836

Tilg H, Moschen AR (2014) Microbiota and diabetes: an evolving relationship. Gut 63(9):1513–1521

Vaziri N, Dure-Smith B, Miller R, Mirahmadi M (1985). Pathology of gastrointestinal tract in chronic hemodialysis patients: an autopsy study of 78 cases. Am J Gastroenterol (Springer Nature) 80(8)

Vaziri ND, Wong J, Pahl M, Piceno YM, Yuan J, DeSantis TZ, Ni Z, Nguyen T-H, Andersen GL (2013) Chronic kidney disease alters intestinal microbial flora. Kidney Int 83(2):308–315

Wen L, Duffy A (2017) Factors influencing the gut microbiota, inflammation, and type 2 diabetes. J Nutr 147(7):1468S-1475S

Yang T, Santisteban MM, Rodriguez V, Li E, Ahmari N, Carvajal JM, Zadeh M, Gong M, Qi Y, Zubcevic J (2015) Gut dysbiosis is linked to hypertension. Hypertension 65(6):1331–1340

Yatim KM, Gosto M, Humar R, Williams AL, Oberbarnscheidt MH (2016) Renal dendritic cells sample blood-borne antigen and guide T-cell migration to the kidney by means of intravascular processes. Kidney Int 90(4):818–827

Yoshifuji A, Wakino S, Irie J, Tajima T, Hasegawa K, Kanda T, Tokuyama H, Hayashi K, Itoh H (2016) Gut Lactobacillus protects against the progression of renal damage by modulating the gut environment in rats. Nephrol Dial Transplant 31(3):401–412

Chapter 9
Gut Microbiota in Elderly Onset Rheumatoid Arthritis

Almagul Kushugulova, Jeannette Kunz, Dimitri Poddighe, and Bayan Ainabekova

Abstract RA is a chronic and systemic autoimmune disease characterized by a significant role in disease development of external (environmental, diet, infection) and internal (genetic, microbiome, hormonal) factors. Currently, the diagnosis of RA is mainly based on the presence and high titer of rheumatoid factor (RF) and anti-citrullinated protein antibodies (ACPAs) or anti-cyclic citrullinated peptide (anti-CCP) antibodies in the plasma. It is known that pathogenesis is associated with pathological processes in the mucous. One significant factor affecting disease development, course, treatment, and patient outcome is the microbiome. Rheumatoid arthritis in aged people is significantly different due to intestinal aging as an aggravating factor, is more challenging to treat, and most often has an unfavorable outcome. Aged people have lower gut immunity and higher leaky gut syndrome, and the constant inflammatory changes in the aged gut are setting more fire on the RA of the elderly. In this chapter, we describe the EORA microbiome and mycobiome features, immune imbalances including data on clinical and genetic characteristics.

Keywords EORA · Age · Microbiome · Rheumatoid arthritis · Inflammatory

A. Kushugulova (✉)
Center for Life Sciences, National Laboratory Astana, Nazarbayev University, Nursultan, Kazakhstan
e-mail: akushugulova@nu.edu.kz

J. Kunz · D. Poddighe
Department of Medicine, School of Medicine, Nazarbayev University, Nursultan, Kazakhstan

B. Ainabekova
Medical University Astana, Nursultan, Kazakhstan

F. Marotta (ed.), *Gut Microbiota in Aging and Chronic Diseases*, Healthy Ageing and Longevity 17, https://doi.org/10.1007/978-3-031-14023-5_9

9.1 Introduction

Geriatric rheumatology (Kobak and Bes 2018) is a branch of medicine studying the changes in joints, muscles, connective tissues, and related diseases in the elderly population (van Lankveld et al. 2005).

In recent decades, numerous studies have reported that the prevalence of autoimmune diseases, which were previously common only in the young population, has significantly increased in older age groups. The physiological and immune system changes in older individuals can be among the fundamental factors responsible for an increase in the risk of morbidity and also cause multi-clinical manifestations of inflammatory processes.

Rheumatoid arthritis (RA) is a multisystem chronic inflammatory disease characterized by destructive synovitis, which leads to joint deformities, disability, and even death if not diagnosed and treated early (Fig. 9.1). Any joints can be involved in the pathological process, but the most pathognomonic are the small joints of the hands and feet. The prevalence of RA is highly variable and depends on patient age and ethnicity, and it shows a significant increase of up to 2% in the geriatric population. The onset of RA after the age of 60 years is referred to as elderly onset RA (EORA). Population studies of EORA incidence and prevalence have demonstrated that the annual incidence of EORA globally can widely vary depending on gender and ethnicity (Gabriel 2001; Kumaradev et al. 2021; Lufkin et al. 2021; Xu and Wu 2021).

The isolation of the elderly onset of rheumatoid arthritis (EORA) is directly related to the aging process. Morphological and functional changes occur in the aging intestine. Age-related changes in the intestine include changes in the microbiota, the immune system, and epithelial functions. Certain alterations, such as the decrease in microvilli number, intestinal lumen defect, and low bacterial diversity in microbiome composition, trigger the aging process (Pimentel). Studies on mice demonstrated that the intestinal alkaline phosphatase (IAP), Sirtuin 1 (SIRT1) (Wellman et al. 2017), and Nicotinamide Adenine Dinucleotide (NAD) (Chini et al. 2017) depletion play a crucial role in aging. Additionally deficiency in spermidine (SPD) (Frank et al. 2018) and urolithin A (Espín et al. 2013), as well as "leaky gut" (Qi et al. 2017) may contribute to gut microbiome aging process and autoimmune intolerance. Moreover, aging and rheumatoid arthritis have common Wnt pathway impairment, which, along with pathological destructive processes in the joints (Miao et al. 2013), initiates aging, affecting intestinal homeostasis (Nalapareddy et al. 2017).

The relevance of the direction is due to the upward trend in the incidence of EORA. Carbonell et al. studied the incidence of RA in Spain per 100,000 people over the age of 60 years and found that the incidence rates were 9.1 and 14.5 among males and females, respectively (Carbonell et al. 2008).

Japanese scientists have observed an increase in patients who develop RA after 70 or even 80 years of age (Jinno et al. 2021). It is assumed that the increase in

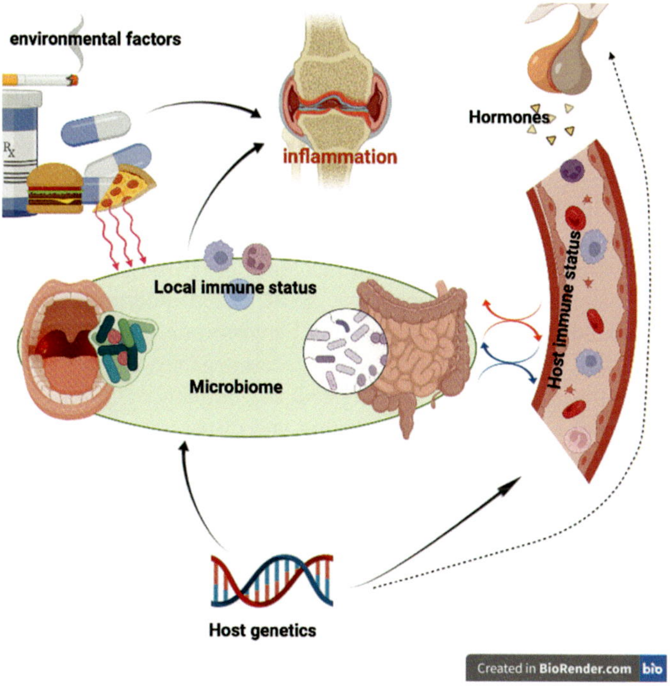

Fig. 9.1 Trigger factors for the development of rheumatoid arthritis. In the body of a genetically susceptible individual, due to dysregulation of the immune response, there is a delay in the resolution of an acute inflammatory process in the joints (clinically overt or latent) caused by any of the trigger factors. Thus, acute inflammation transforms into chronic inflammation, proceeding further along the autoimmune mechanism, in contrast to subjects not predisposed to RA, in which an acute process under conditions of normal immunoregulation ends in complete recovery. No single etiological factor responsible for the development of RA has been identified. The starting factor for the development of the disease can be hypothermia, hyperinsolation (sunlight), intoxication, endocrine pathology, stress, and others

EORA incidence is primarily caused by an increase in the life expectancy of individuals in developed countries. This fact will likely further increase the number of patients with EORA in the coming years.

9.2 Genetics

Genetic factors clearly play a crucial role in the etiopathogenesis of RA. Over 100 risk loci for RA have been identified. Moreover, new loci are being studied and presented to the scientific community (Kwon et al. 2020; Leng et al. 2020; Ha et al. 2021). The largest contributors are genes encoding class II human leukocyte antigens (HLA). However, several studies have demonstrated that genetic loci are specific to different ethnic groups (Achour et al. 2017; Aslam et al. 2020). Ethnic features have been studied, for example, between populations in Japan and China (Suzuki et al. 2019; Wu et al. 2021).

Research findings pertaining to genetic predisposition in EORA are controversial. It has been reported that DRB1 alleles associated with RA show differences between young-onset RA (YORA) and EORA11. Numerous authors studying DRB1 alleles in patients with RA found that YORA is associated with HLA-DRB1*0401, whereas EORA is associated with HLA-DRB1*0110 (Kobak and Bes 2018). In addition, an increased incidence of HLA-DRB1*13*14 has been observed in patients with seronegative EORA and polymyalgia rheumatica. A prospective study reported an association between polymyalgia rheumatica and HLA-DRB1*0101, 0102, and 0401, whereas seronegative EORA was associated with DRB1-0401. Kim et al. investigated the influence of HLA-DRB1 and HLA-DQB1 genes on disease predisposition and disease severity in patients with EORA and YORA (Kobak and Bes 2018). They found that alleles encoding a common epitope were less frequently detected in patients with EORA than in those with YORA (49.2 vs. 66.1%). Moreover, they observed that the effect of the common epitope and HLA-DQ*04 alleles was less significant in relation to EORA. Compared with YORA, EORA is associated with less frequent epitope presence and more minor radiological progression. Hellier et al. also studied the effect of the HLA-DRB1 gene on disease susceptibility and activity in patients with EORA and YORA. The HLA-DRB1/04 alleles were closely associated with YORA, whereas no such associations were observed with EORA. Wu et al. found that the DRB1*04 allele was present in 50% of patients with EORA and 92% of patients with RA onset before the age of 30 years. Oka et al. studying the HLADRB1 gene association with YORA and EORA, found out that the frequencies of DRB104:05/DRB104:06, DRB104:05/DRB112:0, DRB104:05/DRB115:01 genes were higher in YORA than EORA; however, DRB1 ∗ 01:01/DRB1 ∗ 04:05 and DRB1 ∗ 04:05/DRB1 ∗ 04:05 gene frequencies were considerably higher in EORA (Oka et al. 2019). Moreover, the gene dosage effect of the shared epitope alleles was significantly higher in EORA compared to YORA. It was also assumed that DQ molecules might impact the age of RA onset.

According to Mariaselvam et al. HLA-G rs9380142 polymorphism`s AA genotype demonstrated significant frequency in seropositive YORA, however, AG genotype of this gene was more prevalent in seropositive EORA patients. Additionally, HLA-G rs66554220 was considerably less frequent in RF-positive EORA compared to RF-positive YORA (Mariaselvam et al. 2015).

Thus, results related to the genetic predisposition to EORA remain controversial (reference). The lack of research on the genetic predisposition to LORA according to non-HLA genes limits the ability to determine the full genetic picture of this form of the disease.

A number of studies have demonstrated a more severe course and a worse prognosis in patients with EORA compared to YORA (Kobak and Bes 2018; Murata et al. 2019; Ke et al. 2021; Targońska-Stępniak et al. 2021), which in turn suggests that genes associated with severe RA course, such as PADI4, CTLA4,

TRAF1, and ETS1 (Danila et al. 2017), respectively, may be associated with EORA, which in turn determines the need for further studying the genetic markers of EORA.

9.3 Microbiome

Gut microbiota can trigger autoimmune arthritis (Scher et al. 2013; Teng et al. 2016). Several experimental and clinical studies have linked the microbiome of the oral cavity and lower gastrointestinal tract to citrullinization as well as immunological changes that contribute to pathological processes affecting the development and progression of RA. A new area of pharmacomicrobiomics is rapidly developing, which studies the effect of changes in the human microbiome on drugs. Pharmacomicrobial studies have demonstrated that human intestinal microorganisms and their enzyme products affect the bioavailability, clinical efficacy, and toxicity of various drugs via direct and indirect mechanisms. This effect may provide a means to predict treatment response and modulate the microbiome to improve therapy response and reduce drug toxicity. However, there have been contradictory results; some researchers suggested that changes in the microbiome trigger RA pathogenesis, whereas others argued that such changes are secondary to RA. Early changes in the gut microbiome in patients with RA are characterized by an increase in the populations of several genera, including *Lactobacillus*, *Prevotella*, *Clostridium*, *Gordonibacter*, *Eggerthella*, *Collinsella*, *Actinobacteria*, and *Lachnospiraceae*, and a decrease in the populations of proteobacteria, *Firmibacterium*, and *Faecalibacterium*. Bacteria such as *Collinsella aerofaciens* have been found to exacerbate arthritis in mouse models (Horta-Baas et al. 2017).

To date, several mechanisms have been reported for the initiation of RA, which includes the production of citrullinated peptides by *Porphyromonas gingivalis* and the activation of immune responses via Th17 cells by *Prevotella copri*. The first mechanism involves the production of citrulline to promote the pathogenesis of RA. High serum levels of antibodies to *P. gingivalis* have been detected in patients with RA. In addition, *P. gingivalis* was associated with RA; *P. gingivalis* derived peptidylarginine deiminase citrullinates proteins, thus leading to the generation of ACPA targets in the gingival mucosa. *P. copri* is also evidently involved in the pathogenesis of RA. Scher et al. found that compared with healthy patients, the relative abundance of *P. copri* was higher in patients with RA at the first onset. Subsequently, it was demonstrated that antibody responses to *P. copri* were higher in patients with RA compared with those in the control group. In patients with de novo RA, the increased abundance of *Prevotella* in the gastrointestinal tract displaced *Bacteroides fragilis*, a bacterium with marked Treg (Li and Wang 2021) activity. Increased levels of *P. copri* and similar species are associated with low titers of beneficial organisms. This can suppress the immune system and affect the breakdown of vitamins into components that are absorbed into the bloodstream (Maeda and Takeda 2019).

Recently, Japanese scientists have reported that RA is associated with not only *P. copri* but also other *Prevotella* sp. (*P. denticola*, *P. marshii*, *P. disiens*, *P. corporis*, and *P. amnii*) (Kishikawa et al. 2020). Further, the gingival mucosa is closely associated with the pathogenesis of RA through specific bacteria, such as *Aggregatibacter actinomycetemcomitans* and *Cryptobacterium curtum*. *A. actinomycetemcomitans*, found to be part of the oral microbiome, produces leukotoxin, which destroys the cell membrane of neutrophils, leading to hypercitrullination. More recently, a study reported differences in the oral microbiota profile of patients with RA and patients with osteoarthritis and healthy subjects; the oral microbiota profile of patients with RA was enriched with eight genera of bacteria, namely *Prevotella*, *Streptococcus*, *Porphyromonas*, *Haemophilus*, *Rothia*, *Actinomyces*, *Granulicatella*, *Leptotrichia*, and *Lautropia*. Vaahtovuo et al. analyzed the microbiota of patients with untreated premature RA or fibromyalgia via flow cytometry, 16S rRNA hybridization, and DNA staining (Vaahtovuo et al. 2008).

In patients with RA, the number of deliveries of *Bifidobacterium* and *Eubacterium rectale–Clostridium coccoides* from the *B. fragilis* subgroup was reduced, which is consistent with earlier results for individuals with Crohn's disease (Seksik et al. 2003).

According to Abdollahi-Roodsaz et al. at the genus level, the abundance of *Lactobacillus* was lower in patients with RA than in individuals in the control group. *L. bifidus* has been found to cause joint swelling in mice (Abdollahi-Roodsaz et al. 2008). Zhang et al. employed metagenomic shotgun sequencing to report that patients with RA exhibited increased amounts of *L. salivarius* in the intestines, teeth, and saliva (Zhang et al. 2015). However, the abundance of *Haemophilus* species was found to be reduced in all three samples. Furthermore, Liu et al. (Liu et al. 2013) showed that fecal *Lactobacillus* species were more abundant in patients with RA in China than in healthy controls (Danila et al. 2017).

Sun et al. reported differences in microbiome diversity and composition between patients with RA and healthy controls in China (Sun et al. 2019). An increase in the abundance of *Bacteroides* and *Escherichia–Shigella*, along with a decrease in that of *Lactobacillus*, *Alloprevotella*, *Enterobacter*, and *Odoribacter*, was reported to be characteristic of patients with RA in Shanghai, China. These results differ from those of other studies conducted outside China (Sun et al. 2019).

Bacteroides occupy a specific niche in the pathogenesis of RA. In recent years, KEGG pathway analyses have revealed a significant association between the pathophysiology of RA and some substances involved in these pathways:

- Fatty acids are associated with inflammation, and several serum-free fatty acids are reported to be upregulated in patients with RA.
- Terpenoids suppress the signaling of nuclear factor-kB, the main regulator of the pathogenesis of inflammatory diseases. They are a major component of *Tripterygium wilfordii*, a plant used in traditional Chinese medicine, which is reportedly as effective as methotrexate in the treatment of RA.
- Adipocytokines contribute to the proinflammatory state of RA, increasing in abundance in the synovial fluid of patients with RA.

- Glycosaminoglycans are the main components of articular cartilage and other soft connective tissues. A modification in their metabolism has been suggested to play a decisive role in the pathogenesis of RA.

Taken together, functional analyses of pertinent KEGG pathways, evaluation of 16S rRNA gene sequences, and assessment of differences in the microbiome of patients with RA and healthy people revealed the various roles of microbes, including secondary metabolite biosynthesis, xenobiotic degradation, metabolism, and endocrine system regulation and how these functions are altered in RA patients. However, these pathways are evidently depleted in *Bacteroides*, a bacterial genus enriched in RA patients. These data suggest that gut microbiota composition contributes to RA initiation or development, which is attributable to altered microbial function.

Some authors have demonstrated a significant association of the R6FCZ7 gene; the protein encoded by this gene performs a wide range of functions, such as electron transport, redox catalysis, and gene regulation. Moreover, it has been widely reported that reactive oxygen species play a key role in the pathogenesis of RA.

In these data, the metagenomes following R6FCZ7 were associated with the taxonomic reference genomes of *B. uniformis*, *B. rodentium*, and *B. fragilis*, as well as other *Bacteroides* sp. This means that several species belonging to this genus carry this gene. The normal repair function of the microbiome, particularly of *Bacteroides*, seems to play a role in the pathology of RA. Thus, *Bacteroides* plays a crucial role in this process.

To date, it is known that the microbiome in PA (Li and Wang 2021) is composed of Prevotellaceae, *Collinsella aerofaciens*, *Eggerthella lenta* (Scher et al. 2013), *Verrucomicrobia*, *Akkermansia* (Chiang et al. 2019), *Prevotella* (Scher et al. 2013; Lee et al. 2019; Kishikawa et al. 2020; Li and Wang 2021), *Lactobacilli* (Lee et al. 2019), Clostridiaceae (Muñiz Pedrogo et al. 2019), *Bacteroides* (Rodrigues et al. 2019), and *L. salivarius* (Zhang et al. 2015), and *Faecalibacterium* (Lee et al. 2019), *Bacteroides*, *Bifidobacterium*, Bacteroidetes–Firmicutes(Lee et al. 2019), *Clostridium leptum* (Rodrigues et al. 2019), *Eubacterium rectale–Clostridium coccoides* (Vaahtovuo et al. 2008), and *Haemophilus* (Zhang et al. 2015) are found in lower numbers.

Considering the clinical features of RA, the characteristics of intestinal microflora differ between EORA and YORA. Adult gut microbiota composition significantly varies from one individual to another, and changes are encountered during puberty, pregnancy, and menopause. Further, they become more pronounced with the age profile. Physiological changes such as decreased immunity, poor diet, medication, and lifestyle changes cause the microbiota to become less diverse, eventually leading to bowel disease and increased susceptibility to infections. Chronic inflammation in the elderly, associated with several age-related diseases, is caused by an imbalance in gut microbiota. Here, “inflammation” refers to an age-related chronic inflammatory condition associated with limiting stroke, dementia, heart disease, etc.

In one study, the intestinal microbiota of old and young mice was transplanted into young sterile mice, and the immune responses of the spleen, lymph node, and small intestine tissues were analyzed. Further, a genome-wide gene expression analysis in the small intestine was conducted. An imbalance in gut microbiota can be a disease in older adults. Dysbacteriosis can lead to "bad" bacteria becoming dominant, which can have negative consequences and result in the development of various disorders, such as inflammatory bowel disease, obesity, diabetes, cancer, anxiety disorders, and autism. Certain bacterial metabolites serve as environmental triggers, responsible for either the onset or prevention of arthritis. Zhao demonstrated the potential microbiome–metabolite–arthritis links as a guide to the gut–joint axis for future research (Zhao et al. 2021).

No separate studies exist on the microbiome in patients with EORA; as a rule, all patients with EORA and YORA are included in the study group. Therefore, distinguishing distinctive features is difficult. Japanese researchers conducted a study involving 51–69-year-old patients with RA; patients with early RA were not treated with antirheumatic drugs. Among operational taxonomic units classified as *Prevotella* in cluster 4, the majority (79.2%) showed high (0.99%) similarity to *P. copri* (accession no. NR_113411.1), followed by 10.3% to *P. stercorea*. Similar to the results of a previous study in the North American population, *Prevotella*, particularly *P. copri*, was observed in high numbers in gut microbiota in some patients with early RA in Japan (Maeda et al. 2016).

In another study (Wells et al. 2020) that included twins aged 50–66 years, *Prevotella* sp. showed the strongest taxa associations. Furthermore, there evidently exists an association between genetic risk for RA and gut microbiota, as found in two large cohorts. Antibodies to *P. copri* have been shown to be associated with disease severity and immune responses mediated by Th1 and Th17 cells in patients with RA. It is assumed that *P. copri* contributes to the development of the pathology of RA. An increase in the abundance of Ruminococcaceae UCG-14, *Rikenella*, and *Shigella*, in association with RA PRS, has been reported. In addition, an abundance of gut microbiota was found to be associated with a risk genotype for RA, even in the absence of the disease.

The results are very similar to studies of the microbiome of earlier patients with PA (Sun et al. 2019), characterized by α-diversity, as determined using the Chao1 indices, significantly different between the RA and healthy control groups, as well as a moderate decrease in β-diversity. An increase in the population of *Bacteroidetes* has been observed in RA patients, which may be associated with a compensatory mechanism for the suppression of autoimmune reactions. In addition, in patients with RA, a decrease in the abundance of *Lactobacillus*, *Alloprevotella*, *Enterobacter*, *Clostridium*, Ruminococcaceae UCG-014, *Odoribacter*, Rikenellaceae RC9, *Enterococcus*, *Klebsiella*, *Desulfovibrio*, *Citrobacter*, *Akkermansia*, and *Rhodococcus* has been observed. On the contrary, the abundance of *Bacteroides*, *Escherichia–Shigella*, *Parasutterella*, *Flavonifractor*, the *Eubacterium xylanophium* group, *Tyzzerella*, *Sellimonas*, and *Oscillospira* has been found to be significantly higher in patients with RA. Accordingly, these and the *Eubacterium saphenum* group can serve as microbial markers. *Sutterella*,

Megamonas, and *Faecalibacterium* are negatively correlated with RF levels, whereas *Dorea* and *Alloprevotella* are positively correlated with RF and anti-CCP antibody levels. Despite these findings, the microbiome of patients with EORA requires further studies.

9.4 Mycobiome

In addition to bacteria, a few fungi are associated with most parts of the human body, such as the gastrointestinal tract, accounting for 0.001–0.1% of the microbial community. The collection of fungi in the microbiota is called mycobiome. The most common fungi belong to the following genera: *Candida*, *Cladosporium*, *Aureobasidium*, *Saccharomycetes*, *Aspergillus*, and *Fusarium*. Studying the effect of microorganisms on human health has enhanced our understanding of the nature of several diseases. Although the role of the mycobiome in RA is poorly understood, some evidence exists for other autoimmune diseases. For example, *Candida tropicalis*, a representative of the mammalian mycobiome, is capable of exacerbating inflammatory bowel disease when the pattern of the Dectin-1 recognition receptor is deficient. Furthermore, there are other fungi, such as *Saccharomyces boulardii*, that have a capable effect on a set of functional diseases of the gastrointestinal tract.

Various intestinal fungi and their metabolites and toxins have been reported to affect host immunity. Several clinical and epidemiological studies have demonstrated that patients with RA are more prone to opportunistic fungal infections compared with those with other autoimmune diseases (Avasthi et al. 2021). Similar results were reported as part of the Rochester Epidemiology Project (Doran et al. 2002).

Hoarau et al. characterized the gut microbiota and mycobiome in families with multiplex Crohn's disease and healthy relatives, reporting an association between the level of antibodies against *Saccharomyces cerevisiae* (anti-*S. cerevisiae* antibody, a known biomarker of CD) and abundance of *C. tropicalis*. A correlation was determined between *C. tropicalis*, *Escherichia coli*, and *Serratia marcescens* in patients with CD, which was confirmed by in vitro analyses of biofilms (Hoarau et al. 2016).

SKG mice are genetically predisposed to the development of autoimmune arthritis. Japanese scientists demonstrated that under conditions of microbiological purity, despite these mice showing active thymic production of arthritogenic autoimmune T cells and their constancy at the periphery, autoimmune disease does not manifest itself. However, in a clean environment, a single intraperitoneal injection of zymosan, mushroom raw glucan, or purified β-glucans, such as curdlan and laminarin, has been reported to cause severe chronic arthritis in SKG mice but only transient arthritis in normal mice. In addition, antibiotic treatment against fungi can reportedly prevent SKG arthritis in a microbial environment prone to arthritis. Multiple injections of polyinosinic–polycytidylic acid double-stranded RNA have also been found to induce mild arthritis in SKG mice. Thus, certain microbes (e.g., fungi or viruses) can cause autoimmune arthritis, such as RA, by stimulating the innate immunity in people who potentially exhibit the presence of arthritogenic autoimmune T cells owing to genetic abnormalities or variations (Yoshitomi et al. 2005).

9.5 Immunome

The gut is the primary immunological organ involved in the training of immune cells, maturation of the host immune system, and maintenance of the gut barrier function (Alam and Neish 2018; Cheng et al. 2019). Most antigens pass through the intestine.

The microbiota has numerous functions, but its key role is to maintain host immune system homeostasis (Li and Wang 2021a). Genetics seems to influence the uniqueness of the microbiome, particularly the profile of *Prevotella* sp. (Wells et al. 2019; Priya et al. 2021). In patients with early-onset RA, the fecal populations of lactobacilli have been observed to show higher diversity (Diamanti et al. 2016).

Chen et al. reported that compared with healthy controls, individuals with demonstrated impaired microbial variability in the gut was associated with autoantibody titers and duration of disease. Evidently, methotrexate increases the abundance and diversity of microbial species in patients with PA (Chen et al. 2016). Bowel dysbiosis associated with RA is not limited to *Lactobacillus*-related imbalances. In particular, *Haemophilus* populations are depleted, and *P. copri* is more abundant in patients with RA than in healthy controls. Notably, the prevalence of *Haemophilus* sp. is negatively correlated with autoantibody levels, suggesting that bacteria play a protective immunomodulatory role (Lorenzo et al. 2019; Maeda and Takeda 2019). With regard to the immunological spectrum, EORA has its own characteristics; the main antibodies typical of classical RA (RF and ACCP) and included in the ACR/EULAR 2010 classification assessments are often negative or have significantly low titers compared with classical RA. More pronounced destructive changes in the joints and functional characteristics of patients with EORA compared with elderly patients with YORA are observed.

9.6 Inflammatory Cytokine

Among proinflammatory cytokines, interleukin (IL)-1α/β and tumor necrosis factor (TNF)-α trigger the intracellular molecular signaling pathway that is responsible for the pathogenesis of RA, which leads to the activation of mesenchymal cells, recruitment of innate and adaptive immune system cells, and activation of synoviocytes. This, in turn, activates various mediators, including TNF-α, IL-1, IL-6, and IL-8, resulting in synovial inflammation, increased angiogenesis, and decreased lymphangiogenesis. The role of TNF-, IL-1, IL-6, and granulocyte–macrophage colony-stimulating factor in genesis has been confirmed by the effectiveness of therapy with specific antibodies (Fig. 9.2). TNF-α regulates the production of proinflammatory cytokines in rheumatoid synovial tissue and induces the activation of synovial fibroblasts, which then produce a connective tissue growth factor that causes abnormal activation of osteoclasts, resulting in joint destruction. TNF-α also mediates pannus formation, activates endothelial cells, and enhances chemokine production, which facilitates leukocyte migration into the synovial membrane. IL-1 activates T cell activation and is also a potent inducer of matrix metalloproteinases. Granulocyte–macrophage colony-stimulating factor alone may be sufficient to activate the expression of HLA-DR molecules on antigen-presenting cells, which activates T cells (Chantry et al. 1990).

The composition and function of gut microbiota play a key role in regulating the balance of Th17/Treg cells in the lamina propria, influencing host immune responses and tolerance as well as susceptibility to autoimmune diseases, such as PA (Sakkas and Bogdanos 2018; Jiao et al. 2020).

Recent evidence supports the hypothesis that autoimmunity arises in the immune system of the mucous membranes through local autoimmune processes (Maeda et al. 2016; Horta-Baas et al. 2017; Zhong et al. 2018).

The intestinal microbiota can induce cell differentiation through the Th17 pathway, inducing the secretion of various proinflammatory cytokines that are involved in the pathogenesis of RA. The production of short-chain fatty acids, such as butyrate, acetate, and propionate, from the microbial fermentation of dietary fibers is a potential pathway through which gut microbiota influence Treg cell differentiation and systemic inflammation (Silva et al. 2020). It is notable that in the elderly, short-chain fatty acids are reduced.

Another mechanism that may explain the influence of gut microbiota on the occurrence and persistence of inflammatory diseases is gut permeability modulation. Several factors affect the strength and function of the intestinal barrier, including diet, smoking, and the immune system of the mucous membranes. When the mucosal barrier is disrupted, intestinal cells can be invaded, triggering an abnormal immune response (Novella-Navarro et al. 2021).

Thus, the main members of the intestinal microbiota, including *P. copri*, *L. salivarius*, and *Collinsella*, seem to play a key role in the early stage of RA and its pathogenesis.

P. has a noticeable ability to stimulate the production of cytokines associated with Th17 cells, in particular, IL-6 and IL-23. The excess number of species is mediated by the Th17 pathway. This finding is consistent with the marked propensity of *Prevotella* sp. to drive Th17 immune responses in vitro (Li and Wang 2021).

One of the ways *Collinsella* is caused is by increasing intestinal permeability, as evidenced by decreased expression of tight junction proteins. *Collinsella* also affects the release of IL-17A by the epithelium (Chen et al. 2016).

In a study conducted by Sun et al. Spearman's correlation analysis revealed that *Dorea* and *Ruminococcus* were positively correlated with RF-IgA and anti-CCP

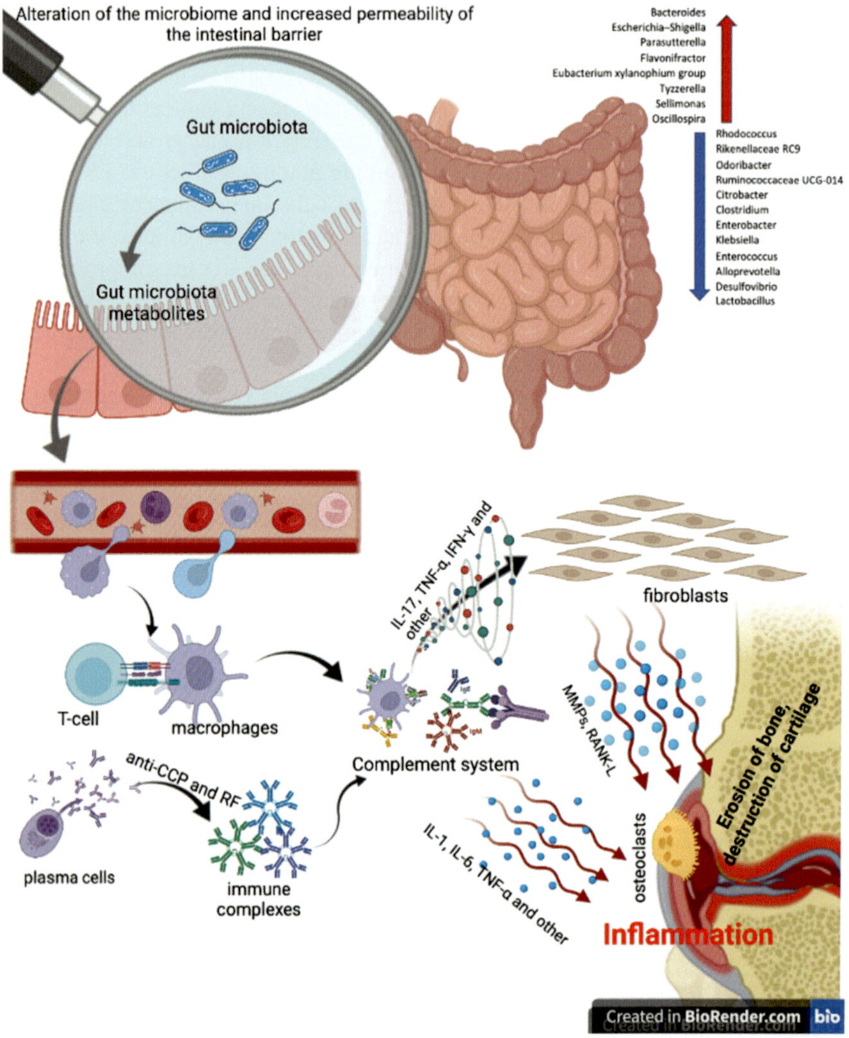

Fig. 9.2 Immunomodulatory role of the intestinal microbiome in the pathogenesis of RA. Persistent changes in the composition of the microbiota are a pathological process characterized by a violation of the intestinal barrier. This promotes the translocation of microorganisms and metabolites of microbial origin from the intestinal lumen to the mucosal layer and intestinal epithelium, which in turn leads to the activation of immune cells (Th1/Th2/Th17) and the production of cytokines. Induction of an inflammatory process by autoreactive Th1 or Th17 T cells, activated autoreactive T cells subsequently activate macrophages and fibroblasts through secretion of the proinflammatory mediators TNF-α, IL-17, IFN-γ, and RANK-L. Activated macrophages secrete proinflammatory cytokines TNF-α, IL-1β, and IL-6. Activated T cells contribute to the production of antibodies to citrulline protein (ACPA) and autoantibodies to rheumatoid factor (RF)

antibodies. Similarly, *Alloprevotella* was positively correlated with rheumatoid factors (RF-IgM, RF-IgA, and RF-IgG) and inflammatory biomarkers (CRP and ESR). The role of these bacteria in RA etiology and pathogenesis warrants further investigations, as the aforementioned study included only a small sample size. Future studies should also evaluate the significance of these changes for diagnosis and prognosis of response to PA drugs (Sun et al. 2019).

Based on these findings, it is evident that dysbiosis influences the levels of proinflammatory cytokines and depletes the levels of protective cytokines, thus playing a role in disease predisposition.

9.7 Solution

The crucial impact of age at the rheumatoid arthritis` debut on prognostic risk factors, disease progression, and pharmacological treatment is undeniable. The patients with EORA are less responsive to traditional treatment methods, such as synthetic DMARDs, biologic DMARDs, and targeted synthetic DMARDs, than those with YORA. Moreover, older age of patients with EORA is associated with many additional factors, such as comorbidities, compared to YORA. The major differences between EORA and YORA are demonstrated in Table 9.1.

The gut microbiota is a critical mediator in maintaining health and is influenced by microbial metabolites of exogenous substrates and endogenous host compounds, as well as immunity. To date, science is aware of the certain factors that affect the homeostasis of the gut microbiota, such as drugs, natural foods, diet, probiotics, and prebiotics, as well as aging. The shifts of the gut microbiome lead to a broad range of pathological processes that lead to autoimmune disorders, including RA, development. It is confirmed with data that gut microbiome shifts occur earlier than RA clinical manifestation. Since aging determines key changes in the immune system and the composition of the intestinal microbiome, pathogenetic differences of RA, depending on the manifestation age, also take place. This explains the differences mentioned above between EORA and YORA.

This knowledge is an excellent tool for correcting intestinal microbiota and suppressing autoimmune advances to achieve stable remission. Animal model studies have demonstrated that drugs, natural foods, diet, probiotics, and prebiotics can alter gut microbiota. In addition, these effects have been associated with the immune response module.

Therapies that alter gut microbiota can significantly affect the preclinical stage of arthritis, as dysbiosis in such cases occurs earlier than in clinical arthritis. The effects of the interventions on gut permeability modulation could not reverse arthritis. In addition, after the development of arthritis, therapies regulating symptom management are limited. Studies on acarbose, probiotics, and prebiotics have suggested that these interventions can reduce disease incidence but not cure it.

Table 9.1 Comparative characteristics of YORA and EORA

	YORA	EORA
Onset Age	30–55	>60
Sex	Women affected 3–5 times more often	Equally men and women
Joint involvement	Small joints are more often affected, including the proximal interphalangeal joints of hands and feet	Large joints are more often affected, including shoulders joints
Articular pattern	symmetrical usually	Asymmetrical more often than YORA
Debut	Typical debut, from symmetrical enduring joint syndrome, mainly, with the small joints defeat	Atypical debut: constitutional symptoms, such as weight loss, myalgia, lymphadenopathy, polymyalgia rheumatic like syndrome may be the only signs in debut, followed by the addition of joint syndrome
Immunological panel	ACCP positive with high titers more often than with EORA	RF and ACCP often negative or low titers
Inflammatory parameters	ESR, CRP have usually lower values than with EORA	ESR, CRP, leukocytosis, and thrombocytosis are usually clearly altered
DAS 28 score	III->5.1 II-5.1–3.2 I-2.6–3.2 0-<2.6	The upper values of the limits indicated for YORA
Radiological patterns	More commonly non-erosive	Often erosive
Complications	Infrequent with timely treatment	Complications develop faster and frequently
Disability	Low risk	High risk
HAQ-score:	0–1.0—minimal disability 1.1–2.0—moderate disability 2.1–3.0—pronounced violations of vital activity	The upper values of the limits indicated for YORA
Response to therapy	Better	Worse

The literature suggests that elderly patients with RA have poorer functional and anatomical performance than younger patients with RA (Abdollahi-Roodsaz et al. 2008). The data from this study confirm that RA is a more severe disease in older compared to younger people.

References

Abdollahi-Roodsaz S et al (2008) Stimulation of TLR2 and TLR4 differentially skews the balance of T cells in a mouse model of arthritis. The J Clin Invest (American Society for Clinical Investigation) 118(1):205–216. https://doi.org/10.1172/JCI32639

Achour Y et al (2017) Analysis of two susceptibility SNPs in HLA region and evidence of interaction between rs6457617 in HLA-DQB1 and HLA-DRB1*04 locus on Tunisian rheumatoid arthritis. J Genet 96(6):911–918. https://doi.org/10.1007/s12041-017-0855-y

Alam A, Neish A (2018) Role of gut microbiota in intestinal wound healing and barrier function. Tissue Barriers (Taylor and Francis) 6(3):1539595. https://doi.org/10.1080/21688370.2018.1539595

Aslam MM et al (2020) Investigating the GWAS-implicated loci for rheumatoid arthritis in the pakistani population. In: Hindawi W-JT (ed) Disease markers. p 1910215. https://doi.org/10.1155/2020/1910215

Avasthi D et al (2021) Disseminated histoplasmosis in an adult with rheumatoid arthritis not on biological immune modulators. Cureus, Inc., Cureus. https://doi.org/10.7759/cureus.15709

Carbonell J et al (2008) The incidence of rheumatoid arthritis in Spain: results from a nationwide primary care registry. Rheumatology 47(7):1088–1092. https://doi.org/10.1093/rheumatology/ken205

Chantry D et al (1990) Granulocyte-macrophage colony stimulating factor induces both HLA-DR expression and cytokine production by human monocytes. Cytokine 2(1):60–67. https://doi.org/10.1016/1043-4666(90)90044-T

Chen J, Wright K, Davis JM et al (2016) An expansion of rare lineage intestinal microbes characterizes rheumatoid arthritis. Genome Med (BioMed Central Ltd) 8(1):43. https://doi.org/10.1186/s13073-016-0299-7

Cheng H-Y et al (2019) Interactions between the gut microbiota and the host innate immune response against pathogens. Front Immunol (Frontiers Media SA) 10:607. https://doi.org/10.3389/fimmu.2019.00607

Chiang HI et al (2019) An association of gut microbiota with different phenotypes in Chinese patients with rheumatoid arthritis. J Clin Med MDPI 8(11). https://doi.org/10.3390/jcm8111770

Chini CCS, Tarragó MG, Chini EN (2017) NAD and the aging process: Role in life, death and everything in between. Mol Cell Endocrinol 455:62–74. https://doi.org/10.1016/j.mce.2016.11.003

Danila MI et al (2017) Dense genotyping of immune-related regions identifies loci for rheumatoid arthritis risk and damage in African Americans. Mol Med (Cambridge, Mass) (Feinstein Institute for Medical Research) 23:177–187. https://doi.org/10.2119/molmed.2017.00081

Diamanti AP et al (2016) Microbiota and chronic inflammatory arthritis: an interwoven link. J Transl Med 14(1):233. https://doi.org/10.1186/s12967-016-0989-3

Doran MF et al (2002) Frequency of infection in patients with rheumatoid arthritis compared with controls: a population-based study. Arthritis Rheum 46(9):2287–2293. https://doi.org/10.1002/art.10524

Espín JC et al (2013) Biological significance of urolithins, the gut microbial ellagic acid-derived metabolites: the evidence so far. Evid-Based Complement Altern Med eCAM (Hindawi Publishing Corporation) 270418. https://doi.org/10.1155/2013/270418

Frank M et al (2018) Spermidine in health and disease. Science (American Association for the Advancement of Science) 359(6374):eaan2788. https://doi.org/10.1126/science.aan2788

Gabriel SE (2001) The epidemiology of rheumatoid arthritis. Rheum Dis Clin N Am 27(2):269–281. https://doi.org/10.1016/S0889-857X(05)70201-5

Ha E, Bae S-C, Kim K (2021) Large-scale meta-analysis across East Asian and European populations updated genetic architecture and variant-driven biology of rheumatoid arthritis, identifying 11 novel susceptibility loci. Ann Rheum Dis 80(5):558. https://doi.org/10.1136/annrheumdis-2020-219065

Hoarau G et al (2016) Bacteriome and mycobiome interactions underscore microbial dysbiosis in familial Crohn's disease. mBio (American Society for Microbiology) 7(5). https://doi.org/10.1128/mBio.01250-16

Horta-Baas G et al (2017) Intestinal dysbiosis and rheumatoid arthritis: a link between gut microbiota and the pathogenesis of rheumatoid arthritis. J Immunol Res (Hindawi) 4835189. https://doi.org/10.1155/2017/4835189

Jiao Y et al (2020) Crosstalk between gut microbiota and innate immunity and its implication in autoimmune diseases. Front Immunol (Frontiers Media SA) https://doi.org/10.3389/fimmu.2020.00282

Jinno S et al (2021) Comparison of the drug retention and reasons for discontinuation of tumor necrosis factor inhibitors and interleukin-6 inhibitors in Japanese patients with elderly-onset rheumatoid arthritis-the ANSWER cohort study. Arthritis Res Ther (BioMed Central) 23 (1):116. https://doi.org/10.1186/s13075-021-02496-w

Ke Y et al (2021) Features and outcomes of elderly rheumatoid arthritis: does the age of onset matter? a comparative study from a single center in China. Rheumatol Ther (Springer Healthcare) 8(1):243–254. https://doi.org/10.1007/s40744-020-00267-8

Kishikawa T et al (2020) Metagenome-wide association study of gut microbiome revealed novel aetiology of rheumatoid arthritis in the Japanese population. Ann Rheum Dis (BMJ Publishing Group) 79(1):103–111. https://doi.org/10.1136/annrheumdis-2019-215743

Kobak S, Bes C (2018) An autumn tale: geriatric rheumatoid arthritis. Ther Adv Musculoskelet Dis (SAGE Publications). 10(1):3–11. https://doi.org/10.1177/1759720X17740075

Kumaradev S et al (2021) Socio-demographic determinants in the evolution of pain in inflammatory rheumatic diseases: results from ESPOIR and DESIR cohorts. Rheumatology. https://doi.org/10.1093/rheumatology/keab562

Kwon Y-C et al (2020) Genome-wide association study in a Korean population identifies six novel susceptibility loci for rheumatoid arthritis. Ann Rheum Dis 79(11):1438. https://doi.org/10.1136/annrheumdis-2020-217663

van Lankveld W, Franssen M, Stenger A (2005) Gerontorheumatology: the challenge to meet health-care demands for the elderly with musculoskeletal conditions. Rheumatology 44 (4):419–422. https://doi.org/10.1093/rheumatology/keh555

Lee JY et al (2019) Comparative analysis of fecal microbiota composition between rheumatoid arthritis and osteoarthritis patients. Genes MDPI AG 10(10). https://doi.org/10.3390/genes10100748

Leng R-X et al (2020) Identification of new susceptibility loci associated with rheumatoid arthritis. Ann Rheum Dis 79(12):1565. https://doi.org/10.1136/annrheumdis-2020-217351

Li M, Wang F (2021a) Role of intestinal microbiota on gut homeostasis and rheumatoid arthritis. J Immunol Res (Ren W-K (ed) Hindawi) 8167283. https://doi.org/10.1155/2021a/8167283.

Liu X et al (2013) Analysis of fecal lactobacillus community structure in patients with early rheumatoid arthritis. Curr Microbiol 67(2):170–176. https://doi.org/10.1007/s00284-013-0338-1

Lorenzo D et al (2019) Oral-gut microbiota and arthritis: is there an evidence-based axis?. J Clin Med MDPI 8(10):1753. https://doi.org/10.3390/jcm8101753

Lufkin L et al (2021) A Bayesian model to analyze the association of rheumatoid arthritis with risk factors and their interactions. Front Public Health (Frontiers Media SA) 9:693830. https://doi.org/10.3389/fpubh.2021.693830

Maeda Y et al (2016) Dysbiosis contributes to arthritis development via activation of autoreactive T cells in the intestine. Arthritis Rheumatol (John Wiley and Sons Inc) 68(11):2646–2661. https://doi.org/10.1002/art.39783

Maeda Y, Takeda K (2019) Host–microbiota interactions in rheumatoid arthritis. Exp Mol Med 51 (12):1–6. https://doi.org/10.1038/s12276-019-0283-6

Mariaselvam CM, Chaaben AB, Salah S (2015) Human leukocyte antigen-G polymorphism influences the age of onset and autoantibody status in rheumatoid arthritis. Tissue Antigens 85 (3):182–189. https://doi.org/10.1111/tan.12521

Miao C et al (2013) Wnt signaling pathway in rheumatoid arthritis, with special emphasis on the different roles in synovial inflammation and bone remodeling. Cell Signal 25(10):2069–2078. https://doi.org/10.1016/j.cellsig.2013.04.002

Muñiz Pedrogo DA et al (2019) An increased abundance of clostridiaceae characterizes arthritis in inflammatory bowel disease and rheumatoid arthritis: a cross-sectional study. Inflamm Bowel Dis (Oxford University Press) 25(5):902–913. https://doi.org/10.1093/ibd/izy318

Murata K et al (2019) Elderly onset of early rheumatoid arthritis is a risk factor for bone erosions, refractory to treatment: KURAMA cohort. Int J Rheum Dis 22(6):1084–1093

Nalapareddy K et al (2017) Canonical Wnt signaling ameliorates aging of intestinal stem cells. Cell Rep 18(11):2608–2621. https://doi.org/10.1016/j.celrep.2017.02.056

Novella-Navarro M et al (2021) Risk factors for developing rheumatoid arthritis in patients with undifferentiated arthritis and inflammatory arthralgia. Front Med (Frontiers Media SA) https://doi.org/10.3389/fmed.2021.668898

Oka S et al (2019) Association of HLA-DRB1 genotype with younger age onset and elder age onset rheumatoid arthritis in Japanese populations. Medicine (Wolters Kluwer Health) 98(48): e18218–e18218. https://doi.org/10.1097/MD.0000000000018218

Priya S et al (2021) Shared and disease-specific host gene-microbiome interactions across human diseases. bioRxiv 2021.03.29.437589. https://doi.org/10.1101/2021.03.29.437589

Qi YF et al (2017) Intestinal permeability biomarker zonulin is elevated in healthy aging. J Am Med Dir Assoc (Elsevier Inc) 18(9):810.e1-810.e4. https://doi.org/10.1016/j.jamda.2017.05.018

Rodrigues GSP et al (2019) Detection of increased relative expression units of bacteroides and prevotella, and decreased clostridium leptum in stool samples from Brazilian rheumatoid arthritis patients: a pilot study. Microorganisms MDPI AG 7(10). https://doi.org/10.3390/microorganisms7100413

Sakkas LI, Bogdanos DP (2018) Multiple hit infection and autoimmunity: the dysbiotic microbiota–ACPA connection in rheumatoid arthritis. Curr Opin Rheumatol 30(4). https://journals.lww.com/co-rheumatology/Fulltext/2018/07000/Multiple_hit_infection_and_autoimmunity__the.17.aspx

Scher JU et al (2013) Expansion of intestinal Prevotella copri correlates with enhanced susceptibility to arthritis. eLife (eLife Sciences Publications, Ltd) 2:e01202–e01202. https://doi.org/10.7554/eLife.01202

Seksik P et al (2003) Alterations of the dominant faecal bacterial groups in patients with Crohn's disease of the colon. Gut. Copyright 2003 by Gut 52(2):237–242. https://doi.org/10.1136/gut.52.2.237

Silva YP, Bernardi A, Frozza RL (2020) The role of short-chain fatty acids from gut microbiota in gut-brain communication. Front Endocrinol (Frontiers Media SA) https://doi.org/10.3389/fendo.2020.00025

Sun Y et al (2019) Characteristics of gut microbiota in patients with rheumatoid arthritis in Shanghai, China. Front Cell Infect Microbiol (Frontiers Media SA) 9: 369. https://doi.org/10.3389/fcimb.2019.00369

Suzuki A, Terao C, Yamamoto K (2019) Linking of genetic risk variants to disease-specific gene expression via multi-omics studies in rheumatoid arthritis. Semin Arthritis Rheum 49(3, Supplement):S49–S53. https://doi.org/10.1016/j.semarthrit.2019.09.007

Targońska-Stępniak B et al (2021) Systemic inflammatory parameters in patients with elderly-onset rheumatoid arthritis (EORA) and young-onset rheumatoid arthritis (YORA)-an observational study. J Clin Med MDPI 10(6):1204. https://doi.org/10.3390/jcm10061204

Teng F et al (2016) Gut microbiota drive autoimmune arthritis by promoting differentiation and migration of Peyer's patch t follicular helper cells. Immunity 44(4):875–888. https://doi.org/10.1016/j.immuni.2016.03.013

Vaahtovuo J et al (2008) Fecal microbiota in early rheumatoid arthritis. J Rheumatol 35(8):1500–1505. http://www.jrheum.org/content/35/8/1500.abstract

Wellman AS et al (2017) Intestinal epithelial sirtuin 1 regulates intestinal inflammation during aging in mice by altering the intestinal microbiota. Gastroenterology 153(3):772–786. https://doi.org/10.1053/j.gastro.2017.05.022

Wells PM et al (2019) 'RA and the microbiome: do host genetic factors provide the link? J Autoimmun (Academic Press) 99:104–115. https://doi.org/10.1016/j.jaut.2019.02.004

Wells PM et al (2020) Associations between gut microbiota and genetic risk for rheumatoid arthritis in the absence of disease: a cross-sectional study. Lancet Rheumatol (Lancet Publishing Group) 2(7):e418–e427. https://doi.org/10.1016/S2665-9913(20)30064-3

Wu C et al (2021) Transcriptome-wide association study identifies susceptibility genes for rheumatoid arthritis. Arthritis Res Ther (Biomed Central) 23(1):38. https://doi.org/10.1186/s13075-021-02419-9

Xu Y, Wu Q (2021) Prevalence trend and disparities in rheumatoid arthritis among US adults, 2005–2018. J Clin Med MDPI 10(15):3289. https://doi.org/10.3390/jcm10153289

Yoshitomi H et al (2005) A role for fungal β-glucans and their receptor Dectin-1 in the induction of autoimmune arthritis in genetically susceptible mice. J Exp Med 201(6):949–960. https://doi.org/10.1084/jem.20041758

Zhang X et al (2015) The oral and gut microbiomes are perturbed in rheumatoid arthritis and partly normalized after treatment. Nat Med 21(8):895–905. https://doi.org/10.1038/nm.3914

Zhao Y et al (2021) Hidden link in gut–joint axis: gut microbes promote rheumatoid arthritis at early stage by enhancing ascorbate degradation. Gut (BMJ Publishing Group). https://doi.org/10.1136/gutjnl-2021-325209

Zhong D et al (2018) The role of gut microbiota in the pathogenesis of rheumatic diseases. Clin Rheumatol 37(1):25–34. https://doi.org/10.1007/s10067-017-3821-4

Chapter 10
Combating the Sustained Inflammation Involved in Aging and Neurodegenerative Diseases with Probiotics

Sabrina Schaly and Satya Prakash

Abstract The gut-brain axis may be a key factor in disease pathologies including colon cancer, irritable bowel syndrome, depression, anxiety, and now neurodegenerative disorders. The multiple signaling pathways, mediators, and immune cells in the GIT create a systemic impact on the body. Aging and inflammation (termed inflamm-aging) are underlying factors in neurodegenerative disorders that may be orchestrated by the gut-brain axis. The sustained inflammation along with inflammatory cytokines and chemokines present in both the gut and brain of people with neurodegenerative disorders reveals a gut-brain dysbiosis. This dysbiosis and low-grade inflammation are present in Alzheimer's, Parkinson's, Huntington's, and other neurodegenerative diseases indicating a potential causative or additive role. Through the supplementation of commensal bacteria, via probiotics, inflammation and age-related damage can be reduced. This may be a beneficial tool to combat neurodegeneration and provide a neuroprotective ability to the body through the gut microbiome.

Keywords Inflammation · Aging · Neurodegenerative disorders · Probiotics · Gut-brain axis · Alzheimer's disease · Parkinson's disease · Huntington's disease

10.1 The Microbiome

Humans have symbiotically evolved together with microorganisms in the gastrointestinal tract (GIT) to form a commensal and essential relationship (Li et al. 2008). The GIT has diversity and recent studies reveal just how crucial the microbiome is for human health. The gut microbiome constitutes the trillions of

S. Schaly · S. Prakash (✉)
Biomedical Technology and Cell Therapy Research Laboratory, Department of Biomedical Engineering, Faculty of Medicine, McGill University, 3775 University Street, Montreal, Quebec H3A 2B4, Canada
e-mail: satya.prakash@mcgill.ca

F. Marotta (ed.), *Gut Microbiota in Aging and Chronic Diseases*, Healthy Ageing and Longevity 17, https://doi.org/10.1007/978-3-031-14023-5_10

archae, viruses, bacteria, and fungi within the gastrointestinal tract (GIT) (Limage et al. 2020). The gut microbiome is predominantly inhabited by four phyla of bacteria; *Bacteroidetes, Firmicutes, Proteobacteria,* and *Actinobacteria* (Villanueva-Millán et al. 2015). These commensal microbes use the gut space, nutrients, and breeding environment and in return they assist the host with several functions (Chang and Kao 2019).

The gut microbiome has been reported to have several key roles in human health. First, these microorganisms assist with metabolism. This includes regulating carbohydrate fermentation, synthesizing vitamins, gut motility, mucosal permeability, mucosal blood flow, gastric secretions, and more (Kau et al. 2011). However, they also act as a first-line defense by creating a physically epithelial barrier and have several mechanisms to prevent pathogenic bacteria from invading. This includes strengthening the mucosal barrier, inhibiting viral attachment, secreting antimicrobial peptides, and regulating the immune system (Dobson et al. 2012; Torres et al. 2013; Schroeder 2019). These antimicrobial peptides direct pathogen killing via opsonization and inflammatory cell recruitment (Boyton and Openshaw 2002). The gut microbes also compete against pathogen binding, colonization, and survival, reducing infection severity. The microorganisms also induce secretion of secretory IgA which prevents bacteria penetration. The commensal bacteria and the secreted mediators also activate the host's innate and adaptive immune system and have an indirect supportive role in the health of the kidneys, liver, heart, and lungs (Enaud et al. 2020). Given the large role of the microbiome in both health and disease, several researchers have worked on ways to improve gut health and overall health.

Although the localized impact of the immune system is being investigated, the human body is interconnected through the immune system, endocrine signals, and neurological signals creating a systemic response. The systemic circulation enables widespread travel of mediators and immune cells as well as immune-to-central nervous system (CNS) communication. To support this, local inflammation in the gut has shown to produce peripheral inflammation. Moreover, antigen-presenting immune cells (dendritic cells and macrophages) reside in the subepithelial lamina propria tissue of the GIT providing direct communication between the external environment (food consumed) and the systemic immune system (Smith et al. 2011). The GIT immune cells also express toll-like receptors (TLRs) and NOD-like receptors (NLRs) that recognize the microbe-associated molecular patterns (MAMPs) from pathogens and trigger a signaling cascade with cytokines (Faden and Loane 2015). This evidence may explain the connection between the gut dysbiosis and seemingly unrelated diseases.

There are significant differences between individual's microbiomes indicating that environment, diet, genetics, and countless other factors all play a role in shaping the microbiome (Qin et al. 2022). The gut microbiome requires a finely regulated homeostasis of commensal and pathogenic bacteria. In diseased states, this balance appears to favor pathogenic bacteria and may further contribute to the disease pathogenesis.

The gut microbiome has been linked not only to gut-related diseases but also to central nervous system disorders. Specifically, continuous low-grade inflammation and immune system activation contribute to the irritable bowel syndrome pathogenesis and generation of symptoms (O'Malley et al. 2011). Similarly, colon cancer in mice has been treated using probiotics to mitigate gut dysbiosis and combat cancer (Kahouli et al. 2017). Meningitis is a viral infection associated with lymphocyte over-activation and cytokine release that may seem unrelated to the gut microbiome (Patarca-Montero et al. 2001). However, gut bacteria release IgA which correlates with the inflammation, immune activation, and symptoms seen in meningitis (Maes et al. 2012). Moreover, gut dysbiosis has been linked to cardiovascular disease, non-alcoholic fatty liver disease, diabetes, and high cholesterol revealing the systemic impact of the microbiome (Bhathena et al. 2013) (Tomaro-Duchesneau et al. 2014). Specifically, commensal bacteria like *Lactobacillus reuteri* and *Lactobacillus fermentum* were shown to help regulate cholesterol levels and thus reduce the risk of cardiovascular disease (Tomaro-Duchesneau et al. 2014). For type 2 diabetes, *Bifidobacterium, Bacteroides, Faecalibacterium, Akkermansia, and Roseburia* were negatively associated while the *Ruminococcus, Fusobacterium,* and *Blautia* were positively associated with the disease (Gurung et al. 2020). Given that the microbiome is the first contact for nutrients and glucose in the GIT and it regulates glucose metabolism, it is logical that it could contribute to the glucose issues seen in diabetes. Given the increasing evidence of the involvement of the gut microbiome in several seemingly unrelated diseases, it is essential to fully understand the mechanism behind it and develop a way to treat gut dysbiosis.

10.2 Gut-Brain Axis

The gut-brain axis is a constantly changing bidirectional system involving neurological connections, endocrine signals, and immunological inputs (Romijn et al. 2008). This bidirectional axis is constantly at work and responsible for many functions such as stimulating satiety (Romijn et al. 2008). The gut-brain axis communicates with intestinal cells, the endocrine nervous system, and the central nervous system via metabolic and neuroendocrine pathways (Forsythe, Bienenstock and Kunze, 2014). The endocrine system is affected by bacterial metabolites, such as propionic acid and acetic acid, which promote the sympathetic nervous system activation and can alter memory and learning processes (Srikantha and Mohajeri, 2019). The microbiota communicates with the intestine which is lined with sympathetic and parasympathetic nerve fibers. These fibers can affect the brain through both the vagus nerve and the intestinal immune system (Filpa et al. 2016). The gut microbiota can also regulate inflammation through the production of mediators (short-chain fatty acids, neurotransmitters, and more) and through the activation of T cells after presenting antigens (Arpaia et al. 2013; Horai et al. 2015). Local inflammation also has an impact on the peripheral immune system. For example, an injection of pre-formed fibril (PFF) α-Syn leads to activation of microglial, astrocytes, and infiltration of B, CD4+T, CD8+T, and natural killer cells. The stimulated T cells further activate microglial causing them to become an inflammatory phenotype (M1) which contributes to the neurodegenerative process (Kustrimovic et al. 2019).

Similarly, the brain can influence the microbiota both directly and indirectly (Rhee et al. 2009). Directly, the brain can alter gastrointestinal motility, secretion, and permeability. Indirectly, the brain can communicate through signaling networks and cells within the gut lumen (enterochromaffin cells, neurons, and immune cells) to initiate change. Communication can involve epithelial to cell, receptor-mediated signals, and direct stimulation of gut cells. Specifically, enterochromaffin cells play an essential role in communication as they are innervated with afferent nerves (vagal). This innervation provides a direct pathway from the enterochromaffin cells to the nervous system and can assist with the immune response, controlling emotions, and other functions (Rhee et al. 2009).

Given this tight communication between the gut and nervous system, it is no surprise that researchers everywhere are discovering the association between gut dysbiosis and neurodegenerative disorders. Interestingly, neuroinflammation correlates with neurodegeneration, behavioral issues, and other neurological deficits which indicates that dysregulation may contribute to pathogenesis (Faden and Loane 2015; Vivekanantham et al. 2015). Gut dysbiosis is reported in several disorders including autism, depression, anxiety, and stress. Recent studies have reported different metabolites and microbiome alterations in autistic people compared to people without autism (Ming et al. 2012; Guo et al. 2021; Yap et al. 2010). Depression is a mood disorder that can also be linked to neuro-psychiatric changes and immune system deregulation (Dantzer et al. 2008). This could explain the presence of inflammatory markers such as C-reactive protein (CRP), IL-1, IL-6, and tumor necrosis factor (TNF-a), which can be risk factors for depression (Dantzer et al. 2008; Rook et al. 2011). Similarly, anxiety and stress are associated with intestinal dysfunctions, indicating a change in gut-brain signals (neurotransmitters and immune factors) (Diamond et al. 2011; O'Malley et al. 2011; Dinan and Cryan 2012).

10.3 Inflamm-Aging

Inflamm-aging is a term used to describe a spectrum of age-related pathologies, including neurodegeneration (Biagi et al. 2010). It has been shown that, with age, there is a delayed immune response, lower antibody production, and decreased immune cell production (Chandra 2000). This can be accompanied by increased inflammation despite the absence of infection (Salanitro et al. 2012). Aging also leads to a reduction in metabolic activity and lowers short-chain fatty acids (important for combatting inflammation). Moreover, cognitive impairment, altered bowel transit, hypertension, diabetes, arthritis, and vitamin D deficiency are also all age-related (Nagpal et al. 2018).

Within the GIT, commensal bacteria shape the host immune system and can alter the auto-reactivity of the host (Berer and Krishnamoorthy 2012; Rook et al. 2011). With age, the gut microbiome matures yet the diversity declines (Corey et al. 2019). Moreover, pathogenic bacteria increase while commensal ones are compromised leading to low-grade inflammation (Vemuri et al. 2018). An imbalance in commensal and pathogenic bacteria, termed dysbiosis, is consistent with impaired cognition, altered behavior, and decreased brain volume (Ticinesi et al. 2018). Gut dysbiosis has already been linked to several other diseases including metabolic, autoimmune, cancer, and now neurodegenerative disorders (Duvallet et al. 2017; Castelli et al. 2020; Iradj et al. 2019). Without the microbiome, for instance, in germ-free mice, memory, learning, physical activity, and anxiety are noted (Sampson et al. 2016). Moreover, there is an impaired hippocampal development, altered BBB permeability, and hormone levels.

The common underlying risk factor between neurodegenerative disorders like Alzheimer's and Parkinson's disease is aging (Castelli et al. 2019; Franceschi et al. 2018). Moreover, a sustained low-grade immune response and consequent inflammation are associated with neurodegenerative disorders like Alzheimer's disease (Ferreira et al. 2014; Grammas 2011). Specifically, Parkinson's patients have a significantly lower amount of butyrate and other short-chain fatty acids which are known to protect dopaminergic neurons (Sharma and Taliyan 2015; Cantu-Jungles et al. 2019). Early signs of Parkinson's disease include altered gut permeability, dopamine production, and a reduction in Lactococcus bacteria (Darby et al. 2019; Houser and Tansey 2017). Counteractively, fecal transplants from healthy donors to people with Parkinson's disease have ameliorated both non-motor and motor symptoms revealing the crucial role of the gut microbiome (Van Laar et al. 2019). Similarly in Alzheimer's disease, certain pathogenic gut bacteria such as *Mycobacterium spp., Salmonella spp., E. coli., Streptococcus spp.*, and *Staphylococcus aureus* appear to contribute to amyloid beta plaques and consequential disease progression (Petra et al. 2015). Together, this suggests that maintaining a healthy gut is key to preserving brain function and combatting age-related diseases.

10.4 Neurodegenerative Diseases

10.4.1 Dementia

Neurodegenerative diseases show common mechanisms and features. Dementia is characterized by a deteriorating cognitive function beyond what is expected by aging. This disease is progressive and affects memory, thinking, orientation, learning, and judgment. A deterioration in emotional and social intelligence is also common. This generates a physical, psychological, social, and economic impact on the individual and those taking care of them. The most common form of dementia is Alzheimer's disease, contributing to 60–70% of cases. There are three stages termed the early, middle, and late stages, with each stage encompassing progressively worse symptoms. Currently, around 50 million people worldwide have dementia (OMS 2020). Of these 50 million people, 60% live in low or middle-income countries. Moreover, cases increase by 10 million every year, expecting to reach 82 million cases by 2030 (OMS 2020). This large increase may be attributed to the growing elderly population as well as other environmental and dietary factors. Dementia, Alzheimer's disease, and Huntington's disease pathology have all been linked with a dysregulation of tryptophan metabolism via its serotonergic and kynurenine routes (Ruddick et al. 2006). They have also been linked to gut dysbiosis and inflammation as discussed later in the chapter.

10.4.2 Alzheimer's Disease (AD)

Alzheimer's disease (AD) can be characterized by two core pathologies, the presence of β-amyloid (Aβ) plaques and neurofibrillary tangles (NFTs) (Kinney et al. 2018). Improper cleavage of the amyloid precursor protein leads to Aβ aggregation and eventually fibrils and plaques (Selkoe 1994). Moreover, people with AD typically have low levels of acetylcholine. Acetylcholine is an important neurotransmitter for the parasympathetic and autonomic nervous system that is secreted by the intestines (Guo et al. 2021). Some symptoms of Alzheimer's disease can be detected and treated early. However, currently, there is no cure for dementia. Due to the lack of treatment, early-stage prevention is critical.

Microglia are resident immune cells in the central nervous system and have been studied extensively for their role in the disease etiology of Alzheimer's disease. The Aβ plaques and other factors lead to the activation of microglia cells which release pro-inflammatory cytokines, reactive oxygen species, and nitric oxide in response. The resident microglia can also phagocytose the Aβ plaques and release more cytokines which recruit more microglia (Wisniewski, Moretz and Lossinsky, 1981; Baik et al. 2016). The end result is a halo of activated microglial and a continuous positive feedback loop (Yates et al. 2000). Over time, the microglia are no longer able to handle the Aβ plaques and a sustained chronic inflammatory reaction persists damaging surrounding neurons (Sheng et al. 1998; Krukowski et al. 2018). Microglial cells also carry certain receptors that mediate signaling pathways and may contribute to the disease pathology. Specifically, TREM2 regulates phagocytic and autophagic pathways; myeloid survival and proliferation; and inflammation. In Alzheimer's disease, researchers have reported that damaged microglia are activated with TREM2 leading to dysregulation and neurodegeneration (Keren-Shaul et al. 2017). Another key mediator involved is CX3CL1, also called fractalkine, which binds to its corresponding receptor on microglia (Harrison et al. 1998). This, in turn, activates microglia leading to neuronal death and neurodegeneration (Zujovic et al. 2000). Finally, to counteract the inflammatory response of activated microglia, gamma Aminobutyric acid (GABA) is released by astrocytes into the extracellular fluid (Lee, Schwab and Mcgeer 2011). When GABA binds to the $GABA_B$ receptor, it prevents microglia and astrocytes from releasing pro-inflammatory cytokines, like TNFα, IL-6, and IL-12 (Kuhn et al. 2004; (Lee Schwab and Mcgeer 2011).

The inflammatory cytokines released from immune cells in the CNS promote AD pathogenesis. Specifically, people with AD have increased levels of TNF-α, IL-1, IL-6, and TGF-B. TNF-α promotes immune cell migration and infiltration into the area (Perry et al. 2001; Chang et al. 2017). Aβ further stimulates this pathway through directly stimulating microglia and activating NFκB signaling leading to TNF-α release from microglia (Combs et al. 2001). Similarly, IL-1 is crucial for β-amyloid plaque deposition through its regulation of amyloid precursor protein and positive feedback activation of microglia (Chong, 1997). The high levels of IL-6 in the cerebrospinal fluid of AD patients also contribute to neuroinflammation and neurodegeneration (Rothaug, Becker-Pauly and Rose-John, 2016). Moreover, the Aβ plaques stimulate the synthesis and release of IL-6 from microglia creating a positive feedback loop. Microglia and astrocyte also release IL-10 in people with AD in an attempt to regulate the pro-inflammatory response (Szczepanik et al. 2001). However, as the disease progress, it has been shown that IL-10 contributes to neuroinflammation and microglial dysfunction (Guillot-Sestier et al. 2015). Finally, TGF-β typically has a neuroprotective role and is also elevated in people with AD (Grammas and Ovase 2002). However, in AD, it appears that elevated levels of TGF-B promote an inflammatory response and the production of IL-1β and TGF-α.

10.4.3 Parkinson's Disease (PD)

Parkinson's disease is a neurological disorder that leads to neuronal death and brain shrinkage. Specifically, the death of dopaminergic neurons decreases dopamine neurotransmission resulting in motor problems like tremor, bradykinesia, and instability (Alexander 2004). Moreover, the loss of neurotransmitters can cause fatigue, irregular blood pressure changes, and trouble with digestion. The nerve damage also manifests into difficulty with balance, coordination, and walking, which worsens with time. This can hinder the individual's mental health as well as bodily functions like eating and using the bathroom.

Gastrointestinal symptoms are common in people with PD indicating a relationship between the inflammation in the gut and brain (Romano et al. 2021). There is evidence of microbial dysbiosis in PD patients through the significantly elevated levels of indican, a bacterial metabolite and marker of intestinal dysbiosis (Gao et al. 2018). Moreover, Calprotectin in the fecal matter has been identified and proposed as a marker of the gut immune system activation in PD patients (Mulak et al. 2019). There is also a significant reduction in anti-inflammatory butyrate-producing bacteria (i.e., Roseburia, Faecalibacterium spp., and

Prevotellaceae) and a clear increase in pro-inflammatory Proteobacteria species (genus Ralstonia) compared to healthy age-matched controls (Zhang et al. 2005). Moreover, levels of *Enterobacteriaceae* are positively associated with postural instability (Scheperjans et al. 2015). The resultant pro-inflammatory state can induce α-synuclein misfolding and contribute to PD pathology (Houser and Tansey 2017). Counteractively, long-term probiotic treatment has shown a neuroprotective effect on dopamine neurons. This is also supported by several PD animal model studies which have shown that daily probiotic supplementation (containing *Bifidobacterium bifidum, Bifidobacterium longum, Lactobacillus rhamnosus, Lactobacillus rhamnosus GG, Lactobacillus Plantarum LP28*, and *Lactococcus lactis subsp. Lactis*) could reduce motor impairments, improve balance, and improve motor coordination (Hsieh et al. 2020). To further support the role of the immune system in PD and its bidirectional communication between the gut and brain, is that inflammatory bowel disease is a risk factor for developing PD as well as anti-inflammatory therapies reduce the risk of developing PD (Peter et al. 2018).

Similar to Alzheimer's disease, PD has been linked to the activation of enteric nerves leading to inflammation and neurodegeneration. Specifically, Pathogen-Associated Molecular Patterns (PAMPs) and Damage-Associated Molecular Patterns (DAMPs) can be activated in the gut or systemically lead to the activation of microglia and astrocytes in the CNS (Tang 2018). An inflammatory response follows including polarization of microglia toward a classical pro-inflammatory M1 phenotype. PD patients' microbiota is rich in pro-inflammatory cytokines due to enhanced intestinal permeability to endotoxins (lipopolysaccharide). Moreover, damaged neurons can release DAMPs (ATP, neuromelanin, m-calpain, matrix metalloprotease 3, TLRs, α-Syn) that favor the M1 microglia phenotype and release pro-inflammatory mediators, chemokines, and cytokines like IL1β, IL-6, and TNFα (Wang et al. 2015). The released TNFα inhibits an isoform of NF-κB, called c-Rel, which usually has a neuroprotective role by inhibiting apoptosis in dopaminergic neurons (Sarnico et al. 2009). Similarly, there is a high expression of chemokines like CXCL12 in people with PD (Shimoji et al. 2009). When CXCL12 binds to its receptor, CXCR4, it activates caspase 3 which leads to apoptotic neural death and thus neurodegeneration.

Similar to M1 microglia, A1 astrocytes are polarized toward an inflammatory phenotype and are more abundant in people with PD (Liddelow et al. 2017). These A1 astrocytes release IL1α, C1q, and TNFα. The released TNFα binds dopaminergic neurons (like TNFR1/2) and activates an apoptotic pathway (HIRSCH et al. 2003). Chronic overexpression of IL1β can lead to many of the symptoms associated with PD such as dopaminergic cell death, akinesia, and glial activation (Ferrari et al. 2006). Moreover, systemic IL1β has been shown to contribute to neuronal loss in animal models through the activation of T cells (Pott Godoy et al. 2008). Moreover, there is evidence that neurons transfer α-Syn to surrounding astrocytes leading to the production of pro-inflammatory cytokines (IL-6, TNFα), intercellular adhesion molecule 1 (ICAM-1), and reactive oxygen species (Klegeris et al. 2006). α-Syn can also activate microglia within the brain and promote inflammation which contributes to the disease phenotype (Sampson et al. 2016).

With time astrocytes also adopt a senescence-associated pro-inflammatory phenotype further contributing to the disorder (Chinta et al. 2018).

It is important to note that both TNFα and IL1β impact the function of endothelial cells comprising the blood–brain barrier (Wu et al. 2017). These endothelial cells express TLRs and can elicit an immune response in response to DAMPs and PAMPs (Nagyőszi et al. 2010). Another common receptor on endothelial cells is VCAM1 which aids in immune cell adhesion and translocation. Interestingly, VCAM1 is highly expressed in patients with PD and correlates with the disease stage and motor impairment (Perner et al. 2019). Moreover, the high levels of cytokines such as TNFα, IL1β, or IFNγ in PD patients appear to increase the blood–brain barrier (BBB) permeability as indicated by a decrease in trans endothelial electrical resistance in a BBB in vitro model (Raasch et al. 2016).

10.4.4 Huntington's Disease

Huntington's disease (HD) is an inherited disease characterized by a progressive degeneration of neurons in the striatum and cortex. Specifically, a mutation in chromosome 4 that leads to an unstable CAG trinucleotide repeat expansion in exon 1 of the Huntingtin gene, creating a mutant form of the huntingtin protein (HTT) (MacDonald et al. 1993). Normally, the HTT protein protects neurons from apoptotic stimuli (Rigamonti et al. 2000). With the mutant HTT, a progressive decline in motor functions, cognition, and an increase in psychiatric disorders are observed. This can also hinder autonomic function and result in GIT, urinary, and cardiovascular issues (Aziz et al. 2010). Animal models of HD suggest that overexpression of TNF-α contributes to inflammation and cardiac dysfunction (Lin et al. 2015).

Inflammation can be detected prior to the onset of HD symptoms which may indicate a causative effect. Moreover, the disease progression correlates with high plasma levels of chemokines eotaxin-3, MIP-1β, MCP-1, and MCP-4 (Wild et al. 2011). Specifically, MCP-1 levels are proportional to declining clinical motor skills and function. Moreover, non-specific systemic markers of inflammation (IL-6 and CRP) are also elevated in HD patients (Sánchez-López et al. 2012). Similarly, high levels of IL-4, IL-10, and TNF-α are reported in people with PD (Björkqvist et al. 2008). This may be attributed to the mutant HTT which is highly expressed in microglia and monocytes (Moscovitch-Lopatin et al. 2010). This induces the NF-kB signaling pathway leading to the release of inflammatory cytokines (IL-6 and IL-8) and chemokines from astrocytes and microglia (Tai et al. 2007; Khoshnan et al. 2004). The activated microglia also correlate with higher plasma levels of IL-1β, IL-6, IL-8, and TNF-α (Politis et al. 2015). Meanwhile, the progressive neuronal death activates more inflammatory pathways leading to a feedback loop of neurodegeneration (Kaushik and Basu 2013). Again microglia appear to play a large role by releasing IL-6, MMP-9, vascular endothelial growth factor, and TGF-β (Valadão et al. 2020).

10.5 Probiotics to Combat Inflammation and Neurodegeneration

Probiotics have been studied extensively for their beneficial impact on cognition and mental state. Specifically, *Bifidobacterium bifidum* BGN4 and *Bifidobacterium longum* BORI supplementation for 12 weeks promotes mental flexibility and alleviates stress in healthy older adults (Kim et al. 2021). Similarly, *Lactobacillus acidophilus* CUL60 (NCIMB 30,157) and CUL21 (NCIMB 30,156), *Bifidobacterium lactis* CUL34 (NCIMB 30,172), and *Bifidobacterium bifidum* CUL20 (NCIMB 30,153) given for 6 weeks at 2.5×10^{10} CFU/capsule improved attention and lowered anxiety levels (Owen et al. 2014). Probiotic intake also significantly reduced depression scores and altered emotional processes. However, they did report that supplementation did not alter salivary cortisol or circulating CRP concentrations (Baião et al. 2022). Counteractively, probiotics have also been shown to improve memory, cognitive function, and biochemical and histological measures in subjects with dementia and cognitive impairment. The researchers also reported that inflammation of the central nervous system is associated with cognitive decline, and an alteration in neurotransmitter levels can be induced via the gut microbiome (Ruiz-Gonzalez et al. 2021). When investigating probiotics and their association with mental disorders like depression, similar beneficial impacts are reported. For example, *Lactobacillus Rhamnosus G* probiotic (1.9×10^9 CFU) significantly decreased depression scores, anxiety scores, hs-CRP, lipopolysaccharide, and TNF-α in comparison to placebo (Moludi et al. 2021). These findings are consistent among several studies involving a total of 526 participants, with probiotics reducing Hamilton Depression Rating Scale, CRP, IL-10, and malondialdehyde levels (Amirani et al. 2020).

Probiotics have been useful for alleviating symptoms associated with PD. For example, *Lactobacillus casei Shirota* (6.5×10^9 CFU) improves stool consistency as well as bowel habits which are typically impaired in PD patients (Castelli et al. 2020). Similarly, fermented milk, containing probiotic strains and prebiotic fiber, improved constipation (Barichella et al. 2016). *Lactobacillus acidophilus, Bifidobacterium bifidum, Lactobacillus reuteri, and Lactobacillus fermentum* improve the Unified Parkinson's Disease Rating Scale (Tamtaji et al. 2019). *Lactobacillus acidophilus, Lactobacillus casei, Lactobacillus lactis, Bifidobacterium infantis,* and *Bifidobacterium longum* improved bowel movements and whole gut transit time (Ibrahim et al. 2020). *Lactobacillus acidophilus, Lactobacillus reuteri, Lactobacillus gasseri, Lactobacillus rhamnosus, Bifidobacterium bifidum, Bifidobacterium longum, Enterococcus faecalis,* and *Enterococcus faecium* increased spontaneous bowel movements, improved stool consistency, and improved the quality of life with relation to constipation (Tan et al. 2021). However, the full potential of alleviating both motor and non-motor Parkinson's symptoms have yet to be shown and could be initiated through an anti-inflammatory pathway provided by probiotics.

In a new approach, probiotics have a large potential to mitigate the inflammation associated with neurodegeneration and aging. Probiotic supplementation

(Lactobacillus, Bifidobacterium, and Saccharomyces) significantly reduced serum concentrations of pro-inflammatory cytokines including, hs-CRP, TNF-a, IL-6, IL-12, and IL-4, but it did not influence IL-1B, IL-8, IFN-g, and IL-17 concentrations (Milajerdi et al. 2020). A significant reduction in TNF-alpha, a major component in inflammation, can be achieved using Lactobacillus. It is important to note that a reduction in CRP due to probiotic supplementation is an indicator of a reduction of systemic inflammation (Mousavi et al. 2020). Similarly, *Lactobacillus plantarum*, *Lactobacillus casei Shirota*, *Bifidobacterium lactis*, and *Lactobacillus casei* when administered for 12 weeks reduced inflammatory biomarkers in the 365 people studies (Tenorio-Jiménez et al. 2020). Probiotics have also been shown to reduce soluble vascular cell adhesion molecule 1 (sVCAM-1), IL-6, TNF-α, vascular endothelial growth factor, and thrombomodulin. Counteractively, probiotics have been shown to significantly increase IL-10 serum levels, an anti-inflammatory cytokine, in a meta-analysis including 42 randomized control trials with 2258 participants (Milajerdi et al. 2020).

Given the extensive communication between the gut and brain, probiotics may also be a key treatment for inflammation and neurodegeneration associated with aging and neurodegenerative disorders. Probiotics can help regulate kynurenine levels that contribute to dementia, Alzheimer's, and Huntington's disease pathologies (Desbonnet et al. 2008). The systemic and local inflammation present in patients with AD, PD, and HD includes immune cell activation (microglia and astrocytes) along with the release of pro-inflammatory cytokines and chemokines. To combat this, probiotics and selenium supplementation have been shown to reduce serum CRP, insulin, and LDL-cholesterol and increase glutathione as an anti-oxidant (Tamtaji et al. 2019; Krüger et al. 2020). Specifically, a reduction in CRP indicates a reduction in systemic inflammation which is beneficial for neurodegenerative disorder treatment (Sánchez-López et al. 2012). Moreover, the resultant regulation of serum LDL-cholesterol and insulin is essential for diabetes and cardiovascular disease prevention. Similarly, *Lactobacillus acidophilus, Lactobacillus casei, Bifidobacterium bifidum*, and *Lactobacillus fermentum* (2×10^9 CFU/g for each) supplementation for 12 weeks significantly improved mini-mental state examination scores in patients with AD (Akbari et al. 2016). Moreover, markers of inflammation linked to neurodegeneration such as plasma malondialdehyde, serum high-sensitivity C-reactive, and insulin resistance were reduced in the probiotic treatment group. Similarly, *Lactobacillus acidophilus, Bifidobacterium bifidum*, and *Bifidobacterium* (2×10^9 CFU/day each) supplementation for 12 weeks significantly reduced serum high-sensitivity CRP, insulin, serum triglycerides, and insulin resistance and increased oxidative capacity (Tamtaji et al. 2019). The researchers concluded that probiotic consumption for 12 weeks positively affects cognitive function and some metabolic statuses in people with AD. Another meta-analysis drew similar conclusions supplementing *Lactobacillus acidophilus, Lactobacillus casei, Bifidobacterium bifidum, Lactobacillus fermentum, Bifidobacterium longum,* and *Bifidobacterium breve A1* (Den et al. 2020). They report a significant improvement in cognition as well as a significant reduction in malondialdehyde and high-sensitivity CRP post-intervention in people with AD. Moreover, the probiotic mechanism is suggested to be through a reduction in inflammatory and oxidative biomarkers. Probiotics also assist with metabolism and vitamin absorption which can be essential given that people with vitamin D deficiency were at a 21% higher risk of developing AD compared to those with normal vitamin D serum levels (Shen and Ji 2015).

Research also suggests that the gut microbiome and consequent systemic inflammatory response depends on the individual's initial state (ex. healthy vs diseased). For example, co-morbidities like diabetes, arthritis, cirrhosis, and more altered an individual's response to probiotic supplementation (Kazemi et al. 2020). *L. acidophilus* LA- 5, *Bifidobacterium* BB- 12, *S. thermophilus* STY-31, and *L. delbrueckii bulgaricus* LBY-2 have been shown to improve levels of malondialdehyde, glutathione reductase, erythrocyte glutathione peroxidase levels, serum high-sensitivity CRP, and TNF-α levels in women with diabetes (Hajifaraji et al. 2018). Similarly, probiotics or symbiotic supplementation significantly decreases hs-CRP levels and malondialdehyde in diabetic patients compared to people receiving placebos (Zheng et al. 2019). In addition, probiotics were found to increase the total anti-oxidant capacity, nitric oxide, and glutathione levels which are essential for maintaining an anti-oxidant environment, promoting neuronal survival, and improving cognition (Zheng et al. 2019).

To combat age-related inflammation present in neurodegenerative disorders commensal bacteria may be beneficial. Probiotic formulations can improve CNS activity via modulation of inflammation and via positive interactions with the commensal gut microbiota (Lin et al. 2015). Probiotics such as *Lactobacillus* and *Bifidobacterium* may promote an anti-oxidant and anti-inflammatory environment (Nowak et al. 2019). Specifically, *L. fermentum* NCIMB 5221, *B. animalis*, and *L. plantarum* NCIMB 8826 rapidly synthesize ferulic acid (Bhathena et al. 2009; Szwajgier and Dmowska 2010; Tomaro-Duchesneau et al. 2012). Ferulic acid has been used for many years in Chinese Medicine for its ability to inhibit reactive oxygen species and for its anti-inflammatory and anti-oxidant properties (Trombino et al. 2013). Similarly, *B. fragilis* PSA has also been shown to reduce the expression of inflammatory cytokines (IL-17), IFNc, and RORct (Ochoa-Reparaz et al. 2010). *L. helveticus* R0052 also decreases inflammation and fecal corticosterone (a stress response hormone) in a mice model on a western diet (Ohland et al. 2013). Other commensal bacteria, like Lactobacillus and Bifidobacterium species, reduce oxidative stress, pro-inflammatory cytokines, and counteracted pathogenic bacterial overgrowth in PD patients (Magistrelli et al. 2019). Bifidobacterium administration is also effective in enhancing non-spatial and spatial memory (Savignac et al. 2015). *Bacillus subtilis* probiotic has been shown to inhibit α-synuclein aggregation and clear pre-formed aggregates in the established Caenorhabditis elegans model of synucleinopathy (Goya et al. 2020). A recent, clinical trial demonstrated that a probiotic formulation (given via a tablet) improved insulin resistance and sensitivity and improved motor scores (Tamtaji et al. 2019). Moreover, long-term probiotic administration (six bacterial strains) has been shown to exert neuroprotective effects on dopaminergic neurons and counteract motor impairments in a genetic PD mouse model (Hsieh et al. 2020). Bonfili et al. demonstrated that SLAB51 probiotic formulation, through modulation of the microbiota, influenced neuroprotective pathways and counteracted the progression of AD (Bonfili et al. 2017). SLAB51 has an anti-inflammatory ability which helps restore pro-survival and neuroprotective pathways. This protected dopaminergic neurons and ameliorated behavioral impairment in PD. These commensal bacteria can also contribute to short-chain fatty acids (SCFAs) production which are well-known for their potent anti-inflammatory effects (Brahe et al. 2013). Lactobacillus, Lactococcus, Streptococcus, Moreover, Pediococcus, and Enterococcus spp. all have the histidine decarboxylase gene and can produce histamine (Landete et al. 2008). Histamine is a biogenic monoamine that is produced by enterochromaffin cells in the GIT and are directly released from certain Lactobacillus spp. In the brain, histamine induces an allergic inflammatory response by increasing the production of pro-inflammatory cytokines such as IL-1a, IL-1b, and IL-6 and various chemokines. However, the action of histamine through a different receptor, namely H4R, induces an anti-inflammatory response which is particularly critical in the CNS [113]. Overall, both inflammatory cytokines and gut metabolic hormones are important therapeutic targets to consider for alleviating and even improving symptoms associated with neurodegenerative disorders, a summary of which is provided in Fig. 10.1.

Neurodegenerative disease	Alzheimer's disease	Parkinson's disease	Huntington's disease	Probiotics*
TNF-α	↑	↑	↑	↓
IL-1	↑	↑	↑	
IL-4			↑	↓
IL-6	↑	↑	↑	↓
ßIL-8			↑	
IL-10	↑			↑
IL-12	↑			↓
TGF-B	↑		↑	
IFNγ		↑		
CRP	↑	↑	↑	↓
Chemokines	↑	↑	↑	↓

Fig. 10.1 Cytokine and chemokine overexpression in neurodegenerative diseases and probiotics that can mitigate this expression. *The probiotics mentioned in this chapter include *Lactobacillus acidophilus, Lactobacilluscasei, Lactobacillus fermentum, Lactobacillus plantarum, Lactobacillus casei Shirota, Lactobacillus casei, Lactobacillus Rhamnosus G, Bifidobacterium bifidum, Bifidobacterium longum* and *Bifidobacterium breve A1, Bifidobacterium, Bifidobacterium lactis,* and Saccharomyces (Sánchez-López et al. 2012; Hajifaraji et al. 2018; Zheng et al. 2019; Den et al. 2020; Milajerdi et al. 2020; Tenorio-Jiménez et al. 2020; Moludi et al. 2021)

10.6 Conclusion or Translational Perspectives

Aging and inflammation (inflamm-aging) appear to be closely correlated and are common underlying factors in neurodegeneration and neurodegenerative disorders. Both inflammation and aging are systemic conditions that may be regulated by the gut microbiome more extensively than previously expected. The sustained inflammation along with inflammatory cytokines and chemokines present in both the gut and brain of people with neurodegenerative disorders reveals a gut-brain dysbiosis. The bidirectional axis allows for the transport of inflammatory mediators and immune cells as well as endocrine and neurological signals that can dysregulate the whole body. This dysbiosis is present in Alzheimer's, Parkinson's, Huntington's, and other neurodegenerative diseases. To counteract this inflammation, probiotics can deliver commensal bacteria that promote an anti-inflammatory state, improve gut function, promote nutrient intake, and produce metabolites that prevent inflammation. Collectively, probiotics have been shown to be beneficial in mitigating neurodegenerative symptoms and disease progression.

References

Akbari E et al (2016) Effect of probiotic supplementation on cognitive function and metabolic status in Alzheimer's disease: a randomized double-blind and controlled trial. Front Aging Neurosci 8:256. https://doi.org/10.3389/fnagi.2016.00256

Alexander GE (2004) Biology of Parkinson's disease: pathogenesis and pathophysiology of a multisystem neurodegenerative disorder. Dialogues Clin Neurosci 6(3):259–280. https://doi.org/10.31887/DCNS.2004.6.3/galexander

Amirani E et al (2020) The effects of probiotic supplementation on mental health, biomarkers of inflammation and oxidative stress in patients with psychiatric disorders: a systematic review and meta-analysis of randomized controlled trials. Complement Ther Med 49:102361. https://doi.org/10.1016/j.ctim.2020.102361

Arpaia N et al (2013) Metabolites produced by commensal bacteria promote peripheral regulatory T-cell generation. Nature 504(7480):451–455. https://doi.org/10.1038/nature12726

Aziz NA et al (2010) Autonomic symptoms in patients and pre-manifest mutation carriers of Huntington's disease. Eur J Neurol 17(8):1068–1074. https://doi.org/10.1111/j.1468-1331.2010.02973.x

Baião R et al (2022) Multispecies probiotic administration reduces emotional salience and improves mood in subjects with moderate depression: a randomised, double-blind, placebo-controlled study. Psychol Med 2022(02/07):1–11. https://doi.org/10.1017/S003329172100550X

Baik SH et al (2016) Microglia contributes to plaque growth by cell death due to uptake of amyloid β in the brain of Alzheimer's disease mouse model. Glia 64(12):2274–2290

Barichella M et al (2016) Probiotics and prebiotic fiber for constipation associated with Parkinson disease: an RCT. Neurology 87(12):1274–1280

Berer K, Krishnamoorthy G (2012) Commensal gut flora and brain autoimmunity: a love or hate affair? Acta Neuropathol 123(5):639–651

Bhathena J et al (2009) Orally delivered microencapsulated live probiotic formulation lowers serum lipids in hypercholesterolemic hamsters. J Med Food 12(2):310–319. https://doi.org/10.1089/jmf.2008.0166

Bhathena J et al (2013) Oral probiotic microcapsule formulation ameliorates non-alcoholic fatty liver disease in Bio F1B Golden Syrian hamsters. PloS one 8(3):e58394. https://doi.org/10.1371/journal.pone.0058394
Biagi E et al (2010) Through ageing, and beyond: gut microbiota and inflammatory status in seniors and centenarians. PLoS ONE 5(5):e10667
Björkqvist M et al (2008) A novel pathogenic pathway of immune activation detectable before clinical onset in Huntington's disease. J Exp Med 205(8):1869–1877. https://doi.org/10.1084/jem.20080178
Bonfili L et al (2017) Microbiota modulation counteracts Alzheimer's disease progression influencing neuronal proteolysis and gut hormones plasma levels. Sci Rep 7(1):1–21
Boyton RJ, Openshaw PJ (2002) Pulmonary defences to acute respiratory infection. Br Med Bull 61(1):1–12. https://doi.org/10.1093/bmb/61.1.1
Brahe LK, Astrup A, Larsen LH (2013) Is butyrate the link between diet, intestinal microbiota and obesity-related metabolic diseases? Obes Rev 14(12):950–959. https://doi.org/10.1111/obr.12068
Cantu-Jungles TM, Rasmussen HE, Hamaker BR (2019) Potential of prebiotic butyrogenic fibers in Parkinson's disease. Front Neurol:663
Castelli V et al (2019) Neuronal cells rearrangement during aging and neurodegenerative disease: metabolism, oxidative stress and organelles dynamic. Front Mol Neurosci 12:132
Castelli V et al (2020) Effects of the probiotic formulation SLAB51 in in vitro and in vivo Parkinson's disease models. Aging 12(5):4641–4659. https://doi.org/10.18632/aging.102927
Chandra RK (2000) Food allergy and nutrition in early life: implications for later health. Proc Nutr Soc 59(2):273–277. https://doi.org/10.1017/s0029665100000306
Chang C-S, Kao C-Y (2019) Current understanding of the gut microbiota shaping mechanisms. J Biomed Sci 26(1):59. https://doi.org/10.1186/s12929-019-0554-5
Chang R, Yee K-L, Sumbria RK (2017) Tumor necrosis factor α inhibition for Alzheimer's disease. J Central Nervous Syst Disease 9:1179573517709278
Chinta SJ et al (2018) Cellular senescence is induced by the environmental neurotoxin paraquat and contributes to neuropathology linked to Parkinson's disease. Cell Rep 22(4):930–940
Chong Y (1997) Effect of a carboxy-terminal fragment of the Alzheimer's amyloid precursor protein on expression of proinflammatory cytokines in rat glial cells. Life Sci 61(23):2323–2333
Combs CK et al (2001) β-Amyloid stimulation of microglia and monocytes results in TNFα-dependent expression of inducible nitric oxide synthase and neuronal apoptosis. J Neurosci 21(4):1179–1188
Corey S et al (2019) Gut microbiome: lactation, childbirth, lung dysbiosis, animal modeling, stem cell treatment, and CNS disorders. CNS Neurol Disorders-Drug Targets (Formerly Current Drug Targets-CNS & Neurological Disorders), 18(9):687–694
Dantzer R et al (2008) From inflammation to sickness and depression: when the immune system subjugates the brain. Nat Rev Neurosci 9(1):46–56
Darby TM et al (2019) Lactococcus Lactis Subsp. cremoris is an efficacious beneficial bacterium that limits tissue injury in the intestine. Iscience 12:356–367
Den H et al (2020) Efficacy of probiotics on cognition, and biomarkers of inflammation and oxidative stress in adults with Alzheimer's disease or mild cognitive impairment a meta-analysis of randomized controlled trials. Aging 12(4):4010–4039. https://doi.org/10.18632/aging.102810
Desbonnet L et al (2008) The probiotic Bifidobacteria infantis: an assessment of potential antidepressant properties in the rat. J Psychiatr Res 43(2):164–174. https://doi.org/10.1016/j.jpsychires.2008.03.009
Diamond B et al (2011) It takes guts to grow a brain: increasing evidence of the important role of the intestinal microflora in neuro-and immune-modulatory functions during development and adulthood. BioEssays 33(8):588–591
Dinan TG, Cryan JF (2012) Regulation of the stress response by the gut microbiota: implications for psychoneuroendocrinology. Psychoneuroendocrinology 37(9):1369–1378

Dobson A et al (2012) Bacteriocin production: a Probiotic Trait? Appl Environ Microbiol 78(1):1–6. https://doi.org/10.1128/AEM.05576-11
Duvallet C et al (2017) Meta-analysis of gut microbiome studies identifies disease-specific and shared responses. Nat Commun 8(1):1–10
Enaud R et al (2020) The Gut-Lung axis in health and respiratory diseases: a place for inter-organ and inter-kingdom crosstalks. Front Cell Infect Microbiol 10(February):1–11. https://doi.org/10.3389/fcimb.2020.00009
Faden AI, Loane DJ (2015) Chronic neurodegeneration after traumatic brain injury: Alzheimer disease, chronic traumatic encephalopathy, or persistent neuroinflammation? Neurother J Am Soc Exp NeuroTher 12(1):143–150. https://doi.org/10.1007/s13311-014-0319-5
Ferrari CC et al (2006) Progressive neurodegeneration and motor disabilities induced by chronic expression of IL-1β in the substantia nigra. Neurobiol Dis 24(1):183–193. https://doi.org/10.1016/j.nbd.2006.06.013
Ferreira ST et al (2014) Inflammation, defective insulin signaling, and neuronal dysfunction in Alzheimer's disease. Alzheimer's and Dementia 10(1):S76–S83
Filpa V et al (2016) Role of glutamatergic neurotransmission in the enteric nervous system and brain-gut axis in health and disease. Neuropharmacology 111:14–33. https://doi.org/10.1016/j.neuropharm.2016.08.024
Forsythe P, Bienenstock J, Kunze WA (2014) Vagal pathways for microbiome-brain-Gut axis communication BT-microbial endocrinology. In: Lyte M, Cryan JF (eds) The Microbiota-Gut-brain axis in health and disease. Springer, New York, pp 115–133. https://doi.org/10.1007/978-1-4939-0897-4_5
Franceschi C et al (2018) The continuum of aging and age-related diseases: common mechanisms but different rates. Front Med:61
Gao J et al (2018) Impact of the gut microbiota on intestinal immunity mediated by tryptophan metabolism. Front Cell Infect Microbiol:13
Goya ME et al (2020) Probiotic Bacillus subtilis protects against α-Synuclein aggregation in C. elegans. Cell Rep 30(2):367-380.e7. https://doi.org/10.1016/j.celrep.2019.12.078
Grammas P (2011) Neurovascular dysfunction, inflammation and endothelial activation: implications for the pathogenesis of Alzheimer's disease. J Neuroinflammation 8(1):1–12
Grammas P, Ovase R (2002) Cerebrovascular transforming growth factor-β contributes to inflammation in the Alzheimer's disease brain. Am J Pathol 160(5):1583–1587
Guillot-Sestier M-V et al (2015) Il10 deficiency rebalances innate immunity to mitigate Alzheimer-like pathology. Neuron 85(3):534–548
Guo Y et al (2021) A diet high in sugar and fat influences neurotransmitter metabolism and then affects brain function by altering the gut microbiota. Transl Psychiatry 11(1):328. https://doi.org/10.1038/s41398-021-01443-2
Gurung M et al (2020) Role of gut microbiota in type 2 diabetes pathophysiology. eBioMedicine:51. https://doi.org/10.1016/j.ebiom.2019.11.051
Hajifaraji M et al (2018) Effect of probiotic supplements in women with gestational diabetes mellitus on inflammation and oxidative stress biomarkers: a randomized clinical trial. Asia Pac J Clin Nutr 27(3):581–591
Harrison JK et al (1998) Role for neuronally derived fractalkine in mediating interactions between neurons and CX3CR1-expressing microglia. Proc Natl Acad Sci 95(18):10896–10901
Hirsch EC et al (2003) The role of glial reaction and inflammation in Parkinson's Disease. Ann N Y Acad Sci 991(1):214–228. https://doi.org/10.1111/j.1749-6632.2003.tb07478.x
Horai R et al (2015) Microbiota-dependent activation of an autoreactive T cell receptor provokes autoimmunity in an immunologically privileged site. Immunity 43(2):343–353. https://doi.org/10.1016/j.immuni.2015.07.014
Houser MC, Tansey MG (2017) The gut-brain axis: is intestinal inflammation a silent driver of Parkinson's disease pathogenesis? npj Parkinson's Disease 3.[Europe PMC free article]
Hsieh T-H, Kuo C-W, Hsieh K-H, Shieh M-J, Peng C-W, Chen Y-C, Chang Y-L, Huang Y-Z, Chen C-C, Chang P-K (2020a) Probiotics alleviate the progressive deterioration of motor functions in a mouse model of Parkinson's disease. Brain Sci 10(4):206

Ibrahim A et al (2020) Multi-strain probiotics (Hexbio) containing MCP BCMC strains improved constipation and gut motility in Parkinson's disease: A randomised controlled trial. PLoS ONE 15(12):e0244680

Iradj S et al (2019) Colorectal cancer-associated microbiota contributes to oncogenic epigenetic signatures. Proc Natl Acad Sci 116(48):24285–24295. https://doi.org/10.1073/pnas.1912129116

Kahouli I et al (2017) Design and validation of an orally administrated active L. fermentum-L. acidophilus probiotic formulation using colorectal cancer Apc (Min/+) mouse model. Appl Microbiol Biotechnol 101(5):1999–2019. https://doi.org/10.1007/s00253-016-7885-x

Kau AL et al (2011) Human nutrition, the gut microbiome and the immune system. Nature 474 (7351):327–336. https://doi.org/10.1038/nature10213

Kaushik DK, Basu A (2013) A friend in need may not be a friend indeed: role of microglia in neurodegenerative diseases. CNS Neurol Disorders Drug Targets 12(6):726–740. https://doi.org/10.2174/18715273113126660170

Kazemi A et al (2020) Effect of probiotic and synbiotic supplementation on inflammatory markers in health and disease status: A systematic review and meta-analysis of clinical trials. Clin Nutr 39(3):789–819. https://doi.org/10.1016/j.clnu.2019.04.004

Keren-Shaul H et al (2017) A unique microglia type associated with restricting development of Alzheimer's disease. Cell 169(7):1276–1290

Khoshnan A et al (2004) Activation of the IkappaB kinase complex and nuclear factor-kappaB contributes to mutant huntingtin neurotoxicity. J Neurosci Off J Soc Neurosci 24(37):7999–8008. https://doi.org/10.1523/JNEUROSCI.2675-04.2004

Kim C-S et al (2021) Probiotic supplementation improves cognitive function and mood with changes in gut microbiota in community-dwelling older adults: a randomized, double-blind, placebo-controlled, multicenter trial. J Gerontol Ser A 76(1):32–40. https://doi.org/10.1093/gerona/glaa090

Kinney JW et al (2018) Inflammation as a central mechanism in Alzheimer's disease. Alzheimer Dementia Transl Res Clin Interv 4:575–590. https://doi.org/10.1016/j.trci.2018.06.014

Klegeris A et al (2006) Alpha-synuclein and its disease-causing mutants induce ICAM-1 and IL-6 in human astrocytes and astrocytoma cells. FASEB J 20(12):2000–2008. https://doi.org/10.1096/fj.06-6183com

Krüger JF et al (2020) Probiotics for dementia: a systematic review and meta-analysis of randomized controlled trials. Nutr Rev. https://doi.org/10.1093/nutrit/nuaa037

Krukowski K et al (2018) Traumatic brain injury in aged mice induces chronic microglia activation, synapse loss, and complement-dependent memory deficits. Int J Mol Sci 19 (12):3753

Kuhn SA et al (2004) Microglia express GABAB receptors to modulate interleukin release. Mol Cell Neurosci 25(2):312–322. https://doi.org/10.1016/j.mcn.2003.10.023

Kustrimovic N, Marino F, Cosentino M (2019) Peripheral immunity, immunoaging and neuroinflammation in parkinson's disease. Curr Med Chem:3719–3753. https://doi.org/10.2174/0929867325666181009161048

Landete JM et al (2008) Updated molecular knowledge about histamine biosynthesis by bacteria. Crit Rev Food Sci Nutr 48(8):697–714. https://doi.org/10.1080/10408390701639041

Lee M, Schwab C, Mcgeer PL (2011) Astrocytes are GABAergic cells that modulate microglial activity. Glia 59(1):152–165. https://doi.org/10.1002/glia.21087

Li M, Wang B, Zhang M, Rantalainen M, Wang S, Zhou H, Zhang Y, Shen J, Pang X, Zhang M, Wei H, Chen Y, Lu H, Zuo J, Su M, Qiu Y, Jia W, Xiao C, Smith LM, Yang S, Holmes E, Tang H, Zhao G, Nicholson JK, Li L, Zhao L (2008) Symbiotic gut microbes modulate human metabolic phenotypes. Proc Nat Acad Sci 105(6):2117–2122. https://doi.org/10.1073/pnas.0712038105

Liddelow SA et al (2017) Neurotoxic reactive astrocytes are induced by activated microglia. Nature 541(7638):481–487. https://doi.org/10.1038/nature21029

Limage R et al (2020) TiO2 nanoparticles and commensal bacteria alter mucus layer thickness and composition in a gastrointestinal tract model. Small 16(21):2000601. https://doi.org/10.1002/smll.202000601

Lin C-L et al (2015) Oral treatment with herbal formula B307 alleviates cardiac failure in aging R6/2 mice with Huntington's disease via suppressing oxidative stress, inflammation, and apoptosis. Clin Interv Aging 10:1173

MacDonald ME et al (1993) A novel gene containing a trinucleotide repeat that is expanded and unstable on Huntington's disease chromosomes. Cell 72(6):971–983. https://doi.org/10.1016/0092-8674(93)90585-E

Maes M et al (2012) Increased IgA responses to the LPS of commensal bacteria is associated with inflammation and activation of cell-mediated immunity in chronic fatigue syndrome. J Affect Disord 136(3):909–917

Magistrelli L et al (2019) Probiotics may have beneficial effects in parkinson's disease: in vitro evidence. Front Immunol 10:969. https://doi.org/10.3389/fimmu.2019.00969

Milajerdi A et al (2020) The effect of probiotics on inflammatory biomarkers: a meta-analysis of randomized clinical trials. Eur J Nutr 59(2):633–649. https://doi.org/10.1007/s00394-019-01931-8

Ming X et al (2012) Metabolic perturbance in autism spectrum disorders: a metabolomics study. J Proteome Res 11(12):5856–5862

Moludi J et al (2021) The effects of co-administration of probiotics and prebiotics on chronic inflammation, and depression symptoms in patients with coronary artery diseases: a randomized clinical trial. Nutr Neurosci:1–10. https://doi.org/10.1080/1028415X.2021.1889451

Moscovitch-Lopatin M et al (2010) Optimization of an HTRF assay for the detection of soluble mutant huntingtin in human buffy coats: a potential biomarker in blood for huntington disease. PLoS Currents 2:RRN1205. https://doi.org/10.1371/currents.RRN1205

Mousavi SN, Saboori S, Asbaghi O (2020) Effect of daily probiotic yogurt consumption on inflammation: A systematic review and meta-analysis of randomized Controlled Clinical trials. Obesity Med 18:100221. https://doi.org/10.1016/j.obmed.2020.100221

Mulak A et al (2019) Fecal calprotectin as a marker of the gut immune system activation is elevated in Parkinson's Disease. Front Neurosci. https://www.frontiersin.org/article/https://doi.org/10.3389/fnins.2019.00992.

Nagpal R et al (2018) Gut microbiome and aging: Physiological and mechanistic insights. Nutr Healthy Aging 4(4):267–285

Nagyőszi P et al (2010) Expression and regulation of toll-like receptors in cerebral endothelial cells. Neurochem Int 57(5):556–564. https://doi.org/10.1016/j.neuint.2010.07.002

Nowak A, Paliwoda A, Błasiak J (2019) Anti-proliferative, pro-apoptotic and anti-oxidative activity of Lactobacillus and Bifidobacterium strains: a review of mechanisms and therapeutic perspectives. Crit Rev Food Sci Nutr 59(21):3456–3467

O'Malley D et al (2011) Do interactions between stress and immune responses lead to symptom exacerbations in irritable bowel syndrome? Brain Behav Immun 25(7):1333–1341. https://doi.org/10.1016/j.bbi.2011.04.009

Ochoa-Repáraz J et al (2010) A polysaccharide from the human commensal Bacteroides fragilis protects against CNS demyelinating disease. Mucosal Immunol 3(5):487–495. https://doi.org/10.1038/mi.2010.29

Ohland CL et al (2013) Effects of lactobacillus helveticus on murine behavior are dependent on diet and genotype and correlate with alterations in the gut microbiome. Psychoneuroendocrinology 38(9):1738–1747. https://doi.org/10.1016/j.psyneuen.2013.02.008

OMS (2020) 'Dementia @ Www.Who.Int. Demencia. https://www.who.int/es/news-room/fact-sheets/detail/dementia

Owen L et al (2014) A double blind, placebo controlled, randomised pilot trial examining the effects of probiotic administration on mood and cognitive function. Proc Nutr Soc. 2014/05/20, 73(OCE1):E29. https://doi.org/10.1017/S0029665114000433

Patarca-Montero R et al (2001) Cytokine and other immunologic markers in chronic fatigue syndrome and their relation to neuropsychological factors. Appl Neuropsychol 8(1):51–64

Perner C et al (2019) Plasma VCAM1 levels correlate with disease severity in Parkinson's disease. J Neuroinflammation 16(1):1–5

Perry RT et al (2001) The role of TNF and its receptors in Alzheimer's disease. Neurobiol Aging 22(6):873–883

Peter I et al (2018) Anti–Tumor Necrosis Factor Therapy and Incidence of Parkinson Disease Among Patients With Inflammatory Bowel Disease. JAMA Neurol 75(8):939–946. https://doi.org/10.1001/jamaneurol.2018.0605

Petra AI et al (2015) Gut-microbiota-brain axis and its effect on neuropsychiatric disorders with suspected immune dysregulation. Clin Ther 37(5):984–995

Politis M et al (2015) Increased central microglial activation associated with peripheral cytokine levels in premanifest Huntington's disease gene carriers. Neurobiol Dis 83:115–121. https://doi.org/10.1016/j.nbd.2015.08.011

Pott Godoy MC et al (2008) Central and systemic IL-1 exacerbates neurodegeneration and motor symptoms in a model of Parkinson's disease. Brain 131(7):1880–1894. https://doi.org/10.1093/brain/awn101

Qin Y, Havulinna AS, Liu Y, Jousilahti P, Ritchie SC, Tokolyi A, Sanders JG, Valsta L, Brożyńska M, Zhu Q, Tripathi A, Vázquez-Baeza Y, Loomba R, Cheng S, Jain M, Niiranen T, Lahti L, Knight R, Salomaa V, Inouye M, Méric G (2022) Combined effects of host genetics and diet on human gut microbiota and incident disease in a single population cohort. Nat Genet 54(2):134–142. https://doi.org/10.1038/s41588-021-00991-z

Raasch M et al (2016) An integrative microfluidically supported in vitro model of an endothelial barrier combined with cortical spheroids simulates effects of neuroinflammation in neocortex development. Biomicrofluidics 10(4):44102

Rhee SH, Pothoulakis C, Mayer EA (2009) Principles and clinical implications of the brain–gut–enteric microbiota axis. Nat Rev Gastroenterol Hepatol 6(5):306–314. https://doi.org/10.1038/nrgastro.2009.35

Rigamonti D et al (2000) Wild-type huntingtin protects from apoptosis upstream of caspase-3. J Neurosci Off J Soc Neurosci 20(10):3705–3713. https://doi.org/10.1523/JNEUROSCI.20-10-03705.2000

Romano S et al (2021) Meta-analysis of the Parkinson's disease gut microbiome suggests alterations linked to intestinal inflammation. npj Parkinson's Disease 7(1):27. https://doi.org/10.1038/s41531-021-00156-z

Romijn JA et al (2008) Gut-brain axis. Curr Opin Clin Nutr Metabolic Care 11(4):518–521. https://doi.org/10.1097/MCO.0b013e328302c9b0

Rook GAW, Lowry CA, Raison CL (2011) Lymphocytes in neuroprotection, cognition and emotion: is intolerance really the answer? Brain Behav Immun 25(4):591–601

Rothaug M, Becker-Pauly C, Rose-John S (2016) The role of interleukin-6 signaling in nervous tissue. Biochimica et Biophysica Acta (BBA)-Molecular Cell Res 1863(6):1218–1227

Ruddick JP et al (2006) Tryptophan metabolism in the central nervous system: medical implications. Expert Rev Mol Med 8(20):1–27

Ruiz-Gonzalez C et al (2021) Effects of probiotics supplementation on dementia and cognitive impairment: A systematic review and meta-analysis of preclinical and clinical studies. Prog Neuropsychopharmacol Biol Psychiatry 108:110189. https://doi.org/10.1016/j.pnpbp.2020.110189

Salanitro AH et al (2012) Inflammatory biomarkers as predictors of hospitalization and death in community-dwelling older adults. Arch Gerontol Geriatr 54(3):e387–e391. https://doi.org/10.1016/j.archger.2012.01.006

Sampson TR et al (2016) Gut microbiota regulate motor deficits and neuroinflammation in a model of Parkinson's disease. Cell 167(6):1469–1480

Sánchez-López F et al (2012) Oxidative stress and inflammation biomarkers in the blood of patients with Huntington's disease. Neurol Res 34(7):721–724. https://doi.org/10.1179/1743132812Y.0000000073

Sarnico I et al (2009) NF-kappaB dimers in the regulation of neuronal survival. Int Rev Neurobiol 85:351–362
Savignac HM et al (2015) Bifidobacteria modulate cognitive processes in an anxious mouse strain. Behav Brain Res 287:59–72
Scheperjans F et al (2015) Gut microbiota are related to Parkinson's disease and clinical phenotype. Mov Disord Off J Mov Disord Soc 30(3):350–358. https://doi.org/10.1002/mds.26069
Schroeder BO (2019) Fight them or feed them: how the intestinal mucus layer manages the gut microbiota. Gastroenterology Report 7(1):3–12. https://doi.org/10.1093/gastro/goy052
Selkoe DJ (1994) Normal and abnormal biology of the beta-amyloid precursor protein. Annu Rev Neurosci 17(1):489–517
Sharma S, Taliyan R (2015) Targeting histone deacetylases: a novel approach in Parkinson's disease. Parkinson's Disease 2015
Shen L, Ji H-F (2015) Vitamin D deficiency is associated with increased risk of Alzheimer's disease and dementia: evidence from meta-analysis. Nutr J 14(1):76. https://doi.org/10.1186/s12937-015-0063-7
Sheng JG et al (1998) Progressive neuronal injury associated with amyloid plaque formation in Alzheimer disease. J Neuropathol Exp Neurol 57(7):714–717
Shimoji M et al (2009) CXCR4 and CXCL12 Expression is Increased in the Nigro-Striatal System of Parkinson's Disease. Neurotox Res 16(3):318–328. https://doi.org/10.1007/s12640-009-9076-3
Smith PD et al (2011) Intestinal macrophages and response to microbial encroachment. Mucosal Immunol 4(1):31–42. https://doi.org/10.1038/mi.2010.66
Srikantha P, Mohajeri MH (2019) The possible role of the microbiota-gut-brain-axis in autism spectrum disorder. Int J Mol Sci. https://doi.org/10.3390/ijms20092115
Szczepanik AM et al (2001) IL-4, IL-10 and IL-13 modulate A beta(1–42)-induced cytokine and chemokine production in primary murine microglia and a human monocyte cell line. J Neuroimmunol 113(1):49–62. https://doi.org/10.1016/s0165-5728(00)00404-5
Szwajgier D, Dmowska K (2010) Novel ferulic acid esterases from bifidobacterium sp. produced on selected synthetic and natural carbon sources [Nowe esterazy kwasu ferulowego produkowane przez bakterie Z rodzaju bifidobacterium Z użyciem wybranych syntetycznych I naturalnych źródeł we{. Acta Scientiarum Polonorum, Technologia Alimentaria 9:305–318
Tai YF et al (2007) Microglial activation in presymptomatic Huntington's disease gene carriers. Brain J Neurol 130(7):1759–1766. https://doi.org/10.1093/brain/awm044
Tamtaji OR, Taghizadeh M et al (2019) Clinical and metabolic response to probiotic administration in people with Parkinson's disease: a randomized, double-blind, placebo-controlled trial. Clin Nutr (Edinburgh, Scotland) 38(3):1031–1035. https://doi.org/10.1016/j.clnu.2018.05.018
Tamtaji OR, Heidari-soureshjani R et al (2019) Probiotic and selenium co-supplementation, and the effects on clinical, metabolic and genetic status in Alzheimer's disease: a randomized, double-blind, controlled trial. Clin Nutr 38(6):2569–2575. https://doi.org/10.1016/j.clnu.2018.11.034
Tan AH et al (2021) Probiotics for constipation in Parkinson disease: a randomized placebo-controlled study. Neurology 96(5):e772–e782
Tang Y (2018) Editorial: microglial polarization in the pathogenesis and therapeutics of neurodegenerative diseases. Front Aging Neurosci:154. https://doi.org/10.3389/fnagi.2018.00154
Tenorio-Jiménez C et al (2020) Effects of probiotics on metabolic syndrome: a systematic review of randomized clinical trials. Nutrients. https://doi.org/10.3390/nu12010124
Ticinesi A et al (2018) Gut microbiota, cognitive frailty and dementia in older individuals: a systematic review. Clin Interv Aging 13:1497
Tomaro-Duchesneau C et al (2012) Probiotic ferulic acid esterase active lactobacillus fermentum NCIMB 5221 APA microcapsules for oral delivery: preparation and in vitro characterization. Pharmaceuticals (basel, Switzerland) 5(2):236–248. https://doi.org/10.3390/ph5020236

Tomaro-Duchesneau C et al (2014) Cholesterol assimilation by Lactobacillus probiotic bacteria: an in vitro investigation. BioMed Res Int:380316. https://doi.org/10.1155/2014/380316
Torres NI et al (2013) Safety, formulation and in vitro antiviral activity of the antimicrobial peptide subtilosin against herpes simplex virus type 1. Probiotics Antimicrob Proteins 5(1):26–35. https://doi.org/10.1007/s12602-012-9123-x
Trombino S et al (2013) Trans-ferulic acid-based solid lipid nanoparticles and their antioxidant effect in rat brain microsomes. Colloids Surf B Biointerfaces 109:273–279. https://doi.org/10.1016/j.colsurfb.2013.04.005
Valadão PAC et al (2020) Inflammation in Huntington's disease: a few new twists on an old tale. J Neuroimmunol 348. https://doi.org/10.1016/j.jneuroim.2020.577380
Van Laar T, Boertien JM, Herranz AH (2019) Faecal transplantation, pro-and prebiotics in Parkinson's disease; hope or hype? J Parkinson's Dis 9(s2):S371–S379
Vemuri R et al (2018) Gut microbial changes, interactions, and their implications on human lifecycle: an ageing perspective. BioMed Res Int
Villanueva-Millán MJ, Pérez-Matute P, Oteo JA (2015) Gut microbiota: a key player in health and disease. A review focused on obesity. J Physiol Biochem 71(3):509–525. https://doi.org/10.1007/s13105-015-0390-3
Vivekanantham S et al (2015) Neuroinflammation in Parkinson's disease: role in neurodegeneration and tissue repair. Int J Neurosci 125(10):717–725. https://doi.org/10.3109/00207454.2014.982795
Wang WY et al (2015) Role of pro-inflammatory cytokines released from microglia in Alzheimer's disease. Ann Transl Med 3(10). https://atm.amegroups.com/article/view/6546
Wild E et al (2011) Abnormal peripheral chemokine profile in Huntington's disease. PLoS Currents 3:RRN1231. https://doi.org/10.1371/currents.RRN1231
Wisniewski HM, Moretz RC, Lossinsky AS (1981) Evidence for induction of localized amyloid deposits and neuritic plaques by an infectious agent. Ann Neurol off J Am Neurol Assoc Child Neurol Soc 10(6):517–522
Wu F, Liu L, Zhou H (2017) Endothelial cell activation in central nervous system inflammation. J Leukoc Biol 101(5):1119–1132
Yap IKS et al (2010) Urinary metabolic phenotyping differentiates children with autism from their unaffected siblings and age-matched controls. J Proteome Res 9(6):2996–3004
Yates SL et al (2000) Amyloid β and amylin fibrils induce increases in proinflammatory cytokine and chemokine production by THP-1 cells and murine microglia. J Neurochem 74(3):1017–1025
Zhang W et al (2005) Aggregated alpha-synuclein activates microglia: a process leading to disease progression in Parkinson's disease. FASEB J Off Publ Fed Am Soc Exp Biol 19(6):533–542. https://doi.org/10.1096/fj.04-2751com
Zheng HJ et al (2019) The effect of probiotic and synbiotic supplementation on biomarkers of inflammation and oxidative stress in diabetic patients: a systematic review and meta-analysis of randomized controlled trials. Pharmacol Res 142:303–313. https://doi.org/10.1016/j.phrs.2019.02.016
Zujovic V et al (2000) Fractalkine modulates TNF-α secretion and neurotoxicity induced by microglial activation. Glia 29(4):305–315

Chapter 11
Lung and Gut Microbiota Interactions with Air Pollution and Aging in Human Chronic Diseases

Afzaal Nadeem Mohammed and Jagjit Singh Yadav

Abstract Human body harbors trillions of microbes that constitute the human Microbiota (aka Microbiome) distributed across various organs. This chapter focuses on gut and lung microbiota. The microbiota diversity plays an essential role in regulating the body's essential functions by sharing a symbiotic relationship with the host. Any changes in the microbial diversity (aka dysbiosis) may disrupt host homeostasis leading to various diseases. Microbiota diversity in both lung and gut may get altered by environmental factors such as air pollution or by natural biological phenomena such as aging. While air pollution directly interacts with lung microbiota, it also impacts gut microbiota conceivably via a lung–gut axis. Epidemiological studies have shown that human populations with underdeveloped microbiota such as that seen in children or older population, are vulnerable to developing chronic diseases such as respiratory, cardiovascular, and metabolic diseases and the effect is compounded in those with higher exposure to air pollution. Mechanistically, aging and air pollution are two common factors that may cause a compromised gut and lung mucosal epithelial barrier resulting in leaking of microbes and microbial products into systemic circulation. This eventually triggers a crosstalk between these and other organs thereby leading to a higher risk of developing chronic diseases.

Keywords Microbiota · Air pollution · Aging · Aging and microbiota · Air pollution and microbiota · Microbiota and pulmonary diseases · Microbiota and chronic diseases

---

A. N. Mohammed · J. S. Yadav (✉)
Pulmonary Pathogenesis and Immunotoxicology Laboratory, Department of Environmental and Public Health Sciences, College of Medicine, University of Cincinnati, Cincinnati, OH 45267-0056, USA
e-mail: yadavjs@ucmail.uc.edu

F. Marotta (ed.), *Gut Microbiota in Aging and Chronic Diseases*, Healthy Ageing and Longevity 17, https://doi.org/10.1007/978-3-031-14023-5_11

11.1 Introduction

Recent advancements in the next-generation sequencing technology have unraveled the enormous diversity in the microbiota, designated as "Microbiome" (genomic nomenclature), associated with the human body and its diverse functions relevant to the host health. The physiological importance of microbiota (interchangeably used with the term Microbiome) can be imagined from the fact that the number of microbial cells in the human body outnumbers the total number of human cells (Bander et al. 2020; Salazar et al. 2020). Though historically neglected for their role in human health and disease, recent clinical, epidemiological, and experimental studies have linked microbiota to a variety of diseases such as metabolic, pulmonary, neurodevelopmental, and other disorders and diseases. The human microbiome consists of trillions of microbial species ("commensals") that share a symbiotic relationship with the host. The lion's share of human microbiota is associated with the gut where they are involved in maintaining host physiology by performing essential functions such as fermentation of dietary fibers into absorbable metabolites, synthesis of essential vitamins, removal of toxic compounds, and protection against infections by out-competing the invading microbes (Heintz-Buschart and Wilmes 2018). Considering such diverse role of microbiota in host defense, microbiome research has been gaining much scientific interest making this topic one of the most extensively studied and cited in the recent literature (Bufard 2020).

Microbial colonization of the human body starts immediately after birth, though some studies have indicated the presence of bacteria in amniotic fluid and umbilical cord (Nagpal et al. 2018), suggesting the initiation of colonization before birth. During infancy, microbiota undergoes a series of changes in composition and functions, reaching a stable pattern at 2–3 years of age (Salazar et al. 2014; Lozupone et al. 2012). A stable microbiota is essentially required for maintaining the normal physiological functions of the body. As the human system grows older, there are gradual changes associated with the composition of microbiota, becoming prominent at senescence (Lozupone et al. 2012). Senescence is the irreversible growth arrest of individual mitotic cells and one of the drivers of age-related tissue dysfunction and numerous age-related diseases (Childs et al. 2015). Globally, older population poses a significant challenge to the healthcare system due to the frequent demand for hospitalization and primary care (Gjestsen et al. 2018). The higher rate of hospitalizations may be associated with several factors such as living conditions, socio-economic status, and/or social deprivation (Landi et al. 2004). In addition, microbial dysbiosis associated with older age may also be a determining factor for aging-related complications and susceptibility to diseases. Indeed, emerging studies have emphasized that microbiome changes at senescence are responsible for higher susceptibility of the older population to developing chronic non-communicable diseases (Bana and Cabreiro 2019; Nagpal et al. 2018; Zhou et al. 2018).

Besides aging, air pollution has been identified as a major factor contributing to the development of various chronic diseases in human population. Globally, human sub-populations residing in the neighborhoods with greater exposure to air pollutants, are at a higher risk of developing non-communicable diseases (Kaufman et al. 2020). In the United States, the National Ambient Air Quality Standards (NAAQS) defined by Clean Air Act have led to a 74% reduction in the emission of the six criteria pollutants (Carbon Monoxide, Lead, Carbon dioxide, Ozone, Sulfur dioxide, and ambient Particulate Matter) between 1970 and 2018 (Kaufman et al. 2020). Despite stringent regulations and improvements in the air quality standards over the last three decades, poor air quality remains one of the main reasons for developing cardiovascular diseases (CVD). There are increasing number of studies emphasizing air pollution as a prominent factor responsible for altering the human microbiota composition leading to allergic, inflammatory, and/or cardiovascular diseases (Bailey et al. 2020; Alderete et al. 2018; Salim et al. 2013). Moreover, air pollution-related cardiovascular complications are more prevalent in elderly population with altered microbiota.

Microbiota being a common target of both air pollution exposure and aging, studying the dynamical changes in the microbiota in these two conditions may provide the knowledgebase necessary for establishing precautionary measures/interventions, policy implementations, and therapeutic innovations. This would reduce the risk of developing chronic diseases and eventually improve the quality of life.

11.2 Gut and Lung Microbiota

Microbiota in the human body is widespread inhabiting various organs, particularly the mucosal surfaces. The gastrointestinal tract is the primary organ harboring most of the body's microbiome on the gut mucosa. The total microbial content in human gut is estimated as about 10^{14} bacterial cells consisting of up to 1000 different species (Salazar et al. 2020). The gut microbiota is predominantly composed of three bacterial divisions namely, Firmicutes, Bacteroidetes, and Actinobacteria comprising more than 90% of the gut microbial population (Gerard 2016). The composition of gut microbiota is critical in defining the health status because of the interaction of microbiota and its metabolites with the host at different levels (Salazar et al. 2020). Previous studies have addressed the role microbiota plays in regulating normal metabolism (Leustean et al. 2018; Young and Kweon 2016). The alterations in the metabolic status of the body due to lifestyle, environmental factors, or natural phenomena are known to drastically increase the risk of developing inflammatory or metabolic disorders including diabetes and cardiovascular diseases (Cani et al. 2007). Another important organ inhabiting diverse mucosal microbiota is lungs. Until recently, lungs were considered as sterile organs. After the discovery of more sensitive and culture-independent techniques, it has been revealed that lung indeed is a microbiome-rich site. However, as compared to the GI tract, the microbiota composition of lungs is very distinct and lower in density (Khatiwada and Subedi 2020). A healthy human lung contains 10–100 bacteria per 1000 lung cells (Enaud et al. 2020). The lung microbial community is majorly composed of bacterial phyla such as Proteobacteria, Bacteroidetes, Cyanobacteria, and Firmicutes with Proteobacteria being the dominant phylum with a total abundance of nearly 90% (Li et al. 2020). In addition to the bacterial species, advanced molecular techniques have also evidenced the presence of non-bacterial organisms such as viruses (virome) and fungi (mycobiome). Since the concept of mycobiome emerged, it has been gaining scientific interest due to several reasons (Huffnagle and Noverr 2013). Lung mycobiome is particularly gaining attention because of its clinical significance in chronic respiratory diseases (Nguyen et al. 2015). The respiratory tract is the main port of entry to numerous airborne particles including fungal spores. During the allergy season, there are 50,000 spores per cubic meter of air giving rise to various inflammatory and allergic disorders (Pashely et al. 2012). The unique composition of microbiota (including mycobiota) in an individual defines the lung function and susceptibility for developing allergic and inflammatory diseases (Nguyen et al. 2015).

11.3 Microbiota in Aging and Aging-Associated Diseases

11.3.1 Microbiome in Aging

Aging is a natural phenomenon characterized by changes in physiological, psychological, environmental, and social behaviors, often leading to clinical manifestations. Statistically, patients at older ages are more frequently diagnosed with multi-morbid conditions compared to patients at young ages (van den Akker et al. 1998). Such co-occurrence of multiple chronic conditions ("multi-morbidity") is receiving increasing attention in current medical and scientific world because the "1-patient 1-disease" paradigm no longer fits the medical necessities in most of the patients (Fabbri et al. 2015). A study conducted to analyze the key risk factors associated with multi-morbidity in the population of south Netherlands reported that up to 78% of the population with multi-morbidity were above the age of 80 years. This emphasized aging as one of the significant factors for multi-morbidity (van den Akker et al. 1998). Another study conducted on a sample population representing 1/3rd of Scottish population found that the proportion of the population with multi-morbidity increases with age as more than half of the study population were multi-morbid at the age of 65 (Barnett et al. 2012).

Aging-associated changes are due to progressive loss of cellular functions resulting in increased susceptibility to various diseases and disorders. Recent advancements in aging research dissecting cellular and molecular mechanisms have identified the hallmarks of aging, including oxidative stress (Barbosa et al. 2019), immunosenescence (Xu et al. 2020), mitochondrial dysfunctions (Sun et al. 2015), and genetic instability and epigenetic dysregulation (Barbosa et al. 2019; Lopez-Otin et al. 2013). These hallmarks are the primary cause of increased cellular damage leading to clinical complications frequently reported in older population. Some of the cellular dysfunctions associated with aging hallmarks are a diminishing ability to initiate the cellular response to the damage, induction of compensatory mechanism, and altered intracellular communication which are ultimately responsible for aging (Bana and Cabreiro 2019).

In addition to the proposed hallmarks of aging, there is increasing evidence that body's microbiota composition also plays a major role in multi-morbid complications associated with aging (Bana and Cabreiro 2019). At present, it is unclear if alterations in the body's microbiota are a cause or consequence of aging. However, microbiota of elder population is different from healthy adults and this altered microbiota profile is strongly associated with multi-morbid clinical complications frequently seen in elderly population (Nagpal et al. 2018). In particular, the microbiota of older population is defined by reduced diversity and lower abundance of bacterial species which are beneficial (Kim and Jazwinski 2018; Biagi et al. 2016; Claesson et al. 2011; Zweilehner et al. 2009). This was demonstrated in a population-based study that analyzed the microbial composition of 38 institutionalized elders. This study reported that microbiota of elderly subjects of a mean age of 84 years had a lower abundance of Faecalibacterium genus, Bacteroides group, and increased abundance of Lactobacillus group as compared to the microbiota of elder adults of a mean age of 62 years. Additionally, higher levels of proinflammatory cytokines and reduced levels of short-chain fatty acids (SCFA) have been reported in elderly subjects (Salazar et al. 2019; Spychala et al. 2018. SCFA (such as acetate, propionate, and butyrate) are produced by fermentation of dietary fibers and starches by colonic bacteria. SCFA are the main energy source for colonic epithelial cells that form a tight junction barrier to separate gut luminal content from host. The dysfunctional epithelial barrier may compromise the gut integrity allowing the free flow of bacterial toxins and inflammatory mediators in host's circulation causing endotoxemia which is commonly associated with inflammatory and metabolic disorders (Kim et al. 2016). More recently, a gut microbial metabolite known as trimethylamine-N-oxide (TMAO) has been linked to an increased risk for developing CVD (Fig. 11.1). Microbiota can metabolize the dietary choline or betaine to trimethylamine (TMA) which is then oxidized by a family of liver enzymes Flavin Monoxygenases (FMO) to produce TMAO. TMAO promotes cholesterol accumulation and foamy cell formation from macrophages inside the wall of the arteries leading to atherosclerotic plaque (Zhu et al. 2020). Another recent study reported a direct link between endotoxemia and TMAO accumulation in systemic circulation (Mouladi et al. 2019). However, a detailed analysis of this aspect remains to be elucidated. There is a growing body of literature supporting the role of microbiota in developing Type 1 and Type II Diabetes (Fig. 11.1). Type 1 diabetes (T1D) is characterized by damage to pancreatic β-cells responsible for producing insulin. In contrast, Type 2 diabetes (T2D) is characterized by insulin resistance leading to a rise in glucose level in the blood. Notably, both T1D and T2D are linked to gut dysbiosis (Qin et al. 2012) and endotoxemia

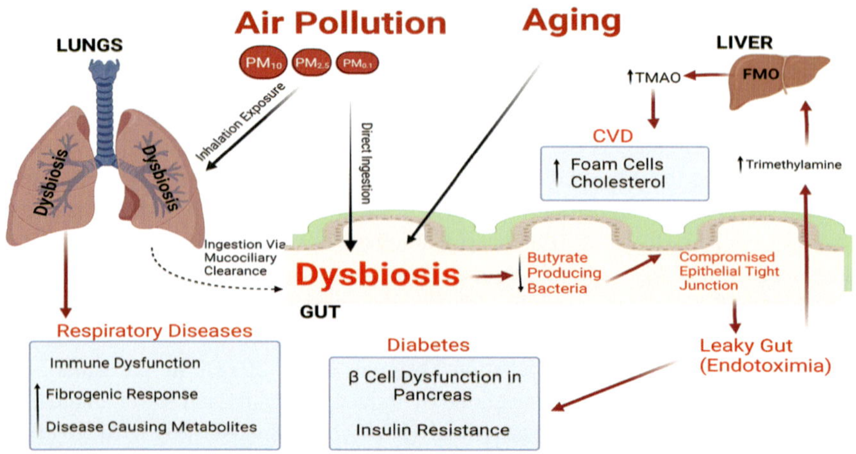

Fig. 11.1 Schematic representation of proposed mechanisms linking air pollutants, aging, and dysbiosis of lung and gut microbiota. Aging and air pollutants cause dysbiosis in both the lungs and the gut. The dysbiosis-induced compromised gut epithelial barrier results in systemic Endotoxemia which may trigger cellular and molecular events leading to Diabetes and Cardiovascular diseases. PM: Particulate Matter with Aerodynamic Diameters 10.0, 2.5, or 0.1 μm; FMO: Flavin Monooxygenase; TMAO: Trimethylamine Oxide; CVD: Cardiovascular diseases

(Blandino et al. 2016). In addition, the compromised gut barrier leads to the release of inflammatory mediators in the systemic circulation which is a potential cause of pancreatic β-cell destruction causing T1D and insulin resistance causing T2D (Li and Atkinson 2015; Cani et al. 2007).

Aging studies have unambiguously identified that compromised gut integrity and level of inflammatory mediators increase with age and are responsible for aging-related complications (Kim et al. 2016). One study specifically reported butyrate as a central SCFA responsible for gut integrity because of its anti-inflammatory action. Microbiome analysis revealed that loss of butyrate-producing bacteria is associated with endotoxemia (Troseid et al. 2020) leading to several other complications (Fig. 11.1). In addition to endotoxemia, decline in the abundance of SCFA-producing microbes gives opportunity to other pathogenic microbes to invade, making an individual susceptible to inflammation and increased risk of cardiovascular events (Lee et al. 2020). In humans, there are conflicting reports about compromised gut permeability and endotoxemia occurring because of dysbiosis. While dysbiosis is a common phenomenon in elderly population, the compromised gut permeability is only seen in the sud-populations who develop inflammatory and/or cardiovascular diseases with age (Ferruci and Fabri 2018).

11.3.2 Lung Microbiota in Aging-Associated Disorders

According to the European Respiratory Society, the status of lung function is one of the predictors of healthy aging. Poor lung function is commonly associated with a higher risk of developing cardiovascular disease and diabetes in elderly population. Conversely, elderly population with robust lung function exhibit higher metabolic rate, better fitness, activity, and physical performance (Saint-Criq et al. 2021). Though the density and diversity of lung microbiota are lower than that of the gut, the lung microbiota is critical for the maintenance of healthy respiratory system. Several recent studies have documented that the composition of lung microbiota plays a causal role in asthma (Di Cicco et al. 2018; Fazlollahi et al. 2018), pulmonary hypertension (Zhang et al. 2020), and other pulmonary diseases such as chronic obstructive pulmonary disease (COPD), cystic fibrosis, and lung cancers (Xue et al. 2020).

Lung microbiota could be a determining factor in the age-related decline in lung health frequently seen in the elderly population (Cho and Stout-Delgado 2020). In this context, elderly population exhibiting reduced lung function and increased arterial stiffness differ in the lung microbiome composition as compared to younger adults. Specifically, the reduced lung function in the elderly has been found to be strongly associated with a higher abundance of Firmicutes and a relative lower abundance of Proteobacteria which predominated in the lungs of healthy adults (Lee et al. 2019). Similarly, another study reported that the upper respiratory tract of the elderly is characterized by an increased abundance of Streptococci. The microbial composition in the upper respiratory tract (URT) is crucially important as it provides defense against the pathogenic microbes from reaching the lower respiratory tract by limiting the nutrient access or secreting growth inhibitors (Khatiwada and Subedi 2020). It has been documented that children of ages below 5 years and elderly people of ages above 65 years have a higher susceptibility to pulmonary infections such as viral and bacterial pneumonia. Interestingly, the URT microbiota profiles in both the age groups are different from those in the healthy adult population resistant to pulmonary infections (Whelan et al. 2014). This can also be emphasized based on a study that evaluated the URT microbiota profiles in elderly pneumonia patients and healthy controls. The profile of elderly pneumonia patients differed significantly from the age-matched healthy controls. The microbiota profile highly correlated with the disease was less diverse and dominated by Lactobacilli, *Rothia* spp, or *Streptococcus pneumoniae* (Piters et al. 2016). Thus, a shift in the microbiome composition may cause opportunistic microbes to invade leading to the disease pathology.

11.3.3 Gut Microbiota in Respiratory Diseases

As gut is the primary organ inhabiting the major proportion of the body's microbiota, compositional and functional changes in the gut microbiota may influence the inflammatory and/or immune status in various organs contributing to various disorders or diseases including but not limited to systemic diseases such as obesity (Cani et al. 2007), diabetes (Blandino et al. 2016), and CVD (Blandino et al. 2016; Li and Atkinson 2015) and organ system-specific disease pathologies like those of the GI system (Chunxi et al. 2020), respiratory system (Dayama et al. 2020), urogenital system (Magruder et al. 2020), CNS (Roy et al. 2019), among others. In this chapter, our focus is on chronic respiratory disorders and diseases associated with the perturbationsin gut microbiota.

Gut microbiota has been extensively studied under various pathological conditions associated with lungs including allergies, asthma, COPD, and cystic fibrosis. Interestingly, these lung diseases have been linked with alterations in the gut microbiota and/or intestinal permeability which is the underlying causal factor for dysbiosis-associated disease pathologies (Zhang et al. 2020).

Asthma: Asthma is a chronic airway inflammatory disease characterized by airflow restriction and airway hyper-responsiveness. The influence of microbiota in this disease can be understood from the 'hygiene hypothesis'. The hygiene hypothesis proposes that early life exposure to certain bacteria is essential for developing immunity against asthma and allergic diseases (Denise 2014). With the advancement in sequencing technologies, it was revealed that the gut microbiota of asthmatic individuals is different from healthy controls (Chunxi et al. 2020). Infancy is the period when the gut microbial composition is a triggering factor for developing asthma (Stokholm et al. 2018). Breastmilk is essentially required for establishing healthy microbiota during infancy. The bacterial diversity is higher in breastfed infants compared to formula-fed infants. Interestingly, the rate of asthma is higher in formula-fed infants compared to breastfed infants (Oddy 2017). Thus, lower bacterial diversity is associated with a higher risk of developing asthma in infants. Another study reported that the lower abundance of bacterial genera such as Bifidobacterium, Ackermansia, and Faecalibacterium in infants is associated with a high risk of developing asthma (Fujimura et al. 2017). In adults, the gut microbiota of asthma patients is characterized by the depletion of gut Faecalibacterium and Bacteroidetes. In addition, the species *F. prausnitzii* which is involved in producing short chain fatty acids (SCFA), declines in asthma patients. SCFA regulates the activation of T-cell subsets which are overactivated in asthma patients (Wang et al. 2018).

Chronic Obstructive Pulmonary Disease (COPD): COPD is a common respiratory disease characterized by persistent airflow limitation and airway inflammation. Like Inflammatory Bowel Disease (IBD), COPD is more prevalent in chronic cigarette smokers compared to non-smokers. Although there is no direct link between the risk of COPD development and microbial dysbiosis, intestinal dysbiosis in chronic smokers is hypothesized to be responsible for a higher incidence of COPD in smokers (Chunxi et al. 2020).

Cystic Fibrosis: Cystic fibrosis (CF) is a common genetic disorder caused by a mutation in cystic fibrosis transmembrane conductor regulatory (CFTR) gene that affects mainly lungs and digestive tract (Li and Somerset 2014; Dayama et al. 2020). Mutation in the CFTR gene is associated with microbiota perturbations in the gut. A murine study demonstrated lower bacterial diversity, evenness, and richness in the CFTR KO mice model. Moreover, CF mice exhibited a lower abundance of protective microbial species such as Acinetobacter. On the other hand, pathogenic bacteria such as *Mycobcaterium* species and *Bacteroidetes fragilis* were abundant in CF mice, both of which have been associated with GI infections and immunomodulation (Lynch et al. 2013). Vernocchi et al. (2013) have proposed a predictive and functional model of gut microbiota enterophenotype of pediatric CF patients of age group 1–6 years. This study reported that the microbiota composition in fecal samples obtained from CF patients was enriched with Propionibacterium, Staphylococcus, and Clostridiaceae. The abundance of Eggerthella, Eubacterium, Ruminococcus, Dorea, *Faecalibacterium prausnitzii*, and Lachnospiraceae was lower in CF patients compared to age-matched healthy controls.

11.3.4 Gut–Lung Axis

As discussed above, there is a difference in the microbiota composition between the gut and the lungs. Gut microbiota is majorly composed of the bacterial phyla Firmicutes, Bacteroidetes, and Actinobacteria, whereas Proteobacteria phylum predominates in lung microbiota. Despite the compositional differences, the anatomical structure and function of mucosal epithelial barrier in both the organs are similar. Due to the similarities in the barrier that restricts microbiota from entering systemic circulation, there is emerging evidence about the vital crosstalk between gut microbiota and lungs giving rise to a concept of the gut–lung axis. The epithelial lining in gut and lung mucosa serves as a barrier for protection against invading pathogenic microbes. This innate protective barrier becomes more permeable in aged individuals or aging animal models. Impaired barrier is the main reason for the inflow and outflow of microbes and microbial products between lungs and gut (Fig. 11.2). Consequently, this leads to alterations in the microbial composition in both the organs (Saint-Criq et al. 2021).

Certain environmental exposures or natural phenomena such as aging are associated with compromised lung epithelial tight junction proteins which maintain the tight junction barrier (Fig. 11.2). Downregulation of these proteins allows infiltration of microbes and inflammatory mediators in the lungs from systemic and lymphatic circulations (Liu et al. 2019a). This seeding of microbes and their products into lungs may lead to the invasion of unwanted microbes primarily responsible for pathologies of pulmonary diseases (Hauptmann and Schaible 2016).

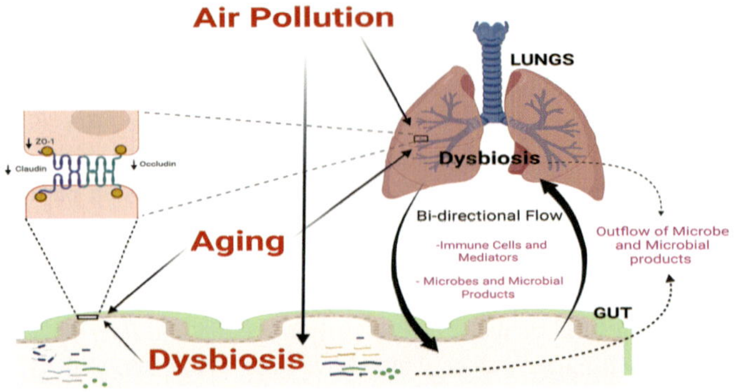

Fig. 11.2 Schematic representation of bidirectional lung–gut Axis. Both aging and air pollutants cause damage to the cell tight-junction proteins in the lung and gut mucosa leading to a compromised epithelial barrier in these organs. The disrupted barriers result in an outflow of microbes and microbial products into systemic circulation allowing bidirectional flow of inflammatory mediators and immune cells between the lungs and the gut. ZO-1: Zona Occludens-1

The concept of gut–lung axis was first introduced by Cooke et al. (2000) who demonstrated that idiopathic pneumonia syndrome is directly associated with the release of gut toxins. Controlling the release of inflammatory mediators from gut can significantly reduce the risk of developing idiopathic pneumonia in bone marrow transplanted mouse model. The gut–lung axis allows the flow of microbial products, cytokines, inflammatory mediators, and hormones from gut into the bloodstream which can then infiltrate into the lungs. Conversely, inflammatory status in lungs has been associated with changes in gut microbiota suggesting a bidirectional nature of gut–lung axis (Dumas et al. 2018). Recent animal studies, including our own, have shown that respiratory exposure to air pollution particulate matter (PM), cigarette smoke (CS), or carbon nanotubes (CNT) not only alters the microbial composition in lungs but also perturbs the microbial diversity in gut (Bhattacharya et al. 2022; Liu et al. 2021; Ran et al. 2021; Fitch et al. 2020; Chen et al. 2019). Our studies further showed that lung inflammatory response due to inhalation exposure to CNT or CS was attenuated in mice with depleted gut microbiota, demonstrating a causal role of gut microbiota in the gut–lung crosstalk (Bhattacharya et al. 2022). Likewise, one of the consequences seen in chronic smokers is intestinal dysbiosis (Lee et al. 2018) which may be one of the causes of increased prevalence of COPD and IBD in chronic cigarette smokers (Erickson et al. 2012). The intestinal dysbiosis in smokers can be significantly improved upon smoking cessation (Biedermann et al. 2013). Another study reported a higher abundance of Streptococcus in fecal samples obtained from children diagnosed with the lung disease cystic fibrosis. As Streptococcus primarily inhabits lungs, a higher abundance of this bacterium in cystic fibrosis patient guts validated the bidirectional nature of gut–lung axis (Enaud et al. 2019).

In the current era, gut–lung axis has been implicated in the severity of the novel lung disease COVID-19 which is caused by SARS-CoV-2 commonly known as

Coronavirus. Emerging studies have indicated altered microbiome composition in the gut and lungs in COVID-19 patients, implying the role of SARS-CoV-2, which is primarily a respiratory virus, in influencing the microbiome composition in both the organs. The dysbiosis in COVID-19 patients was characterized by a relative higher abundance of opportunistic pathogens and depletion of beneficial commensals (Zou et al. 2020; Khatiwada and Subedi 2020).

In a meta-analysis study, gastrointestinal symptoms were analyzed in 4243 patients diagnosed with COVID-19. This study found that 17.6% of the patients with severe COVID-19 symptoms also had gastrointestinal problems. In addition, 48.1% of the total fecal samples obtained from COVID-19 patients were positive for virus RNA and 70% of these stool samples were collected after the recovery from disease (Cheung et al. 2020). Several lines of evidence support that microbiota in lung's mucosal surfaces prevents the virus attachment thereby preventing the invasion of virus. Moreover, lung microbiota may also activate the immunity of the host to fight against the infection (Khatiwada and Subedi 2020). However, to date, no study has unambiguously identified the role of lung microbiome in the COVID-19 disease severity. In this context, the severity of the disease was found to be directly proportional to the abundance of opportunistic microbes (Zou et al. 2020).

11.4 Microbiota in Air Pollution-Associated Chronic Diseases

Human health is strongly influenced by the quality of air we breathe. Poor air quality has long been associated with health problems. However, it has not been recognized as a significant risk factor in human health until historical disasters related to air pollution came into the picture (Anderson et al. 2012). The 1930 Meuse Valley fog in Belgium killed 60 people and several thousand suffered acute pulmonary disorders mainly due to the release of sulfur dioxide (SO_2) and hydrofluoric acid (HF) (Nemery et al. 2001). In 1952, the accumulation of dense smog in London contributed to more than 3000 deaths over a period of three weeks (Bell and Davis 2001). These incidents served as a catalyst for scientists to study the impact of air pollution on human health. Several countries implement stringent regulation for the controlled release of pollutants in air. In 1970, the United States set the limits on emissions and defined the national ambient air quality standard (NAAQS) under clean air act. These standards set the limit on the release of the six primary air pollutants named in the Introduction section (Anderson et al. 2012). Despite global efforts for controlling the air quality by setting the stringent regulatory standards, air pollution continues to be a major risk factor in global burden of disease (GBD) and increased death rates (Cohen et al. 2017). According to the global assessment of exposure and burden of diseases by the World Health Organization (WHO), ambient air pollution affects the people of all regions, settings, socioeconomic strata, and age groups. More than 91% of the world's population lives in areas where air pollution exceeds the WHO limits. This contributes to an estimated seven million deaths every year making it to rank among the top five risk factors for death and disability worldwide (Health Effect Institute 2020: Special report Boston, MA). With the global implementation of WHO-recommended air quality guidelines (AQGs), 80% of deaths related to $PM_{2.5}$ can be avoided in countries with large population and high concentration of $PM_{2.5}$. This will not only reduce the global disparities in air pollution but also contribute towards reducing the global burden of diseases (WHO 2021).

11.4.1 Air Pollution: A Cause of Non-communicable Diseases in Elderly Population

As our knowledge about the air pollution continues to evolve, the current understanding is that poor air quality can affect our health and quality of life in a variety of ways. In addition to developing chronic lung diseases and lung infections, poor air quality is well documented to be a responsible factor for developing non-communicable systemic diseases such as CVD and diabetes. In this context, the first scientific statement about linking air pollution and cardiovascular disease was published in 2004 by the American Heart Association. The statement was based on the growing incidences of hospitalization and mortalities due to cardiovascular diseases associated with exposure to environmental air pollutants including Nitrogen oxides, Sulfur dioxide, Ozone, Lead, and Ambient particulate matter (Brook et al. 2004). Of all the air pollutants, coarse particles (particulate matter in air with aerodynamic diameter between 2.5 and 10 micron or PM_{10}), fine particles ($PM_{2.5}$), and ultrafine particles ($PM_{0.1}$) have been of greater concern because of the fact that the highest proportion of adverse health effects related to air pollution are due to particulate matter in air.

A major Multi-Ethnic Study of Atherosclerosis and Air Pollution (MESA) in the United States (Kaufman et al. 2016) reported that human sub-populations living in areas with a higher concentration of atmospheric PM had a higher rate of CVD-associated hospitalizations as later corroborated in Iran (Dastoorpoor et al. 2019) and China (Yang et al. 2019). Moreover, in sub-populations with already existing cardiovascular complications, exposure to PM is significantly associated with a higher rate of mortalities and a decrease in life expectancies (Brook et al. 2010). Long-term exposure to $PM_{2.5}$ at levels close to NAAQS standards is known to accelerate the calcification of coronary arteries leading to atherosclerosis (Kaufman et al. 2016). Interestingly, all these studies have reported that the major proportion of population affected due to exposure to PM belonged to the age group of 65 years and above. This may suggest that elderly population is at a higher risk of developing cardiovascular complications due to PM exposures.

Recent statistical surveys indicated that exposure to PM is one of the prominent factors responsible for the increase in the prevalence of diabetes across the world

(Rao et al. 2015; Meo et al. 2015). In a prospective study, elderly American population of age 57 and above with long-term exposure to $PM_{2.5}$ was evaluated for their metabolic health. This study found a statistically significant association between exposure to $PM_{2.5}$ and elevated level of glycosylated hemoglobin in blood samples. The level of glycosylated hemoglobin was similar in healthy participants and participants with pre-existing diabetic conditions. This suggested that long-term $PM_{2.5}$ exposure is a key risk factor in abnormal glucose metabolism in elderly population (Honda et al. 2017). Similar findings about PM being a prominent cause of the increased risk of diabetes have been frequently reported in population-based studies in Canada (Brook 2008) and Germany (Kramer et al. 2010; Weinmayr et al. 2015).

11.4.2 Air Pollution Enhances Age-Related Disease Susceptibility Via Microbial Dysbiosis

11.4.2.1 Air Pollution-Induced Gut Dysbiosis

Certain families of pollutants/toxicants can alter the composition of microbiome which in turn contributes to the pollutant/toxicant-induced toxicity and/or vulnerability of an individual for developing disease (Jin et al. 2017; Claus et al. 2016). Conversely, microbiome can bioactivate the toxicants entering the body to their toxic metabolites capable of inducing toxicity at the target site thereby leading to disease susceptibility (Weile et al. 2005). In recent studies, it was revealed that air pollution-associated (Fouladi et al. 2020) inhaled PM can affect the microbial diversity in gut eventually leading to an enhanced risk of developing CVD (Liu et al. 2021; Ran et al. 2020; Chen et al. 2019). According to some reports, inhaled PM may potentially get indirect access to gut due to translocation in part from the upper parts of the respiratory tract (oropharynx and nasopharynx) via mucocilliay clearance. The translocated material can then be swallowed allowing PM to reach the gut where it can disrupt the homeostasis such as the epithelial tight junction barrier (Mutlu et al. 2018). This was demonstrated in an atherosclerotic mouse model that exhibited lower bacterial diversity and altered Firmicutes/Bacteroidetes (F/B) ratio due to exposure to wood smoke or gasoline vehicle exhaust. Reduced bacterial diversity was also accompanied by downregulation of tight junction proteins and increased inflammatory markers in gut (Fitch et al. 2020) which contribute to endotoxemia (Kim et al. 2016). As discussed earlier, the balanced microbial composition is required to maintain epithelial tight junction barrier. SCFA-producing bacteria that provide nourishment to the colonocytes to maintain the integrity of mucosal barrier prevent the leakage of gut contents into systemic circulation. Recent studies evidenced that abundance of SCFA-producing bacteria declines due to exposure to ambient PM (Fig. 11.1). In addition to reduced abundance of SCFA-producing bacteria, studies also reported downregulation of tight junction proteins such as Zonula Occludens-1 (ZO-1), Claudin-1, and Occludin-1 in gut (Kish et al. 2013; Mutlu et al. 2011). These proteins are required to maintain integrity and proper functioning of epithelial barrier. Interestingly, compromised gut integrity and biochemical markers associated with the dysbiosis are also seen in aging-related complications.

Experimental animal studies such as the above have shown that the inhaled air particulate matter influences the gut microbiome eventually impacting the metabolic and cardiovascular health. However, currently, epidemiological studies addressing the microbial dysbiosis associated with air pollution are limited in terms of establishing a direct link between air pollutants, dysbiosis, and disease occurrence. In a recent study conducted at the University of Southern California (Alderete et al. 2018), 43 overweight adolescents were selected from a Meta-AIR study to analyze the impact of traffic-related air pollution (TRAP) on the gut microbiome. The level of TRAP exposure was estimated based on the residential addresses. This study found correlation between TRAP exposure and microbial richness and diversity characterized by 16S rRNA sequencing on fecal samples. They further reported that TRAP exposure was negatively correlated with the relative abundance of phylum Proteobacteria and positively with the relative abundance of phylum Actinobacteria and certain families such as Bacteroidaceae, Tissierellaceae, Corynebacteriaceae, and Coriobacteriaceae. Also, there was a positive correlation between TRAP and fasting glucose level. Some of the bacterial families such as Bacteroidaceae and Coriobacteriaceae which are positively associated with TRAP are also linked with insulin resistance and obesity (Dujardin et al. 2020; Alderete et al. 2018). A similar finding was reported in another population-based study that examined the microbiome status of 5828 adults recruited from the Southern part of China. This study revealed exposure to $PM_{2.5}$ and PM_1 is positively correlated with the abundance of microbial taxa such as Firmicutes, Proteobacteria, and Verrucomicrobia. Additionally, PM exposure was also associated with elevated fasting glucose level which was partially mediated by gut microbiome (Liu et al. 2019b).

11.4.2.2 Air Pollution-Induced Lung Dysbiosis

There is a growing literature addressing the impact of inhaled pollutants on the respiratory microbiome. Inhaled air pollutants can alter the lung microenvironment by changing the pH, oxidative stress, and inflammatory status (Vignal et al. 2017). These changes are not favorable for internal lung flora thus reducing the abundance of beneficial microbes in lungs. Conversely, these changes may damage the epithelial barrier in lungs allowing entry and invasion of pathogenic microbes from bloodstream (Adar et al. 2016). In experimental animal studies, it has been shown that inhaled air pollutants may become a cause of pulmonary dysbiosis which may lead to pollution-related lung diseases (Li et al. 2017a). Higher bacterial burden and diversity were seen in the lungs of rats receiving particulate matter in the form of motor vehicle exhaust (MVE) every day for 4 weeks compared to rats receiving clean air for the same duration (Li et al. 2017a). In another study, 16S rRNA sequencing on the BAL fluid obtained from mice intratracheally administered with

$PM_{2.5}$ revealed that the $PM_{2.5}$ exposure contributes to the changes in the microbial composition. The lung microbial community is primarily composed of phyla such as Proteobacteria, Bacteroidetes, Cyanobacteria, and Firmicutes with Proteobacteria being a dominant phylum with the total abundance of nearly 90%. Exposure to $PM_{2.5}$ alters this composition by decreasing the abundance of Proteobacteria and increasing the abundance of Bacteroidetes (Li et al. 2020). In humans, microbiome status was analyzed in the nasal samples of 40 healthy volunteers to correlate with the exposure to PM_1 and $PM_{2.5}$. The exposure assessment was done by using a personal sampler worn for 24 h prior to sampling. This study found that short-term exposure to PM_1 and $PM_{2.5}$ is associated with lower microbial diversity suggesting an altered homeostasis in the bacterial communities (Mariani et al. 2020). In this context, an earlier study had correlated the presence of microbial communities in the lungs with inhaled air microflora (which contained approximately 100 pathogens/m^3) and microbes dispersed from oropharynx (Adar et al. 2016).

Air pollution research has explained that PM can reach deep into the lung where the microenvironment facilitates the release of the PM-associated toxins that can damage the airway epithelial cells and destroy the pulmonary microbes. Consequently, the pathogenic bacteria and metabolites can translocate into the circulation and thus freely travel to the distal organs including intestine. This may trigger a bidirectional flow of microbes, cytokines, and other mediators across gut-lung axis (Fig. 11.2) (Crawford et al. 2021). This was demonstrated in an animal study that investigated the inflammatory response in lungs and colons of mice exposed to ambient PM. Mice exposed to urban coarse PM in ventilated inhalation chambers for 15 days exhibited an increased oxidative stress and immune cell infiltration in lungs. This was in parallel with increased level of inflammatory cytokines in colons. This low-grade inflammation in lungs and colon was reversed by the administration of the antioxidant N-acetyl cysteine (Vignal et al. 2017). In another study, inhalation exposure to $PM_{2.5}$ in mice downregulated the tight junction proteins in lungs such as claudin, occludin, and ZO-1 leading to the dissemination of bacterial species and metabolites in systemic circulation (Liu et al. 2019a). The epithelial barrier acts as a first line of defense against the environmental pathogens by expressing pattern recognition receptors, secreting antimicrobial peptides, and eliciting anti-inflammatory response through activation of MAP Kinases (Dai et al. 2012). Thus, damage to lung epithelium by any factor such as exposure to PM exposure facilitates the translocation of lung microbes and metabolites to distant organs (Liu et al. 2019a, b). However, it is still unknown if the compromised lung barrier is a cause or consequence of dysbiosis.

11.5 Conclusions and Future Directions

Microbiome plays a crucially important role in normal functioning state as well as disease states of human health. Though the gut-associated microbiota has been the most extensively studied in terms of the microbiome role in health and disease, microbial commensals colonizing other niches in the human body also play crucial functions. Particularly, lung microbiome is receiving increasing attention in understanding the cause–effect relationship in the context of both aging- and air pollution-associated chronic lung diseases and even severe infections including COVID-19.

Any changes in the diversity of the microbiome community (microbial dysbiosis) due to external factors such as exposure to environmental chemicals, antibiotics, lifestyle habits, or natural biological phenomena such as aging can influence susceptibility to diverse health conditions and diseases. In this context, it is well documented that aging and air pollution are risk factors for serious respiratory diseases (asthma, COPD). However, the causal role of microbiome dysbiosis (including gut and lung microbiomes) in these chronic diseases is beginning to emerge. The instability of microbiome with aging is perceived as a central risk factor is aging-related clinical complications and diseases. Deeper mechanistic insights into this risk factor would provide an understanding of the causal relationship between microbiome and disease and help design intervention strategies. In this regard, nutritional interventions such as probiotics and prebiotics are promising options being pursued to replenish the microbial diversity which is lost as a consequence of aging (Salazar et al. 2013). In majority of the studies, the linking of air pollution with chronic diseases particularly COPD, CVD, and metabolic disorders has also been reported for the elderly population. This implies that both aging and air pollution can directly influence the microbiome (gut and/or lung) and trigger or exacerbate these disorders. With the developing hypothesis that microbiome is one of the causal factors for developing these chronic diseases/disorders, microbiome composition and activity are increasingly being targeted as a driver or modulator of the underlying mechanistic pathways. On the other hand, a crosstalk between gut microbiome and distant organs and vice versa has been a topic of much interest in dissecting these modulatory mechanisms. Gut–lung crosstalk is one such mechanism receiving attention to elucidate the role of gut microbiome in respiratory diseases caused by aging or pollutants and systemic diseases/disorders caused by inhaled pollutants. Conceivably, pollutant-impacted microbiomes (dysbiosis) may cause leaky epithelial barriers releasing mediators that may enter circulation leading to lung or systemic disease pathologies. However, little if any is known about the specific metabolites or factors mediating this crosstalk and whether they are a cause or consequence of dysbiosis. Considering the fact that microbiota is the main target of aging and air pollution, future studies should focus to fill this knowledge gap so personalized intervention strategies could be designed for clinical applications.

Acknowledgements Funding support (to JSY) from the University of Cincinnati Center for Environmental Genetics through the NIH/NIEHS award **P30ES006096** is acknowledged.

References

Adar SD, Huffnagle GB, Curtis JL (2016) The respiratory microbiome: an underappreciated player in the human response to inhaled pollutants? Ann Epidemiol 26(5):355–359

Alderete TL, Jones RB, Chen Z, Kim JS, Habre R, Lurmann F et al (2018) Exposure to traffic-related air pollution and the composition of the gut microbiota in overweight and obese adolescents. Environ Res 161:472–478

Anderson JO, Thundiyil JG, Stolbach A (2012) Clearing the air: a review of the effects of particulate matter air pollution on human health. J Med Toxicol 8(2):166–175

Bailey MJ, Naik NN, Wild LE, Patterson WB, Alderete TL (2020) Exposure to air pollutants and the gut microbiota: a potential link between exposure, obesity, and type 2 diabetes. Gut Microbes 11(5):1188–1202

Bana B, Cabreiro F (2019) The microbiome and aging. Annu Rev Genet 53:239–261

Bander ZA, Nitert MD, Mousa A, Naderpoor N (2020) The gut microbiota and inflammation: an overview. Int J Environ Res Public Health 17(20):1–22

Barbosa MC, Grosso RA, Fader CM (2019) Hallmarks of aging: an autophagic perspective. Front Endocrinol 10:1–13

Barnett K, Mercer SW, Norbury M, Watt G, Wyke S, Guthrie B (2012) Epidemiology of multimorbidity and implications for health care, research, and medical education: a cross-sectional study. Lancet 380(9836):37–43

Bell ML, Davis DL (2001) Reassessment of the lethal London Fog of 1952: novel indicators of acute and chronic consequences of acute exposure to air pollution. Environ Health Perspect 109:389–394

Bhattacharya SS, Yadav B, Rosen L, Nagpal R, Yadav H, Yadav JS (2022) Crosstalk between gut microbiota and lung inflammation in mouse toxicity models of respiratory exposure or co-exposure to carbon nanoparticles and cigarette smoke extract. Toxicol Appl Pharmacol 447:116066

Biagi E, Franceschi C, Rampelli S, Severgnini M, Ostan R, Turroni S et al (2016) Gut microbiota and extreme longevity. Curr Biol 26(11):1480–1485

Biedermann L, Zeitz J, Mwinyi J, Sutter-Minder E, Rehman A, Ott SJ et al (2013) Smoking cessation induces profound changes in the composition of the intestinal microbiota in humans. PLoS ONE 8:e59260

Blandino G, Inturri R, Lazzara F, Rosa MD, Malaguarnera L (2016) Impact of gut microbiota on diabetes mellitus. Diabetes Metablsm 42:303–315

Brook RD, Franklin M, Cascio W, Hong Y, Howard G, Lipsett M, Leupker R, Mittleman M, Samet J, Smith SC, Tager I (2004) Air pollution and cardiovascular diseases. Circulation 109:2655–2671

Brook RD (2008) Cardiovascular effects of air pollution. Clin Sci 115:175–187

Brook RD, Rajagopalan S, Pope CA, Brook JR, Bhatnagar A, Diez-Roux AV et al (2010) Particulate matter air pollution and cardiovascular disease: an update to the scientific statement from the American heart association. Circulation 121(21):2331–2378

Bufard TW (2020) The gut microbiome and aging. J Gerontol 75:1229–1231

Cani PD, Amar J, Iglesias A, Poggi M, Knauf C, Bastelica D, Neyrinck AM, Fava F, Touhy KM, Chabo C, Waget A, Delmee E, Cousin B, Sulpice T, Chamontin B, Ferrieres J, Tanti J-F, Gibson GR, Casteilla L, Delzennn MN, Alessi CM, Burceline R (2007) Metabolic endotoxemia initiates obesity and insulin resistance. Am Diabetes Assoc 57:1761–1772

Chen D, Xiao C, Jin H, Yang B, Niu J, Yan S et al (2019) Exposure to atmospheric pollutants is associated with alterations of gut microbiota in spontaneously hypertensive rats. Exp Ther Med 18(5):3484–3492

Cheung KS, Hung IFN, Chan PY, Lung KC, Tso E, Raymond L, Ng YY et al (2020) Gastrointestinal manifestations of SARS-CoV-2 infection and virus load in fecal samples from a Hong Kong Cohort: systematic review and meta-analysis. Gastroenterology 159(1):81–95

Childs BG, Durik M, Baker DJ, van Deursen JM (2015) Cellular senescence in and aging-related disease: from mechanism to therapy. Nat Med 2(12):1424–1435

Chunxi L, Haiyue L, Yanxia L, Jiangbing P, Jin S (2020) The gut microbiota and respiratory diseases: new evidence. J Immunol Res 20:2340670

Cho SJ, Stout-Delgado HW (2020) Aging and lung disease. Annu Rev Physiol 82:433–459

Claesson MJ, Cusack S, O'Sullivan O, Greene-Diniz R, De Weerd H, Flannery E et al (2011) Composition, variability, and temporal stability of the intestinal microbiota of the elderly. Proc Natl Acad Sci USA 108:4586–4591

Claus SP, Guillou H, Ellero-Simatos S (2016) The gut microbiota: a major player in the toxicity of environmental pollutants? Biofilms Microbes 2:16003

Cohen AJ, Brauer M, Burnett R, Anderson HR, Frostad J, Estep K, Balakrishnan K, Brunkreef B, Dandona L, Dandona R, Feigin V, Freedman G, Hubbel B, Jobling A, Kan H, Knibbs L, Liu Y, Martin R, Morawska L, Pope III CA, Shin H, Straif K, Shaddik G, Thomas M, Dingenen R, Donkelaar AV, Vos T, Murray CL, Forouzanfart MH (2017) Estimates and 25-years trend of global burden of diseases attributable to ambient air pollution: an analysis of data from global burden of disease study 2015. Lancet 389:1907–1917

Cooke KR, Hill GR, Gerbitz A, Kobzik L, Martin TR, Crawfard JM, Brewer JP, Ferrara JLM (2000) Hyporesponsiveness of donor cells to lipopolysaccharide stimulation reduces the severity of experimental idiopathic pneumonia syndrome: potential role for a gut-lung axis of inflammation. J Immunol 165:6612–6619

Crawford MS, Nordgren TM, McCole FD (2021) Every breath you take: impacts of environmental dust exposure on intestinal barrier function-from the gut-lung axis to COVID 19. Am J Physiol: Gastrointest Liver Physiol 320:586–600

Dai C, Zhao D-H, Jiang M (2012) Probiotics regulate the intestinal epithelial barrier in vivo and in vitro via P38 and ERK signaling pathways. Int J Mol Med 29:202–208

Dastoorpoor M, Sekhavatpour Z, Masoumi K, Mohammadi MJ, Aghababaeian H, Khanjani N et al (2019) Air pollution and hospital admissions for cardiovascular diseases in Ahvaz, Iran. Sci Total Environ 652:1318–1330

Dayama G, Priya S, Niccum DE, Khoruts A, Blekhman R (2020) Interaction between the gut microbiome and host gene regulation in cystic fibrosis. Genome Med 12:12

Denise D (2014) The evolution of hygiene hypothesis. The role of early-life exposure to viruses and microbes and their relationship to asthma and allergic diseases. Curr Opin Allergy Clin Immunol 14:390–396

Di Cicco M, Pistello M, Jacinto T, Ragazzo V, Piras M, Freer G et al (2018) Does lung microbiome play a causal or casual role in asthma? Pediatr Pulmonol 53(10):1340–1345

Dujardin CE, Mars RAT, Manemann SM, Kashyap PC, Clements NS, Hassett LC, Roger VL (2020) Impact of air quality on gastrointestinal microbiome. A review. Environ Res 186:109485

Dumas A, Bernard L, Poquet Y, Lugo-Villarino G, Neyrolles O (2018) The role of the lung microbiota and the gut–lung axis in respiratory infectious diseases. Cell Microbiol 20:e12966

Erickson AR, Cantarel BL, Lamendella R, Darzi Y, Mongodin EF, Pan C et al (2012) Integrated metagenomics/metaproteomics reveals human host-microbiota signatures of Crohn's disease. PLoS ONE 7:e4913

Enaud R, Hooks KB, Barre A, Barnetche T, Hubert C, Massot M et al (2019) Intestinal inflammation in children with cystic fibrosis is associated with Crohn's-like microbiota disturbances. J Clin Med 8:645–657

Enaud R, Prevel R, Ciarlo E, Beaufils F, Wieers G, Guery B, Delhaes L (2020) The gut-lung axis in health and respiratory diseases: A place for inter-organ and inter-kingdom crosstalk. Front Cell Infect Microbiol 10:9

Fabbri E, An Y, Schrack JA, Gonzalez-Freire M, Zoli M, Simonsick EM et al (2015) Energy metabolism and the burden of multimorbidity in older adults: results from the baltimore longitudinal study of aging. J Gerontol-Ser Biol Sci Med Sci 70(11):1297–1303

Fazlollahi M, Lee TD, Andrade J, Oguntuyo K, Chun Y, Grishina G et al (2018) The nasal microbiome in asthma. J Allergy Clin Immunol 142(3):834–843

Feroci L, Fabri E (2018) Inflammaging: chronic inflammation in aging, cardiovascular disease and frailty. Nat Rev 15(9):505–522

Fitch MN, Phillippi D, Zhang Y, Lucerro J, Panday RS, June L, Brower J, Allen MS, Campen MJ, McDonald JD, Lund MK (2020) Effect of inhaled air pollution on markers of integrity, inflammation, and microbiota profiles of intestines in Apo-/- mice. Environ Ress 181:108913

Fouladi F, Bailey MJ, Patterson WB, Sioda M, Blakley IC, Fodor AA et al (2020) Air pollution exposure is associated with the gut microbiome as revealed by shotgun metagenomic sequencing. Environ Int 138:105604

Fujimura KE, Sitarik AR, Havstad S, Lin DL, Levan S, Fadrosh D, Panzer AR et al. (2017) Neonatal gut microbiota associates with childhood multi–sensitized atopy and T–cell differentiation. Nat Med 22(10):1187–91

Gerard P (2016) Gut microbiota and obesity. Cell Mol Life Sci 73:147–162

Gjestsen MT, Brønnick K, Testad I (2018) Characteristics and predictors for hospitalizations of home-dwelling older persons receiving community care: a cohort study from Norway. BMC Geriatr 18(1):1–9

Hauptmann M, Schaible UE (2016) Linking microbiota and respiratory diseases. FEBS Lett 590:3721–3738

Huffnagle GB, Noverr MC (2013) The emerging world of the fungal microbiome. Trends Microbiol 21(7):334–341

Health effect institute (2020) State of global air, special report. Boston, MA

Heintz-Buschart A, Wilmes P (2018) Human gut microbiome: function matters. Cell Press: Trends Microbiol 26:563–574

Honda T, Pun VC, Manjourides J, Suh H (2017) Association between long-term exposure to air pollution, glycosylated hemoglobin and diabetes. Int J Hyg Environ Health 220(7):1124–1132

Jin Y, Wu S, Zeng Z, Fu Z (2017) Effects of environmental pollutants on gut microbiota. Environ Pollut 222:1–9

Kaufman JD, Adar SD, Barr RG, Budoff M, Burke GL, Curl CL et al (2016) Association between air pollution and coronary artery calcification within six metropolitan areas in the USA (the Multi-Ethnic Study of Atherosclerosis and Air Pollution): a longitudinal cohort study. Lancet 388(10045):696–704

Kaufman JD, Elkind SLV, Bhatnagar A, Koehler K, Balmes RJ, Sidney S, Pena MSB, Dockery DW, Hou L, Brook RD, Laden F, Rajagopalan S, Kendrick KB, Turner JR (2020) Guidance to reduce the cardiovascular burden of ambient air pollutants: a policy statement from the American Heart Association. Circulation 142:432–447

Khatiwada S, Subedi A (2020) Lung microbiome and coronavirus disease 2019 (COVID-19): possible link and implications. Hum Microbiome J 17:10073

Kim KA, Jeong JJ, Yoo SY, Kim DH (2016) Gut microbiota lipopolysaccharide accelerates inflamm-aging in mice. BMC Microbiol 16(1):1–9

Kim S, Jazwinski SM (2018) The gut microbiota and healthy aging. Gerontology 64(6):513–520

Kish L, Hotte N, Kaplan GG, Vincent R, Tso R, Gänzle M et al (2013) Environmental particulate matter induces murine intestinal inflammatory responses and alters the gut microbiome. PLoS ONE 8(4):e62220

Krämer U, Herder C, Sugiri D, Strassburger K, Schikowski T, Ranft U, Rathmann W (2010) Traffic-related air pollution and incident type 2 diabetes: results from the SALIA cohort study. Environ Health Perspect 118(9):1273–1279

Landi F, Graziano O, Matteo C, Christian B, Fabrizia L, Pier UC, Roberto B (2004) Comorbidity and social factors predicted hospitalization in frail elderly patients. J Clin Epidemiol 57 (8):832–836

Leustean AM, Manuela C, Anca S, Claudia FC, Mariana F, Claudia CT, Daniela MT (2018) Implications of the intestinal microbiota in diagnosing the progression of diabetes and the presence of cardiovascular complications. J Diabetes Res 2018:5205126

Lee SH, Yun Y, Kim SJ, Lee E-J, Chang Y, Ryu S, Shin H, Kim H-L, Kim H-N, Lee JH (2018) Association between cigarette smoking status and composition of gut microbiota: population-based cross-sectional study. J Clin Med 7:282–297

Lee SY, MacAogain A, Famk D, Chia KL, Binte Ali NA, Yap MMC, Eric PH, Sanjay H, Chin C, Lim L (2019) Airway microbiome composition correlates with lung function and arterial stiffness in an age-dependent manner. PLOS ONE 14(11):e0225636

Lee J, Aigle J, Atadja J, Quaicoe V, Honarpisheh P, Ganesh BP, Hasan A, Graf J, Patrosino J, Putluri N, Zhu L, Durgan DJ, Bryan Jr RM, McCullough LD, Venna VR (2020) Gut microbiota derived short term fatty acids promotes post stroke recovery in aged mice. Cir Res 127(4):453–465

Li X, Atkinson MA (2015) Role of gut permeability in pathogenesis of type I diabetes—a solid or leaky concept. Pediatr Diabetes 16(7):485–492

Li L, Somerset S (2014) The clinical significance of the gut microbiota in cystic fibrosis and the potential for dietary therapies. Clin Nutr 33:571–580

Li N, He F, Liao B, Zhou Y, Li B, Ran P (2017a) Exposure to ambient particulate matter alters the microbial composition and induces immune changes in rat lung. Respir Res 18(1):1–10

Li J, Hu Y, Liu L, Wang Q, Zeng J, Chen C (2020) PM2.5 exposure perturbs lung microbiome and its metabolic profile in mice. Sci Total Environ 721:137432

Liu J, Chen X, Dou M, He H, Ju M, Ji S et al (2019a) Particulate matter disrupts airway epithelial barrier via oxidative stress to promote pseudomonas aeruginosa infection. J Thorac Dis 11:2617–2627

Liu T, Chen X, Xu Y, Wu W, Tang W, Chen Z, Ji G, Peng J, Jiang Q, Xiao J, Li X, Zeng W et al (2019b) Gut microbiota partially mediates the effects of fine particulate matter on type 2 diabetes: evidence from population based epidemiological study. Environ Int 130:104882

Liu Y, Wang T, Si B, Du H, Liu Y, Waqas A et al (2021) Intratracheally instillated diesel PM2.5 significantly altered the structure and composition of indigenous murine gut microbiota. Ecotoxicol Environ Saf 210:111903

López-Otín C, Blasco MA, Partridge L, Serrano M, Kroemer G (2013) The hallmarks of aging. Cell 153:1194

Lozupone CA, Stombaugh JI, Gordon JI, Jansson JK, Knight R (2012) Diversity, stability and resilience of the human gut microbiota. Nature 489(7415):220–230

Lynch SV, Goldfarb KC, Wild YK, Kong W, De Lisley RC, Brodie EL (2013) Cystic fibrosis transmembrane conductance regulator knockout mice exhibit aberrant gastrointestinal microbiota. Gut Microbes 4:41–47

Magruder M, Emmanuel E, Lisa Z, Shady A, Michael JS, Lars FW, Line M et al (2020) Gut commensal microbiota and decreased risk for enterobacteriaceae bacteriuria and urinary tract infection. Gut Microbes 12(1):1805281

Mutlu EA, Engen PA, Soberanes S, Urich D, Forsyth CB, Nigdelioglu R et al (2011) Particulate matter air pollution causes oxidant-mediated increase in gut permeability in mice. Part Fiber Toxicol 8:1–13

Mutlu EA, Comba IY, Cho T, Engen PA, Yazıcı C, Soberanes S et al (2018) Inhalational exposure to particulate matter air pollution alters the composition of the gut microbiome. Environ Pollut 240:817–830

Mariani J, Favero C, Carugno M, Pergoli L, Ferrari L, Bonzini M et al (2020) Nasal microbiota modifies the effects of particulate air pollution on plasma extracellular vesicles. Int J Environ Res Public Health 17:611

Meo SA, Memon AN, Sheik SA, Rouq FA, Usmani AM, Hassan A, Arain SA (2015) Effect of environmental air pollution on type 2 diabetes mellitus. Eur Rev Med Pharmacol Sci 19:123–128

Mouladi J, Malek V, Jafri-Vayghyan H, Vager-Mehrabani E, Alizadeh M (2019) Metabolic endotoxemia and cardiovascular disease: a systemic review about potential roles of prebiotics and probiotics. Clin Exp Pharmacol Physiol 47:927–936

Nagpal R, Mainali R, Ahmadi S, Wang S, Singh R, Kavanagh K, Kitzman DW, Almagul Kushugulova A, Marotta F, Yadav H (2018) Gut microbiome and aging: physiological and mechanistic insights. Nutr Health Aging 4:267–285

Nemery B, Hoet PHM, Nemmar A (2001) The Meuse valley fog of 1930: an air pollution disaster. Lancet 357:704–708

Nguyen TLA, Vieira-Silva S, Liston A, Raes J (2015) How informative is the mouse for human gut microbiota research? Dis Model Mech 8:1–16

Oddy W (2017) Breastfeeding in childhood asthma, and allergic diseases. Ann Nutr Metab 70:26–36

Pashely CH, Fairs A, Free RC, Warldlaw AJ (2012) DNA analysis of outdoor air reveals a high degree of fungal diversity, temporal variability, and genera not seen by spore morphology. Fungal Biol 116:214–224

Piters de WAS, Huijskens EGW, Wyllie AL, Biesbroek G, van den Bergh MR, Veenhoven RH, Xinhui Wang X, Trzcinski K, Bonten5 MJ, Rossen JWA, Sanders EAM, Bogaert D (2016) Dysbiosis of upper respiratory tract microbiota in elderly pneumonia patients. ISME J 10:97–108

Qin J, Li Y, Cai Z, Li S, Zhu J, Zhang F, Liang S, Zhang W, Guan Y, Shen D, Peng Y, Zhang D, Jie Z, Wu W et al (2012) A metagenome-wide association study of gut microbiota in type 2 diabetes. Nature 490:55–60

Ran Z, An Y, Zhou J, Yang J, Zhang Y, Yang J et al (2021) Subchronic exposure to concentrated ambient PM2.5 perturbs gut and lung microbiota as well as metabolic profiles in mice. Environ Pollut 272:115987

Rao X, Patel P, Puett R, Rajagopalan S (2015) Air pollution as a risk factor for type 2 diabetes. Toxicol Sci 143(2):231–241

Roy SS, Banerjee S (2019) Gut microbiota in neurodegenerative disorders. J Neuroimmunol 328:98–104

Saint-Criq V, Lugo-Villarino G, Thomas M (2021) Dysbiosis, malnutrition and enhanced gut-lung axis contribute to age-related respiratory diseases. Aging Res Rev 66:101235

Salazar N, López P, Valdés L, Margolles A, Suárez A, Patterson ÁM et al (2013) Microbial targets for the development of functional foods accordingly with nutritional and immune parameters altered in the elderly. J Am Coll Nutr 32(6):399–406

Salazar N, Arboleya S, Valdés L, Stanton C, Ross P, Ruiz L et al (2014) The human intestinal microbiome at extreme ages of life. Dietary intervention as a way to counteract alterations. Front Genet 5:1–9

Salazar N, Arboleya S, Fernández-Navarro T, de los Reyes-Gavilán CG, Gonzalez S, Gueimonde M (2019) Age-associated changes in gut microbiota and dietary components related with the immune system in adulthood and old age: a cross-sectional study. Nutrients 11 (8):1–11

Salazar N, González S, Nogacka AM, Rios-Covián D, Arboleya S, Gueimonde M, de los Reyes-Gavilán CG (2020) Microbiome: effects of ageing and diet. Curr Issues Mol Biol 36:33–62

Salim SY, Kaplan GG, Madsen KL (2013) Air pollution effects on the gut microbiota: a link between exposure and inflammatory disease. Gut Microbes 5(2):215–217

Spychala MS, Venugopal RV, Michal J, Sarah JD, David JD, Bhanu PG, Nadim JA et al (2018) Age-related changes in the gut microbiota influence systemic inflammation and stroke outcome. Ann Neurol 84(1):23–36

Stokholm J, Blaser MJ, Thorsen J, Rasmussen MA, Waage J, Vinding RK et al (2018) Maturation of the gut microbiome and risk of asthma in childhood. Nat Commun 9:1–10

Sun N, Jeanho Y, Jie L, Daniela M, Chengyu L, Ilsa IR, Kira M, Holmström et al (2015) Measuring in vivo mitophagy. Mol Cell 60(4):685–96

Trøseid M, Andersen GØ, Broch K, Hov JR (2020) The gut microbiome in coronary artery disease and heart failure: current knowledge and future directions. Bio-Medicine 52:102649

van der Akker M, Buntinx F, Metsemaker JF, Roos S, Knottnerus A (1998) Multimorbidity in general practice:prevalence, incidence, and determinants of co-occurring chronic and recurrent diseases. J Clin Epidemiol 51:367–375

Vernocchi P, Del Chierico F, Majo F, Rossitto M, Casadei L, La Storia A, De Filoppi F, Rizzo C, Manetti C, Paci P, Ercolini D, Marini F, Fiscarelli EV, Dallapiccola B, Lucidi V, Miccheli A, Pitingnani L (2013) Gut microbiota signatures in cystic fibrosis: loss of host CFTR function drives the microbiota enterophenotype. PLOS ONE 13:e0208171

Vignal C, Pichavant M, Alleman LY, Djouina M, Dingreville F, Perdrix E, Waxin C, Alami AO, Gower-Rouseau C, Desreumaux P, Body-Malapel M (2017) Effects of urban coarse particles inhalation on oxidative and inflammatory parameters in the mouse lung and colon. Part Fiber Toxicol 14 (1):46

Wang Q, Li F, Liang B, Liang Y, Chen S, Mo X, Ju Y, Zhao H, Jia H, Spector TD, Xie H, Guo R (2018) A metagenome-wide association study of gut microbiota in asthma in UK adults. BMC Microbiol 18:114–121

Weinmayr G, Hennig F, Fuks K, Nonnemacher M, Jakobs H, Möhlenkamp S et al (2015) Long-term exposure to fine particulate matter and incidence of type 2 diabetes mellitus in a cohort study: Effects of total and traffic-specific air pollution. Part Fiber Toxicol 14:1–8

Weile T, Van De, Vanhaecke L, Boeckart C, Peru K, Headly J, Verstraete W, Siciliano S (2005) Human colon microbiota transform policyclic aromatic hydrocarbons to estrogenic metabolites. Environ Health Prospect 113:6–10

Whelan FJ, Verschoor CP, Stearns JC, Rossi L, Luinstra K, Loeb M, Smieja M, Johnstone5 J, Surette MG, Bowdish DME (2014) The loss of topography in the microbial communities of the upper respiratory tract in the elderly. Annu Am Thorac Soc 4:513–521

World Health Organization (2021) WHO global air quality guidelines: particulate matter (PM2.5 and PM10), ozone, nitrogen dioxide, sulfur dioxide and carbon monoxide. World Health Organization. ISBN 9789240034228 (electronic version)

Xu W, Glenn W, You-Yi H, Anis L (2020) The Untwining of Immunosenescence and Aging. Semin Immunopathol 42(5):559–572

Xue Y, Chu J, Li Y, Kong X (2020) The influence of air pollution on respiratory microbiome: a link to respiratory disease. Toxicol Lett 334:14–20

Yang BY, Guo Y, Markevych I, Qian Z (min), Bloom MS, Heinrich J et al (2019) Association of long-term exposure to ambient air pollutants with risk factors for cardiovascular disease in China. JAMA Netw Open 2(3):1–15

Young JY, Kweon MN (2016) the gut microbiota: a key regulator of metabolic diseases. BMB Rep 49(10):536–541

Zhang C, Zhang T, Lu W, Duan X, Luo X, Liu S et al (2020) Altered airway microbiota composition in patients with pulmonary hypertension. Hypertension 76:1589–1599

Zhu Y, Li Q, Jiang H (2020) Gut microbiota in atherosclerosis: focus on trimethylamine N-oxide. J Pathol Microbiol Immunol 128:353–366

Zhou X, Jing L, Junli G, Bin G, Wenjie J, Qian Z, Jinlong L et al (2018) Gut-dependent microbial translocation induces inflammation and cardiovascular events after ST-elevation myocardial infarction. Microbiome 6(1):66

Zou T, Zhang F, Lui CYG, Yeoh YK, Li YLA, Zhan H, Yatin W, Chung CKA, Cheung CP, Chen N, Lai KCC, Chen Z, Tso EYK, Fung KSC, Chan V, Ling L, Joynt G, Hui DSC, Chan FKL, Chan PKS, Siew Ng C (2020) Alterations in the gut microbiota of patients with COVID-19 during the time of hospitalization. Gastroenterology 159(3):944-955

Zweilehner J, Liszt K, Handschur M, Lassi C, Lapin A, Haslberger AG (2009) Combined PCR-DGGE fingerprinting and quantitative-PCR indicates shift in fecal population sizes and diversity of Bacteroidetes, Bifidobacteria and Clostridium cluster IV in institutionalized elderly. Exp Gerontol 44:440–446

Chapter 12
Aging Gut Microbiota and Sarcopenia Signaling

Samatha M. Jain, Aparimita Das, Antara Banerjee, Manuel J. Castillo, Francesco Marotta, Arun Kumar Radhakrishnan, and Surajit Pathak

Abstract Recent advances in technology have allowed us to delve into the world of genetics in a way that could barely be conceived a couple of decades ago. Now besides the human genome, we can also study of the commensal microbiota present within the human gut and discover the link it shares to the functioning of the human body. Previously, researchers believed in the notion that the impact of gut microbiota extended only as far as the digestive system. Now it has been discovered that the gut microbiota has the potential to impact nearly every system in the body and possibly plays a pivotal role in aging and related sarcopenia. By directly influencing anabolic and catabolic processes, it may have the ability to alter muscle protein metabolism. While a large extent of this has been demonstrated in the laboratory on rodent models, it is yet to be fully demonstrated in human trials. It has been found that the transfer of gut microbiota can significantly alter metabolic processes in mice models. The microbiome as such plays an important role in signaling pathways at the molecular level, largely through hormone secretion. Significant differences have been observed between the gut commensal composition of a younger

Samatha M. Jain and Aparamita Das are Equal contributing authors.

S. M. Jain · A. Banerjee · S. Pathak (✉)
Department of Medical Biotechnology, Faculty of Allied Health Sciences, Chettinad Academy of Research and Education (CARE), Chettinad Hospital and Research Institute (CHRI), Chennai, India
e-mail: drsurajitpathak@care.edu.in; surajit.pathak@gmail.com

A. Das
Chettinad Hospital and Research Institute (CHRI), Chennai, India

M. J. Castillo
Department of Medical Physiology, School of Medicine, University of Granada, Granada, Spain

F. Marotta
ReGenera R&D International for Aging Intervention, Milano, Italy

A. K. Radhakrishnan
Department of Pharmacology, Chettinad Academy of Research and Education (CARE), Chettinad Hospital and Research Institute (CHRI), Chennai, India

F. Marotta (ed.), *Gut Microbiota in Aging and Chronic Diseases*, Healthy Ageing and Longevity 17, https://doi.org/10.1007/978-3-031-14023-5_12

population as opposed to an older one, which indicates that there exists a relationship between them and aging in the host. With an increase in the number of clinical trials, we may yet determine other roles played by gut commensals, and this could hold the key to newer clinical approaches to known diseases and change our perspective on aging. This chapter reviews the existing information about gut microbiota and its functioning with special emphasis on its role in sarcopenia.

Keywords Gut microbiota · Sarcopenia · Aging · Gut-muscle · Endocrine system

12.1 Introduction to Gut Microbiota

The human gastrointestinal (GI) tract is one of the greatest interfaces (250–400 m^2) in the human body between the host, environmental variables, and antigens. In a typical lifetime, around 60 tonnes of food travel through the human GI tract, coupled with a plethora of bacteria from the environment, posing a significant danger to gut integrity (Thursby and Juge 2017). As soon as humans and animals are born, huge numbers of bacteria invade their intestines. Anaerobic bacteria dominate this community, which consists of at least 10^{13} members and 500–1000 species. In ecology, the "gut microbiota" is an ensemble of bacteria, archaea, and eukarya that colonize and interact with the GI tract over thousands of years in a mutually advantageous manner (Neish 2009). The term "microbiota" refers to the whole population of microorganisms that inhabit a certain region, which includes not only bacteria but also fungus, archaea, viruses, and protozoans. Firmicutes and Bacteroidetes are the phyla that make up the majority of the healthy gut microbiota. *Actinobacteria* and *Verrucomicrobia* are the next two phyla (Bäckhed 2011). Furthermore, bacterial cells outnumber human cells in the human body. The gut microbiota has a lot of inter-individual variation, yet it's generally stable within individuals over time. Various human hosts, independent of gender, geographic region, or age, appear to have a functioning core microbiome supplied by numerous bacterial species. As people become older, their gut flora becomes more varied and unpredictable (Kim and Jazwinski 2018). It has long been assumed that the gut microbiota plays an important role in sustaining gut health in normal people and human health in general. The microbiota provides several benefits to the host through a variety of physiological activities, including gut integrity and epithelial shape, energy harvesting, pathogen protection, and host immunity regulation. These processes, however, may be disturbed as a result of dysbiosis, which is characterized by a change in microbial makeup. The innate immune response and persistent low-grade inflammation can be triggered by gut dysbiosis, leading to a variety of age-related degenerative diseases and unhealthy aging. Various chemicals, nutrition signaling-independent routes, and epigenetic processes are used by the gut microbiota to interact with the host. The disruption of these communications caused by age-related gut dysbiosis can have a negative impact on the host's health and longevity. This might explain how the gut microbiota affects health and aging.

It has been assumed that the microbiota develops at birth; however, some investigations have found microorganisms in tissues in the womb including the placenta, which calls this assumption into question (Rodríguez et al. 2015). The GI tract is rapidly colonized after birth, with life events such as illness, antibiotic treatment, and dietary changes producing chaotic fluctuations in the microbiota. The microbiota is often limited in variety in the early stages of development and is dominated by two major phyla, Actinobacteria and Proteobacteria (Bäckhed 2011). The microbiota composition converges towards a distinct adult-like microbial profile throughout the first year of life, with temporal patterns that are unique to each newborn. Despite the fact that the gut microbiota's composition is largely stable throughout adulthood, life experiences can nevertheless disrupt it (Dethlefsen and Relman 2011).

The human gut microbiota is dynamic and sensitive to dietary changes, which might have a significant impact on gut microbial composition. The microbiota composition in the GI tract is stratified on both a transverse and longitudinal axis and reflects the physiological characteristics of a specific area. Chemical, nutritional, and immunological gradients in the gut impact the density and composition of the microbiota. Most understanding regarding the adult human gut microbiota came from labor-intensive culture-based approaches approximately a decade ago. A common strategy is to target the bacterial 16S ribosomal RNA (rRNA) gene (Poretsky et al. 2014). Because this gene is found in all bacteria and archaea and has nine highly variable sections (V1–V9), it can be used to differentiate between species. Recent studies in mice colonized with a diverse pathogen-free microbiota revealed that the large intestine's outer mucus forms a distinct microbial niche, with bacterial species present in the mucus exhibiting differential proliferation and resource utilization compared to the same species in the intestinal lumen (Li et al. 2015). Host and environmental selection forces shape the microbiome makeup (Hooper and Macpherson 2010).

However, microbiota with different compositions may have some functional redundancy, resulting in a comparable protein or metabolite profile. Inter-individual variance arises as a result of factors that impact the gut environment, which in turn contribute to the unique gut microbial community.

Dietary habits can lead to some strains dominating the gut more than others, resulting in unique phenotypic features in the host. Dietary changes will increase the availability of different nutrients, resulting in a shift in which strains dominate, adding to the microbiota's dynamic character (Chen et al. 2014). Furthermore, ethnographic studies show that culture influences the general types of food ingested, and hence collective enterotypes, and so the effect of nutrition on gut microbiota. This may be seen in Asian people that consume a lot of starchy foods like rice; their enterotypes are marked by a lot of *Bifidobacterium*, a species renowned for producing a lot of starch-metabolizing enzymes (Al Bander et al. 2020).

12.2 Endocrine System in Gut Microbiota

The gut microbiota and its wide influence in regulation and production of hormones have resulted in it being considered a full-fledged endocrine organ (Clarke et al. 2014). Of late, with better understanding of human microbial ecology, the importance of gut microbiota and its involvement with various bodily systems have come to the fore. Gut microbiota has been found to greatly influence host metabolism and immunity. While its direct impact on the digestive system is more obvious to researchers, it has been recently discovered that gut microbiota also plays a crucial role in the modulation of hormone secretion (Diamant et al. 2011). Microbial colonization of the intestines since birth seems to have a role in host immunity via hormone secretion and bidirectional cross-talk between hormones secreted by commensal bacteria and the human host (IEκANU et al. 2020).

Bacteria have been found to perform quorum sensing, which revolves around sensing and producing auto-inducer molecules (Papenfort and Bassler 2016). This has further strengthened the belief that hormonal communication between microbes and hosts not only exists but helps to organize bacterial growth, virulence, and motility. These may also modulate host cell signal transduction and in effect activate signaling pathways (Table 12.1).

Table 12.1 Hormones and their impact on gut microbiota

Host factors	Hormone	Effect	Mechanism
Diet, exercise, stress, gender	Norepinephrine	Bacterial growth, biofilm growth, increased virulence	1. Binding with Ferric iron; Eg: *H. Pylori* 2. Formation of transferrin-norepinephrine complexes
	Epinephrine	Bacterial growth, biofilm growth, decreased resistance to host, increased virulence	1. Promoting cell adhesion 2. Induction of locus of enterocyte effacement (LEE) gene that impacts virulence
	Progesterone	Decreases bacterial growth	1. Upregulation of occludin results in decreased gut permeability
	Estradiol	Bacterial growth, decreased virulence	1. Induction of Intestinal Alkaline Phosphatase (IAP) by up-regulating KLF4 transcription factor 2. Lowering of Lipopolysaccharides (LPS) 3. Decreasing gut permeability
	Estriol	Decreased virulence	1. Decrease in gut permeability 2. Lowering of LPS
	Dopamine	Bacterial growth, biofilm growth	1. Converts to norepinephrine to exert effect 2. Other mechanisms are under investigation

Gut microbiota and the power it may possess to regulate the endocrine system can hold the key to understanding diseases ranging from irritable bowel syndrome to autism. This is essentially referred to as the "gut-brain axis, which is mediated mainly by the vagus nerve and hormonal modulations (Sun et al. 2020). Neurohormones and stress hormones can initiate physiological changes in the body, preparing the host for a certain response, which impacts blood pressure, heart rate, and metabolism (Revhaug et al. 1988). One of the most important neurotransmitters, serotonin, is mainly found in the intestines. This is a key component of the gut-brain axis and is regulated by gut microbiota, impacting host behavior (Jenkins et al. 2016). Certain bacteria such as *Streptococcus and Escherichia* can also produce serotonin (O'Mahony et al. 2015).

Stress, modulated via cortisone and adrenocorticotropic hormone, has been found to be elevated to a greater extent in germ-free mice in response to mild stress as opposed to other mice. In a study, *L. helveticus* and *B. longum* were found to reduce stress hormones and anxiety-like behavior in humans (Messaoudi et al. 2011).

Gender plays an important role in gut microbiota composition and function, especially given the alteration in hormone function and metabolism. "Microgenderome" is the term coined to encompass gender-specific immune changes and disease processes that may be a result of gut microbiota based on its differential composition in different genders (Vemuri et al. 2019).

Mating behavior has been observed to be influenced by commensal gut bacteria. In post-menopausal women, there is an established association between the gut microbiota diversity and ratio of estrogen metabolites in urine (Fuhrman et al. 2014).

The latest information linking gut microbiota and the endocrine system comes from research involving gastric bypass surgery. Roux-en-Y bypass performed on mice showed compositional changes in microbiota and transfer of microbiota from these mice to germ-free mice resulted in weight loss and decreased fat mass (Fouladi et al. 2019). This was correlated to the abundance of specific groups of bacteria, compounded with reduced food intake and bile acid reduction. Intestinal hormones such as GLP1 and glucose-dependent insulinotropic polypeptide (GIP) and insulin levels are also altered, adding to the effect.

It has been demonstrated that the development of diabetes could be prevented in mice using androgen treatment (Fox 1992). Increased levels of testosterone in male mice were found to be protective against Type 1 diabetes. The same effect was observed in immature female mice when microbiome transplant was done from the former to the latter.

Gut microbiota is also reported to play a role in thyroid disorders, particularly affecting women. Gut microbiota was found to vary in patients with thyroid cancer as opposed to healthy controls. In the former, *Neisseria* and *Streptococcus* were found to be elevated and *Lactobacilli* were fewer than usual. This supports the hypothesis that the composition of gut microbiota can alter and be altered in response to thyroid disorders (Knezevic et al. 2020).

By extrapolation of the above-mentioned findings, it has been hypothesized that the growth and development of host may be indirectly affected by gut microbiota via hormone regulation. Therefore, numerous studies link gut microbiota with hormonal regulation, which in turn affects every system in the host's body. A better understanding could hold the key to clinical approach in a number of diseases. Microbiota can indirectly control the hosts' metabolism, appetite, reproduction and immune response via hormonal regulation and crosstalk between host and bacterial hormones.

12.3 Aging and Gut Microbiota

12.3.1 Role of Aging in Gut Microbiota

A vast range of physiological, genetic, metabolic, and immunological activities are affected by the process of aging, making it a very complex process. As we age, our organs and associated physical functions begin to deteriorate. Frailty is a phrase used to describe a person's vulnerability to age-related health problems, such as falls, impairment, institutionalization, and mortality (Maffei et al. 2017). Characteristics of frailty include anorexia and sarcopenia as well as declining cognition, relative inactivity, and dependence on others to conduct everyday duties. The capacity of the host to maintain a robust and effective immune response and metabolic health has been known for decades as part of the aging process (Nagpal et al. 2018). The effect of these age-related changes on the host's health state, as well as the effect of these age-related changes on the gut microbiota composition, is gaining a lot of interest, especially in terms of enhancing the wellness of older individuals by manipulating the gut microbiota balance. Particular variables such as age-related changes in lifestyle, diet, frailty, and inflammation have a significant impact on the pace of age-related degradation of a mutualistic relationship between the host and intestinal microbiota (Shen et al. 2018). The stability and diversity of a person's gut microbiota diminish with age, as do their health. Microbiota composition generally preserves the stability and compositional make-up of a healthy younger adult if health is preserved. Physiological changes, food choices and malnutrition, living environment (community-dwelling, hospitalized, or long-term care), and use of antibiotics and other prescription medications are the most common age-related variables impacting the microbial population of the gut (Voreades et al. 2014). Biagi et al. found that the microbial makeup of young adults and individuals in their seventies is quite similar, but that it differs substantially from that of centenarians (Biagi et al. 2010). The gut microbiota of people who lived over 100 years showed a shift in the Firmicutes population and an enrichment in Proteobacteria, including opportunistic pro-inflammatory bacteria ("pathobiobacteria"). Tongeren et al. investigated the link between microbial diversity and frailty in older people. Individuals with high frailty scores had a significantly lower

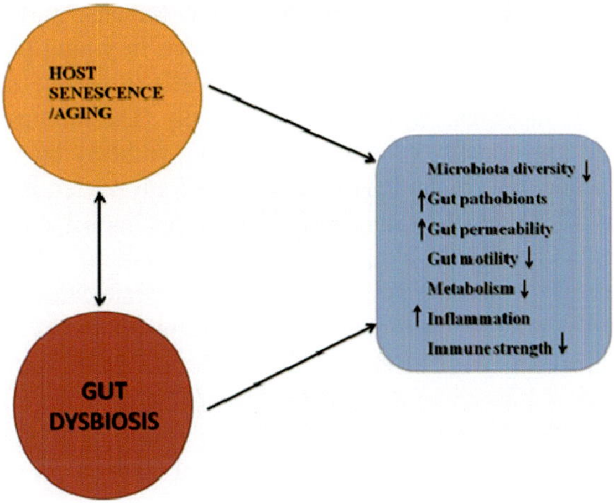

Fig. 12.1 Effects of ageing and gut dysbiosis on gut microbiota

number of lactobacilli, Bacteroides/Prevotella, and Faecalibacterium prausnitzii, and a significantly higher proportion of *Ruminococcus, Atopobium*, and *Enterobacteriaceae* (van Tongeren et al. 2005). A centenarian study found that genes controlling short-chain fatty acid synthesis and saccharolytic capability decreased with age in the human intestinal microbiome. Older people had a greater number of proteolytic functions within their intestinal metagenomes. Elderly people have few commensals like bacteroides and *bifidobacteria*, which are normally found in the gut, and the variety of their gut microbiota is often reduced. Despite this, the levels of opportunists such as *enterobacteria*, *Clostridium perfringens*, and *Clostridium difficile* are elevated (Rampelli et al. 2013). The results of these studies suggest that aging-related changes in gut bacteria may vary according to geographical location, since different results have been found in older people from different regions. The root cause or effect of gut dysbiosis in aging and senescence-related inflammatory diseases is unknown. However, the gut microbiota is believed to contribute to healthy aging as well as restorative medicine (Fig. 12.1).

12.3.2 Age-Related Alteration in Human Gut Microbiota

It's no surprise that the gut microbial ecology is garnering a lot of interest as a possible target for creating innovative techniques for healthy aging and older welfare, given the recent discovery of the link between aging and the intestinal microbiota. The gut microbiota has the potential to influence a number of age-related factors directly or indirectly, and it is an area of research that has yet to be fully investigated. Increased abundance of Gram-negative bacteria, such as Enterobacteriaceae and other pathogens, is one of the most serious consequences of age-related changes in gut microbiota composition. The amount of short chain fatty acid (SCFAs) in the gut of elderly persons, particularly acetate, butyrate, and propionate, is thought to be lower than in young people (Biagi et al. 2010). This may cause the intestinal epithelial cells to secrete mucins, allowing infections to enter the intestinal mucosa more easily. The immune system can be hampered by the age-related reduction in SCFA synthesis by the gut microbiota of elderly people, which can lead to a variety of gut-related illnesses in humans. The decline in SCFA levels with age may have an impact on the host's immunological and inflammatory health, leading to a variety of gut-related disorders in the elderly. Another research found that butyrate causes epigenetic changes in immune cells, such as macrophages, which decreases the release of inflammatory molecules like IL-6 (Nagpal et al. 2016). The permeability of the mucosal membrane is reduced as a result, reducing the influx of toxic chemicals such as lipopolysaccharides from the colon. Additional study is needed in this field to address the information gaps between age-related changes in the gut microbial ecology and human health (Kumar et al. 2016).

According to the study, maintaining a healthy gut microbiota through fiber/ microbiota-accessible carbohydrate (MAC)-rich diets or taking the right pro/ prebiotics might help prevent and treat age-related pathophysiological illnesses, hence extending lifespan (Kumar et al. 2016).

It is thought that changes in the gut microbiota caused by aging offer a favorable environment for pathogen development in the gut. Prebiotics and probiotics may help to maintain a favorable diversity of gut microorganisms, which may help to avoid or postpone *C. difficile* infections in older people. Such research will provide more precise treatment targets for treating or delaying gut-related illnesses in the elderly (Kumar et al. 2016).

12.4 Relationship Between Skeletal Muscle and Microbiome

The link between health evolution through time and gut microbiota (GM) and skeletal muscle metabolic and immunological activities is now well understood. The effects of gut microbiota on skeletal muscle, however, have only been studied in a few researches. Skeletal muscles in mammals contract to allow the body to move more easily, resulting in energy consumption. Intestinal bacteria are thought to regulate muscle tissue development and function in humans and animals, according to several studies. Germ-free (GF) mice showed muscular atrophy, lower insulin-like growth factor (IGF) production, and lower expression of genes involved in skeletal muscle development and mitochondrial function, according to a recent study (Qi et al. 2021).

Endocrine organs such as skeletal muscle and adipose tissue release myokines and adipokines, which affect skeletal muscle. Gut bacteria are linked to skeletal muscle metabolism, according to a growing number of researches in recent years.

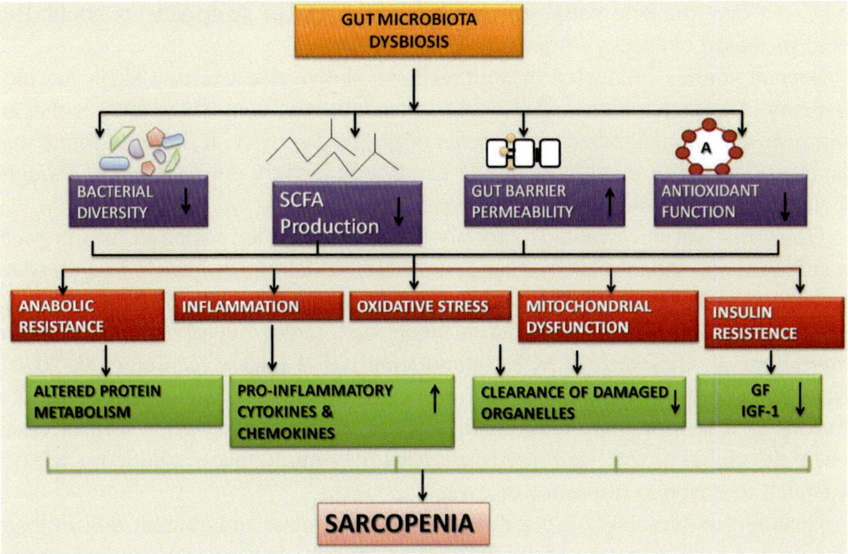

Fig. 12.2 Relationship between gut microbiota and sarcopenia

The composition and function of skeletal muscle may be influenced by gut bacteria's health is explained in Fig. 12.2. Patients and even the elderly suffer from gastrointestinal microbial diseases, which reduce their quality of life. As a result, understanding the involvement of the gut microbiota in skeletal muscle metabolism is critical for us to discover an effective intervention strategy.

12.5 Potential Role of Gut Microbiota in Gut-Muscle Axis

Sarcopenia, a loss of muscle mass, is mainly a geriatric syndrome that is often exacerbated by acute or chronic conditions. The gut-muscle axis may be a strong influencer of onset and progression of sarcopenia. The gut-muscle axis depends on nutrition and exercise, which in turn affect inflammation, insulin sensitivity, and anabolism. Thus inappropriate nutrition or a lack of exercise are the stepping stones to the onset of sarcopenia. Diagnostic criteria for sarcopenia involve, reduced appendicular lean mass, reduced skeletal muscle index, analyzed via bioimpedance analysis or DXA (Ticinesi et al. 2017). A study showed that species diversity in feces of older subjects is inversely related to physical activity. Research has even successfully established an association between certain bacterial taxa and frailty via the Rockwood Clinical Frailty Scale. This can be of great value clinically, where reduced chances of survival can be associated with gut microbiota dysbiosis and increased opportunistic pathogens present in the gut of such patients could also yield increased chances of infection.

Several studies conducted in animals have shown that exercise alters the biodiversity of gut microbiota and it works in tandem with diet, which implies that an improper diet can neutralize any benefits of physical activity. It has been found that the alteration in gut microbiota due to exercise is slightly different to that brought about by a healthy diet, making it unique.

Given that gut microbiota is strongly influenced by diet, which in turn impacts metabolism and inflammation, a suggestion of an existing gut-muscle axis presents itself. Germ-free mice have been seen to persist in showing a lean phenotype even on high-fat diets. Gut microbiota is, therefore, largely involved in anabolism through metabolites such as SCFA stimulating IGF-1 release (Jensen et al. 2020). Transplantation of feral microbiota from malnourished African children to mice resulted in a loss of body mass in these mice whereas this wasn't seen in cases where the children were well nourished. This may imply that dysbiotic microbiota is related to anabolic resistance or even catabolism.

Healthy gut microbiota is capable of producing folate and vitamin B12 in large quantities, which may prevent oxidative stress-related damage and promote muscle anabolism. Among other things, gut microbiota can synthesize amino acids such as tryptophan, which furthers expression of genes resulting in myofibrillar synthesis. Betaine derived from diet also impacts skeletal muscle cells (Banerjee et al. 2021; Rowland et al. 2018).

The most documented mediators include SFCA, derived from commensal metabolism, through diet. They mainly target mitochondria within skeletal muscles. These enter the hosts's systemic circulation and are taken up by skeletal muscles where they ultimately regulate glucose metabolism, promote insulin sensitivity and also affect mitochondrial biogenesis. Butyrate seems to be one of the most active SFCAs, playing an important role in inflammatory processes and activation of a number of regulatory pathways that encompass ATP production and prevention of apoptosis. It possibly provides some degree of protection against muscle protein catabolism and subsequent loss of muscle mass, as seen in a study performed on aging mice who were given butyrate. While butyrate is considered beneficial, as far as sarcopenia is concerned, another SFCA, acetate is considered detrimental with effects opposite to that of butyrate (Amedei and Morbidelli 2019).

Multiple studies have found that an important class of polyphenols, ellagitannins, is metabolized by commensal bacteria to form urothilin, which prevents accumulation of dysfunctional mitochondria and has been seen to improve exercise capacity in mice upon consumption of urolithin A (Toney et al. 2019). So far most studies involving the gut-muscle axis have been done in animal models. In mouse models with cancer, *Lactobacillus Reuters* was administered, and it was found that cachexia could be prevented and muscle mass was preserved. In other mouse models, reduced inflammation and improved liver anabolism was noted upon administration of *Faecalibacterium prausnitzii* (Crovesy et al. 2020). Older mice, upon *Spirulina* administration showed reduced effects of oxidative stress at the cellular level. Out of 10 clinical and 26 pre-clinical trials analyzed in a study, 3 showed a relationship between commensals and muscle, 3 others focused on germ-free mice that successfully replicated the muscle phenotype of their donors post fecal transplantation (Liu et al. 2021). This homeostasis, maintained by gut microbiota, can be disturbed by antibiotic administration, as demonstrated by multiple animal studies ultimately increasing muscle protein catabolism and promoting sarcopenia.

Even though such studies are limited in humans, an important conclusion drawn from a series of 6 studies involving more than 600 elderly people and more than 200 young adults was the association that could be shown between the muscle and gut microbiota. Prebiotic administration in 60 old patients was carried out in one randomized controlled trial as opposed to placebo and showed improved handgrip strength and reduced muscle exhaustion, giving way to hope for future manipulation of gut microbiota in humans to reverse or delay age-related sarcopenia.

Of all the studies, mechanisms through which gut microbiota impacts muscles have been narrowed down to the metabolism of glucose, protein, and lipids, alteration in inflammatory pathways, at the neuromuscular junctions and modulation of mitochondrial function. All these mechanisms may be manipulated via supplements such as probiotics, bacteria depletion, and fecal transplantation.

12.6 Gut Microbiota—A Novel Player in Sarcopenia

Skeletal muscle mass is almost entirely regulated by muscle protein turnover, involving synthesis and breakdown. Sarcopenia, understood as age related decay in this muscle mass, can be categorized into a primary form and a secondary form. Primary sarcopenia involves loss of muscle mass mainly as a result of chronic subclinical inflammation and cellular oxidative stress coupled with a lack of exercise and poor nutrition, even though the exact cause hasn't been determined. Secondary sarcopenia is a direct result of a distinct disease process such as COPD, cirrhosis, or cancer (Aryana et al. 2018). Sarcopenia must also be distinguished from physical frailty, which is an outcome of a stressor event, measured using the Fried criteria or the Rockwood frailty index (Bone et al. 2017). Sarcopenia is associated with poor clinical outcomes, including increase in hospital admissions and morbidity. Loss in muscle mass with old age could be natural fallout of "anorexia of aging or disease-related malnutrition combined with declining physical activity (Morley 1997).

All the factors leading up to sarcopenia seem to be controlled by the gut commensals via various molecular mechanisms and metabolic pathways. Although more animal studies have been done in this area as compared to human studies, it is clear that a relationship between gut microbiota and muscle aging does exist. It has also been concluded that timely interventions could potentially reverse sarcopenia, making gut microbiota a novel target in preventing muscle wasting and dysfunction.

From previous studies, we know that mice with microbial dysbiosis at a young age went on to age with decreased physical function and muscle mass as shown in Fig. 12.3. It has been established that decreased gut biodiversity can directly induce muscle atrophy. Germ-free mice were observed to be more prone to fatigue and generally had lower muscle function and mass. Studies suggest that recipient mice often adapt the muscle phenotype, grip strength, and endurance of their donors after microbiome transplantation (Liu et al. 2021). However, one study also correlates microbiome depletion to better locomotor activity (Chaudhari et al. 2017).

Based on these studies, further trials were conducted to test appropriate interventions targeting the gut commensals. These found that beneficial probiotic supplements include, *Lactobacillus salivarius, Lactobacillus casei, Lactobacillus paracasei, Lactobacillus plantarum, Bifidobacterium longum* and *Saccharomyces boulardii* (Holzapfel et al. 2001). SFCAs involved were butyrate, acetate, and propionate, which work to maintain muscle mass and strength (Kelder et al. 2014). Prebiotics containing fructooligosaccharides and inulin showed direct impact on grip strength and endurance capacity in older subjects (Ticinesi et al. 2019). In younger subjects, the same showed reduced fatiguability and increase in muscle mass. It was also confirmed that antibiotics cause gut dysbiosis. Other key findings include; correlation between gut microbiome and physical fitness as well as muscle mass in young women.

Although the gut microbiota undergoes rapid changes for the first couple of years of life, it stagnates after a certain age possibly due to aging. These changes include a reduced number of butyrate producers. Older individuals have been seen to have a larger Bacteroidetes population and fewer *Ruminococcus* and *Blautia* species (Lippert et al. 2017). *Escherichia* is also seen to be more abundant in the elderly. A lower Firmicutes/Bacteroides ratio (F/B ratio) has been reported in the aging population (Bibbò et al.2017).

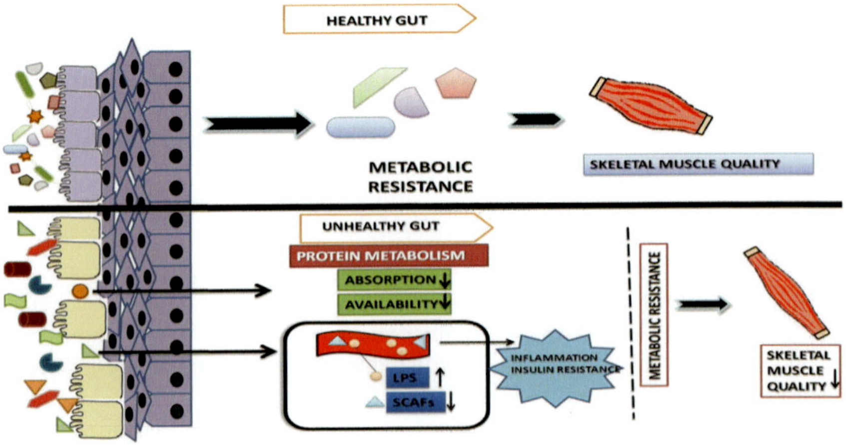

Fig. 12.3 Representation of skeletal muscle quality influenced by gut microbiota

On the whole, gut microbiota plays a pivotal role in maintaining the balance between pro- and anti-inflammatory responses as well protein synthesis and breakdown. Both of these processes are directly related to aging. With increasing age, the gut commensals seem less capable of counteracting pathogens and reduced SFCA levels trigger mucin secretion, which ultimately allow for pathogens to enter the intestinal mucosa. LPS leakage from intestine to circulation is the fallout of increased intestinal permeability with age. All of this contributes to triggering innate immunity, resulting in a state of chronic inflammation. Therefore, chronic inflammation may form the backbone of sarcopenia (Grosicki et al. 2018).

The ease of conducting studies in this field has resulted from the advent of sequencing technologies. The next-generation sequencing (NGS) enables the characterization of microbial species including pathogens and has helped us to understand microbial diversity (Greay et al. 2018). This has paved the way for metagenomics, which is even more advanced in its approach to analysis. Now larger sample sizes can be studied, and a reference catalog is being assembled for future reference. Sequences are assigned to respective taxonomic groups called operational taxonomic unit (OTU) in what is known as the binning process (He et al. 2015). All of this leaves scope for more research, especially among human subjects but is enough to establish gut microbiota as the key to a better understanding of sarcopenia and its underlying processes.

12.7 Gut-Related Mechanisms Leading to Sarcopenia

Sarcopenia when studied with a view of the gut-muscle axis involves specific mechanisms via which muscle breakdown and dysfunction are perpetuated. As a result of animal studies, it has been found that complete absence of gut microbiota resulted in breakdown of branched-chain amino acids in muscle. This pathway was typified by decreased expression of branched-chain-keto acid dehydrogenase kinase and increased expression of branched-chain-keto acid dehydrogenase, branched-chain aminotransferase 2 and Kruppel-like factor 15 (Nie et al. 2018).

Extensive research conducted on germ-free mice showed higher quantities of alanine and glycine along with increased phosphorylation of the adenosine 5′-monophosphate-activated protein kinase (AMPK), which prevented obesity and increased fatty acid oxidation (Nie et al. 2018; Lahiri et al. 2019). Upregulation of forkhead box O3 (FoxO3), Atrogin-1 and muscle RING-finger protein-1 (Murf-1) was observed but reversed on microbiota transplantation. FoxO3 and other genes related to muscle atrophy were found to be higher in GF mice treated with antibiotics. These mice also showed mitochondrial dysfunction and decreased mtDNA along with dampened succinate dehydrogenase (SDH) activity (Waddell et al. 2008).

Studies conducted to improve the understanding of muscle function alteration, assessed serum lactate, ammonia, creatine kinase (CK), and blood urea nitrogen (BUN), which are markers of fatigue and found that exercise combined with LP10, kefir, curcumin, and other supplements, reduced these markers considerably (Hsu et al. 2018).

Further analysis indicated that muscle atrophy was related to the breakdown of creatine in aging mice. Antibiotic administration even caused the shortening of myofibres. Other studies also showed that germ-free mice demonstrate increased fatty acid metabolism involving AMPK, which regulate oxidative metabolism and mitochondrial content, in effect, offering some degree of protection against diet-induced obesity. Other factors detrimental to muscle growth in GF mice include low levels of insulin-like growth factor 1 gene, myoblast determination protein 1, and myogenin. SFCAs cause the effect through Atrogin-1 and MyoD.39 (Suriano et al. 2020).

Another possible mechanism resulting in sarcopenia involves the "leaky gut hypothesis" where a breakdown of the intestinal barrier allows toxins that promote inflammation into the blood. This could result in low-grade chronic inflammation, eventually contributing to anabolic resistance. In animal models, the leaky gut has been observed to be the fallout of endotoxemia due to gut bacteria. It is possible that restricting gut permeability to toxic bacterial products could better muscle function by preventing insulin resistance, low-grade inflammation and glucose intolerance. This was achieved in mice through probiotic supplementation (de Marco Castro et al. 2021; DeJong 2020). SFCAs, mainly produced by probiotic bacteria work to enhance the function of muscle nutrient sensors. Studies show a strong association between SFCA concentration and muscle mass and strength even though the mechanism is not entirely clear.

"Inflamm-aging is a process that is of late discovered to be driven by diet-derived micro RNA (inflamma-miRs). Diet-derived exogenous miRs (referred as "xenomiRs") can affect circulating miRs homeostasis that is altered in many diseases. This is done by the modulation of common signaling pathways (Bacalini et al. 2014). One study took up three miRs, miR-155, miR-146a, and miR-21. These were seen to be regulating some pathways such as fatty acid lengthening and lysine degradation. These miRs were found in cow fat and cow milk, which suggests the possible roles these play in the inflammatory pathways (Teodori et al. 2019). Another such pathway is the kynurenine pathway and serotonin synthesis, which are parts of tryptophan metabolism. Some studies have found that gut microbiota may also be responsible for the control of tryptophan metabolism, modulating neuroendocrine function (Kennedy et al. 2017). Antibiotic treatment, also widely studied alongside gut microbiota, showed that modulation of gut commensals, using the same, reduced TLR4 pathway activation and JNK activation and IκBα degradation by lowering circulating LPS levels. A consequence of this was increase in specific proteins through insulin induced phosphorylation (Boulangé et al. 2016).

12.8 Conclusion

Sarcopenia, through various studies, has been strongly linked to gut microbiota. The concept of "healthy aging" is evolving and the need to understand the workings of gut commensals and their modulation is on the rise. We are yet to definitively establish a direct impact of human gut commensals on muscle protein metabolism, but evidence provided by animal studies suggest a causal relationship between the gut microbiota and muscle protein synthesis. However, data do point to correlation between the two on the basis of improved outcomes in the elderly population on modulation of gut microbiota through probiotic supplementation. In such population, inflammation and metabolic endotoxemia were seen to be reducing as a result of better cell wall integrity and reduced anabolic resistance. On filling the lacunae in research in human subjects we may better understand the relationship between gut microbiota and sarcopenia and more importantly use them as potential future targets for manipulation with a view to enhancing muscle function and activity even with progressing age. This will keep on improving our understanding of nutrition together with efficient and tailored physical activity (Castillo-Garzón et al. 2006) to favor healthy aging strategies.

References

Al Bander Z, Nitert MD, Mousa A, Naderpoor N (2020) The gut microbiota and inflammation: an overview. Int J Environ Res Public Health 17(20):7618

Amedei A, Morbidelli L (2019) Circulating metabolites originating from gut microbiota control endothelial cell function. Molecules 24(21):3992

Aryana S, Lestari AW, Putrawan IB, Purnami NKR, Astika IN, Kuswardhani RT (2018) The relationship between IL-6 and CRP with Sarcopenia in indigenous elderly population at Pedawa Village, Buleleng, Bali Indonesia. Health Sci J Indonesia 9(1):37–44

Bäckhed F (2011) Programming of host metabolism by the gut microbiota. Ann Nutr Metab 58 (Suppl. 2):44–52

Biagi E, Nylund L, Candela M, Ostan R, Bucci L, Pini E, Nikkïla J, Monti D, Satokari R, Franceschi C, Brigidi P (2010) Through ageing, and beyond: gut microbiota and inflammatory status in seniors and centenarians. PLoS ONE 5(5):e10667

Banerjee A, Marotta F, Sriramulu S, Chabria Y, Hari S, Catanzaro R, Barbagallo M, Balakrishnan B, He F, Nandagopal Radha RK, Pathak S (2021) Beyond Physical Exercise: The Role of Nutrition, Gut Microbiota and Nutraceutical Supplementation in Reducing Age-Related Sarcopenia. Curr Aging Sci 14(2):94–104

Bone AE, Hepgul N, Kon S, Maddocks M (2017) Sarcopenia and frailty in chronic respiratory disease: Lessons from gerontology. Chron Respir Dis 14(1):85–99

Bibbò S, Dore MP, Pes GM, Delitala G, Delitala AP (2017) Is there a role for gut microbiota in type 1 diabetes pathogenesis? Ann Med 49(1):11–22

Bacalini MG, Friso S, Olivieri F, Pirazzini C, Giuliani C, Capri M, Santoro A, Franceschi C, Garagnani P (2014) Present and future of anti-ageing epigenetic diets. Mech Ageing Dev 136:101–115

Boulangé CL, Neves AL, Chilloux J, Nicholson JK, Dumas ME (2016) Impact of the gut microbiota on inflammation, obesity, and metabolic disease. Genome Medicine 8(1):1–12

Chen J, He X, Huang J (2014) Diet effects in gut microbiome and obesity. J Food Sci 79(4):R442–R451

Clarke G, Stilling RM, Kennedy PJ, Stanton C, Cryan JF, Dinan TG (2014) Minireview: gut microbiota: the neglected endocrine organ. Mol Endocrinol 28(8):1221–1238

Crovesy L, Masterson D, Rosado EL (2020) Profile of the gut microbiota of adults with obesity: a systematic review. Eur J Clin Nutr 74(9):1251–1262

Chaudhari SN, McCurry MD, Devlin AS (2021) Chains of evidence from correlations to causal molecules in microbiome-linked diseases. Nat Chem Biol 17(10):1046–1056

Castillo-Garzón MJ, Ruiz JR, Ortega FB, Gutiérrez Á (2006) Anti-aging therapy through fitness enhancement. Clin Interv Aging 1(3):213

Dethlefsen L, Relman DA (2011) Incomplete recovery and individualized responses of the human distal gut microbiota to repeated antibiotic perturbation. Proc Natl Acad Sci 108(supplement_1):4554–4561

Diamant M, Blaak EE, De Vos WM (2011) Do nutrient–gut–microbiota interactions play a role in human obesity, insulin resistance and type 2 diabetes? Obes Rev 12(4):272–281

de Marco Castro E, Murphy CH, Roche HM (2021) Targeting the gut microbiota to improve dietary protein efficacy to mitigate sarcopenia. Front Nutr:331

DeJong E (2020) Investigating the role of the intestinal microbiota in unhealthy aging, Doctoral dissertation

Fuhrman BJ, Feigelson HS, Flores R, Gail MH, Xu X, Ravel J, Goedert JJ (2014) Associations of the fecal microbiome with urinary estrogens and estrogen metabolites in postmenopausal women. J Clin Endocrinol Metab 99(12):4632–4640

Fouladi F, Brooks AE, Fodor AA, Carroll IM, Bulik-Sullivan EC, Tsilimigras MC, Sioda M, Steffen KJ (2019) The role of the gut microbiota in sustained weight loss following Roux-en-Y gastric bypass surgery. Obes Surg 29(4):1259–1267

Fox HS (1992) Androgen treatment prevents diabetes in nonobese diabetic mice. J Exp Med 175 (5):1409–1412

Grosicki GJ, Fielding RA, Lustgarten MS (2018) Gut microbiota contribute to age-related changes in skeletal muscle size, composition, and function: biological basis for a gut-muscle axis. Calcif Tissue Int 102(4):433–442

Greay TL, Gofton AW, Paparini A, Ryan UM, Oskam CL, Irwin PJ (2018) Recent insights into the tick microbiome gained through next-generation sequencing. Parasit Vectors 11(1):1–14

Hooper LV, Macpherson AJ (2010) Immune adaptations that maintain homeostasis with the intestinal microbiota. Nat Rev Immunol 10(3):159–169

Holzapfel WH, Haberer P, Geisen R, Björkroth J, Schillinger U (2001) Taxonomy and important features of probiotic microorganisms in food and nutrition. Am J Clin Nutr 73(2):365s–373s

He Y, Caporaso JG, Jiang XT, Sheng HF, Huse SM, Rideout JR, Edgar RC, Kopylova E, Walters WA, Knight R, Zhou HW (2015) Stability of operational taxonomic units: an important but neglected property for analyzing microbial diversity. Microbiome 3(1):1–10

Hsu YJ, Huang WC, Lin JS, Chen YM, Ho ST, Huang CC, Tung YT (2018) Kefir supplementation modifies gut microbiota composition, reduces physical fatigue, and improves exercise performance in mice. Nutrients 10(7):862

Ieкanu MARA, Zahiu CDM, Georgescu IA, Zăgrean L, Zăgrean A (2020) The microbiota-gut-brain axis-a complex system cross-talk. Fiziologia-Physiol 30(1)

Jenkins TA, Nguyen JC, Polglaze KE, Bertrand PP (2016) Influence of tryptophan and serotonin on mood and cognition with a possible role of the gut-brain axis. Nutrients 8(1):56

Jensen EA, Young JA, Mathes SC, List EO, Carroll RK, Kuhn J, Onusko M, Kopchick JJ, Murphy ER, Berryman DE (2020) Crosstalk between the growth hormone/insulin-like growth factor-1 axis and the gut microbiome: a new frontier for microbial endocrinology. Growth Hormon IGF Res 53:101333

Kim S, Jazwinski SM (2018) The gut microbiota and healthy aging: a mini-review. Gerontology 64(6):513–520

Knezevic J, Starchl C, Tmava Berisha A, Amrein K (2020) Thyroid-gut-axis: how does the microbiota influence thyroid function? Nutrients 12(6):1769

Kumar M, Babaei P, Ji B, Nielsen J (2016) Human gut microbiota and healthy aging: recent developments and future prospective. Nutr Healthy Aging 4(1):3–16

Kelder T, Stroeve JHM, Bijlsma S, Radonjic M, Roeselers G (2014) Correlation network analysis reveals relationships between diet-induced changes in human gut microbiota and metabolic health. Nutr Diabetes 4(6):e122–e122

Kennedy PJ, Cryan JF, Dinan TG, Clarke G (2017) Kynurenine pathway metabolism and the microbiota-gut-brain axis. Neuropharmacology 112:399–412

Li H, Limenitakis JP, Fuhrer T, Geuking MB, Lawson MA, Wyss M, Brugiroux S, Keller I, Macpherson JA, Rupp S, Stolp B (2015) The outer mucus layer hosts a distinct intestinal microbial niche. Nat Commun 6(1):1–13

Liu C, Cheung WH, Li J, Chow SKH, Yu J, Wong SH, Ip M, Sung JJY, Wong RMY (2021) Understanding the gut microbiota and sarcopenia: a systematic review. J Cachexia Sarcopenia Muscle 12(6):1393–1407

Lippert K, Kedenko L, Antonielli L, Kedenko I, Gemeier C, Leitner M, Kautzky-Willer A, Paulweber B, Hackl E (2017) Gut microbiota dysbiosis associated with glucose metabolism disorders and the metabolic syndrome in older adults. Benef Microbes 8(4):545–556

Lahiri S, Kim H, Garcia-Perez I, Reza MM, Martin KA, Kundu P, Cox LM, Selkrig J, Posma JM, Zhang H, Padmanabhan P (2019) The gut microbiota influences skeletal muscle mass and function in mice. Sci Transl Med 11(502):5662

Messaoudi M, Violle N, Bisson JF, Desor D, Javelot H, Rougeot C (2011) Beneficial psychological effects of a probiotic formulation (Lactobacillus helveticus R0052 and Bifidobacterium longum R0175) in healthy human volunteers. Gut Microbes 2(4):256–261

Maffei VJ, Kim S, Blanchard E IV, Luo M, Jazwinski SM, Taylor CM, Welsh DA (2017) Biological aging and the human gut microbiota. J Gerontol Ser Biomed Sci Med Sci 72 (11):1474–1482

Morley JE (1997) Anorexia of aging: physiologic and pathologic. Am J Clin Nutr 66(4):760–773

Neish AS (2009) Microbes in gastrointestinal health and disease. Gastroenterology 136(1):65–80

Nagpal R, Mainali R, Ahmadi S, Wang S, Singh R, Kavanagh K, Kitzman DW, Kushugulova A, Marotta F, Yadav H (2018) Gut microbiome and aging: Physiological and mechanistic insights. Nutr Healthy Aging 4(4):267–285

Nagpal R, Kumar M, Yadav AK, Hemalatha R, Yadav H, Marotta F, Yamashiro Y (2016) Gut microbiota in health and disease: an overview focused on metabolic inflammation. Benef Microbes 7(2):181–194

Nie C, He T, Zhang W, Zhang G, Ma X (2018) Branched chain amino acids: beyond nutrition metabolism. Int J Mol Sci 19(4):954

O'Mahony SM, Clarke G, Borre YE, Dinan TG, Cryan JF (2015) Serotonin, tryptophan metabolism and the brain-gut-microbiome axis. Behav Brain Res 277:32–48

Poretsky R, Rodriguez-R LM, Luo C, Tsementzi D, Konstantinidis KT (2014) Strengths and limitations of 16S rRNA gene amplicon sequencing in revealing temporal microbial community dynamics. PLoS ONE 9(4):e93827

Papenfort K, Bassler BL (2016) Quorum sensing signal–response systems in Gram-negative bacteria. Nat Rev Microbiol 14(9):576–588

Qi R, Sun J, Qiu X, Zhang Y, Wang J, Wang Q, Huang J, Ge L, Liu Z (2021) The intestinal microbiota contributes to the growth and physiological state of muscle tissue in piglets. Sci Rep 11(1):1–14

Rodríguez JM, Murphy K, Stanton C, Ross RP, Kober OI, Juge N, Avershina E, Rudi K, Narbad A, Jenmalm MC, Marchesi JR (2015) The composition of the gut microbiota throughout life, with an emphasis on early life. Microb Ecol Health Dis 26(1):26050

Revhaug A, Michie HR, Manson JM, Watters JM, Dinarello CA, Wolff SM, Wilmore DW (1988) Inhibition of cyclo-oxygenase attenuates the metabolic response to endotoxin in humans. Arch Surg 123(2):162–170

Rampelli S, Candela M, Turroni S, Biagi E, Collino S, Franceschi C, O'Toole PW, Brigidi P (2013) Functional metagenomic profiling of intestinal microbiome in extreme ageing. Aging (Albany NY) 5(12):902

Rowland I, Gibson G, Heinken A, Scott K, Swann J, Thiele I, Tuohy K (2018) Gut microbiota functions: metabolism of nutrients and other food components. Eur J Nutr 57(1):1–24
Sun LJ, Li JN, Nie YZ (2020) Gut hormones in microbiota-gut-brain cross-talk. Chin Med J 133 (07):826–833
Suriano F, Van Hul M, Cani PD (2020) Gut microbiota and regulation of myokine-adipokine function. Curr Opin Pharmacol 52:9–17
Thursby E, Juge N (2017) Introduction to the human gut microbiota. Biochem J 474(11):1823–1836
Ticinesi A, Lauretani F, Milani C, Nouvenne A, Tana C, Del Rio D, Maggio M, Ventura M, Meschi T (2017) Aging gut microbiota at the cross-road between nutrition, physical frailty, and sarcopenia: is there a gut–muscle axis? Nutrients 9(12):1303
Toney AM, Fan R, Xian Y, Chaidez V, Ramer-Tait AE, Chung S, Urolithin A (2019) A gut metabolite, improves insulin sensitivity through augmentation of mitochondrial function and biogenesis. Obesity 27(4):612–620
Ticinesi A, Nouvenne A, Cerundolo N, Catania P, Prati B, Tana C, Meschi T (2019) Gut microbiota, muscle mass and function in aging: a focus on physical frailty and sarcopenia. Nutrients 11(7):1633
Teodori L, Petrignani I, Giuliani A, Prattichizzo F, Gurău F, Matacchione G, Olivieri F, Coppari S, Albertini MC (2019) Inflamm-aging microRNAs may integrate signals from food and gut microbiota by modulating common signalling pathways. Mech Ageing Dev 182:111127
Vemuri R, Sylvia KE, Klein SL, Forster SC, Plebanski M, Eri R, Flanagan KL (2019) The microgenderome revealed: sex differences in bidirectional interactions between the microbiota, hormones, immunity and disease susceptibility. In Seminars in immunopathology (vol 41, no 2, pp 265–275). Springer Berlin Heidelberg
van Tongeren SP, Slaets JP, Harmsen HJM, Welling GW (2005) Fecal microbiota composition and frailty. Appl Environ Microbiol 71(10):6438–6442
Voreades N, Kozil A, Weir TL (2014) Diet and the development of the human intestinal microbiome. Front Microbiol 5:494
Waddell DS, Baehr LM, Van Den Brandt J, Johnsen SA, Reichardt HM, Furlow JD, Bodine SC (2008) The glucocorticoid receptor and FOXO1 synergistically activate the skeletal muscle atrophy-associated MuRF1 gene. Am J Physiol Endocrinol Metab 295(4):E785–E797

Part III
Gut Microbiota in Neurodegernative Diseases and Cancer

Chapter 13
Gut Microbiota Changes and Parkinson's Disease: What Do We Know, Which Avenues Ahead

Masaaki Hirayama and Kinji Ohno

Abstract PD is caused by a decrease in the number of dopaminergic neurons in the substantia nigra, which are caused by the abnormal aggregation of alpha-synuclein fibers (Lewy bodies) in neurons. PD is caused by a decrease in the number of dopaminergic neurons in the substantia nigra. Constipation, idiopathic rapid eye movement sleep behavior disorder (iRBD), and depression can be frequently observed about 20, 10, and 5 years before the onset of motor symptoms in PD, and Braak et al. reported that α-synuclein fibers start in the dorsal vagal nucleus and migrate upward to the nucleus accumbens and substantia nigra, consistent with the progression of symptoms. Our nonparametric meta-analysis of the gut microbiota of PD in five countries, including our own case, revealed an increase in mucinolytic Akkermansia and a decrease in short-chain fatty acid (SCFA)-producing bacteria in PD. As a result, it increases the exposure of the enteric plexus to toxins such as lipopolysaccharides (LPS) and pesticides. In addition, it can cause abnormal aggregation of alpha-synuclein fibers. The gut microbiota of REM sleep behavior disorder, which is considered to be a precursor stage of Parkinson's disease, is an intermediate enterotype between healthy individuals and Parkinson's disease. A 2-year longitudinal study revealed that the pathological development of early Parkinson's disease patients can be defined by three types of gut bacteria. Butyric acid, a type of short-chain fatty acid, is mainly consumed by the colonic mucosa and plays an important role in the maintenance of mucosal production. Short-chain fatty acids are similarly decreased in the advanced stages of the disease. These

M. Hirayama (✉)
Department of Pathophysiological Laboratory Sciences, Nagoya University Graduate School of Medicine, Nagoya, Japan
e-mail: hirasan@met.nagoya-u.ac.jp

K. Ohno
Department of Pathophysiological Laboratory Sciences, Division of Neurogenetics, Center for Neurological Diseases and Cancer, Nagoya University Graduate School of Medicine, Nagoya, Japan
e-mail: ohnok@med.nagoya-u.ac.jp

F. Marotta (ed.), *Gut Microbiota in Aging and Chronic Diseases*, Healthy Ageing and Longevity 17, https://doi.org/10.1007/978-3-031-14023-5_13

results suggest that therapeutic intervention for Parkinson's disease is possible. Biochemical, cellular, and animal studies are needed to further elucidate the direct relationship between gut microbiota and PD.

Keywords Parkinson's disease · Small intestinal bacterial growth syndrome · Colon microbiota · Short-chain fatty acid · Polyamine

13.1 Parkinson's Disease Gastrointestinal Symptoms and Gastrointestinal Abnormalities

Gastrointestinal symptoms of PD are among the most common non-motor symptoms, among which constipation is present in about 80% (Maeda et al. 2017). The onset of constipation usually precedes the motor symptoms of PD and worsens as the disease progresses. Prolonged small and large bowel transit times have been reported. In addition, the onset of constipation in PD is independent of age and physical activity and is considered to be unique to PD (Abbott et al. 2001).

Because of this early damage to the intestinal tract, it is thought that intestinal biopsy may be useful in diagnosing Parkinson's disease. However, the detection of α-synuclein in biopsiable tissues is not sufficiently useful as a biomarker for PD diagnosis. According to a systematic review of studies using biopsies of skin, olfactory, and gastrointestinal tissues, tissues of the gastrointestinal tract and salivary glands (13 postmortem, 13 biopsy) showed higher sensitivity and specificity than skin (3 postmortem, 8 biopsy) and olfactory mucosa and olfactory bulb (6 postmortem) (Schneider et al. 2015) In contrast to phosphorylated α-synuclein, α-synuclein was also detectable in peripheral tissues of healthy controls (Schneider et al. 2015).

Although the gastrointestinal tract and salivary glands appear to be the most promising candidate tissues for peripheral biopsy, the sensitivity and specificity of phosphorylated α-synuclein have not yet been accepted for use as a reliable peripheral biomarker of PD in clinical routine. In most reports, biopsy is performed in the submucosal or mesenteric plexus. Although submucosal nerves can be harvested by endoscopy, there are risks such as perforation in actual clinical practice. In addition, it is still unclear which nerves in the submucosal sample are most likely to be involved in PD.

13.2 Evidence that Parkinson's Disease Progresses from the Intestinal Tract

PD is commonly recognized as a movement disorder characterized by muscle rigidity, postural instability, gait disturbance, bradykinesia, and resting tremor. However, PD patients often present with non-motor symptoms, including mild-to-severe cognitive impairment. Braak et al. studied PD patients without cognitive symptoms and healthy subjects, focusing on the appearance of Lewy bodies, and found that Lewy bodies are distributed ascending from the dorsal nucleus of the vagus to the Locus coeruleus (LC) and substantia nigra (SN). As a result, the ascending hypothesis was proposed (Braak et al. 2002). Later, the double-hit hypothesis was proposed. The hypothesis was proposed that a pathogen that appears to be a virus may travel through two pathways: one from the olfactory bulb to the brain via olfactory cells, and the other from the gastrointestinal tract and the gastrointestinal nervous system (ENS) to the brain (Hawkes et al. 2007).

Although there is still no consensus on the olfactory bulb-mediated pathway, constipation, REM sleep behavior abnormalities, and depression have been observed before the onset of motor symptoms, and the hypothesis of alpha-synuclein ascending through the brainstem is consistent with pathology and symptom progression (Cersosimo and Benarroch 2012).

Inflammatory bowel disease (IBD), such as ulcerative colitis, increases the risk of developing Parkinson's disease by 22–35% of the incidence of PD (Weimers et al. 2019). IBD is known to have a genetic susceptibility. Genome-wide association studies have reported that the leucine-rich repeat kinase 2 (LRRK2) gene is one of the major loci associated with increased susceptibility to Crohn's disease (Franke et al. 2010). This LRRK2 is also a gene associated with familial and sporadic PD and multiple system atrophy (MSA) (Cookson 2010).

LRRK2 has been reported to be involved in the regulation of inflammatory responses and α-synuclein clearance, suggesting that it may be responsible for α-synuclein accumulation. However, ulcerative colitis increased the risk of developing PD, while for MSA there was a trend in the risk of developing PD, but it was not significant. In Crohn's disease, there was no significant association between the risk of developing both PD and MSA (Villumsen et al. 2019).

It has been reported that appendicitis may be involved in the etiology of PD. The appendix is associated with immune surveillance, microbial storage, and the lymphatic system, and has been reported to be rich in α-synuclein aggregates (Killinger et al. 2018; Stokholm et al. 2016). Some reports have found early removal of the appendix to be associated with a decreased risk of developing PD (Mendes et al. 2015) and others with no association (Palacios et al. 2018).

These findings are consistent with the idea that increased intestinal permeability is causally related to PD. Epidemiological studies have shown that truncal vagotomy reduces the probability of developing PD (Svensson et al. 2015; Liu et al. 2017). Enteroendocrine cells have direct synapses to vagal afferents, and the vagal nucleus has synaptic pathways to the SN and striatum (Kaelberer et al. 2018).

Animal models also show prion-like transmission of alpha-synuclein fibrils through the vagus nerve (Sacino et al. 2014). Furthermore, vagus nerve stimulation after administration of mouse LPS was reported to reduce inflammation-induced production of TNF-α and IL-6 by microglia, but this effect is no longer observed in vagotomy mice. These results support the vagus nerve hypothesis as the pathway for alpha-syn propagation and eventual microglial activation (Forsythe et al. 2014). It is possible that intestinal bacteria may be involved as a factor that causes inflammation in the gastrointestinal tract and promotes abnormal protein aggregation. Intraperitoneal administration of LPS increases intestinal permeability and accumulation of phosphorylated α-synuclein in the intestinal mucosa and dorsal vagal nucleus (Clairembault et al. 2015). LPS promotes the formation of α-synuclein fibrils in vitro. Dyskinesia and constipation in the germ-free mouse model of PD are milder than in the model mice without the specific pathogen. Fecal transplants from PD patients exacerbate dyskinesia and constipation in a sterile PD mouse model compared to fecal transplants from healthy individuals (Sampson et al. 2016).

The expression of Lewy bodies in the gastrointestinal plexus was reported by Qualman in 1984 (Qualman et al. 1984). Later, Lewy bodies were found in the Auerbach's plexus, an intrinsic gastrointestinal plexus with vasoactive intestinal peptide (VIP) (Wakabayashi et al. 1990). The neurons of the submucosal plexus (SMP) innervate the goblet cells by releasing neurotransmitters such as acetylcholine (ACh) and vasoactive internal peptide (VIP) (Alcolado et al. 2014). In addition, Lewy bodies have already been found in exogenous nerve fibers originating from the dorsal nucleus of the vagus nerve and the sacral medulla in a case of PD after colon biopsy (Shannon et al. 2012). The accumulation of α-synuclein in the gastrointestinal tract of PD has been reported to be more frequent in the upper gastrointestinal tract than in the lower gastrointestinal tract (Beach et al. 2010).

However, this was reported in autopsies and the frequency of occurrence in biopsies is not clear (Hilton et al. 2014). Therefore, it is difficult to assert that the upper gastrointestinal tract is the preferred site of Lewy bodies. As the disease progresses, the number of cells accumulating in Lewy bodies decreases. In addition, at autopsy, the intestinal mucosa has already melted, making it difficult to stain histopathology of the colon. For these reasons, the apparent incidence of the disease may be reduced. In animal experiments, oxidative stress on enteric nerves resulted in the appearance of α-synuclein (Paillusson et al. 2010). Furthermore, in an animal model of Parkinson's disease in which LPS was administered intraperitoneally, α-synuclein appeared less in the small intestine and increased in the large intestine, and increased intestinal permeability was observed in the large intestine (Kelly et al. 2014). In a PET study that used the parasympathetic marker donepezil to observe the progression of Parkinson's disease neuropathy in the colon, small intestine, and pancreas (Fedorova et al. 2017), the damage progressed from colon to small intestine to pancreas in parallel with the progression of the disease. Therefore, it is possible that the deposition of α-synuclein in the colon is rather reduced due to a high degree of cell destruction at autopsy. PD is a multifactorial disease with strong environmental factors, and less than 10% of PD is hereditary. Environmental exposure to herbicides and pesticides increases the risk of developing PD (Chen et al. 2013). Inhibition of mitochondrial function and induction of oxidative stress are said to be involved (Tanner et al. 2011). In animal experiments, rotenone, which inhibits mitochondrial complex I, causes alpha-synuclein to be seen first in the enteric nervous system and then in the pars compacta of the substantia nigra. The effects of rotenone inhibited the retrograde progression of alpha-synuclein to the dorsal motor nucleus of the vagus nerve, which is also pathologically affected by vagotomy (Pan-Montojo et al. 2010). Pesticides such as rotenone have been shown to promote the progression of Parkinson's disease through the intestinal tract.

13.3 The Relationship Between Small Intestinal Bacteria and Parkinson's Disease

In general, intestinal bacteria often refer to bacteria in the large intestine, but there have been reports on their association with disorders of the small intestine, including drug absorption.

13.3.1 Helicobacter Pylori Infection

The presence of spiral-shaped bacteria in the stomach has been known, but in 1983 they were successfully cultured and newly recognized as H. pylori (HP) (Warren and Marshall 1983). HP is considered to be a risk factor for various diseases such as gastric ulcer, duodenal ulcer, chronic gastritis, and MALT lymphoma. In PD, a Danish epidemiological study reported that chronic infection increases the risk of developing the disease by 45% (Nielsen et al. 2012). However, some reports suggest that HP infection does not differ between healthy individuals and PD patients (Fasano et al. 2013). In Japan, the incidence of HP infection is higher than in Western countries, but the incidence of PD is not high (Yamawaki et al. 2009). Whether HP infection is a factor in the pathogenesis of PD needs to be further investigated. l-dopa, a drug used to treat PD, is easily soluble in acidic conditions. In fact, taking l-dopa with lemon water in PD patients has been shown to increase blood levels and improve motor function. In clinical practice, taking l-dopa with lemon water in PD patients has been shown to increase blood levels and improve motor function. HP produces ammonia by urease activity and decreases the acidity of gastric juice, which may inhibit l-dopa absorption. Although some reports suggest that HP infection does not affect changes in blood levels of l-dopa (Narozanska et al. 2014; Pierantozzi et al. 2001), in a double-blind study of PD patients with motor fluctuations, removal of HP infection significantly increased l-dopa absorption and ON duration significantly increased (Pierantozzi et al. 2006). HP eradication reduced the OFF time and delayed-on of PD symptoms (Lee et al. 2008).

13.3.2 Small Intestinal Bacterial Growth Syndrome

Normally, most bacteria in the small intestine are killed by the strong acid in the stomach, and the number of bacteria is low due to the action of bile and pancreatic juice. The cause of small intestinal bacterial overgrowth (small intestinal bacterial growth syndrome (SBO)) is well known to be organic changes in the intestine (Grace et al. 2013). These include conditions after small bowel diverticulosis, surgical appendicectomy, and gastrectomy. However, diabetic neuropathy, systemic scleroderma, amyloidosis, and hypothyroidism can also result in poor clearance of bacteria. In addition, in the elderly, achlorhydria and idiopathic changes in intestinal motility may lead to SBO (Grover et al. 2008). In addition, chronic use of antacids in recent years has been thought to potentially increase SBO (Lo and Chan 2013). Bacterial growth is thought to consume nutrients such as carbohydrates and vitamin B12, leading to caloric deficiency and vitamin B12 deficiency. Bacteria deconjugate bile salts, leading to micelle formation failure and subsequent fat malabsorption. Severe intestinal bacterial abnormalities can also damage the intestinal mucosa. Many patients are asymptomatic and may exhibit weight loss and malnutrition. Gastrointestinal symptoms include abdominal discomfort, diarrhea, bloating, and excessive flatulence. SBO is also found to be more common in PD (Gabrielli et al. 2011). In a report of expiratory gas test with glucose tolerance test by Gabrielli et al., significantly more SBO was found in 54.17% compared to 8.33% in healthy subjects (Gabrielli et al. 2011) However, a similar study conducted with lactulose tolerance test showed that SBO was less common in patients with prolonged small intestinal transit time (Davies et al. 1996, 672), and SBO was more common in patients with shorter disease duration and was associated with motor symptoms, not gastrointestinal symptoms. It has been reported that SBO is associated with duration of disease, Yahr, severity of UPDRS-III, and severity of motor complications, with higher rates of SBO complications with higher drug doses (Gabrielli et al. 2011). However, when adjusted for duration of illness, some reports have found no association with motor complications, but only with severity (Tan et al. 2014). The clinical manifestations of SBO are unremarkable, and it is difficult to distinguish the symptoms of SBO from those of PD, since many of the non-motility symptoms of PD and anti-PD drugs have an inhibitory effect on intestinal motility. However, it has been reported that the presence of SBO may induce peripheral inflammation of the intestinal tract (Quigley and Quera 2006), which may contribute to increased intestinal permeability and trigger an immune response that activates glial cells, exacerbating the neurodegenerative process (Tan et al. 2014). SBOs may also alter the composition of the intestinal environment as a result of mucosal damage, fermentation of sugars, and deconjugation reactions of bile acids (Devos et al. 2013).

13.3.3 The Relationship Between Colon Bacteria and Parkinson's Disease

The gut microbiota has been associated with a variety of diseases, but the greatest impact has been reported with fecal transplantation therapy for refractory clostridium difficile infection (van Nood et al. 2013). It has been reported that gut bacteria are independent factors for obesity (Ridaura et al. 2013) and diabetes (Hartstra et al. 2015), and that antibiotic exposure alters the diversity of gut bacteria, especially fetal antibiotic exposure induces obesity (Cox and Blaser 2015) In neuropsychiatric disorders, autism (Hsiao et al. 2013), irritable bowel syndrome (Collins 2014), and multiple sclerosis (Miyake et al. 2015) have also been reported. Many of these mechanisms are still unknown, but they are gradually being elucidated, including the effects of changes in the composition and reduced diversity of the bacterial flora, the bidirectional action of the flora and the immune system, the production of signaling molecules by the vagus nerve, changes in tryptophan metabolism and microglial activation by products of gut bacteria, and the effects of cell wall components. Unlike other neurodegenerative diseases, the development of PD is greatly influenced by environmental factors. A history of smoking reduces the risk of developing PD by about 36–50%, and coffee consumption reduces the risk by 30% (Wirdefeldt et al. 2011). Unlike other neurodegenerative diseases, the development of PD is more influenced by environmental factors than by genetic abnormalities. A history of smoking reduces the risk of developing PD by about 36–50%, and coffee consumption reduces the risk by 30% (Wirdefeldt et al. 2011). These are thought to be due to the effects of nicotine in cigarettes (Quik et al. 2012) and polyphenols in coffee (Prediger 2010). Smokers have a higher ratio of Bacteroides/Prevotella in their feces, but after quitting smoking, this ratio decreases along with Proteobacteria, while the amount of Firmicutes and Actinobacteria increases (Biedermann et al. 2014). Coffee has also been reported to activate gallbladder contraction and colonic motility, and to cause an increase in Bifidobacterium and a decrease in Clostridium (Khokhlova et al. 2012). Coffee consumption has been negatively correlated with the prevalence of constipation (Murakami et al. 2006). Constipation is more frequent in PD patients. Constipation itself may be a cause of altered intestinal microbiota, increased mucosal permeability, and inflammation. In severe constipation, Staphylococcus aureus and Escherichia coli increase in the intestine, and serum antibody titers to Staphylococcus aureus and Escherichia coli also increase (Khalif et al. 2005).

In our report, the degree of constipation in PD patients correlated with a decrease in LPS-binding protein (LBP) (Hasegawa et al. 2015). LPS is present in the outer membrane of the cell wall of bacteria such as E. coli and is known to induce inflammation; LBP is a protein that traps these, and chronic entry of LPS through the intestinal tract is thought to lead to a decrease in LBP. In PD, intestinal permeability is increased, coliform bacteria and nitrotyrosine in the intestinal mucosa are increased, and α-synuclein staining is enhanced (Kelly et al. 2014).

In addition, reports have been seen that short-chain fatty acids (SCFA) produced by intestinal bacteria are involved in the immune system in PD. PD model mice (ASO), in which human α-synuclein was overexpressed in neurons, showed worse motor function and lower intestinal peristalsis compared to wild-type mice. However, when the mice were made sterile, there was no decrease in motor function or intestinal peristalsis in ASO mice. When germ-free ASO mice were treated with SCFA, microglia, immune cells in the central nervous system, were activated and abnormal accumulation of α-synuclein occurred, resulting in worsened motor function and intestinal peristalsis. Furthermore, when germ-free mice were administered stool from healthy individuals and stool from PD patients, respectively, symptoms were worse in the stool from PD patients (Sampson et al. 2016). They thought that SCFAs crossed the blood–brain barrier (BBB) and that the increased activation state of microglia in the CNS and the increased production of cytokines before inflammation were involved in the development of symptoms. However, SCFAs, especially butyrate, are considered to be a regulator of the immune system (Morrison and Preston 2016), and further investigation is needed because butyrate production in the gut is decreased in PD patients, as discussed below.

13.4 Intestinal Microbiota in PD

Abnormalities in the gut microbiota of PD vary widely from report to report (Table 13.1). The classification of the bacteria depends on the database used for the analysis, and the names of the strains also differ from report to report. SILVA is the most commonly used classification database, but three other classifications are also in use. We report 16S ribosomal RNA gene sequencing of fecal samples from PD patients and controls, and performed a meta-analysis of our dataset along with four previously reported datasets (Nishiwaki et al. 2020c).

The results showed that the enterobacteria that increased in genus level in PD were Akkermansia and Catabacter, and the bacteria that decreased were Roseburia, Faecalibacterium, and Lachnospiraceae ND3007. Akkermansia increased with the progression of PD, and Faecalibacterium and Roseburia decreased with the progression of PD. We also performed pseudo-functional analysis with PICRUSt2, which predicts functional gene dosage, pathway, etc. from compositional data of bacterial flora. KO generated by PICRUSt2 were assigned to gene expression profiles, and KEGG orthology set enrichment analysis was performed. The results showed that butyrate and propionate metabolism was altered in the gut microbiota in PD. This is consistent with a decrease in SCFAs-producing bacteria. Furthermore, network analysis using SparCC revealed two enterobacterial networks: one consisting of the Akkermansia and Christensenellaceae R-7 groups, which are increasing in PD, and the other consisting of decreasing bacteria such as Faecalibacterium, Roseburia, and Fusicatenibacter. Christensenellaceae R-7 was the hub of the increased subcluster and was associated with Akkermansia, suggesting

Table 13.1 Gut microbiota in Parkinson's disease

Dataset	First author	[a]City, Country	[b]Date	# PD	# Controls	Sequencing	Primers
1	Scheperjans et al. (2015)	Helsinki, Finland	Dec. 2014	72	72	16S rRNA V1-3	27F/530R
2	Keshavarzian et al. (2015)	Chicago, IL, USA	July 2015	38	36	16S rRNA V4	515F/806R
3	Hasegawa et al. (2015)	Nagoya, Japan	Nov. 2015	52	36	qRT-PCR	Not applicable
4	Unger et al. (2016)	Homburg, Germany	Sep. 2016	34	34	qPCR	Not applicable
5	Petrov et al. (2017)	Tomsk, Russia	Dec. 2016	89	66	16S rRNA V3-4	Not specified
6	Hill-Burns et al. (2017)	Birmingham, AL, USA	Feb. 2017	197	130	16S rRNA V4	515F/806R
7	Bedarf et al. (2017)	Bonn, Germany	Apr. 2017	31	28	Shotgun	Not applicable
8	Hopfner et al. (2017)	Kiel, Germany	May 2017	29	29	16S rRNA V1-2	Not specified
9	Li et al. (2017)	Beijing, China	May 2017	24	14	16S rRNA (region not specified)	Not specified
10	Qian et al. (2018)	Shanghai, China	Mar. 2018	45	45	16S rRNA V3-4	341F/806R
11	Lin et al. (2018)	Guangzhou, China	May 2018	75	45	16S rRNA V4	V4F/V4R
12	Heintz-Buschart et al. (2018)	[c]Kassel, Germany	Aug. 2018	26	38	16S rRNA V4	515F/805R
13	Tan	Kuala Lumpur, Malaysia	Oct. 2018	104	96	16S rRNA V3-4	Not specified
14	Barichella et al. (2019)	Milan, Italy	Mar 2019	193	113	16S rRNA V3–V4	341F/785R

(continued)

Table 13.1 (continued)

Dataset	First author	[a]City, Country	[b]Date	# PD	# Controls	Sequencing	Primers
15	Aho et al. (2019)	Helsinki, Finland	Jun. 2019	64	64	16S rRNA V3-4	341F/785R
16	Li	Changchun, China	July 2019	51	48	16S rRNA V4	515F/806R
17	Pietrucci et al. (2019)	Rome, Italy	Oct. 2019	80	72	16S rRNA V3–V4	Not specified
18	Nishiwaki et al. (2020c)	Nagoya, Japan	Jun. 2020	223	137	16S rRNA V3-4	341F/805R
19	Cristea et al. (2020)	British Columbia, Canada	July 2020	197	103	16S rRNA V4	515F/806R
20	Wallen et al. (2020)	Birmingham, AL,USA	Jun. 2020	323	184	16S rRNA V4	515F/806R
21	Nishiwaki et al. (2020a)	Nagoya, Japan	Dec 2020	PD:223,RBD 26	137	6S rRNA V3–V4	341F/805R

[a] The first author's address. [b] Date of online publication. [c] The first author's address was Esch-sur-Alzette, Luxembourg, but the samples were collected at Kassel, Germany. We reported datasets 3. 21 and 18

that Akkermansia and Christensenellaceae R-7 groups may play an important role in the onset and progression of the disease. These results suggest that the networks of the two major microbial groups are altered. When dietary fiber is deficient, Akkermansia degrades the intestinal mucosal layer and increases intestinal permeability; Faecalibacterium and Roseburia, which are reduced in PD, are butyrate-producing bacteria; butyrate is an energy source for colonic mucosal cells and is important for the maintenance of intestinal epithelium. These results suggest that the networks of the two major microbial groups are altered. When dietary fiber is deficient, Akkermansia degrades the intestinal mucosal layer and increases intestinal permeability; Faecalibacterium and Roseburia, which are reduced in PD, are butyrate-producing bacteria; butyrate is an energy source for colonic mucosal cells and is important for the maintenance of intestinal epithelium. It is also reported to inhibit histone deacetylase, which promotes Foxp3 gene expression and differentiation of naïve T cells into regulatory T cells, and to act on macrophage and dendritic cell receptors, promoting cytokine secretion, Treg activation, and anti-inflammatory effects. There is also an increase in mucinolytic Akkermansia and LPS-producing Gammaproteobacteria. Both are associated with increased intestinal permeability, exposing the enteric plexus to toxins such as LPS and pesticides (Fig. 13.1, left). The association of pesticides and herbicides with PD has been repeatedly reported in epidemiological studies and reviews, and these can cause aggregation of the ASN. Aggregation of the ASN propagates via the vagus nerve to the substantia nigra (Fig. 13.2, right (vagal pathway)). In addition, LPS generates chronic inflammation. SCFAs transported to the brain hematogenously and immune system responses activate microglia in the brain, resulting in increased inflammation in the brain (Fig. 13.1, right (blood pathway)). As a result, it is thought that ASN aggregates in the substantia nigra and dopaminergic neurons are lost. We call this pathophysiology the double-hit pathophysiology hypothesis of Parkinson's disease caused by intestinal bacteria. It is known that about 90% of RBD patients develop neurodegenerative diseases, most of which are PD. Using a topic model, we clustered the gut microbiota of PD, RBD, and controls and found four enterotypes(Nishiwaki et al. 2020b). Akkermansia in RBD was significantly increased compared to controls. On the other hand, as for SCFAs-producing bacteria, Faecalibacterium tended to decrease in the RBD, but Roseburia did not differ significantly from the control group. Decreases in Faecalibacterium and Roseburia may be a biomarker to predict that RBD will lead to PD (Fig. 13.2). Probiotics/prebiotics to increase the production of SCFAs and administration of SCFAs to the colon are expected to inhibit the progression of alpha-synucleinopathy in the RBD stage. Therefore, we divided PD patients, HY stages 1–3 (n = 165), 1 (n = 24), 2 (n = 85), and 3 (n = 56) at year 0 into worsening and stable groups to see if gut bacteria were involved in the progression of Parkinson's disease.

The deteriorated group had an increased HY stage, whereas the stable group had an unchanged or decreased HY stage.

We made random forest models to differentiate the deteriorated and stable groups for HY stages 1–3, 1, 2, and 3 at year 0 using bacterial features, and compared the models with those using clinical features (Nishiwaki et al. 2022). We

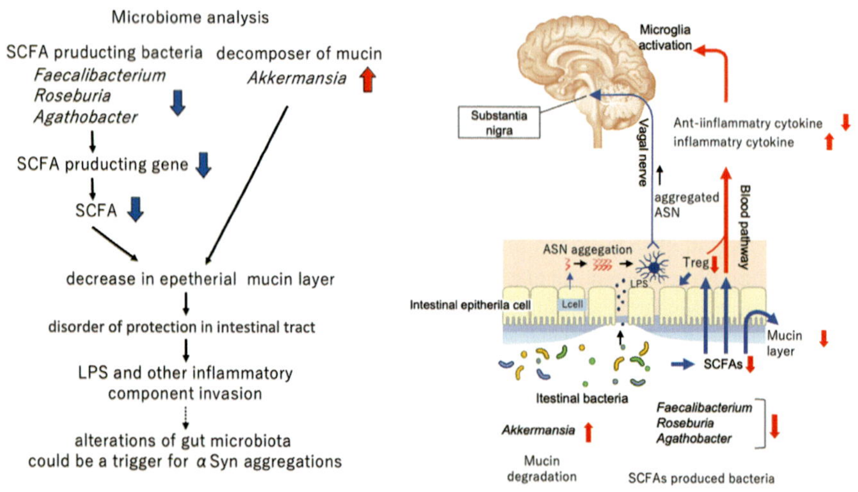

Fig. 13.1 The double-hit pathophysiology hypothesis of Parkinson's disease caused by intestinal bacteria

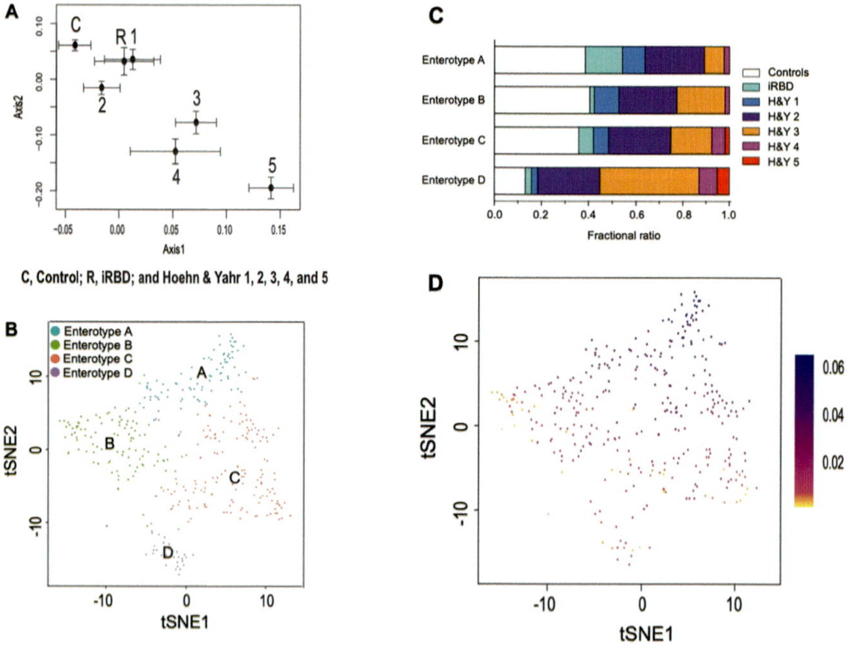

Fig. 13.2 The four enterotypes of gut microbiota among PD, RBD, and controls

first examined the validity of our modeling strategy by nested cross-validation with recursive feature elimination (RFE), which should have no leakage (type 1 circularity) between the training and test datasets. The nested cross-validation for HY stages 1–3 at year 0 yielded AUROCs of 0.548 (95% confidence interval: 0.456–0.641) in bacterial feature-based models (red line in Fig. 13.3a) and 0.559 (0.464–0.654) in clinical feature-based models (blue line in Fig. 13.3a). Thus, both bacterial and clinical features failed to make decent models. We thus constructed models for each HY stage. For HY stage 1 at year 0, the nested cross-validation yielded AUROCs of 0.799 (95% confidence interval: 0.615–0.982) in bacterial feature-based models (red line in Fig. 13.3d) and 0.549 (0.307–0.791) in clinical feature-based models (blue line in Fig. 13.3d). In the nested cross-validation, the bacterial feature-based model yielded 75% sensitivity and 83% specificity, whereas the clinical feature-based model failed to make a decent model (Fig. 13.3d). Seven statistical measures including sensitivities and specificities to indicate the performance of each model were all better with the bacterial feature-based models than with the clinical feature-based models. As the nested cross-validation did not provide us with essential features that determined the progression of PD, we made models by recursively eliminating features and evaluated them by cross-validation. Recursive feature elimination with cross-validation showed that maximum AUROCs were 0.868 (0.719–1.000) at 2 genera (Fusicatenibacter and Faecalibacterium) in bacterial feature-based models (red line in Fig. 13.3e, f) and 0.639 (0.402–0.876) at 2 features (BMI and age) in clinical feature-based models (blue line in Fig. 13.3e, f). AUROC was as high as 0.861 (0.707–1.000) even when a model was generated using genus Fusicatenibacter alone (green dot in Fig. 13.3e and green line in Fig. 13.3f). The bacterial feature-based model with Fusicatenibacter and Faecalibacterium yielded 83% sensitivity and 83% specificity, and Fusicatenibacter-based model yielded 92% sensitivity and 75% specificity (Fig. 13.3f). In contrast, the clinical feature-based model with BMI and age yielded 75% sensitivity and 67% specificity (Fig. 13.3f). For HY stage 2 at year 0, the nested cross-validation yielded AUROCs of 0.705 (0.592–0.818) in bacterial feature-based models (red line in Fig. 13.3g) and 0.719 (0.602–0.835) in clinical feature-based models (blue line in Fig. 13.3g). Seven statistical measures to indicate the performance of each model were slightly better with the clinical feature-based models than the bacterial feature-based models except for specificity. Recursive feature elimination with cross-validation showed that maximum AUROCs were 0.793 (0.692–0.895) at 7 genera (Lactobacillus, Blautia, Fusicatenibacter, Anaerostipes, Ruminococcus gnavus group, Akkermansia, Bifidobacterium) in bacterial feature-based models (red line in Fig. 13.3h, i) and 0.719 (0.602–0.835) at 10 features in clinical feature-based models (blue line in Fig. 13.3h, i). For HY stage 3 at year 0, the nested cross-validation yielded AUROCs of as low as 0.509 (0.301–0.719) in bacterial feature-based models (red line in Fig. 13.3j) and 0.772 (0.619–0.925) in clinical feature-based models (blue line in Fig. 13.3j). Thus, bacterial feature-based models were dependable for the early stage of PD, but clinical feature-based models became reliable with the advancement of PD. Plots of the four genera, Fusicatenibacter, Faecalibacterium, Blautia, and Akkermansia, at

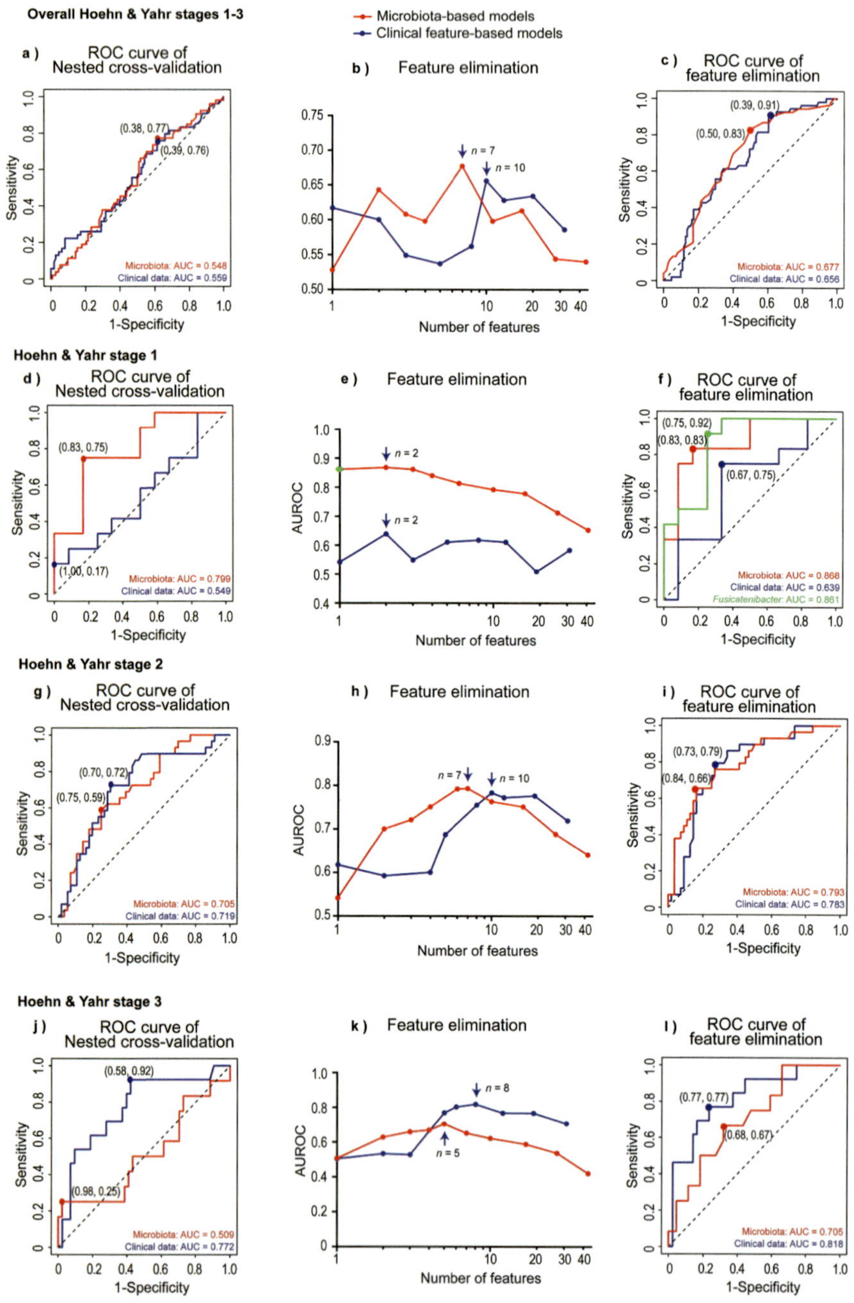

Fig. 13.3 Validation of modeling strategy by determination of essential features by cross-validation

years 0 and 2 in the deteriorated group showed that all the four genera remained unchanged in 2 years. Additionally, plots of these four genera in the course of progression of α-synucleinopathy showed that Akkermansia becomes higher with disease progression, whereas Fusicatenibacter, Faecalibacterium, and Blautia become lower with disease progression, with statistically significant monotonous trends for all the four genera. No change of the four genera even in the deteriorated group and the change of the four genera with disease progression suggest that patients harboring dysbiosis of the four genera progress more rapidly than those without dysbiosis.

13.5 Effect of Gut Bacterial Metabolites on Parkinson's Disease

Intestinal bacteria produce SCFA as well as vitamins B2, B6, B12, K, folic acid, pantothenic acid, biotin (B7), serotonin, and polyamines. Serotonin promotes intestinal peristalsis. Intestinal SCFAs are decreased in PD (Unger et al. 2016; Hasegawa et al. 2015). We found that in PD, SCFA not only decreased, but BCFA increased (Hirayama et al. 2019). The main products of microbial fermentation activities in the gut are SCFAs, especially acetic acid, propionic acid, and butyric acid. However, when there is a lack of fiber as a source of fermentation, the microorganisms switch to less energy-efficient sources, such as amino acids from food or endogenous proteins, or dietary fat. Although SCFAs are sometimes produced from proteins, in most cases, branched-chain fatty acids (BCFAs) such as isobutyric acid, 2-methylbutyric acid, and isovalerate are produced. These are mostly derived from the branched-chain amino acids valine, isoleucine, and leucine. When dietary fiber is degraded by microorganisms in the intestine, three major SCFAs are synthesized: acetic acid, propionic acid, and butyric acid. Butyrate is an important energy source for colon cells and is consumed locally, while other SCFAs are drained into the portal vein. Propionic acid is metabolized in the liver and has low concentration in the peripheral circulation, and the most abundant SCFA is said to be acetic acid. Acetic acid also crosses the blood–brain barrier and reduces appetite through central homeostatic mechanisms. Acetic acid and propionic acid in the intestine stimulate GPR41 and GPR43 to release PYY and GLP-1, affecting satiety and intestinal transit; butyric acid in Luminal exerts anti-inflammatory effects by binding to GPR109A and also by inhibiting histone deacetylase. In addition, propionic acid is converted to glucose by glycogenesis in the intestine, which leads to satiety and decreases hepatic glucose production (Koh et al. 2016).

Immunosuppressive mechanisms are essential for maintaining intestinal homeostasis. In addition to its effects on intestinal epithelial cells, butyrate can also modulate the activity of the enteric nervous system; GPR41 is widely expressed in the peripheral nervous system, including sympathetic ganglia, vagus ganglia, dorsal root ganglia, and trigeminal ganglia (Kimura et al. 2011).

They can act on distal sites, such as the brain, to modulate permeability, neurogenesis, and host behavior. Therefore, a decrease in intestinal SCFAs may be sufficient to promote alpha-synuclein-mediated worsening of PD symptoms. Further characterization of the relationship between microbiota and microglia is expected. Serum polyamines are also reduced in PD (Saiki et al. 2019). Polyamines (PAs) are defined as low molecular weight polycations with a wide range of biological functions, including gene regulation, stress tolerance, cell proliferation, and cell differentiation. Ingested food is the main source of PAs in the lumen, and the majority of these compounds are absorbed in the upper part of the intestine. Therefore, the main source of PAs levels in the lower part of the intestine should be the gut microbiota (Kibe et al. 2014). PAs in the colonic lumen are transferred to the bloodstream through the colonic mucosa. The availability of intracellular PAs contributes to tissue homeostasis, epithelial cell division, and apoptosis in the gastrointestinal mucosa by regulating the expression of various growth-related genes (Timmons et al. 2012). Putrescine and spermidine are the most common PAs produced by bacteria. To the best of our knowledge, no bacterium has an enzyme that converts spermidine to spermine. Little is known about the production and degradation of biogenic amines, especially PAs, by enteric bacteria. The genus Bacteroides possesses a CASDC homolog that is essential for spermidine biosynthesis (Sakanaka et al. 2016). However, many E. coli species do not have a complete synthetic pathway for the production of PA (Qin et al. 2010). Arginine is converted in bacteria to ornithine, which is further converted to putrescine. In another pathway, arginine is converted to agmatine, which is further converted to putrescine. The acid-tolerant system of E. coli produces agmatine from arginine in a low pH environment; Enterococcus faecalis produces putrescine from agmatine via the agmatine/putrescine antiporter of the agmatine deiminase pathway (Matsumoto and Benno, 2004). Bacteria such as Bifidobacterium, which produce acid in the gut, create a favorable environment for the production of putrescine.

13.6 Conclusion

The intestinal bacteria in the small intestine have many implications for therapy. In addition, intestinal bacteria in the large intestine not only have the potential to cause the onset of Parkinson's disease, but are also associated with the progression of the disease. In the future, it will be necessary to study the metabolic products of the intestinal bacteria and to study the intestinal bacteria using sterile animals and other methods.

References

Abbott RD, Petrovitch H, White LR, Masaki KH, Tanner CM, Curb JD, Grandinetti A, Blanchette PL, Popper JS, Ross GW (2001) Frequency of bowel movements and the future risk of parkinson's disease. Neurology 57:456–462

Aho VTE, Pereira PAB, Voutilainen S, Paulin L, Pekkonen E, Auvinen P, Scheperjans F (2019) Gut microbiota in parkinson's disease: temporal stability and relations to disease progression. EBioMedicine 44:691–707

Alcolado NG, Conrad DJ, Poroca D, Li M, Alshafie W, Chappe FG, Pelis RM, Anini Y, Xu Z, Hamidi S, Said SI, Chappe VM (2014) Cystic fibrosis transmembrane conductance regulator dysfunction in vip knockout mice. Am J Physiol Cell Physiol 307:C195-207

Barichella M, Severgnini M, Cilia R, Cassani E, Bolliri C, Caronni S, Ferri V, Cancello R, Ceccarani C, Faierman S, Pinelli G, de Bellis G, Zecca L, Cereda E, Consolandi C, Pezzoli G (2019) Unraveling gut microbiota in parkinson's disease and atypical parkinsonism. Mov Disord 34:396–405

Beach TG, Adler TG, Sue LI, Vedders L, Lue L, White III CL, Akiyama H, Caviness JN, Shill HA, Sabbagh MN, Walker DG, Arizona Parkinson's Disease C (2010) Multi-organ distribution of phosphorylated alpha-synuclein histopathology in subjects with lewy body disorders Acta Neuropathol 119 689 702

Bedarf JR, Hildebrand F, Coelho LP, Sunagawa S, Bahram M, Goeser F, Bork P, Wullner U (2017) Functional implications of microbial and viral gut metagenome changes in early stage l-dopa-naive parkinson's disease patients. Genome Med 9:39

Biedermann L, Brulisauer K, Zeitz J, Frei P, Scharl M, Vavricka SR, Fried M, Loessner MJ, Rogler G, Schuppler M (2014) Smoking cessation alters intestinal microbiota: insights from quantitative investigations on human fecal samples using fish. Inflamm Bowel Dis 20:1496–1501

Braak H, Del Tredici K, Bratzke H, Hamm-clement J, Sandmann-keil D, Rub U (2002) Staging of the intracerebral inclusion body pathology associated with idiopathic Parkinson's disease (preclinical and clinical stages). J Neurol 249 Suppl 3:III/1–5

Cersosimo MG, Benarroch EE (2012) Pathological correlates of gastrointestinal dysfunction in Parkinson's disease. Neurobiol Dis 46:559–564

Chen H, Burton EA, Ross GW, Huang X, Savica R, Abbott RD, Ascherio A, Caviness JN, Gao X, Gray KA, Hong JS, Kamel F, Jennings D, Kirshner A, Lawler C, Liu R, Miller GW, Nussbaum R, Peddada SD, Rick AC, Ritz B, Siderowf AD, Tanner CM, Troster AI, Zhang J (2013) Research on the premotor symptoms of Parkinson's disease: clinical and etiological implications. Environ Health Perspect 121:1245–1252

Cirstea MS, Yu AC, Golz E, Sundvick K, Kliger D, Radisavljevic N, Foulger LH, Mackenzie M, Huan T, Finlay BB, Appel-Cresswell S (2020) Microbiota composition and metabolism are associated with gut function in Parkinson's disease. Mov Disord 35:1208–1217

Clairembault T, Leclair-Visonneau L, Coron E, Bourreille A, le Dily S, Vavasseur F, Heymann MF, Neunlist M, Derkinderen P (2015) Structural alterations of the intestinal epithelial barrier in Parkinson's disease. Acta Neuropathol Commun 3:12

Collins SM (2014) A role for the gut microbiota in IBS. Nat Rev Gastroenterol Hepatol 11:497–505

Cookson MR (2010) The role of leucine-rich repeat kinase 2 (LRRK2) in Parkinson's disease. Nat Rev Neurosci 11:791–797

Cox LM, Blaser MJ (2015) Antibiotics in early life and obesity. Nat Rev Endocrinol 11:182–190

Davies KN, King D, Billington D, Barrett JA (1996) Intestinal permeability and orocaecal transit time in elderly patients with Parkinson's disease. Postgrad Med J 72:164–167

Devos D, Lebouvier T, Lardeux B, Biraud M, Rouaud T, Pouclet H, Coron E, Bruley des varannes S, Naveilhan P, Nguyen JM, Neunlist M, Derkinderen P (2013) Colonic inflammation in Parkinson's disease Neurobiol Dis 50:42–48

Fasano A, Bove F, Gabrielli M, Petracca M, Zocco MA, Ragazzoni E, Barbaro F, Piano C, Fortuna S, Tortora A, di Giacopo R, Campanale M, Gigante G, Lauritano EC, Navarra P, Marconi S, Gasbarrini A, Bentivoglio AR (2013) The role of small intestinal bacterial overgrowth in Parkinson's disease. Mov Disord 28:1241–1249

Fedorova TD, Seidelin LB, Knudsen K, Schacht AC, Geday J, Pavese N, Brooks DJ, Borghammer P (2017) Decreased intestinal acetylcholinesterase in early Parkinson disease: an (11)C-donepezil PET study. Neurology 88:775–781

Forsythe P, Bienenstock J, Kunze WA (2014) Vagal pathways for microbiome-brain-gut axis communication. Adv Exp Med Biol 817:115–133

Franke A, McGovern DP, Barrett JC, Wang K, Radford-Smith GL, Ahmad T, Lees CW, Balschun T, Lee J, Roberts R, Anderson CA, Bis JC, Bumpstead S, Ellinghaus D, Festen EM, Georges M, Green T, Haritunians T, Jostins L, Latiano A, Mathew CG, Montgomery GW, Prescott NJ, Raychaudhuri S, Rotter JI, Schumm P, Sharma Y, Simms LA, Taylor KD, Whiteman D, Wijmenga C, Baldassano RN, Barclay M, Bayless TM, Brand S, Buning C, Cohen A, Colombel JF, Cottone M, Stronati L, Denson T, de Vos M, D'Inca R, Dubinsky M, Edwards C, Florin T, Franchimont D, Gearry R, Glas J, van Gossum A, Guthery SL, Halfvarson J, Verspaget HW, Hugot JP, Karban A, Laukens D, Lawrance I, Lemann M, Levine A, Libioulle C, Louis E, Mowat C, Newman W, Panes J, Phillips A, Proctor DD, Regueiro M, Russell R, Rutgeerts P, Sanderson J, Sans M, Seibold F, Steinhart AH, Stokkers PC, Torkvist L, Kullak-Ublick G, Wilson D, Walters T, Targan SR, Brant SR, Rioux JD, D'Amato M, Weersma RK, Kugathasan S, Griffiths AM, Mansfield JC, Vermeire S, Duerr RH, Silverberg MS, Satsangi J, Schreiber S, Cho JH, Annese V, Hakonarson H, Daly MJ, Parkes M (2010) Genome-wide meta-analysis increases to 71 the number of confirmed Crohn's disease susceptibility loci. Nat Genet 42:1118–1125

Gabrielli M, Bonazzi P, Scarpellini E, Bendia E, Lauritano EC, Fasano A, Ceravolo MG, Capecci M, Rita Bentivoglio A, Provinciali L, Tonali PA, Gasbarrini A (2011) Prevalence of small intestinal bacterial overgrowth in Parkinson's disease. Mov Disord 26:889–892

Grace E, Shaw C, Whelan K, Andreyev HJ (2013) Review article: small intestinal bacterial overgrowth–prevalence, clinical features, current and developing diagnostic tests, and treatment. Aliment Pharmacol Ther 38:674–688

Grover M, Kanazawa M, Palsson OS, Chitkara DK, Gangarosa LM, Drossman DA, Whitehead WE (2008) Small intestinal bacterial overgrowth in irritable bowel syndrome: association with colon motility, bowel symptoms, and psychological distress. Neurogastroenterol Motil 20:998–1008

Hartstra AV, Bouter KE, Backhed F, Nieuwdorp M (2015) Insights into the role of the microbiome in obesity and type 2 diabetes. Diabetes Care 38:159–165

Hasegawa S, Goto S, Tsuji H, Okuno T, Asahara T, Nomoto K, Shibata A, Fujisawa Y, Minato T, Okamoto A, Ohno K, Hirayama M (2015) Intestinal dysbiosis and lowered serum lipopolysaccharide-binding protein in Parkinson's disease. PLoS ONE 10:e0142164

Hawkes CH, del Tredici K, Braak H (2007) Parkinson's disease: a dual-hit hypothesis. Neuropathol Appl Neurobiol 33:599–614

Heintz-Buschart A, Pandey U, Wicke T, Sixel-Doring F, Janzen A, Sittig-Wiegand E, Trenkwalder C, Oertel WH, Mollenhauer B, Wilmes P (2018) The nasal and gut microbiome in Parkinson's disease and idiopathic rapid eye movement sleep behavior disorder. Mov Disord 33:88–98

Hill-Burns EM, Debelius JW, Morton JT, Wissemann WT, Lewis MR, Wallen ZD, Peddada SD, Factor SA, Molho E, Zabetian CP, Knight R, Payami H (2017) Parkinson's disease and Parkinson's disease medications have distinct signatures of the gut microbiome. Mov Disord 32:739–749

Hilton D, Stephens M, Kirk L, Edwards P, Potter R, Zajicek J, Broughton E, Hagan H, Carroll C (2014) Accumulation of alpha-synuclein in the bowel of patients in the pre-clinical phase of Parkinson's disease. Acta Neuropathol 127:235–241

Hirayama M, Hamaguchi T, Ito M, Ishida T, Kuki M, Maeda T, Kashihara K, Tsuboi Y, Ohno K (2019) Relationship between intestinal bacteria, bacterial metabolites and residual pesticides in Parkinson's disease. Mov Disord 34:S362–S362

Hopfner F, Kunstner A, Muller SH, Kunzel S, Zeuner KE, Margraf NG, Deuschl G, Baines JF, Kuhlenbaumer G (2017) Gut microbiota in Parkinson disease in a northern German cohort. Brain Res 1667:41–45

Hsiao EY, McBride SW, Hsien S, Sharon G, Hyde ER, McCue T, Codelli JA, Chow J, Reisman SE, Petrosino JF, Patterson PH, Mazmanian SK (2013) Microbiota modulate behavioral and physiological abnormalities associated with neurodevelopmental disorders. Cell 155:1451–1463

Kaelberer MM, Buchanan KL, Klein ME, Barth BB, Montoya MM, Shen X, Bohorquez DV (2018) A gut-brain neural circuit for nutrient sensory transduction. Science:361

Kelly LP, Carvey PM, Keshavarzian A, Shannon KM, Shaikh M, Bakay RA, Kordower JH (2014) Progression of intestinal permeability changes and alpha-synuclein expression in a mouse model of Parkinson's disease. Mov Disord 29:999–1009

Keshavarzian A, Green SJ, Engen PA, Voigt RM, Naqib A, Forsyth CB, Mutlu E, Shannon KM (2015) Colonic bacterial composition in Parkinson's disease. Mov Disord 30:1351–1360

Khalif IL, Quigley EM, Konovitch EA, Maximova ID (2005) Alterations in the colonic flora and intestinal permeability and evidence of immune activation in chronic constipation. Dig Liver Dis 37:838–849

Khokhlova EV, Smeianov VV, Efimov BA, Kafarskaia LI, Pavlova SI, Shkoporov AN (2012) Anti-inflammatory properties of intestinal Bifidobacterium strains isolated from healthy infants. Microbiol Immunol 56:27–39

Kibe R, Kurihara S, Sakai Y, Suzuki H, Ooga T, Sawaki E, Muramatsu K, Nakamura A, Yamashita A, Kitada Y, Kakeyama M, Benno Y, Matsumoto M (2014) Upregulation of colonic luminal polyamines produced by intestinal microbiota delays senescence in mice. Sci Rep 4:4548

Killinger BA, Madaj Z, Sikora JW, Rey N, Haas AJ, Vepa Y, Lindqvist D, Chen H, Thomas PM, Brundin P, Brundin L, Labrie V (2018) The vermiform appendix impacts the risk of developing Parkinson's disease. Sci Transl Med 10

Kimura I, Inoue D, Maeda T, Hara T, Ichimura A, Miyauchi S, Kobayashi M, Hirasawa A, Tsujimoto G (2011) Short-chain fatty acids and ketones directly regulate sympathetic nervous system via G protein-coupled receptor 41 (GPR41). Proc Natl Acad Sci U S A 108:8030–8035

Koh A, de Vadder F, Kovatcheva-Datchary P, Backhed F (2016) From dietary fiber to host physiology: short-chain fatty acids as key bacterial metabolites. Cell 165:1332–1345

Lee WY, Yoon WT, Shin HY, Jeon SH, Rhee PL (2008) Helicobacter pylori infection and motor fluctuations in patients with Parkinson's disease. Mov Disord 23:1696–1700

Li W, Wu X, Hu X, Wang T, Liang S, Duan Y, Jin F, Qin B (2017) Structural changes of gut microbiota in Parkinson's disease and its correlation with clinical features. Sci China Life Sci 60:1223–1233

Lin A, Zheng W, He Y, Tang W, Wei X, He R, Huang W, Su Y, Huang Y, Zhou H, Xie H (2018) Gut microbiota in patients with Parkinson's disease in southern China. Parkinsonism Relat Disord 53:82–88

Liu B, Fang F, Pedersen NL, Tillander A, Ludvigsson JF, Ekbom A, Svenningsson P, Chen H, Wirdefeldt K (2017) Vagotomy and Parkinson disease: a swedish register-based matched-cohort study. Neurology 88:1996–2002

Lo WK, Chan WW (2013) Proton pump inhibitor use and the risk of small intestinal bacterial overgrowth: a meta-analysis. Clin Gastroenterol Hepatol 11:483–490

Maeda T, Shimo Y, Chiu SW, Yamaguchi T, Kashihara K, Tsuboi Y, Nomoto M, Hattori N, Watanabe H, Saiki H, Group JF (2017) Clinical manifestations of nonmotor symptoms in 1021 Japanese Parkinson's disease patients from 35 medical centers. Parkinsonism Relat Disord 38:54–60

Matsumoto M, Benno Y (2004) Consumption of Bifidobacterium lactis LKM512 yogurt reduces gut mutagenicity by increasing gut polyamine contents in healthy adult subjects. Mutat Res 568:147–153

Mendes A, Goncalves A, Vila-Cha N, Moreira I, Fernandes J, Damasio J, Teixeira-Pinto A, Taipa R, Lima AB, Cavaco S (2015) Appendectomy may delay Parkinson's disease Onset. Mov Disord 30:1404–1407

Miyake S, Kim S, Suda W, Oshima K, Nakamura M, Matsuoka T, Chihara N, Tomita A, Sato W, Kim SW, Morita H, Hattori M, Yamamura T (2015) Dysbiosis in the gut microbiota of patients with multiple sclerosis, with a striking depletion of species belonging to clostridia XIVa and IV Clusters. PLoS ONE 10:e0137429

Morrison DJ, Preston T (2016) Formation of short chain fatty acids by the gut microbiota and their impact on human metabolism. Gut Microbes 7:189–200

Murakami K, Okubo H, Sasaki S (2006) Dietary intake in relation to self-reported constipation among Japanese women aged 18–20 years. Eur J Clin Nutr 60:650–657

Narozanska E, Bialecka M, Adamiak-Giera U, Gawronska-Szklarz B, Soltan W, Schinwelski M, Robowski P, Madalinski MH, Slawek J (2014) Pharmacokinetics of levodopa in patients with Parkinson disease and motor fluctuations depending on the presence of Helicobacter pylori infection. Clin Neuropharmacol 37:96–99

Nielsen HH, Qiu J, Friis S, Wermuth L, Ritz B (2012) Treatment for Helicobacter pylori infection and risk of Parkinson's disease in Denmark. Eur J Neurol 19:864–869

Nishiwaki H, Hamaguchi T, Ito M, Ishida T, Maeda T, Kashihara K, Tsuboi Y, Ueyama J, Shimamura T, Mori H, Kurokawa K, Katsuno M, Hirayama M, Ohno K (2020a) Short-chain fatty acid-producing gut microbiota is decreased in Parkinson's disease but not in rapid-eye-movement sleep behavior disorder. mSystems:5, epub

Nishiwaki H, Hamaguchi T, Ito M, Ishida T, Maeda T, Kashihara K, Tsuboi Y, Ueyama J, Shimamura T, Mori H, Kurokawa K, Katsuno M, Hirayama M, Ohno K (2020b) Short-chain fatty acid-producing gut microbiota is decreased in Parkinson's disease but not in rapid-eye-movement sleep behavior disorder. mSystems 5

Nishiwaki H, Ito M, Hamaguchi T, Maeda T, Kashihara K, Tsuboi Y, Ueyama J, Yoshida T, Hanada H, Takeuchi I, Katsuno M, Hirayama M, Ohno K (2022) Short chain fatty acids-producing and mucin-degrading intestinal bacteria predict the progression of early Parkinson's disease. NPJ Parkinsons Dis 8:65

Nishiwaki H, Ito M, Ishida T, Hamaguchi T, Maeda T, Kashihara K, Tsuboi Y, Ueyama J, Shimamura T, Mori H, Kurokawa K, Katsuno M, Hirayama M, Ohno K (2020c) Meta-Analysis of Gut Dysbiosis in Parkinson's Disease. Mov Disord 35:1626–1635

Paillusson S, Tasselli M, Lebouvier T, Mahe MM, Chevalier J, Biraud M, Cario-Toumaniantz C, Neunlist M, Derkinderen P (2010) alpha-Synuclein expression is induced by depolarization and cyclic AMP in enteric neurons. J Neurochem 115:694–706

Palacios N, Hughes KC, Cereda E, Schwarzschild MA, Ascherio A (2018) Appendectomy and risk of Parkinson's disease in two large prospective cohorts of men and women. Mov Disord 33:1492–1496

Pan-Montojo F, Anichtchik O, Dening Y, Knels L, Pursche S, Jung R, Jackson S, Gille G, Spillantini MG, Reichmann H, Funk RH (2010) Progression of Parkinson's disease pathology is reproduced by intragastric administration of rotenone in mice. PLoS ONE 5:e8762

Petrov VA, Saltykova IV, Zhukova IA, Alifirova VM, Zhukova NG, Dorofeeva YB, Tyakht AV, Kovarsky BA, Alekseev DG, Kostryukova ES, Mironova YS, Izhboldina OP, Nikitina MA, Perevozchikova TV, Fait EA, Babenko VV, Vakhitova MT, Govorun VM, Sazonov AE (2017) Analysis of gut microbiota in patients with Parkinson's disease. Bull Exp Biol Med 162:734–737

Pierantozzi M, Pietroiusti A, Brusa L, Galati S, Stefani A, Lunardi G, Fedele E, Sancesario G, Bernardi G, Bergamaschi A, Magrini A, Stanzione P, Galante A (2006) Helicobacter pylori eradication and l-dopa absorption in patients with PD and motor fluctuations. Neurology 66:1824–1829

Pierantozzi M, Pietroiusti A, Sancesario G, Lunardi G, Fedele E, Giacomini P, Frasca S, Galante A, Marciani MG, Stanzione P (2001) Reduced L-dopa absorption and increased clinical fluctuations in Helicobacter pylori-infected Parkinson's disease patients. Neurol Sci 22:89–91

Pietrucci D, Cerroni R, Unida V, Farcomeni A, Pierantozzi M, Mercuri NB, Biocca S, Stefani A, Desideri A (2019) Dysbiosis of gut microbiota in a selected population of Parkinson's patients. Parkinsonism Relat Disord 65:124–130

Prediger RD (2010) Effects of caffeine in Parkinson's disease: from neuroprotection to the management of motor and non-motor symptoms. J Alzheimers Dis 20(Suppl 1):S205–S220

Qian Y, Yang X, Xu S, Wu C, Song Y, Qin N, Chen SD, Xiao Q (2018) Alteration of the fecal microbiota in Chinese patients with Parkinson's disease. Brain Behav Immun 70:194–202

Qin J, Li R, Raes J, Arumugam M, Burgdorf KS, Manichanh C, Nielsen T, Pons N, Levenez F, Yamada T, Mende DR, Li J, Xu J, Li S, Li D, Cao J, Wang B, Liang H, Zheng H, Xie Y, Tap J, Lepage P, Bertalan M, Batto JM, Hansen T, le Paslier D, Linneberg A, Nielsen HB, Pelletier E, Renault P, Sicheritz-Ponten T, Turner K, Zhu H, Yu C, Li S, Jian M, Zhou Y, Li Y, Zhang X, Li S, Qin N, Yang H, Wang J, Brunak S, Dore J, Guarner F, Kristiansen K, Pedersen O, Parkhill J, Weissenbach J, Meta HITC, Bork P, Ehrlich SD, Wang J (2010) A human gut microbial gene catalogue established by metagenomic sequencing. Nature 464:59–65

Qualman SJ, Haupt HM, Yang P, Hamilton SR (1984) Esophageal Lewy bodies associated with ganglion cell loss in achalasia similarity TO Parkinson's disease. Gastroenterology 87:848–856

Quigley EM, Quera R (2006) Small intestinal bacterial overgrowth: roles of antibiotics, prebiotics, and probiotics. Gastroenterology 130:S78-90

Quik M, Perez XA, Bordia T (2012) Nicotine as a potential neuroprotective agent for Parkinson's disease. Mov Disord 27:947–957

Ridaura VK, Faith JJ, Rey FE, Cheng J, Duncan AE, Kau AL, Griffin NW, Lombard V, Henrissat B, Bain JR, Muehlbauer MJ, Ilkayeva O, Semenkovich CF, Funai K, Hayashi DK, Lyle BJ, Martini MC, Ursell LK, Clemente JC, van Treuren W, Walters WA, Knight R, Newgard CB, Heath AC, Gordon JI (2013) Gut microbiota from twins discordant for obesity modulate metabolism in mice. Science 341:1241214

Sacino AN, Brooks M, Thomas MA, McKinney AB, Lee S, Regenhardt RW, McGarvey NH, Ayers JI, Notterpek L, Borchelt DR, Golde TE, Giasson BI (2014) Intramuscular injection of alpha-synuclein induces CNS alpha-synuclein pathology and a rapid-onset motor phenotype in transgenic mice. Proc Natl Acad Sci U S A 111:10732–10737

Saiki S, Sasazawa Y, Fujimaki M, Kamagata K, Kaga N, Taka H, Li Y, Souma S, Hatano T, Imamichi Y, Furuya N, Mori A, Oji Y, Ueno SI, Nojiri S, Miura Y, Ueno T, Funayama M, Aoki S, Hattori N (2019) A metabolic profile of polyamines in Parkinson disease: a promising biomarker. Ann Neurol 86:251–263

Sakanaka M, Sugiyama Y, Kitakata A, Katayama T, Kurihara S (2016) Carboxyspermidine decarboxylase of the prominent intestinal microbiota species Bacteroides thetaiotaomicron is required for spermidine biosynthesis and contributes to normal growth. Amino Acids 48:2443–2451

Sampson TR, Debelius JW, Thron T, Janssen S, Shastri GG, Ilhan ZE, Challis C, Schretter CE, Rocha S, Gradinaru V, Chesselet MF, Keshavarzian A, Shannon KM, Krajmalnik-Brown R, Wittung-Stafshede P, Knight R, Mazmanian SK (2016) Gut microbiota regulate motor deficits and neuroinflammation in a model of Parkinson's disease. Cell 167(1469–1480):e12

Scheperjans F, Aho V, Pereira PA, Koskinen K, Paulin L, Pekkonen E, Haapaniemi E, Kaakkola S, Eerola-Rautio J, Pohja M, Kinnunen E, Murros K, Auvinen P (2015) Gut microbiota are related to Parkinson's disease and clinical phenotype. Mov Disord 30:350–358

Schneider SA, Boettner M, Alexoudi A, Zorenkov D, Deuschl G, Wedel T (2015) Can we use peripheral tissue biopsies to diagnose Parkinson's disease? A review of the literature. Eur J Neurol

Shannon KM, Keshavarzian A, Dodiya HB, Jakate S, Kordower JH (2012) Is alpha-synuclein in the colon a biomarker for premotor Parkinson's disease? Evidence from 3 cases. Mov Disord 27:716–719

Stokholm MG, Danielsen EH, Hamilton-Dutoit SJ, Borghammer P (2016) Pathological alpha-synuclein in gastrointestinal tissues from prodromal Parkinson disease patients. Ann Neurol 79:940–949

Svensson E, Horvath-Puho E, Thomsen RW, Djurhuus JC, Pedersen L, Borghammer P, Sorensen HT (2015) Vagotomy and subsequent risk of Parkinson's disease. Ann Neurol 78:522–529

Tan AH, Mahadeva S, Thalha AM, Gibson PR, Kiew CK, Yeat CM, Ng SW, Ang SP, Chow SK, Tan CT, Yong HS, Marras C, Fox SH, Lim SY (2014) Small intestinal bacterial overgrowth in Parkinson's disease. Parkinsonism Relat Disord 20:535–540

Tanner CM, Kamel F, Ross GW, Hoppin JA, Goldman SM, Korell M, Marras C, Bhudhikanok GS, Kasten M, Chade AR, Comyns K, Richards MB, Meng C, Priestley B, Fernandez HH, Cambi F, Umbach DM, Blair A, Sandler DP, Langston JW (2011) Rotenone, paraquat, and Parkinson's disease. Environ Health Perspect 119:866–872

Timmons J, Chang ET, Wang JY, Rao JN (2012) Polyamines and gut mucosal homeostasis. J Gastrointest Dig Syst 2

Unger MM, Spiegel J, Dillmann KU, Grundmann D, Philippeit H, Burmann J, Fassbender K, Schwiertz A, Schafer KH (2016) Short chain fatty acids and gut microbiota differ between patients with Parkinson's disease and age-matched controls. Parkinsonism Relat Disord 32:66–72

van Nood E, Vrieze A, Nieuwdorp M, Fuentes S, Zoetendal EG, de Vos WM, Visser CE, Kuijper EJ, Bartelsman JF, Tijssen JG, Speelman P, Dijkgraaf MG, Keller JJ (2013) Duodenal infusion of donor feces for recurrent Clostridium difficile. N Engl J Med 368:407–415

Villumsen M, Aznar S, Pakkenberg B, Jess T, Brudek T (2019) Inflammatory bowel disease increases the risk of Parkinson's disease: a Danish nationwide cohort study 1977–2014. Gut 68:18–24

Wakabayashi K, Takahashi H, Ohama E, Ikuta F (1990) Parkinson's disease: an immunohistochemical study of Lewy body-containing neurons in the enteric nervous system. Acta Neuropathol 79:581–583

Wallen ZD, Appah M, Dean MN, Sesler CL, Factor SA, Molho E, Zabetian CP, Standaert DG, Payami H (2020) Characterizing dysbiosis of gut microbiome in PD: evidence for overabundance of opportunistic pathogens. NPJ Parkinsons Dis 6:11

Warren JR, Marshall B (1983) Unidentified curved bacilli on gastric epithelium in active chronic gastritis. Lancet 1:1273–1275

Weimers P, Halfvarson J, Sachs MC, Saunders-Pullman R, Ludvigsson JF, Peter I, Burisch J, Olen O (2019) Inflammatory bowel disease and Parkinson's Disease: a nationwide swedish cohort study. Inflamm Bowel Dis 25:111–123

Wirdefeldt K, Adami HO, Cole P, Trichopoulos D, Mandel J (2011) Epidemiology and etiology of Parkinson's disease: a review of the evidence. Eur J Epidemiol 26(Suppl 1):S1-58

Yamawaki M, Kusumi M, Kowa H, Nakashima K (2009) Changes in prevalence and incidence of Parkinson's disease in Japan during a quarter of a century. Neuroepidemiology 32:263–269

Chapter 14
Gut Microbiota, Alzheimer and Psychiatric Diseases: Unveiling the Relationships and Treatment Options

Hsin-Yi Lai and Ting Shen

Abstract The gastrointestinal tract is colonized by approximately a hundred trillion gut microorganisms consisting of more than thousands of species, including bacteria, viruses, and fungi, which are called gut microbiota. It is not only involved in food digestion, nutrient acquisition and energy regulation, but also participants in regulating brain activity and behavior via the gut-microbiota-brain axis. Accumulating evidence has revealed the association between gut dysbiosis and a series of neuropsychiatric diseases including neurological diseases such as Alzheimer's disease, psychiatric disorders, such as schizophrenia, depression, and anxiety. The gut microbiota system modulates the bidirectional communication between gut and brain through the neuroanatomical, neuroendocrine, inflammatory, and metabolic pathways, as well as the barrier system consisted of gut mucosal barrier and blood–brain barrier. Mechanisms of gut microbiota mediated pathophysiological processes have indicated potential therapeutic targets for treating neuropsychiatric diseases. Treatment methods including the probiotics supplements, dietary intervention, fecal microbial transplantation, and microbial metabolism have been applied in both animal models and humans. This review focused on gut dysbiosis related neuropsychiatric diseases, roles of the microbiota-gut-brain axis, as well as the effectiveness of related treatment approaches.

Keywords Brain-Gut bidirectionality · Neuroendocrine system · Fecal microbial transplantation · Neuropsychiatric diseases · Dietary modification · Probiotic supplementation

---

H.-Y. Lai (✉) · T. Shen
Kexuelou, Hwajiachi Campus, Zhejiang University, Hangzhou City 310029, Zhejiang Province, China
e-mail: laihy@zju.edu.cn

F. Marotta (ed.), *Gut Microbiota in Aging and Chronic Diseases*, Healthy Ageing and Longevity 17, https://doi.org/10.1007/978-3-031-14023-5_14

Abbreviations

GI	Gastrointestinal
AD	Alzheimer's disease
MCI	Mild cognitive impairment
FMT	Fecal microbial transplantation
MGBA	Microbiota-gut-brain axis
ENS	Enteric nervous system
CNS	Central nervous system
Aβ	Amyloid-beta
CSF	Cerebrospinal fluid
p-tau	Phosphorylated tau
RA	Relative abundance
SCFAs	Short-chain fatty acids
AA	Acetic acid
PPA	Propionic acid
BTA	Butyrate
VA	Valeric acid
HPA	Hypothalamus-pituitary-adrenal
ARNTL	Arylhydrocarbon receptor nuclear translocator-like gene
MAOA	Monoamine oxidase A
MDD	Major depression disorder
GAD	Generalized anxiety disorder
SZ	Schizophrenia
BBB	Blood brain barrier
ANS	Autonomic nervous system
IL	Interleukin
5-HT	5-Hydroxytryptamine
ACTH	Adrenocorticotropic hormone
TNF-α	Tumor necrosis factor-α
GALT	Gut-associated lymphoid tissue
Th	T helper
IgA	Immunoglobulin A
TLR	Toll-like receptor
IFN-γ	Interferon-γ
PGE2	Prostaglandin E2
DA	Dopamine
NE	Norepinephrine
GABA	Gamma-aminobutyric acid
LPS	Lipopolysaccharide
ApoE	Apolipoprotein E
ABCA7	ATP-binding cassette A7
GMB	Gut mucosal barrier
BDNF	Brain-derived neurotrophic factor
CFU	Colony forming units

ABA	Applied behavior analysis
RBANS	Repeatable battery for assessment of neuropsychological status
6-GSI	Six-item gastrointestinal severity index
BDI	Beck depression inventory
HAMA	Hamilton rating scale for anxiety
PANSS	Positive and negative syndrome scale
hs-CRP	High-sensitivity C-reactive protein
HOMA-IR	Homeostasis model of assessment insulin resistance
CDI	Clostridioides difficile infection
TD	Typically developing
GPR40	G-protein coupled receptor 40

14.1 Gut Microbiota and Neurological Diseases

The gut microbiota is a complex system in the enteric nervous system (ENS), also considered as the human "second brain", which bidirectionally communicates with the central nervous system (CNS). Beneficial gut microbiota produces many sorts of protective compounds and acts as a barrier against pathogens. Healthy gut microbiota composition might be an integral part of functional pathways that modulates the development of brain structure and function, the regulation of immune function, and influences mood and behavior, etc. Gut microbiota dysbiosis can trigger the occurrence of neuropsychiatric diseases, including neurodegenerative and psychiatric diseases. As a new perspective, the crucial role of the microbiota-gut-brain axis (MGBA) has brought about widespread attention.

14.1.1 Alzheimer's Disease

AD is one of the most common neurodegenerative causes of dementia, characterized by cognitive decline, abnormal behavior, and altered personalities. In addition to the well known impairment of cognitive function in AD, neuropsychiatric symptoms are commonly seen across whole disease duration, such as depression, apathy, hallucinations, and aggression (Lyketsos et al. 2011). The main neuropathological hallmarks of AD are the accumulation of extracellular amyloid-beta (Aβ) plaques and intracellular neurofibrillary tangles in the brain, which lead to neuronal death (Bloom 2014). Recently, growing evidence has suggested that these neuropathological structures might form in the gut and enter the brain via the gut-brain connecting vagus nerve (Chen et al. 2021a). As an environmental factor, the gut dysbiosis might interact with genetic risk factors to cause the occurrence of AD pathology. For example, three months old APP/PS1 mice received the donor fecal matter from the 16 months old APP/PS1 mice for 7 consecutive days led to a significant increase in Aβ plaques and suppressed astrocyte activation around Aβ plaques (Wang et al. 2021a). Risk factors such as unhealthy diet, aging, stress, obesity, and those lead to gut dysbiosis might also related to the neurodegenerative process. Thus, the gut-brain axis plays an important role in pathological mechanism of AD.

Recent studies reported decreased microbial diversity and gut microbiota structural alterations (Vogt et al. 2017; Cattaneo et al. 2017; Liu et al. 2019a; Zhuang et al. 2018; Haran et al. 2019; Li et al. 2019), as shown in Table 14.1. Although the directionality of gut microbiota alterations was divergent, enriched *Escherichia/Shigella*, depleted *Clostridium* and *Lachnoclostridium* at genus level were observed consistently in most studies (Borsom et al. 2020; He et al. 2020). The abundance of certain microbiota changed along with the progression of AD, which was demonstrated by reported correlations between gut microbiota and clinical characteristics, including the severity of cognitive impairment and disease duration (Liu et al. 2019a; Li et al. 2019). Therefore, the disturbance of gut microbiota is one of the potential risk factors for AD, which appears to affect disease development and symptoms. Growing evidence has indicated that the abnormality of the gut microbiota system might trigger the formation of amyloidosis (Liu et al. 2021b; Leblhuber et al. 2021). The gut microbiota whose abundance was altered in AD correlated with cerebrospinal fluid (CSF) biomarkers including the Aβ42/Aβ40 ratio, phosphorylated tau (p-tau) and p-tau/Aβ42 ratio (Vogt et al. 2017), suggesting that the gut microbiota disturbance might be associated with the deposition of Aβ and tau proteins. Significant alterations of gut microbiota involving metabolic pathways were further found in AD patients, such as glutamate metabolism (Chang et al. 2020), lipid metabolism (Mahmoudian Dehkordi et al. 2019), metaorganismal metabolism (Vogt et al. 2018), glycan biosynthesis and metabolism (Liu et al. 2019a), and energy metabolism (Liu et al. 2019a), which were also found to contribute to the development of AD. In addition, the role of the inflammatory pathway in the pathological mechanism of AD was widely discussed and accepted previously. With respect to the gut-brain axis, dysregulated pathways of gut microbiota that relate to the immune system was also observed in AD patients (Liu et al. 2019a). Meanwhile, correlations of pro-inflammatory cytokines with inflammatory and anti-inflammatory gut microbiota was further detected in AD (Cattaneo et al. 2017). These findings revealed the possible role of the MGBA in the pathological process of AD.

14.1.2 Dementia

Gut microbiota participates in modulating brain plasticity, regulating neurotransmitters, influencing behavior and cognitive functions through MGBA. In addition to AD, cognitive decline due to other diseases might also exhibit altered gut microbial communities. Mild cognitive impairment (MCI) is a stage between

Table 14.1 Alterations of gut microbiota composition in AD patients

References	Patient description	Richness and diversity	Alterations of RA at different taxonomic levels		Predicted functional pathways	Correlation analysis
Vogt et al. (2017)	25 AD cases	↓ ACE, Chao1, Shannon index, Faith's PD	↓ p_Firmicutes P_Actinobacteria f_Ruminococcaceae f_Turicibacteraceae f_Peptostreptococcaceae f_Clostridiaceae f_Mogibacteriaceae f_Bifidobacteriaceae g_Dialister g_Clostridium g_Turicibacter g_Bifidobacterium g_Adlercreutzia	↑ p_Bacteroidetes f_Gemellaceae f_Bacteroidaceae f_Rikenellaceae g_Blautia g_Phascolarctobacterium g_Gemella g_Bacteroides g_Alistipes g_Bilophila	–	–
Cattaneo et al. (2017)	73 cognitively impaired cases	-	↓ g_Bacteroides fragilis G_Eubacterium rectale	↑ g_Escherichia/Shigella	–	+correlation between pro-inflammatory cytokines IL-1, NLRP3 and CXCL2 with RA of Escherichia/Shigella -correlation with RA of Eubacterium rectale
Zhuang et al. (2018)	43 AD cases	-	↓ p_Bacteroidetes p_Verrucomicrobia c_Negativicutes c_Bacteroidia o_Bacteroidales o_Selenomonadales f_Lanchnospiraceae f_Bacteroidaceae	↑ p_Actinobacteria c_Actinobacteria c_Bacilli o_Lactobacillales f_Ruminococcaceae f_Enterococcaceae f_Lactobacillaceae g_Subdoligranulum	–	–

(continued)

Table 14.1 (continued)

References	Patient description	Richness and diversity	Alterations of RA at different taxonomic levels		Predicted functional pathways	Correlation analysis
			f_Veillonellaceae g_Bacteroides g_Lachnoclostridium			
Liu et al. (2019a, b)	33 AD cases, 32 MCI cases	↓ Shannon index, Simpson index	↓ p_Firmicutes c_Clostridia o_Clostridiales f_Clostridiaceae f_Lachnospiraceae f_Ruminococcaceae g_Blautia	↑ p_Proteobacteria c_Gammaproteobacteria o_Enterobacteriales f_Enterobacteriaceae g_Ruminococcus	↓ Transcription Energy metabolism ↑ Signal transduction Protein folding and associated processing Glycan biosynthesis and metabolism	+correlation between CDR and RA of f_Enterobacteriaceae +correlation between MMSE and RAs of p_Bacteroidetes and f_Ruminococcaceae +correlation between MoCA and RA of f_Clostridiaceae -correlation between MoCA and RA of f_Veillonellaceae -correlation between MMSE and MoCA and RAs of p_Proteobacteria, c_Gammaproteobacteria, o_Enterobacteriales and f_Enterobacteriaceae
Haran et al. (2019)	24 AD cases	–	↓ g_Eubacterium g_Lachnoclostridium g_Roseburia	↑ g_Bacteroides g_Alistipes g_Odoribacter g_Barnesiella g_Collinsella	–	–
Li et al. (2019)	30 AD cases, 30 MCI cases	↓ Faith's PD	↓ g_Parabacteroides g_Alistipes g_Bacteroides	↑ g_Lactobacillus g_Akkermansia g_Dorea	↓ Metabolisms of amino acid Carbohydrate	+ correlation between AD duration and RA of g_Megamonas

(continued)

Table 14.1 (continued)

References	Patient description	Richness and diversity	Alterations of RA at different taxonomic levels		Predicted functional pathways	Correlation analysis
			g_Alloprevotella g_Haemophilus g_Paraprevotella g_Succinivibrio, g_Sutterella g_Prevotella g_Barnesiella g_Butyricimonas	g_Bifidobacterium g_Streptococcus g_Acinetobacter g_Blautia	Energy metabolism Glycan Lipid metabolism cofactors and vitamins, nucleotide terpenoids polyketides	+correlation between MMSE and RAs of g_Faecalibacterium, g_Butyricicoccus and g_Hungatella -correlation between MMSE and RAs of g_Fusicatenibacter, g_Blautia and g_Dorea Correlation between ApoE genotype and RA of g_Anaerotruncus
Guo et al. (2021)	18 AD cases, 20 MCI cases	↑ beta-diversity	↓ g_Bacteroides g_Lachnospira g_Ruminiclostridium 9	↑g_Prevotella	–	+correlation between MMSE and RAs of g_Bacteroides, g_Lachnospira and g_Ruminiclostridium_9 -correlation between MMSE and RA of g_Prevotella
Zhou et al. (2021)	60 AD cases	↓ Observed species	↓ f_Eubacteriaceae g_Odoribacter g_Eubacterium g_Anaerobacterium g_Papillibacter	↑ c_Actinobacteria o_Lactobacillales f_Lactobacillaceae f_Bifidobacteriales g_Bifidobacterium g_Sphingomonas g_Bilophila g_Neorhizobium g_Flavobacterium g_Moheibacter g_Lactobacillus	↓Metabolism of cofactors and vitamins Energy metabolism Immune system Folding/sorting and degradation	+correlation between MMSE and RAs of g_Odoribacter, g_Eubacterium and g_Anaerobacterium +correlation between CDR and RAs of g_Sphingomonas, g_Clostridium XIVa and g_Bilophila -correlation between MMSE and RAs of g_Sphingomonas,

(continued)

Table 14.1 (continued)

References	Patient description	Richness and diversity	Alterations of RA at different taxonomic levels		Predicted functional pathways	Correlation analysis
				g_Weissella g_Blautia g_Clostridium XIVa g_Solobacterium	Environmental adaptation Cell growth and death ↑ Xenobiotics biodegradation and metabolism Membrane transport Lipid metabolism	g_Clostridium XIVa, g_Bilophila, g_Weissella and g_Moheibacter -correlation between CDR and RAs of g Odoribacter, g_Eubacterium and g_Anaerobacterium
Xi et al. (2021)	21 AD cases	–	↓ g_Tyzzerella g_Erysipelatoclostridium	↑ g_Agathobacter g_Eubacterium g_Lachnospiraceae g_Coprococcus g_Faecalibacterium g_Ruminococcaceae g_Alloprevotella g_Atopobium g_Parvimonas g_Cloacibacillus g_Solobacterium g_Pseudomonas	↓ Steroid hormone biosynthesis N-Acyl amino acid metabolism Piperidine metabolism	Correlation between ApoE genotype and RAs of g_Parvimona, g_Agathobacter, g_Lachnospiraceae and g_Faecalibacterium - correlation between MMSE and RAs of g_Parvimona, g_Atopobium and g_Coprococcus

AD: Alzheimer's disease; MCI: mild cognitive impairment; RA: relative abundance; ACE: Abundance-based coverage estimator; Faith's PD: Faith's phylogenetic diversity; p: phylum; c: class; o: order; f: family; g: genus; IL: interleukin; NLRP3: NOD-like receptors protein 3; CXCL2: CXC motif chemokine ligand 2; MMSE: mini mental state examination; MoCA: Montreal cognitive assessment; CDR: clinical dementia rating; ApoE: apolipoprotein E cognitive changes during normal aging and serious cognitive decline of dementia, which is characterized by impaired cognitive ability loss in one or more cognitive domains, such as memory, language, or visual/spatial perception (Geda 2012). MCI patients might have a similar pattern of alterations in gut microbiota as AD patients (Li et al. 2019; Guo et al. 2021), such as trend in altered abundances of *Escherichia*, *Lactobacillus* and *Bacteroides*. Negative correlation between increased abundance of *Prevotella* and better cognitive functions were found in MCI (Guo et al. 2021). Furthermore, MCI patients showed a specific correlation pattern between gut microbiota and brain spontaneous activities. For example, the abundance of *Ruminococcaceae* was negatively correlated with brain activities in the left lenticular nucleus and pallidum, while the *Veillonellaceae* was positively correlated with brain activity of left precentral gyrus (Liu et al. 2021a). In Parkinson's disease, patients with MCI showed significantly altered gut microbiota compared to patients without MCI, manifesting in enriched *Rikenellaceae* and *Ruminococcaceae* families, as well as *Alistipes*, *Barnesiella*, *Butyricimonas,* and *Odoribacter* genera (Ren et al. 2020). The abundances of *Butyricimonas*, *Barnesiella*, *Alistipes*, *Odoribacter,* and *Ruminococcus* genera were inversely related to cognition ability (Ren et al. 2020). Vascular cognitive impairment is a term that refers to cognitive disorders attributed to cerebrovascular diseases (Dichgans and Leys 2017). Patients with post-stroke cognitive impairment exhibited significantly enriched *Enterococcus*, *Bacteroides* and *Escherichia/Shigella*, and depleted *Faecalibacterium* compared to stroke patients without cognitive impairment (Huang et al. 2021b). Another group of patients with post-stroke cognitive impairment and depression were observed with decreased levels of several short-chain fatty acids (SCFA)-producing gut microbiota, indicating possible impairment of gut microbial metabolic pathway (Ling et al. 2020). In the light of this, different types of cognitive decline might be accompanied by specific gut microbial disturbances.

14.2 Gut Microbiota and Psychiatric Diseases

The gut microbiota system is put forward as an important factor related to a spectrum of psychiatric diseases, such as depression, schizophrenia, and anxiety. Psychiatric diseases usually share common risk factors, features, and pathological mechanisms. Both clinical and preclinical studies have provided growing evidence that alterations of gut microbiota might contribute to the occurrence and development of mental illness through the bidirectional MGBA. Thus, the role of the MGBA might also be a component of shared pathophysiology of psychiatric diseases. In this section, we describe the current findings in depression, generalized anxiety disorder (GAD) and schizophrenia.

14.2.1 Depression

Depression is a common but serious mood disorder. Patients suffer from persistent feelings of sadness or hopelessness, inability to feel pleasure in hobbies and activities, slowed movements and speech, lowered self-esteem, apathy, fatigue, insomnia, and even suicide if more serious. Risk factors such as the biochemistry, genetics, environmental factors, and personalities contribute to the development of depression through the hypothalamus–pituitary–adrenal (HPA) axis of neuroendocrine system, regulation of neurotransmission and neuroinflammation response of immune system (Li et al. 2021a). Growing evidence has shown alteration of gut microbiota patterns in depression patients (Table 14.2), meanwhile the MGBA also participates in modifying the function of neurotransmission, neuroinflammation, and neuroendocrine systems in depression.

Based on a meta-analysis study carried out on the genus level, the *Oscillibacter, Blautia, Holdemania, Clostridium XIX, Anaerostipes, Anaerofilum, Streptococcus, Gelria, Turicibacter, Parabacteroides, Eggerthella, Klebsiella, Paraprevotella, Veillonella, Clostridium IV, Erysipelotrichaceae incertae sedis, Eubacterium, Parvimonas, Desulfovibrio, Parasutterella, Actinomyces, Asaccharobacter, Atopobium,* and *Olsenella* were enriched, while the *Coprococcus, Lactobacillus, Escherichia/Shigella, Clostridium XlVa, Dialister, Howardella, Pyramidobacter,* and *Sutterella* were decreased in patients with depression (Barandouzi et al. 2020). The butyrate (BTA)-producing *Faecalibacterium* and *Coprococcus* bacteria were associated with the mental quality of life (Valles-Colomer et al. 2019). The *Faecalibacterium* was also correlated with the severity of depressive symptoms (Jiang et al. 2015). Moreover, the species of *Lactobacillus* was related to positive self-judgement, cognitive depression, and lower affective empathy in depression, indicating that the protective effect from beneficial gut microbiota on depression might be impaired along with its depletion (Heym et al. 2019). The gut microbiota system also interacted with host genetic activities. Lower diversity and evenness of gut microbiome were linked to increased methylation level of the CPG position in the clock gene, *Arylhydrocarbon receptor nuclear translocator-like gene (ARNTL)*, which might lead to reduction of monoamine oxidase A (MAOA) activity and increased neurotransmitter levels (Bengesser et al. 2019). Specific gut microbiota produces enzymes involved in the production of depression-related metabolites (Averina et al. 2020). SCFAs are considered to influence emotional state and contribute to the pathogenesis of depression through the MGBA. Levels of the SCFAs including acetic acid (AA), propionic acid (PPA) and BTA were decreased in depression patients, and further associated with disease severity (Skonieczna-Zydecka et al. 2018; Averina et al. 2020). Thus, the disturbance of MGBA was evident in depression patients.

Table 14.2 Alterations of gut microbiota composition in patients with depression

Reference	Patient description	Richness and diversity	Alterations of RA at the different taxonomic levels		Predicted functional pathways	Correlation analysis
Naseribafrouei et al.(2014)	37 depressed cases	–	↓ f_Lachnospiraceae	↑ p_Bacteroidetes g_Alistipes g_Oscillibacter	–	–
Jiang et al. (2015)	29 active-MDD cases	↑ Shannon index	↓ p_Actinobacteria p_Firmicutes f_Veillonellaceae f_Prevotellaceae f_Lachnospiraceae f_Erysipelotrichaceae f_Bacteroidaceae g_Ruminococcus g_Prevotella g_Faecalibacterium g_Dialister g_Bacteroides	↑ p_Fusobacteria p_Proteobacteria p_Bacteroidetes f_Rikenellaceae f_Porphyromonadaceae f_Fusobacteriaceae f_Enterobacteriaceae f_Acidaminococcaceae g_Roseburia g_Oscillibacter g_Phascolarctobacterium g_Parasutterella g_Parabacteroides g_Megamonas g_Lachnospiracea-incertae sedis g_Clostridium XIX g_Blautia g_Alistipes	–	–
Jiang et al. (2015)	17 responsed-MDD cases	–	↓ p_Actinobacteria p_Fusobacteria p_Firmicutes f_Veillonellaceae f_Ruminococcaceae f_Lachnospiraceae	↑ p_Proteobacteria p_Bacteroidetes f_Bacteroidaceae f_Rikenellaceae f_Porphyromonadaceae f_Enterobacteriaceae	–	–

(continued)

Table 14.2 (continued)

Reference	Patient description	Richness and diversity	Alterations of RA at the different taxonomic levels		Predicted functional pathways	Correlation analysis
			g_Ruminococcus g_Prevotella g_Faecalibacterium g_Dialister g_Oscillibacter g_Escherichia/Shigella	f_Acidaminococcaceae g_Bacteroides g_Roseburia g_Phascolarctobacterium g_Parabacteroides g_Alistipes		
Aizawa et al. (2016)	43 MDD cases	–	↓ g_Bifidobacterium	–	–	–
Zheng et al. (2016)	58 MDD cases	–	↓ p_Bacteroidetes f_Bacteroidaceae f_Rikenellaceae f_Lachnospiraceae f_Acidaminococcaceae f_Veillonellaceae f_Sutterellaceae	↑ p_Actinobacteria f_Actinomycineae f_Coriobacterineae f_Lactobacillaceae f_Streptococcaceae, f_Clostridiales incertae sedis XI f_Eubacteriaceae f_Lachnospiraceae f_Ruminococcaceae f_Erysipelotrichaceae incertae sedis	–	–
Lin et al. (2017)	10 MDD cases		↓ p_Bacteroidetes	↑ p_Firmicutes g_Prevotella g_Klebsiella g_Streptococcus g_Clostridium XI	–	–
Chen et al. (2018a, b)	10 MDD cases		↓ p_Proteobacteria p_Bacteroidetes f_Enterobacteriaceae	↑ p_Actinobacteria p_Firmicutes f_Actinomycetaceae	Altered KEGG pathways:	–

(continued)

Table 14.2 (continued)

Reference	Patient description	Richness and diversity	Alterations of RA at the different taxonomic levels		Predicted functional pathways	Correlation analysis
			f_Sutterellaceae f_Oscillospiraceae f_Chitinophagaceae f_Mariniabiliaceae f_Prevotellaceae	f_Nocardiaceae f_Streptomycetaceae f_Bifidobacteriaceae f_Erysipelotrichaceae f_Clostridiaceae f_Lachnospiraceae f_Ruminococcaceae f_Rikenellaceae f_Porphyromonadaceae	Energy production and conversion Cell cycle control, cell division, chromosome partitioning Amino acid transport and metabolism Nucleotide transport and metabolism Carbohydrate transport and metabolism Coenzyme transport and metabolism Lipid transport and metabolism Translation, ribosomal structure, and biogenesis Transcription Replication, recombination and repair Cell wall/ membrane/envelope biogenesis Cell motility	

(continued)

Table 14.2 (continued)

Reference	Patient description	Richness and diversity	Alterations of RA at the different taxonomic levels		Predicted functional pathways	Correlation analysis
					Posttranslational modification, protein turnover, chaperones Inorganic ion transport and metabolism Secondary metabolites biosynthesis, transport, and catabolism General function prediction only Signal transduction mechanisms Intracellular trafficking, secretion, and vesicular transport	
Chen et al. (2018a, b)	24 female MDD cases	–	↓ g_Howardella g_Sutterella g_Pyramidobacter	↑ g_Actinomyces g_Bifidobacterium g_Asaccharobacter, g_Atopobium g_Eggerthella g_Gordonibacter g_Olsenella g_Eubacterium		-correlation between HDRS and RAs of g_Clostridium XIVa, g_Erysipelotrichaceae incertae sedis and g_Streptococcus

(continued)

Table 14.2 (continued)

Reference	Patient description	Richness and diversity	Alterations of RA at the different taxonomic levels		Predicted functional pathways	Correlation analysis
				g_Anaerostipes g_Blautia g_Roseburia g_Faecalibacterium g_Desulfovibrio		
Chen et al. (2018a, b)	20 male MDD cases	–	↓ g_Anaerovorax g_Gordonibacter g_Pyramidobacter	↑ g_Bacteroides g_Erysipelotrichaceae incertae sedis g_Veillonella g_Atopobium		+correlation between HDRS and RA of g_Collinsella -correlation between HDRS and RAs of g_Veillonella
Huang et al. (2018)	27 MDD cases	↓ ACE, Chao1, Shannon index, Faith's PD	↓ f_Lachnospiraceae f_Ruminococcaceae f_Clostridiaceae g_Faecalibacterium g_Coprococcus g_Blautia g_Dorea	↑ g_Oxalobacter g_Parvimonas g_Bulleidia g_Peptostreptococcus g_Gemella	↓ Sporulation Bacterial motility proteins Bacterial chemotaxis Nitrotoluene degradation Germination Synthesis and degradation of ketone bodies ↑ Lipopolysaccharide biosynthesis Ubiquinone and other terpenoid-quinone biosynthesis	–

(continued)

Table 14.2 (continued)

Reference	Patient description	Richness and diversity	Alterations of RA at the different taxonomic levels		Predicted functional pathways	Correlation analysis
					Glycosaminoglycan degradation Glycosphingolipid biosynthesis Toluene degradation Cellular antigens Protein digestion and absorption Steroid hormone biosynthesis Lipoic acid metabolism	
Painold et al. (2019)	32 BD cases	–	↓ f_Ruminococcaceae g_Faecalibacterium	↑ p_Actinobacteria c_Coriobacteria o_Coriobacteriales f_Coriobacteriaceae	–	-correlation between disease duration and observed species
Rong et al. (2019)	31 MDD cases	↓ Chao1, G_m coefficient	↓ p_Bacteroidetes g_Bacteroides g_Tannerella	↑ p_Firmicutes p_Actinobacteria g_Clostridium g_Bifidobacterium g_Oscillibacter g_Streptococcus g_Odoribacter g_Selenomonas g_Megasphaera g_Acidaminococcus g_Veillonella	–	–

(continued)

Table 14.2 (continued)

Reference	Patient description	Richness and diversity	Alterations of RA at the different taxonomic levels		Predicted functional pathways	Correlation analysis
Rong et al. (2019)	30 BPD cases	↓ G_m coefficient	↓ p_Bacteroidetes g_Bacteroides g_Tannerella	↑ p_Firmicutes p_Actinobacteria g_Escherichia g_Clostridium g_Bifidobacterium g_Oscillibacter g_Klebsiella g_Streptococcus g_Odoribacter g_Selenomonas g_Megasphaera g_Acidaminococcus	–	+correlation between HAMA and G_m coefficient and -correlation between HAMA and Shannon index
Hu et al. (2019)	52 untreated BD cases	↓ Obs, Chao1, ICE index	↓ p_Firmicutes g_Roseburia g_Faecalibacterium g_Ruminococcus g_Gemmiger g_Parasutterella g_Coprococcus	↑ p_Bacteroidetes g_Parabacteroides g_Bacteroides g_Weissella g_Halomonas	–	+correlation between disease duration and RA of g_Allisonella +correlation between MADRS and RAs of g_Acinetobacter and g_Cronobacter -correlation between disease duration and RAs of g_Escherichia/Shigella, g_Flavonifractor and g_Staphylococcus -correlation between MADRS and RAs of g_Acetanaerobacterium, g_Stenotrophomonas, g_Anaerotruncus and g_Raoultella

(continued)

Table 14.2 (continued)

Reference	Patient description	Richness and diversity	Alterations of RA at the different taxonomic levels		Predicted functional pathways	Correlation analysis
Chen et al. (2020)	25 young MDD cases		↓ p_Firmicutes f_Clostridiaceae 1 f_Peptostreptococcaceae f_Ruminococcaceae g_Clostridium sensu stricto g_Clostridium XI g_Clostridium XVIII	↑ g_Bacteroidetes	–	–
Chen et al. (2020)	45 middle-aged MDD cases	–	↓ p_Bacteroidetes f_Sutterellaceae f_Acidaminococcaceae f_Bacteroidaceae f_Prevotellaceae g_Anaerovorax g_Sutterella g_Oscillibacter g_Phascolarctobacterium g_Coprococcus g_Bacteroides g_Faecalibacterium	↑ p_Actinobacteria f_Lachnospiraceae f_Coriobacteriaceae f_Strcptococcaccac f_Eubactenaceae f_Actinomycetaceae g_Blautia g_Clostridium XVIII g_Anaerostipes g_Veillonella g_Streptococcus g_Eggenhella g_Rosebuna g_Eubactcrium g_Actinomyces	–	–
Fontana et al. (2020)	34 MDD cases	–	↓ p_Nitrospirae f_Paenibacillaceae	↑ p_Elusimicrobia f_Flavobacteriaceae g_Fenollaria	–	–
Liskiewicz et al. (2021)	16 MDD cases	–	–	–	–	+correlation between HAMD and RAs of p_Firmicutes, c_Clostridia,

(continued)

Table 14.2 (continued)

Reference	Patient description	Richness and diversity	Alterations of RA at the different taxonomic levels		Predicted functional pathways	Correlation analysis
						c_Alphaproteobacteria, o_Clostridiales and g_Faecalibacterium -correlation between HAMD and RAs of p_Bacteroidetes, c_Gammaproteobacteria, c_Bacteroidia, o_Bacteroidales and g_Paraprevotella
Lai et al. (2021)	25 BPD cases	↓ Fisher index ↑ Shannon index	↓ g_Bacteroides g_Tannerella g_Paludibacter g_Sphingobacterium	↑ g_Escherichia g_Bifidobacterium g_Lachnoclostridium g_Megasphaera g_Clostridium g_Oscillibacter g_Selenomonas g_Acidaminococcus g_Streptococcus	↓ K00658 and K00837 in Tryptophan metabolism ↑ K01696 in Tryptophan biosynthesis K00382, K00626, K01667, K03781 in Tryptophan metabolism	–
Zhang et al. (2021)	36 MDD cases	–	↓ o_Pasteurellales f_Prevotellaceae f_Clostridiaceae f_Peptostreptococcaceae f_Pasteurellaceae g_Prevotella g_Haemophilus g_Weissella	↑ o_Actinomycetae f_Bacteroidaceae f_Porphyromonadaceae f_Rikenellaceae G_Bacteroides g_Parabacteroides g_Escherichia g_Alistipes g_Alloprevotella g_Tyzzerella	–	+correlation between HAMD and RAs of g_Butyricicoccus and g_Peptococcus -correlation between HAMD and RA of g_Dorea -correlation between PSQI and RAs of g_Blautia, g_Coprococcus, g_Dorea and g_Intestinibacter

(continued)

Table 14.2 (continued)

Reference	Patient description	Richness and diversity	Alterations of RA at the different taxonomic levels		Predicted functional pathways	Correlation analysis
				g_Paraprevotella g_Flavonifractor g_Anaerotruncus g_Hungatella g_Intestinimonas g_Faecalitalea g_Holdemania g_Eggerthella		-correlation between ISI and RA of g_Intestinibacter
Dong et al. (2021)	23 MDD cases	–	↓ g_Sutterella g_Fusicatenibacter	–	–	-correlation between HAMD factor score (Hopelessness) and RA of g_Christensenellaceae_R7 -correlation between FT4 levels and RA of g_Fusicatenibacter

RA: relative abundance; MDD: major depressive disorder; BD: bipolar depression; BPD: bipolar disorder with current major depressive episode; ACE: Abundance-based coverage estimator; Faith's PD: Faith's phylogenetic diversity; G_m coefficient: Gini coefficient of microbiota; ICE: incidence-based coverage estimators; p: phylum; c: class; o: order; f: family; g: genus; KEGG: Kyoto encyclopedia of genes and genomes; HDRS: Hamilton sepression rating scale; MADRS: Montgomery-Asberg depression rating scale; HAMD: Hamilton depression rating scale; PSQI: Pittsburgh sleep quality index; ISI: insomnia severity index; FT4: free thyroxine 4

14.2.2 Generalized Anxiety Disorder

Generalized anxiety disorder (GAD) is a common mood disorder, characterized by excessive and uncontrollable worry and fear about events, and which causes symptoms such as panic, insomnia, shortness of breath, heart palpitations, tense muscles, as well as gastrointestinal (GI) problems. The causative factors of GAD involve combination of genetics, environmental, and neurobiological factors, life experiences and medical conditions. Since the MGBA plays a key role in modifying emotional status and mood disorders, it also involves in the pathophysiology of anxiety.

The dysbiosis of gut microbiota system was reviewed and summarized in Table 14.3. Gut microbial richness and diversity were decreased in patients with anxiety (Jiang et al. 2018; Chen et al. 2019; Dong et al. 2021). Higher anxiety level was correlated with decreased gut microbiome diversity (Huang et al. 2021a). SCFA-producing bacteria were depleted, while *Escherichia/Shigella*, *Fusobacterium,* and *Ruminococcus* were enriched (Jiang et al. 2018; Chen et al. 2019). Higher anxiety severity was correlated with increased abundances of *Bacteroides* and *Escherichia/Shigella*, as well lower abundances of *Eubacterium*, *Ruminococcaceae*, *Prevotella* (Chen et al. 2019), and *Christensenellaceae* (Dong et al. 2021). Levels of anxiety reduction after treatment were positively associated with abundances of *Eubacterium*, *Ruminococcaceae,* and *Prevotella*, which could be regarded as biomarkers to monitor the effectiveness of treatment (Chen et al. 2019). Correlation between anxiety levels and brain grey matter volumes of the right dorsolateral prefrontal cortex was reported. This relationship was mediated by the gut microbiome diversity (Huang et al. 2021a), which provided evidence of interactions between gut and brain structure through the MGBA. The biosynthetic pathway of indoles, such as the production of indoxyl sulfate, was mainly found in the gut microbiota system (Agus et al. 2018). Gut microbiota-derived uremic toxins were correlated with psychic anxiety levels (Brydges et al. 2021). Through modifying the functional connectivities of brain regions including the cingulate cortex, anterior insula, and premotor areas that are involved in processing aversive stimuli, gut microbiota might contribute to a neurophysiological progress of anxiety (Brydges et al. 2021).

14.2.3 Schizophrenia

Schizophrenia is a serious and lifelong neurodevelopmental disorder characterized by psychiatric symptoms and cognitive impairments. Psychiatric symptoms include positive symptoms, such as delusions and hallucinations, and negative symptoms, such as alogia, avoilition, social withdrawal, and attentional deficits. It has been widely accepted that a combination of genetic, neurochemical, and environmental factors contribute to the development of schizophrenia. Both genetic and

Table 14.3 Alterations of gut microbiota composition in GAD patients

References	Patient description	Richness and diversity	Alterations of RA at the different taxonomic levels		Predicted functional pathways	Correlation analysis
Jiang et al. (2018)	40 GAD cases	↓ Observed OTUs	↓ p_Firmicutes g_Bacteroidetes g_Ruminococcus gnavus g_Fusobacterium	↑ p_Fusobacteria p_Bacteroidetes g_Faecalibacterium g_Eubacterium rectale g_Sutterella g_Lachnospira g_Butyricicoccus	–	–
Chen et al. (2019)	36 GAD cases	↓ ACE	↓ p_Firmicutes, p_Tenericutes c_Mollicutes f_Prevotellaceae f_Muribaculaceae g_Prevotella 9 g_Dialister g_Subdoligranulum g_Megamonas g_Agathobacter g_Coprococcus 1 g_Coprococcus 3 g_Lachnospiraceae NK4A136 g_Buchnera g_Ruminococcaceae NK4A214 g_Mollicutes_RF39_norank g_Acinetobacter	↑ c_Betaproteobacteriales o_Enterobacteriales f_Burkholderiaceae f_Enterobacteriaceae f_Bacteroidaceae g_Tyzzerella 3 g_Hungatella g_Escherichia/Shigella g_Bacteroides	–	+correlation between GAF scores and RAs of g_Prevotellaceae UCG-001, g_Succinivibrio, g_Mollicutes_RF39_norank, g_Prevotella 9 and g_Ruminococcaceae_NK4A214 +correlation between anxiety/ depression scores and RAs of g_Bacteroides and Escherichia/ Shigella -correlation between anxiety/ depression scores and RAs of f_g_Prevotellaceae UCG-001, g_Mollicutes_RF39_norank, g_Succinivibrio, g_Mitsuokella, g_Prevotella 9, g_Subdoligranulum, g_Ruminococcaceae_NK4A214 and g_Ruminococcaceae_UCG-014

(continued)

Table 14.3 (continued)

References	Patient description	Richness and diversity	Alterations of RA at the different taxonomic levels		Predicted functional pathways	Correlation analysis
Dong et al. (2021)	21 GAD cases	↓ ACE, Chao1, Shannon index ↑ Simpson index	↓ g_Fusicatenibacter g_Christensenellaceae_R7_group	–	–	-correlation between PTC levels and RA of g_Faecalibacterium

GAD: generalized anxiety disorder; RA: relative abundance; OTU: operational taxonomic units; ACE: abundance-based coverage estimator; p: phylum; c: class; o: order; f: family; g: genus; GAF: global assessment of functioning; PTC: plasma total cortisol environmental risk factors might also trigger comorbid GI inflammation, and cause schizophrenia through the gut-brain axis.

The pattern of gut microbiota composition is altered in schizophrenia (Table 14.4). Although alterations of many microbiota were inconsistent across different studies, the *Lactobacillus*, *Prevotella,* and *Bifidobacterium* genera were more abundant in patients with schizophrenia in most studies (Castro-Nallar et al. 2015; Shen et al. 2018; Ma et al. 2020). A diagnostic classifier based on gut microbial species was established, and its performance reached 0.896 (Zhu et al. 2020), indicating the effective diagnosis function of this biomarker. Furthermore, associations between the abundance of *Ruminococcaceae* family and lower severity of negative symptoms, abundance of *Bacteroides* and worse depressive symptoms were observed in schizophrenia patients (Nguyen et al. 2019). Neuroimaging studies have further revealed the interaction between gut microbiota and brain structural signatures. Correlations of the grey matter volume of right middle frontal gyrus to schizophrenia-related microbiota, such as the genus *Actinobacillus* and family *Veillonellaceae* (Ma et al. 2020). Functional pathways of gut microbiota including SCFAs synthesis, tryptophan metabolism, and synthesis/degradation of neurotransmitters were different in schizophrenia patients compared to healthy controls (Zhu et al. 2020). Gut microbiota that are involved in the production of serotonin were also correlated with higher risk of developing schizophrenia (Zhuang et al. 2020). The dysbiosis of gut microbiota system led to disturbance in its roles in the regulation of neurotransmission, immunological response, and brain development, which further promotes the pathogenesis of schizophrenia (Yuan et al. 2019).

14.3 Role of Microbiota-Gut-Brain Axis

The role of gut microbiota goes far beyond digestion. Accumulating evidence has shown that the gut microbiota contributes to immune defense, production of vitamins, anti-inflammatory compounds, and neurotransmitters. In general, the gut microbiota in the enteric nervous system (ENS) bidirectionally interacts with brain through neuroanatomical, neuroendocrine, inflammatory, and metabolic pathways, as well as the barrier system consisted of gut mucosal barrier (GMB) and blood–brain barrier (BBB). These components interact with each other to maintain host homeostasis. A group of risk factors including diet, obesity, psychological stress, antibiotics, and medical conditions can disturb the healthy gut microbiota system and lead to dysbiosis. Subsequently, an imbalanced gut microbiome contributes to trigger the pathogenesis of neuropsychiatric diseases. In the following section, we review these impaired pathways related to gut microbiota dysbiosis (Fig. 14.1).

Table 14.4 Alterations of gut microbiota composition in schizophrenia patients

References	Patient description	Richness and diversity	Alterations of RA at the different taxonomic levels		Predicted functional pathways	Correlation analysis
Shen et al. (2018)	64 SZ cases	–	↓ p_Firmicutes g_Blautia g_Coprococcus g_Roseburia g_Streptococcus	↑ p_Proteobacteria p_Fusobacteria g_Succinivibrio g_Megasphaera g_Collinsella g_Clostridium g_Klebsiella g_Methanobrevibacter g_Prevotella g_Fusobacterium g_Lactobacillus g_Megasphaera g_Acidaminococcus g_Citrobacter g_Phascolarctobacterium g_Desulfovibrio	↓ Cytoskeleton proteins Base excision repair; RNA transport Lysine biosynthesis Flavone and flavonol biosynthesis; Phenylpropanoid biosynthesis Starch and sucrose metabolism Methane metabolism; Photosynthesis; Photosynthesis proteins Linoleic acid metabolism; Secondary bile acid biosynthesis Cyanoamino acid metabolism Bisphenol degradation; Polycyclic aromatic hydrocarbon degradation ↑ Cell cycle; Phagosome Bacterial secretion system; Secretion system; mTOR signaling pathway Sulfur relay system; mRNA surveillance pathway Cysteine and methionine metabolism; Tryptophan metabolism; Valine, leucine and isoleucine degradation Ascorbate and aldarate metabolism Lipopolysaccharide biosynthesis proteins alpha-Linolenic acid metabolism; Biosynthesis of unsaturated fatty acids; Fatty acid metabolism; Synthesis and degradation of ketone bodies Ubiquinone and other terpenoid-quinone biosynthesis; Vitamin B6 metabolism beta-Alanine metabolism; Glutathione metabolism; Taurine and hypotaurine metabolism Biosynthesis of siderophore group nonribosomal peptides; Geraniol degradation Benzoate degradation; Caprolactam degradation; Fluorobenzoate degradation; Styrene degradation	
Nguyen et al. (2019)	25 SZ cases	–	↓ p_Proteobacteria g_Haemophilus g_Sutterella g_Clostridium	↑ g_Anaerococcus	-	+correlation between onset age and RA of p_Cyanobacteria +correlation between severity of depressive symptoms and RA of g_Bacteroides +correlation between overall self-reported mental well-being and RA of p_Verrucomicrobia -correlation between negative symptoms and RA of f_Ruminococcaceae

(continued)

Table 14.4 (continued)

References	Patient description	Richness and diversity	Alterations of RA at the different taxonomic levels		Predicted functional pathways	Correlation analysis
Xu et al. (2020)	84 SZ cases	↓ Chao1	↓ o_Rhodocyclales f_Alcaligenaceae f_Enterococcaceae f_Leuconostocaceae f_Rhodocyclaceae f_Rikenellaceae g_Enterococcus	↑ p_Actinobacteria c_Deltaproteobacteria o_Actinomycetales o_Sphingomonadales f_Sphingomonadaceae g_Eggerthella g_Megasphaera,	↓ Alanine, aspartate, and glutamate metabolism Alanine, aspartate, and glutamate metabolism D-Glutamine and D-glutamate metabolism ↑ Phenylalanine, tyrosine and tryptophan biosynthesis/ Biosynthesis of amino acids Alanine, aspartate and glutamate metabolism	–
Ma et al. (2020)	40 FSZ cases,	↓ Shannon index, Chao1	↓ f_Pasteurellaceae f_Turicibacteraceae f_Peptostreptococcaceae f_Veillonellaceae f_Succinivibrionaceae g_Actinobacillus g_Fusobacterium g_Megasphaera	↑ f_Christensenellaceae f_Enterobacteriaceae f_Victivallaceae g_Escherichia	–	+correlation between right middle frontal gyrus volume and RAs of f_Veillonellaceae and g_Actinobacillus
	85 TSZ cases	↓ Shannon index, Chao1	↓ f_Pasteurellaceae f_Turicibacteraceae	↑ f_Christensenellaceae f_Enterobacteriaceae f_Enterococcaceae f_Lactobacillaceae g_Escherichia g_Enterococcus g_Lactobacillus g_Shigella g_Streptococcus g_Veillonella	–	–
Zhu et al. (2020)	90 SZ cases	↑ alpha diversity	–	–	↓ Pectin degradation Lipopolysaccharide biosynthesis Autoinducer-2 transport system Glutamate/aspartate transport system Beta-carotene biosynthesis ↑ Methanogenesis Gamma-aminobutyrate (GABA) shunt Transport system of manganese, zinc and iron	–

(continued)

Table 14.4 (continued)

References	Patient description	Richness and diversity	Alterations of RA at the different taxonomic levels		Predicted functional pathways	Correlation analysis
Li et al. (2020)	82 SZ cases	–	↓ p_Firmicutes g_Adlercreutzia g_Anaerostipes g_Ruminococcus g_Faecalibacterium	↑ p_Actinobacteria g_Collinsella g_Lactobacillus g_Succinivibrio g_Mogibacterium g_Corynebacterium	↓ Polyketide sugar unit biosynthesis Valine, Leucine and Isoleucine biosynthesis Pantothenate and CoA biosynthesis C5-Branched dibasic acid metabolism Phenylpropanoid biosynthesis ↑ Ascorbate and aldarate metabolism Nucleotide metabolism Propanoate metabolism	+correlation between severity of SZ symptoms and RA of g_Succinivibrio -correlation between increased negative symptoms and g_Corynebacterium
Nguyen et al. (2021)	48 SZ cases	–	–	↑ f_Lachnospiraceae	↓ Trimethylamine-N-oxide reductase Kdo2-lipid A biosynthesis ↑ Glycerol degradation to 1,3-propanediol	+correlation between Framingham risk score and log-ratios of Trimethylamine-N-oxide reductase/bottom-20 -correlation between IL-10 level and log-ratio of Kdo2-lipid A/ bottom-20
Yuan et al. (2021)	107 SZ cases	↓ Shannon index, Simpson index ↑ PD whole tree index	↓ g_Alistipes g_Blautia g_Bifidobacterium g_Romboutsia g_Fusobacterium g_Anaerostipes g_Streptococcus g_Dorea g_Fusicatenibacter g_Lactococcus g_Weissella g_Terrisporobacter g_Haemophilus g_Enterococcus g_Citrobacter	↑ g_Lachnoclostridium	–	+correlation between PANSS-FSNS and RA of p_Firmicutes and g_Lachnoclostridium +correlation between changes in the PANSS-FSNS and changes in RA of g_Lachnoclostridium - correlation between changes in the PANSS-FSNS and changes in RA of g_Romboutsia
Zhu et al. (2021)	42 SZ cases in acute stage	–	↓ g_Haemophilus g_Faecalibacterium g_Coprococcus g_Lachnospira	↑ o_Fusobacteriales o_Turicibacterales o_Chthoniobacterales g_Succinivibrio, g_Megasphaera g_Actinomyces g_Turicibacter g_Clostridium	–	+correlation between negative psychiatric symptoms and RA of g_Haemophilus +correlation between RA of g_Haemophilus and symptoms including cognition, excitement and depression -correlation between negative psychiatric symptoms and RA of g_Coprococcus

(continued)

Table 14.4 (continued)

References	Patient description	Richness and diversity	Alterations of RA at the different taxonomic levels		Predicted functional pathways	Correlation analysis
	40 SZ cases in remission stage	–	↓ g_Coprococcus g_Lachnospira g_Haemophilus	↑ f_Succinivibrionaceae f_Lactobacillaceae g_Desulfovibrio g_Mitsuokella g_Lactobacillus g_Succinivibrio, g_Megasphaera g_Megamonas g_Clostridium		
Li et al. (2021a, b, c)	38 SZ cases	-	↓ g_Ruminococcus g_Roseburia	↑ g_Veillonella	–	+ correlation between Faith_PD/observed species and the GMV of bilateral insula and right postcentral gyrus + correlation between Faith_PD and the GMV of left inferior operculum frontal cortex +correlation between Shannon index and the ReHo of bilateral calcarine cortex, bilateral lingual gyrus, left superior occipital cortex and right superior parietal cortex +correlation between evenness index and the ReHo of the right cuneus lobe, bilateral fusiform gyrus, left postcentral gyrus and left superior parietal cortex
Miao et al. (2021)	100 SZ cases	–	↓ p_Firmicutes c_Clostridia o_Clostridiales f_Ruminococcaceae g_Subdoligranulum g_Faecalibacterium	↑ p_Bacteroidetes c_Bacteroidia o_Bacteroidales f_Peptostreptococcaceae f_Bacteroidaceae g_Bacteroides g_Blautia	–	+correlation between serum folic acid level and RA of g_Bifidobacterium +correlation between speed of processing and RA of g_Bifidobacterium

SZ: schizophrenic; FSZ: first-episode drug-naïve SZ: TSZ, treated SZ; RA: relative abundance; p: phylum; c: class; o: order; f: family; g: genus; PANSS-FSNS: Positive and negative syndrome scale-factor scores for negative symptoms; GMV: grey matter volume

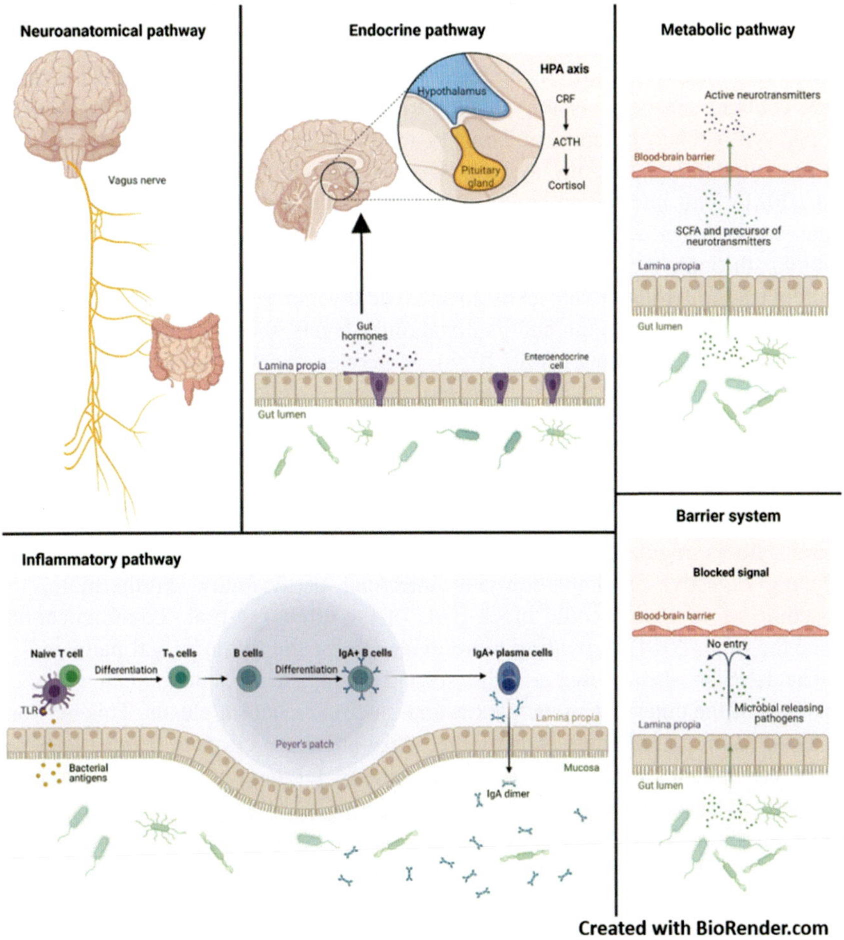

Fig. 14.1 Schematic outlining the bidirectional pathways of microbiota-gut-brain axis

14.3.1 Neuroanatomical Pathways

From a neuroanatomic perspective, the communication pathway between the gut and brain consists of the vagus nerve and autonomic nervous system (ANS) passing through the spinal cord, as well as the ENS in the gut (Makris et al. 2021). The ANS is part of the peripheral nervous system that regulates the function of internal organs and controls involuntary physiologic processes. It comprises three branches including the sympathetic, parasympathetic and enteric nervous systems (Waxenbaum et al. 2021). Of particular concern is that the ANS plays an important part in modulating and coordinating GI motility, secretion, and immune function (Martinez et al. 2012; Olsson and Holmgren 2011). Correspondingly, gut microbiota and SCFAs also alter the parasympathetic outflow to modulate gut motility (Olsson and Holmgren 2011).

The vagus nerve pathway is an important part of the gut-brain axis. The vagus nerve contains sensory, gustatory, motor and autonomous fibers (parasympathetic). Its parasympathetic component belongs to the ANS, which makes up 75% of the parasympathetic nervous system (Olsson and Holmgren 2011, Waxenbaum et al. 2021). Gut microbiota could impact the brain activity and behavior via the vagus nerve. Vagal afferent fibers recognize indirect signals produced by gut microbiota, through the diffusion of the gut microbiota-derived compounds or its metabolites. Enteroendocrine cells located in the gut epithelium also detect the alterations of gut microbiota and its metabolites levels, thereby indirectly affect the vagal afferent fibers (Bonaz et al. 2018). Moreover, receptors of several cytokines such as interleukin (IL)-1 are expressed by paraganglia cells in afferent nerves (Thayer and Sternberg 2010), thus the immune signals generated by gut microbiota might also be received by the vagus nerve. The afferent nerves convey these mechanical and chemical sensory information, as well as immune signals that reflect the state of the gut to the CNS (Bonaz et al. 2018). The activated vagal efferent fibers regulate the gut microbiota composition and function through reducing digestive inflammation and intestinal permeability. Furthermore, the procedure of vagotomy could block parts of the direct gut-brain communications (Forsythe et al. 2014), providing more evidence for the role of vagal pathway.

The ENS is also known as "the second brain", a separate branch of ANS, consisted of the outer myenteric plexus and inner submucosal plexus. This network of 200–600 million neurons is distributed throughout the GI tract. It innervates the whole GI system, and participates in regulating GI motility, intestinal immune response, nutrient absorption, intestinal endocrine secretion, and blood flow (De Vadder et al. 2018). The gut microbiota system cooperates with the immune system to help in regulating the maturation of the ENS after birth. A study reported that the mechanism of gut microbiota regulating the maturation of neuroanatomical pathway was through initiation of serotonin release and activation of the 5-hydroxytryptamine (5-HT) receptor (De Vadder et al. 2018). Gut microbiota bidirectionally communicates with the brain through the MGBA. On one hand, gut microbiota can modulate brain activities and behaviors by stimulating the ENS and afferent signaling via the vagus nerve, and then transmits signals to the CNS. On the other hand, the CNS sends regulatory information to the intestinal microenvironment via the ENS, the neuroendocrine pathway, and subsequently reshapes the gut microbiota community (Zhao et al. 2018).

14.3.2 Neuroendocrine Pathways

The neuroendocrine system is defined as a series of cells that synthesize and secrete neurohormones or neuropeptides after receiving neuronal signals to control and regulate the physiological processes of body. The hypothalamus–pituitary–adrenal

(HPA) axis that consists of the hypothalamus, pituitary gland and adrenal cortex is the main part of the neuroendocrine system and participates in mediating the gut-brain axis (Makris et al. 2021). In response to certain stressors such as stress and infections, the HPA axis, which is regulated by the corticotropin-releasing factor (CRF), will induce the release of adrenocorticotropic hormone (ACTH), followed by the secretion of glucocorticoids. The glucocorticoids, such as cortisol in humans, regulate the body and ensures an adequate response to stressors. The HPA axis is also crucial for emotion regulation. Its dysfunction has been observed in many mood disorders such as anxiety and depression (Makris et al. 2021).

The gut microbiota interacts with brain partially via the HPA axis. Stress-induced activation of the HPA axis can impact the GI functions including permeability, motility and secretion, and trigger intestinal inflammation. The cortisol level as a marker of the HPA axis was also found to be associated with certain gut bacterial taxa (Aizawa et al. 2018). The increased cortisol level could be ameliorated by probiotics treatment (Gareau et al. 2007). Thus, the HPA axis activities might lead to a reshape of gut microbial community. On the contrary, the gut microbiota system also contributes to regulation of the HPA axis. Healthy gut microbiota system plays an important role in the maintaining the normal activities of HPA axis. In contrast to specific pathogen free mice, germ-free mice models showed an increased corticosterone levels, indicating an overreaction of HPA axis to restraint stress (Neufeld et al. 2011). Gut microbiota-derived peptidoglycan is a constituent of bactieral wall, which could be translocated into the brain and recogniced by the innate immune system (Arentsen et al. 2017). Gut dysbiosis could regulate a cascade of inflammatory responses, especially stimulating the release of pro-inflammatory cytokines such as IL-1β, IL-6 and tumor necrosis factor-α (TNF-α), and subsequently activated the HPA axis (Misiak et al. 2020). Furthermore, SCFAs, the motabolites of gut microbiota might attenuate the increased cortisol level and HPA axis response to psychosocial stress (Dalile et al. 2020). Thus, the association between the gut microbiota and HPA axis works from both directions.

14.3.3 Inflammatory Pathways

The GI tract is the largest immunological organ within which the gut-associated lymphoid tissue (GALT) comprises 70–80% of all immunoglobulin-producing cells (Vighi et al. 2008). Gut microbiota colonizes the gut and regulates protective gut immunity against pathogens, thereby modifying the communications between the immune system and the nervous system. Under normal circumstances, the immune system can distinguish invading pathogens from symbiotic bacteria and activate immune responses to eliminate the pathogen, while inducing immunological tolerance to symbiotic bacteria (Takiishi et al. 2017). Meanwhile, the gut immune system specifically regulates the functional microbial fitness in the gut, which might fine-tune host-microbial mutualism.

It has long been recognized that the gut microbiota plays a crucial role in the development of the immune system and maintenance of normal immune functions. The GI tract of newborns is sterile at birth, and the immune system is not fully developed. With the colonization and establishment of gut microbiota, normal development and gradual maturation of the host immune system is promoted. Gut dysbiosis might disrupt the development of host immune system for defense mechanisms against pathogens in the gut and increase the susceptibility of immune associated diseases (Walker 2017). Certain gut microbiota-dereived components were able to stimulate the proliferation of lymphoblasts and differentiation of T cell such as T helper 1 (Th1), Th2, Th17, and regulatory T cells (Gaboriau-Routhiau et al. 2009). Only specific members of gut microbiota were capable of efficiently shaping the development of the host immune system. For example, the *Clostridia*-related species that well-known as potent inducer of mucosal immunoglobulin A (IgA), played a unique role in the postnatal maturation of gut immune functions (Gaboriau-Routhiau et al. 2009). The *Bacteroides fragilis* strain mediated the conversion of CD4 + T cells into Foxp3 + regulatory T cells and and induced anti-inflammatory cytokine production (Round and Mazmanian 2010). Germ-free mice models showed developmental defects in the immune system and function, which appeared as smaller mesenteric lymph nodes, Peyer's patches and less IgA-producing cells that led to decreased pathogen resistance (Sommer and Backhed 2013).

Neuroinflammation is also an important way to inference the crosstalk between gut and brain, regulating the structure and function of each other. When facing the challenge of gut dysbiosis, the immune system in the gut can transmit the activity information of gut microbiota to the CNS and respond accordingly. Gut dysbiosis induces the production of pro-inflammatory cytokines, which crosses the BBB and enters into the brain. These pro-inflammatory cytokines activate the vagal pathway and the HPA axis to affect brain activity and behavior. Gut microbiota-derived components can be recognized by toll-like receptors (TLRs) expressed in both ENS and CNS (Bonfili et al. 2021). Thus the alterations of gut microbiota can be directly detected by macrophage-like cells in the periventricular apparatus and the choroid plexus of the brain and subsequently induces production of cytokines. Furthermore, the neuroprotective metabolites of gut microbiota, SCFAs are important mediums of the MGBA, which contribute to maintain the gut immunological function and the permeability of gut barrier. The SCFAs was regarded as anti-inflammatory mediators by regulating the migration of immune cells, inhibitng production of cytokines including TNF-α, interferon (IFN)-γ, IL-1β, IL-2, IL-6 and IL-10, regulating the maturation and function of microglia, as well as inducing production of prostaglandin E2 (PGE2) (Cox et al. 2009; Vinolo et al. 2011). The gut microbiota is speculated to play a mediational role in the gut-brain crosstalk through these inflammatory pathways, and involved in the regulation of CNS activities such as brain development, stress response, anxiety, depression, and cognitive functions.

These studies have shown aberrant gut microbiota-inflammation-brain axis in some neuropsychiatric disorders such as AD(Goyal et al. 2021), depression (Makris et al. 2021), anxiety (Evrensel et al. 2020), and schizophrenia (Nguyen et al. 2021).

14.3.4 Metabolic Pathways

The gut microbiota is also involved in the synthesis and metabolism of neurotransmitters and metabolic by-products. The gut can exchange information with the brain through these substances, thereby affecting brain activity and behavior. Neurotransmitters are chemical messengers that conduct signal transmission from one neuron to the another target neuron, muscle cell or gland cell. A wide range of neurotransmitters are generated by gut microbiota, including dopamine (DA), norepinephrine (NE), serotonin, and gamma-aminobutyric acid (GABA) (Strandwitz 2018). Especially, approximately 95% of 5-HT and 50% DA are synthesized in the gut (Chen et al. 2021c). DA and NE belong to the catecholamine class, which were reported to be produced by *Bacillus*, *Escherichia*, *Proteus* and *Serratia* strains (Strandwitz 2018). DA and NE play essential roles in the brain and body in functions such as movement, motivation, reward, emotion, memory, learing, and attention. The neuroendocrine transmitter serotonin is mainly synthesized by the enterochromaffin cells in the gut, and also related with several microbial communities such as *Escherichia*, *Bacillus*, *Lactococcus*, *Lactobacillus*, *Streptococcus* (Strandwitz 2018). The essential amino acid tryptophan is the precursor for biosynthesis of serotonin, which is believed to be linked to mood disorders including depression and anxiety. Gut microbiota can regulate 5-HT signaling that modulates neuroendocrine and gut immune responses by mediating the tryptophan metabolism (Gao et al. 2020). Several gut microbiota such as the *Bifidobacterium* and *Lactobacillus* are related with the production of the inhibitory neurotransmitter GABA (Strandwitz 2018). GABA levels and GABAergic activity were found to be associated with gut motility, sensation, and secretion (Strandwitz 2018), as well as neuropsychiatric diseases such as AD, depression and anxiety (Chen et al. 2021c).

SCFAs are the main metabolites of gut microbiota, including AA, PPA, BTA, and valeric acid (VA). A group of gut microbiota such as the *Archaea*, *Acetobacter*, *Bacteroides*, *Bifidobacterium*, *Butyrivibrio*, *Clostridium*, *Collinsella*, *Coprococcus*, *Desulfovibrio*, *Eubacterium*, *Faecalibacterium*, *Lactobacillus*, *Prevotella*, *Propionibacterium*, and *Roseburia* were found to participant in producing and releasing SCFAs(Strandwitz, 2018; Oleskin and Shenderov, 2016; Shen et al. 2021). Accumulating evidence has pointed to an important role for the SCFAs in regulating the gut physiological, metabolic, inflammatory and behavioral processes (Wang et al. 2021b; Oleskin and Shenderov 2016). It has been considered to directly or indirectly influence the brain activies via activating the vagal pathways (Cook et al. 2021), inducing synthesis and metabolism of other neuroactive compounds (such as hormones and neurotransmitters) (Bonfili et al. 2021), as well as regulating cytokine production and recruitment of immune cells (such as lipopolysaccharide (LPS) or TNFα induced inflammatory responses) (Li et al. 2018).

The changes of gut microbiota are clustered in other metabolism pathways, such as energy metabolism, and the primary and secondary bile acid metabolism processes of lipid metabolism pathways are also found in neuropsychiatric disorders. Gut microbiota help to digest and harvest energy from food. In addition to being signaling molecules, SCFAs also serve as energy source in energy metabolisim. The MGBA plays an essential role in regulating energy balance, which might be linked to obesity and cognitive decline. Regarding the energy metabolisim pathway, its abnormalities have been found to relate with depression and anxiety symptoms, which could be improved by probiotic supplement (Yamamura et al. 2021). SCFAs form the core of multiple functional pathways of the gut microbiota system. Except for being used as precursors for producing neurotransmitters and SCFAs, the healthy gut microbiota system provides beneficial lipids that contribute to cell membrane fluidity, sources of energy metabolism and cell signaling pathways (Ooi et al. 2021). Dysregulation of lipid metabolism pathways might lead to cognitive impairment or even AD (Mahmoudian Dehkordi et al. 2019; Baloni et al. 2020). It was reported that specific gut microbiota such as *Prevotellaceae* and *Ruminococcaceae* were associated with the ApoE genotype (Tran et al. 2019), which has been linked to lipid metabolism and transport in brain (El Gaamouch et al. 2016). Thus, the genetics of host might influence and interact with gut microbiota to affect brain activity and behavior. Moreover, altered lipid metabolism pathways have also been observed in mood disorders, which could be ameliorated by antidepressant treatment, suggesting potential therapeutic targets (Liu et al. 2021c). In general, different gut microbiota involving pathways cooperate with each other to exert normal functions in the gut-brain axis.

14.3.5 Barrier System

The barrier system including GMB and BBB is indispensable to the gut-brain communicantion. The GMB serves as a primary defense barrier against enteric invading pathogens, and plays a vital role in the digestive and absorptive functions of the gut (Kedinger and Freund 2006) that contributes to maintenance of gut homeostasis (Desai et al. 2016). The gut epithelium consists of epithelial cells that are tightly interconnected with intercellular junctional complexes. Disrupted and unbalanced gut microbiota system induces impaired GMB, which leads to activation of inflammatory, neural, neuroendocrine, and metabolic pathways. Several neuropsychiatric conditions exhibit breakdown of the GMB and increased gut permeability, which subsequently caused increased susceptibility to pathogen and inflammation (Gwak and Chang 2021). The BBB comprises capillary endothelial cells sealed by tight junctions, astrocytic end-feet, and pericytes. The BBB is a diffusion barrier that selectively restricts the passage and exchange of soluble substances between the blood and brain parenchyma. Once the GMB is broken down, gut microbiota-derived compounds and its metabolites, as well as released cytokines induced by gut dysbiosis enter into the blood circulatory system (Gwak and Chang 2021). These substances reach the brain and interact with brain at the BBB level, which alters the integrity and transport functions of the BBB via the activation of microglia and brain local inflammation, and allows the entrance of harmful substances into the brain. In general, the barrier system acts as gatekeepers for fighting off foreign invaders. Growing evidence has shown that the structural and functional disruption of the barrier system contributes to the occurrence and development of neuropsychiatric diseases such as AD (Liu et al. 2021b), traumatic brain injury (Amoo et al. 2021), multiple sclerosis (Logsdon et al. 2018), stress-related psychiatric disorders (Kelly et al. 2015), which could be served as a potential therapeutic target.

14.4 Potential Therapeutic Approaches

Current strategies to treat these neuropsychiatric diseases are still dominated by drug treatment. Unfortunately, it is difficult for these drugs to enter the appropriate brain regions at the desired concentrations. The drugs might accumulate in the gut by gut microbiota and some species of gut microbiota also produce drug metabolizing enzymes, which play important roles in the biotransformation of drugs and affect their bioavailability, effectiveness, and side effects. Modifying the gut microbiota components of MGBA might be easier than applying treatment to the brain directly. Thus the gut microbiota might be potential therapeutical targets in treatment of neuropsychiatric diseases. The effectiveness of treatment methods, such as probiotics, dietary intervention, fecal microbial transplantation (FMT), and small molecule regulators targeting gut microbial pathways has been proven in preclinical and clinical studies (Fig. 14.2).

14.4.1 Probiotics

Probiotics are live, non-pathogenic microorganisms that have health benefits to the host, such as *Lactobacillus*, *Bifidobacterium*, *Lactococcus*, *Streptococcus*, *Bacillus*, and *Enterococcus* (Liu et al. 2015; Oleskin and Shenderov, 2019). Psychobiotics is a term defined as subgroup of probiotics, which confers a mental health benefit by modulating psychiatric abnormalities. Probiotics impact the moderating role of MGBA on brain activity and behavior, via neuroendocrine, neural, inflammatory, metabolic, and autonomic pathways. A growing body of research has provided evidence for the effectiveness and mechanisms of probiotics supplement treatment.

Animal models treated with *Lactobacillus rhamnosus* showed altered GABA receptor expression in the cingulate, prelimbic regions, prefrontal cortex,

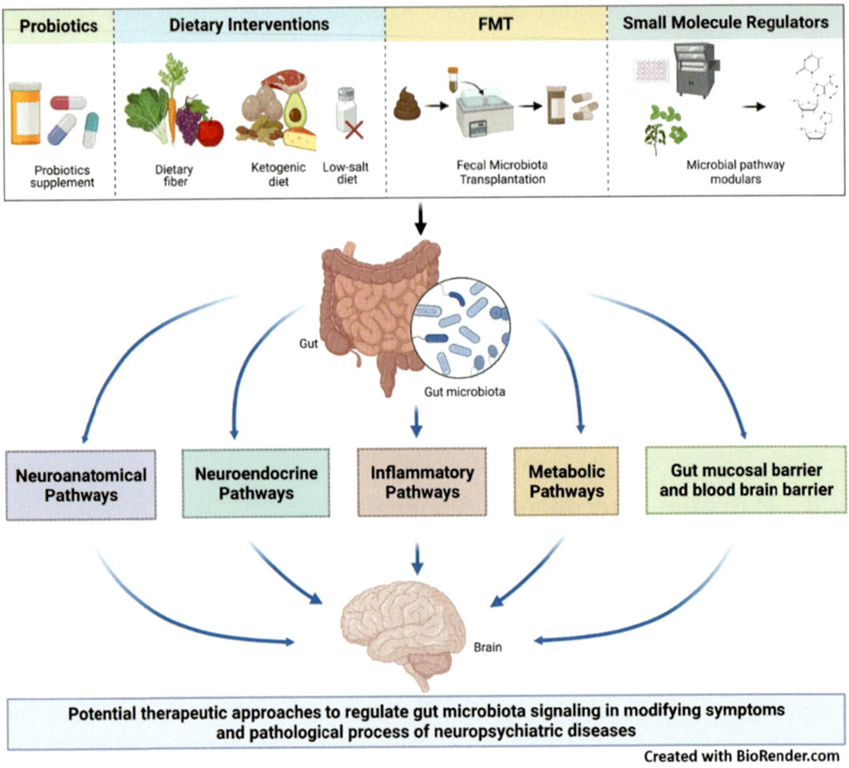

Fig. 14.2 Potential therapeutic approaches of neuropsychiatric diseases

hippocampus, amygdala, and locus coeruleus, as well as reduced anxiety- and depression-related behaviors (Bravo et al. 2011). However, these emotion regulation functions of probiotics might be interupted by vagotomy, indicating the *Lactobacillus rhamnosus* worked through vagus nerve pathway (Bravo et al. 2011). A probiotic combination of *Lactobacillus rhamnosus* and *Lactobacillus helveticus* strains ameliorated gut functional abnormalities and regulating neuroendocrine pathway via reducing the elevated corticosterone levels in maternal separation rat pups (Gareau et al. 2007). Rat models that were chronically supplemented with *Bifidobacteria infantis* exhibited an attenuation of pro-inflammatory immune responses and an elevation of tryptophan, a serotonergic precursor (Desbonnet et al. 2008). These results have demonstrated the effects of probiotics in the inflammatory and neural pathways.

Several probiotic strains and combinations have been evaluated in human studies (Table 14.5). Healthy participants treated with probiotic *Bifidobacterium longum* were observed to have reduced stress levels and subtly improved hippocampus-dependent visuospatial memory that might function via modulating neural activities in the hippocampus, fusiform, temporal, frontal, and cingulate cortex (Allen et al. 2016; Wang et al. 2019a). However, short-term (one week) use of *Bifidobacterium longum* didn't obtain similar results (Siegel and Conklin 2020). It is noteworthy that long-term intake of *Bifidobacterium longum* could improve depressive symptoms, and the improvements were associated with changes in limbic reactivity (Pinto-Sanchez et al. 2017). Another probiotic *Lactobacillus rhamnosus* strain also alleviated depression and anxiety severity in the postpartum period (Slykerman et al. 2017). Intervention of probiotic *Lactobacillus plantarum P8* reduced stress and anxiety levels, as well as improving memory and cognition functions in stressed adults (Lew et al. 2019). The probiotic *Lactobacillus Plantarum 299v* led to reduction of kynurenine concentration, which played an crucial part in the pathophysiology of psychiatric disorders and cognitive decline (Rudzki et al. 2019). As expected, this intervention could improve cognitive functions in patients with major depression.

In addition, the probiotic combinations of several strains were further developed. Although with inconsistent results, the probiotic combination of *Lactobacillus helveticus* and *Bifidobacterium longum* associated with the reduction of the depression scores in patients with major depression. Meanwhile, decreased serum kynurenine/tryptophan ratios and increased serum tryptophan/isoleucine ratios were found after treatment of this combined probiotic, indicating that this probiotic combination improved depression symptom via increasing the supply of tryptophan, a precursor of serotonin (Kazemi et al. 2019). In addtion, this combined probiotic contributes to maintainance of gut immune homeostasis in aging subjects by modulating immune cells (Finamore et al. 2019), and supplements increased level of brain-derived neurotrophic factor (BDNF) (Heidarzadeh-Rad et al. 2020). Another probiotic combination of *Lactobacillus acidophilus*, *Lactobacillus casei*, and *Bifidobacterium bifidum* improved depression symptoms of patients with major depression while altering the function of the insulin pathway (Akkasheh et al. 2016). These results proved that the probiotics improved mental health through multiple signaling pathways, although its function might be limited. It could be more benificial to take probiotics supplements as a preventative approach or an adjuvant treatment.

14.4.2 Dietary Intervention

Dietary habit is one of the most important influence factors in modulating the composition and function of the gut microbiota system, thereby affecting brain activity and structure. An intervention of healthy diet combined with drug treatment or psychotherapy might achieve better improvements in mental therapy, while an unhealthy diet was well known as a risk factor for neuropsychiatric diseases. This might be due to the close correlations between diet and the MGBA. Gut microbiota-targeted dietary interventions have been used as powerful means to maintain and improve human health, through enhancing, introducing or eliminating specific functions or species of gut microbiota.

Table 14.5 The effects of probiotic supplements on patients with neuropsychiatric diseases

References	Sample description	Sample size (intervention/control)	Type of probiotics	Intervention	Control	Duration	Observation summary
Akbari, et al. (2016)	AD patients	30/30	Lactobacillus acidophilus, Lactobacillus casei, Bifidobacterium bifidum and Lactobacillus fermentum	Oral probiotics (2×10^9 CFU/g for each strain)	Not mentioned	12 weeks	Significant improvement in MMSE Changes in plasma malondialdehyde, serum high-sensitivity C-reactive protein, homeostasis model of assessment-estimated insulin resistance, Beta cell function, serum triglycerides, and quantitative insulin sensitivity check index
Agah et al. (2018)	AD patients	25/23	Capsule one: Lactobacillus acidophilus, Lactobacillus casei, Lactobacillus fermentum and Bifidobacterium bifidum Capsule two: Lactobacillus acidophilus, Bifidobacterium bifidum, and Bifidobacterium longum	Oral probiotics (3×10^9 CFU/day)	Placebo packets (500 mg maltodextrine)	12 weeks	No significant difference in cognitive test Not pronounced change percent was found in the serum levels of oxidant/antioxidant and inflammatory/anti-inflammatory biomarkers
Leblhuber et al. (2018)	AD patients	28/25	Lactobacillus helveticus R0052 and Bifidobacterium longum R0175; CEREBIOME	Oral probiotics ($\geq 10 \times 10^9$ CFU/5 g sachet)	Prebiotic (galactalogosaccharide powder) Placebo (excipients)	8 weeks	Significant increased BDNF levels

(continued)

Table 14.5 (continued)

References	Sample description	Sample size (intervention/ control)	Type of probiotics	Intervention	Control	Duration	Observation summary
Tamtaji et al. (2019)	AD patients	27/26	Lactobacillus acidophilus, Bifidobacterium bifidum and Bifidobacterium longum	Oral probiotics (2×10^9 CFU/ day) + selenium (0.2 mg/day)	Placebo	12 weeks	Significant reduction in serum hs-CRP, HOMA-IR, LDL-cholesterol and total-/ HDL-cholesterol ratio Significant increase in total glutathione and quantitative insulin sensitivity check index
Ton et al. (2020)	AD patients	13/-	Acetobacter aceti, Acetobacter sp , Lactobacillus delbrueckii, Lactobacillus fermentum, Lactobacillus fructivorans, Enterococcus faecium, Leuconostoc spp., Lactobacillus kefiranofaciens	Oral probiotics (2 mL/kg.bw/day)	–	3 months	Improvement in memory, visual-spatial/abstraction abilities and executive/language functions Decrease in several cytokine markers of inflammation and oxidative stress markers
Hwang et al. (2019)	MCI cases	50/50	Lactobacillus plantarum C29-fermented soybean	Oral probiotics (1×10^{10} CFU/ day)	Placebo (cellulose and manufactured)	12 weeks	Greater improvements in the combined cognitive functions Increased serum BDNF levels
Kobayash et al. (2019)	MCI cases	61/60	Bifidobacterium breve A1	Oral probiotics (2×10^{10} CFU/ day)	Placebo (corn starch)	12 weeks	Significant difference in subscale "immediate memory" of RBANS and MMSE total score in the subjects with low RBANS total score at baseline No significant differences in terms of blood parameters
Xiao et al. (2020)	MCI cases	40/40	Bifidobacterium breve	Oral probiotics (2×10^{10} CFU/ day)	Placebo (maize starch)	16 weeks	Significant increase in RBANS total score Significant improvement in domain scores of immediate

(continued)

Table 14.5 (continued)

References	Sample description	Sample size (intervention/ control)	Type of probiotics	Intervention	Control	Duration	Observation summary
							memory, visuospatial/ constructional, and delayed memory
Akkasheh et al. (2016)	MDD patients	20/20	Lactobacillus acidophilus, Lactobacillus casei and Bifidobacterium bifidum	Oral probiotics (2×10^9 CFU/g for each strain)	Placebo	8 weeks	Significantly decreased BDI total scores, serum insulin levels, and hs-CRP concentrations
Bambling et al. (2017)	Treatment resistant depression patients	12/-	Lactobacillus acidophilus, Bifidobacterium bifidum, Streptoccocus thermophiles	Oral probiotics (2×10^{10} CFUs, bid) + magnesium orotate (1600 mg, bid)	-	8 weeks	Significant reduction in the depression scores and quality of life
Miyaoka et al. (2018)	Treatment resistant MDD patients	20/20	Clostridium butyricum MIYAIRI 588	Oral probiotics (60 mg/ d) + antidepressants	Control + antidepressants	8 weeks	Relief of depressive and anxiety symptoms. No serious adverse events occurred
Rudzki et al. (2019)	MDD patients	30/30	Lactobacillus Plantarum 299v	Oral probiotics (10×10^9 CFU)	Placebo (crystalline cellulose powder)	8 weeks	Significant decrease in kynurenine concentration, increase in 3-hydroxykynurenine/ kynurenine ratio No significant changes of TNF-α, IL-6 and IL-1b and cortisol concentrations
Chahwan et al. (2019)	Depressed participants	34/37	Bifidobacterium bifidum W23, Bifidobacterium lactis W51 and W52, Lactobacillus acidophilus W37, Lactobacillus brevis W63,	Oral probiotics (1×10^{10} CFU/ day)	Placebo packets (freezedried maize-starch and maltodextrins)	8 weeks	No greater reduction in depressive symptoms compared to placebo group

(continued)

Table 14.5 (continued)

References	Sample description	Sample size (intervention/control)	Type of probiotics	Intervention	Control	Duration	Observation summary
			Lactobacillus casei W56, Lactobacillus salivarius W24, Lactococcus lactis W19 and W58				No significantly alteration of the microbiota composition
Kazem et al. (2019)	MDD patients	28/26	Lactobacillus helveticus R0052 and Bifidobacterium longum R0175	Oral probiotics (10×10^9 CFU/5 g sachet)	Placebo sachet (dispersible powder)	8 weeks	Significant decrease in BDI score compared to placebo group Significant increasement in tryptophan/isoleucine ratio compared to placebo group
Eskandarzadeh et al. (2021)	GAD patients	24/24	Bifidobacterium longom, Bifidobacterium bifidum, Bifidobacterium lactis and Lactobacillus acidophilus bacteria	Oral probiotics (18×10^9 CFU/day) + Sertraline	Placebo + Sertraline	8 weeks	Greater reduction in HAMA compared to placebo group
Dickerson, et al. (2014)	SZ patients	33/32	Lactobacillus rhamnosus strain GG and Bifidobacterium animalis subsp. lactis strain Bb12	Oral probiotics (1×10^9 CFU for each strain)	Placebo	14 weeks	No significant differences in the PANSS total score
Severance, et al. (2017)	SZ male patients	22/15	Lactobacillus rhamnosus strain GG and Bifidobacterium animalis subsp. lactis Bb12	Oral probiotics (1×10^9 CFU/day for each strain)	Placebo	14 weeks	No significant differences in the PANSS total score Significantly reduced Candida albicans antibodies levels in male
Okubo, et al. (2019)	SZ patients	29/-	Bifidobacterium breve A-1	Oral probiotics (5×10^{10} CFU, bid)	-	4 weeks	12 responders and 17 non-responders Responders have fewer negative symptoms, reduced intake of dairy products, and higher RA of g_Parabacteroides

AD: Alzheimer's disease; MCI: mild cognitive impairment; MDD: major depression disorder; GAD: generalized anxiety disorder; SZ: schizophrenia; CFU: colony forming units; ABA: applied behavior analysis; GI: gastrointestinal; RBANS: repeatable battery for assessment of neuropsychological status; ATEC: autism treatment evaluation checklist; 6-GSI: six item gastrointestinal severity index; BDI: Beck depression inventory; HAMA: Hamilton rating scale for anxiety; PANSS: positive and negative syndrome scale; hs-CRP: high-sensitivity C-reactive protein; HOMA-IR: homeostasis model of assessment insulin resistance

As an important part of a healthy diet, dietary fiber with physiological health benefits is very essential for human health. Dietary fiber has been demonstrated to contribute to regulating the gut microbial community and metabolism. Fermentation of dietary fiber occurs mainly in cecum and colon by gut microbiota, which is the main source of SCFAs including AA, PPA, and BTA (Zmora et al. 2019). Increasing dietary fiber intake could restructure the gut microbiota composition and obtain higher levels of SCFAs, and thus reduce the level of pro-inflammatory cytokines, indicating its anti-inflammatory role (Kuo 2013). An animal study found that a high-fiber diet could significantly attenuate cognitive deficits and depressive-like behaviors that upregulated the formation of neurotransmitters including 5-HT and NE, and helped in mood modulation (Liu et al. 2020). Prebiotics refer mainly to fiber substances as food for probiotics, such as fructooligosaccharides and galactooligosaccharides, which is beneficial to the growth and function promotion of gut microbiota (Barbosa and Vieira-Coelho 2020). In addition to improving the depression-like and anxiety-like behaviors in mice models, intake of fructooligosaccharides and galactooligosaccharides prebiotics was also found to increase the expressions of BNDF and GABA receptor genes in the hippocampus and hypothalamus, increase AA and PPA concentrations, and reduce stress-induced elevations in corticosterone and pro-inflammatory cytokine levels (Burokas et al. 2017). Thus, high-fiber diets could result in beneficial effects through metabolic, neural, endocrine and inflammatory pathways.

The ketogenic diet is a very low carbohydrate and high-fat diet that has been regarded as an effective strategy to treat certain diseases such as refractory epilepsy (Sampaio 2016), obesity (Paoli 2014), AD (Rusek et al. 2019), and other cognitive decline (Nagpal et al. 2019). This type of diet has potential impact on modulating gut microbiota expressing as a reproducible depletion of the *Bifidobacteria* and reduced levels of pro-inflammatory Th17 cells (Ang et al. 2020). In addition to reducing inflammation, the ketogenic diet also regulated mitochondrial function and reduced the expression of apoptotic mediators to slow the progress of the disease (Rusek et al. 2019). Furthermore, since high-salt intake inhibits the production of beneficial bacteria and SCFAs, the high intake of salt subsequently increases vulnerability to inflammation (Miranda et al. 2018). More interestingly, parents who have a high-salt diet result in gut microbial abnormalities and autism-like behaviors in their offspring (Afroz et al. 2021). Moreover, a high-fat diet also promoted the growth of pro-inflammatory gut microbiota, indicating a harmful effect (Bibbo et al. 2016). Thus, a low-salt and low-fat diet could also be regarded as an adjunctive therapy.

14.4.3 Fecal Microbiota Transplantation

The concept of FMT was first introduced by a traditional Chinese medical doctor named Ge Hong approximately 1,700 years ago. FMT has obtained new attention recent years, and already commonly applied in treatment of GI diseases such as

Clostridium difficile infection (CDI), inflammatory bowel disease (IBD), Crohn's disease, and ulcerative (Gupta et al. 2016). Acuumulating evidence has also demonstriated the therapeutic role of FMT in non-GI disorders including neurological diseases such as AD, Parkinson's disease, multiple sclerosis and amyotrophic lateral sclerosis, as well as psychiatric diseases such as depression (Ruggiero, 2016). The so-called FMT is a process that includes extracting a solution of functional gut microbiota that from fecal sample of healthy donor, and administrating the fecal matter into the receiver's gut (Evrensel and Ceylan 2016). Increasing the diversity of gut microbiota and restoring a healthy microbial community are key points in the treatment of neuropsychiatric diseases.

The effectiveness of the FMT in neuropsychiatric diseases has been tested in a series of studies. Transplantation of microbiota from young donors into aged recipient mice reversed aging-associated impairments in cognitive behavior, as well as brain alterations, such as peripheral and brain immunity (Boehme et al. 2021). After treatment with FMT, the APP/PS1 transgenic mice showed reconstructed gut microbial community and SCFAs levels, ameliorated cognitive deficits, increased synaptic plasticity, and improved pathologic changes including reduced brain deposition of Aβ and phosphorylation of tau protein (Sun et al. 2019). Thus, FMT may be a potential therapeutic strategy for AD. Regarding neurodevelopmental disorders, FMT was also a promising therapeutic strategy. FMT treatment ameliorated depression-like behaviors through modulating inflammatory pathways, such as suppressing the activation of glial cells and inhibiting NOD-like receptor protein 3 (NLRP3) inflammation (Rao et al. 2021). FMT treatment also has been applied and evaluated in clinical studies. The rapid reversal of symptoms in one AD patient following FMT treatment was reported. GI symptoms were resolved and mental acuity, affect and memory function were improved (Hazan 2020). With respect to mood disorders, anxiety or depression symptoms in GI disorder patients could be alleviated after FMT treatment (Huang et al. 2019; Mizuno et al. 2017). In general, these studies revealed that FMT could be considered as a promising therapeutic strategy to treat neuropsychiatric diseases.

14.4.4 Small Molecule Regulators

Molecules that target key elements of the gut microbial functional pathways provide another therapeutic strategy for neuropsychiatric disorders via regulating the functions of gut microbiota. Neuroinflammation and oxidative stress has been recognized as crucial factors that contribute to AD progression. As mentioned above, gut microbiota is one of the important links of the inflammatory pathways. Polyphenols have anti-inflammatory and antioxidant properties, which could alleviate gut microbiota imbalance through altering abundance of bacteria associated with oxidative stress and inflammatory, such as genera *Alistipes*, *Helicobacter*, *Rikenella*, *Desulfovibrio*, and *Faecalibaculum* (Li et al. 2021b; Grinan-Ferre et al. 2021). Moreover, it also effectively inhibited Aβ deposition in hippocampus and improved cognitive function (Li et al. 2021b). From the perspective of another pathway, the G-protein coupled receptor 40 (GPR40), which is well acknowledged as a receptor for long chain fatty acids, is involved in the neuropathological processes of AD (Khan and He, 2017). The GPR40 agonist might reduce the risk of developing AD through downregulating levels of pro-inflammatory cytokines, as well as upregulating the levels of gut-brain peptides in both serum and hypothalamus via the gut-brain axis to ameliorate neuroinflammation (Chen et al. 2021b). Furthermore, a carbohydrate-based anti-AD drug, the sodium oligomannate also reshaped the gut microbial community and related neuroinflammation to inhibit the progression of AD and improved cognitive function (Wang et al. 2019b).

With respect to psychiatric disorders, various kinds of small molecule regulators targeting gut microbiota may also be regarded as potential treatments. Vitamin A plays a crucial part in regulating the development of the CNS, which is also involved in modulating gut immunity and maintaining gut mucosal epithelial integrity (Cassani et al. 2012). The level of vitamin A was found to be associated with the proportion of *Bacteroidetes/Bacteroidales* and improved social behavioral symptoms (Liu et al. 2017; Yang et al. 2020). Coniferyl ferulate, a phenolic acid compound, was found to ameliorate inflammation in the colon, lower levels of pro-inflammatory cytokines such as IL-6, IL-1β and TNF-α, restructure gut microbiome and related metabolism functions, so as to attenuate depression-like and anxiety-like behaviors in mice models (Hao et al. 2021b).

Traditional Chinese medicines with wide therapeutic spectrum also serve as remedies for neuropsychiatric disorders through targeting gut microbiota. Treatment with the Huang-Lian-Jie-Du-Decoction formula attenuated learning and memory deficiencies and changed gut microbial structure in AD mice model (Liu et al. 2019b). A compound formula, Xiaoyaosan, could improve anxious and depressive behaviors, regulate LPS-producing gut microbiota, and inhibit immoderate activation of the NLRP3 pathway in colon (Hao et al. 2021a). The Tiansi liquid altered the abundance of *Lachnospiraceae_NK4A136*, and modulated the tryptophan-kynurenine metabolism pathway, and resulted in ameliorating the depressive symptoms (Cheng et al. 2018). In general, exploring new medical resources targeting the crosstalk between gut microbiota and brain in neuropsychiatric disorders might be a promising research direction.

14.5 Conclusion

In this chapter, we have reviewed the roles the MGBA and the effectiveness of related treatment approaches in neuropsychiatric diseases. A series of clinical studies have been designed and are about to be carried out. How to accelerate the clinical transformation and expand drug development has become the focus of attention. Although the therapeutic strategy based on gut microbiota regulation has proven to have positive effects in modifying crosstalk pathways between the gut and brain, there are still many obstacles from microbial target validation to safe clinical application. The therapeutic methods that have entered the stage of clinical research mainly focus on reshaping the gut microbial community towards healthy populations. However, the therapeutic effect is still not strong and stable enough. Current studies of the gut-brain axis are still on superficial levels. Many aspects of fundamental knowledge of the gut-brain axis remained to be explored, such as mapping the interaction between gut microbiota and brain, investigating the preferred brain regions of specific microbiota, elaborating in-depth mechanisms of host-gut microbiota interaction. The answers to these questions will lay the foundation for understanding the pathophysiological mechanisms of neuropsychiatric disorders and developing new therapeutic strategies.

The MGBA plays important roles in maintaining healthy homeostasis and regulating pathophysiological processes via neuroanatomical, neuroendocrine, inflammatory, metabolic, and neural pathways. With in-depth studies of the pathophysiological mechanism of gut-microbiota-brain axis in neuropsychiatric diseases, one potential avenue for realizing the promises of precision medicine can be provided in future clinical applications. A series of key circuits of the gut-microbiota-brain axis could serve as new diagnostic biomarkers and therapeutic strategies in a wide range of therapeutic indications.

Acknowledgements We would like to thank Dr. Yi-Ching Lynn Ho from the Centre for Population Health Research & Implementation, Singapore Health Services for her especially valuable edits and comments. This work was supported by grants from National Natural Science Foundation of China (82101323).

References

Afroz KF, Reyes N, Young K, Parikh K, Misra V, Alvina K (2021) Altered gut microbiome and autism like behavior are associated with parental high salt diet in male mice. Sci Rep 11:8364

Agahi A, Hamidi GA, Daneshvar R, Hamdieh M, Soheili M, Alinaghipour A, Esmaeili Taba SM, Salami M (2018) Does severity of alzheimer's disease contribute to its responsiveness to modifying gut microbiota? a double blind clinical trial. Front Neurol 9:662

Agus A, Planchais J, Sokol H (2018) Gut microbiota regulation of tryptophan metabolism in health and disease. Cell Host Microbe 23:716–724

Aizawa E, Tsuji H, Asahara T, Takahashi T, Teraishi T, Yoshida S, Koga N, Hattori K, Ota M, Kunugi H (2018) Bifidobacterium and lactobacillus counts in the gut microbiota of patients with bipolar disorder and healthy controls. Front Psychiatry 9:730

Aizawa E, Tsuji H, Asahara T, Takahashi T, Teraishi T, Yoshida S, Ota M, Koga N, Hattori K, Kunugi H (2016) Possible association of bifidobacterium and lactobacillus in the gut microbiota of patients with major depressive disorder. J Affect Disord 202:254–257

Akbari E, Asemi Z, Daneshvar Kakhaki R, Bahmani F, Kouchaki E, Tamtaji OR, Hamidi GA, Salami M (2016) Effect of probiotic supplementation on cognitive function and metabolic status in Alzheimer's disease: a randomized, double-blind and controlled trial. Front Aging Neurosci 8:256

Akkasheh G, Kashani-Poor Z, Tajabadi-Ebrahimi M, Jafari P, Akbari H, Taghizadeh M, Memarzadeh MR, Asemi Z, Esmaillzadeh A (2016) Clinical and metabolic response to probiotic administration in patients with major depressive disorder: a randomized, double-blind, placebo-controlled trial. Nutrition 32:315–320

Allen AP, Hutch W, Borre YE, Kennedy PJ, Temko A, Boylan G, Murphy E, Cryan JF, Dinan TG, Clarke G (2016) Bifidobacterium longum 1714 as a translational psychobiotic: modulation of stress, electrophysiology and neurocognition in healthy volunteers. Transl Psychiat 6:E939

Amoo M, O'halloran, P. J., Henry, J., Husien, M. B., Brennan, P., Campbell, M., Caird, J. & Curley, G. F. (2021) Permeability of the blood-brain barrier after traumatic brain injury; radiological considerations. J Neurotrauma 39:20–34

Ang QY, Alexander M, Newman JC, Tian Y, Cai J, Upadhyay V, Turnbaugh JA, Verdin E, Hall KD, Leibel RL, Ravussin E, Rosenbaum M, Patterson AD, Turnbaugh PJ (2020) Ketogenic diets alter the gut microbiome resulting in decreased intestinal Th17 cells. Cell 181 (1263–1275):E16

Arentsen T, Qian Y, Gkotzis S, Femenia T, Wang T, Udekwu K, Forssberg H, Diaz Heijtz R (2017) The bacterial peptidoglycan-sensing molecule Pglyrp2 modulates brain development and behavior. Mol Psychiatr 22:257–266

Averina OV, Zorkina YA, Yunes RA, Kovtun AS, Ushakova VM, Morozova AY, Kostyuk GP, Danilenko VN, Chekhonin VP (2020) Bacterial Metabolites Of Human Gut Microbiota Correlating With Depression. Int J Mol Sci 21:9234

Baloni P, Funk CC, Yan J, Yurkovich JT, Kueider-Paisley A, Nho K, Heinken A, Jia W, Mahmoudiandehkordi S, Louie G, Saykin AJ, Arnold M, Kastenmuller G, Griffiths WJ, Thiele I, Metabolomics AD, C., Kaddurah-Daouk, R. & Price, N. D. (2020) Metabolic network analysis reveals altered bile acid synthesis and metabolism in Alzheimer's disease. Cell Rep Med 1:100138

Bambling M, Edwards SC, Hall S, Vitetta L (2017) A combination of probiotics and magnesium orotate attenuate depression in a small ssri resistant cohort: an intestinal anti-inflammatory response is suggested. Inflammopharmacology 25:271–274

Barandouzi ZA, Starkweather AR, Henderson WA, Gyamfi A, Cong XS (2020) Altered composition of gut microbiota in depression: a systematic review. Front Psychiatry 11:541

Barbosa RSD, Vieira-Coelho MA (2020) Probiotics and prebiotics: focus on psychiatric disorders - a systematic review. Nutr Rev 78:437–450

Bengesser SA, Morkl S, Painold A, Dalkner N, Birner A, Fellendorf FT, Platzer M, Queissner R, Hamm C, Maget A, Pilz R, Rieger A, Wagner-Skacel J, Reininghaus B, Kapfhammer HP, Petek E, Kashofer K, Halwachs B, Holzer P, Waha A, Reininghaus EZ (2019) Epigenetics of the molecular clock and bacterial diversity in bipolar disorder. Psychoneuroendocrino 101:160–166

Bibbo S, Ianiro G, Giorgio V, Scaldaferri F, Masucci L, Gasbarrini A, Cammarota G (2016) The role of diet on gut microbiota composition. Eur Rev Med Pharmacol Sci 20:4742–4749

Bloom GS (2014) Amyloid-beta and tau: the trigger and bullet in alzheimer disease pathogenesis. Jama Neurol 71:505–508

Boehme M, Guzzetta KE, Bastiaanssen TFS, Van De Wouw M, Moloney GM, Gual-Grau A, Spichak S, Olavarría-Ramírez L, Fitzgerald P, Morillas E, Ritz NL, Jaggar M, Cowan CSM, Crispie F, Donoso F, Halitzki E, Neto MC, Sichetti M, Golubeva AV, Fitzgerald RS, Claesson MJ, Cotter PD, O'leary, O. F., Dinan, T. G. & Cryan, J. F. (2021) Microbiota from young mice counteracts selective age-associated behavioral deficits. Nature Aging 1:666–676

Bonaz B, Bazin T, Pellissier S (2018) The vagus nerve at the interface of the microbiota-gut-brain axis. Front Neurosci 12:49

Bonfili L, Cecarini V, Gogoi O, Gong C, Cuccioloni M, Angeletti M, Rossi G, Eleuteri AM (2021) Microbiota modulation as preventative and therapeutic approach in Alzheimer's disease. Febs J 288:2836–2855

Borsom EM, Lee K, Cope EK (2020) Do the bugs in your gut eat your memories? relationship between gut microbiota and Alzheimer's disease. Brain Sci 10:814

Bravo JA, Forsythe P, Chew MV, Escaravage E, Savignac HM, Dinan TG, Bienenstock J, Cryan JF (2011) Ingestion of lactobacillus strain regulates emotional behavior and central gaba receptor expression in a mouse via the vagus nerve. Proc Natl Acad Sci U S A 108:16050–16055

Brydges CR, Fiehn O, Mayberg HS, Schreiber H, Dehkordi SM, Bhattacharyya S, Cha J, Choi KS, Craighead WE, Krishnan RR, Rush AJ, Dunlop BW, Kaddurah-Daouk R, Medicine MDP, C. (2021) Indoxyl sulfate, a gut microbiome-derived uremic toxin, is associated with psychic anxiety and its functional magnetic resonance imaging-based neurologic signature. Sci Rep 11:21011

Burokas A, Arboleya S, Moloney RD, Peterson VL, Murphy K, Clarke G, Stanton C, Dinan TG, Cryan JF (2017) Targeting the microbiota-gut-brain axis: prebiotics have anxiolytic and antidepressant-like effects and reverse the impact of chronic stress in mice. Biol Psychiat 82:472–487

Cassani B, Villablanca EJ, De Calisto J, Wang S, Mora JR (2012) Vitamin A and immune regulation: role of retinoic acid in gut-associated dendritic cell education, immune protection and tolerance. Mol Aspects Med 33:63–76

Castro-Nallar E, Bendall ML, Perez-Losada M, Sabuncyan S, Severance EG, Dickerson FB, Schroeder JR, Yolken RH, Crandall KA (2015) Composition, taxonomy and functional diversity of the oropharynx microbiome in individuals with schizophrenia and controls. PeerJ 3:E1140

Cattaneo, A., Cattane, N., Galluzzi, S., Provasi, S., Lopizzo, N., Festari, C., Ferrari, C., Guerra, U. P., Paghera, B., Muscio, C., Bianchetti, A., Volta, G. D., Turla, M., Cotelli, M. S., Gennuso, M., Prelle, A., Zanetti, O., Lussignoli, G., Mirabile, D., Bellandi, D., Gentile, S., Belotti, G., Villani, D., Harach, T., Bolmont, T., Padovani, A., Boccardi, M., Frisoni, G. B. & Group, I.-F (2017) Association of brain amyloidosis with pro-inflammatory gut bacterial taxa and peripheral inflammation markers in cognitively impaired elderly. Neurobiol Aging 49:60–68

Chahwan B, Kwan S, Isik A, Van Hemert S, Burke C, Roberts L (2019) Gut feelings: a randomised, triple-blind, placebo-controlled trial of probiotics for depressive symptoms. J Affect Disord 253:317–326

Chang CH, Lin CH, Lane HY (2020) D-Glutamate and gut microbiota in Alzheimer's disease. Int J Mol Sci 21:2676

Chen C, Zhou Y, Wang H, Alam A, Kang SS, Ahn EH, Liu X, Jia J, Ye K (2021a) Gut inflammation triggers c/ebpbeta/delta-secretase-dependent gut-to-brain propagation of abeta and tau fibrils in Alzheimer's disease. Embo J 40:E106320

Chen J, Li Q, Zhu J, Yuan Z, Wang T, Song J (2021b) Gpr40 agonist ameliorate pathological neuroinflammation of Alzheimer's disease via the modulation of gut microbiota and immune system, a mini-review. Neurotox Res 39:2175–2185

Chen JJ, He S, Fang L, Wang B, Bai SJ, Xie J, Zhou CJ, Wang W, Xie P (2020) Age-specific differential changes on gut microbiota composition in patients with major depressive disorder. Aging (albany Ny) 12:2764–2776

Chen JJ, Zheng P, Liu YY, Zhong XG, Wang HY, Guo YJ, Xie P (2018a) Sex differences in gut microbiota in patients with major depressive disorder. Neuropsychiatr Dis Treat 14:647–655

Chen Y, Xu J, Chen Y (2021c) Regulation of neurotransmitters by the gut microbiota and effects on cognition in neurological disorders. Nutrients 13:2099

Chen YH, Bai J, Wu D, Yu SF, Qiang XL, Bai H, Wang HN, Peng ZW (2019) Association between fecal microbiota and generalized anxiety disorder: severity and early treatment response. J Affect Disord 259:56–66

Chen Z, Li J, Gui S, Zhou C, Chen J, Yang C, Hu Z, Wang H, Zhong X, Zeng L, Chen K, Li P, Xie P (2018b) Comparative metaproteomics analysis shows altered fecal microbiota signatures in patients with major depressive disorder. NeuroReport 29:417–425

Cheng D, Chang H, Ma S, Guo J, She G, Zhang F, Li L, Li X, Lu Y (2018) Tiansi liquid modulates gut microbiota composition and tryptophan(-)kynurenine metabolism in rats with hydrocortisone-induced depression. Molecules 23:2832

Cook TM, Gavini CK, Jesse J, Aubert G, Gornick E, Bonomo R, Gautron L, Layden BT, Mansuy-Aubert V (2021) Vagal neuron expression of the microbiota-derived metabolite receptor, free fatty Acid receptor (Ffar3), is necessary for normal feeding behavior. Mol Metab 54:101350

Cox MA, Jackson J, Stanton M, Rojas-Triana A, Bober L, Laverty M, Yang X, Zhu F, Liu J, Wang S, Monsma F, Vassileva G, Maguire M, Gustafson E, Bayne M, Chou CC, Lundell D, Jenh CH (2009) Short-chain fatty acids act as antiinflammatory mediators by regulating prostaglandin E(2) and cytokines. World J Gastroenterol 15:5549–5557

Dalile B, Vervliet B, Bergonzelli G, Verbeke K, Van Oudenhove L (2020) Colon-delivered short-chain fatty acids attenuate the cortisol response to psychosocial stress in healthy men: a randomized, placebo-controlled trial. Neuropsychopharmacol 45:2257–2266

De Vadder F, Grasset E, Manneras Holm L, Karsenty G, Macpherson AJ, Olofsson LE, Backhed F (2018) Gut microbiota regulates maturation of the adult enteric nervous system via enteric serotonin networks. P Natl Acad Sci Usa 115:6458–6463

Desai MS, Seekatz AM, Koropatkin NM, Kamada N, Hickey CA, Wolter M, Pudlo NA, Kitamoto S, Terrapon N, Muller A, Young VB, Henrissat B, Wilmes P, Stappenbeck TS, Nunez G, Martens EC (2016) A dietary fiber-deprived gut microbiota degrades the colonic mucus barrier and enhances pathogen susceptibility. Cell 167(1339–1353):E21

Desbonnet L, Garrett L, Clarke G, Bienenstock J, Dinan TG (2008) The probiotic bifidobacteria infantis: an assessment of potential antidepressant properties in the rat. J Psychiatr Res 43:164–174

Dichgans M, Leys D (2017) Vascular cognitive impairment. Circ Res 120:573–591

Dickerson FB, Stallings C, Origoni A, Katsafanas E, Savage CL, Schweinfurth LA, Goga J, Khushalani S, Yolken RH (2014) Effect of probiotic supplementation on schizophrenia symptoms and association with gastrointestinal functioning: a randomized, placebo-controlled trial. Prim Care Companion Cns Disord 16

Dong Z, Shen X, Hao Y, Li J, Li H, Xu H, Yin L, Kuang W (2021) Gut microbiome: a potential indicator for differential diagnosis of major depressive disorder and general anxiety disorder. Front Psychiatry 12:651536

El Gaamouch F, Jing P, Xia J, Cai D (2016) Alzheimer's disease risk genes and lipid regulators. J Alzheimers Dis 53:15–29

Eskandarzadeh S, Effatpanah M, Khosravi-Darani K, Askari R, Hosseini AF, Reisian M, Jazayeri S (2021) Efficacy of a multispecies probiotic as adjunctive therapy in generalized anxiety disorder: a double blind, randomized, placebo-controlled trial. Nutr Neurosci 24:102–108

Evrensel A, Ceylan ME (2016) Fecal microbiota transplantation and its usage in neuropsychiatric disorders. Clin Psychopharmacol Neurosci 14:231–237

Evrensel A, Unsalver BO, Ceylan ME (2020) Immune-kynurenine pathways and the gut microbiota-brain axis in anxiety disorders. Adv Exp Med Biol 1191:155–167

Finamore A, Roselli M, Donini L, Brasili DE, Rami R, Carnevali P, Mistura L, Pinto A, Giusti A, Mengheri E (2019) Supplementation with bifidobacterium longum Bar33 and lactobacillus helveticus bar13 mixture improves immunity in elderly humans (over 75 years) and aged mice. Nutrition 63–64:184–192

Fontana A, Manchia M, Panebianco C, Paribello P, Arzedi C, Cossu E, Garzilli M, Montis MA, Mura A, Pisanu C, Congiu D, Copetti M, Pinna F, Carpiniello B, Squassina A, Pazienza V (2020) Exploring the role of gut microbiota in major depressive disorder and in treatment resistance to antidepressants. Biomedicines 8:311

Forsythe P, Bienenstock J, Kunze WA (2014) Vagal pathways for microbiome-brain-gut axis communication. Adv Exp Med Biol 817:115–133

Gaboriau-Routhiau V, Rakotobe S, Lecuyer E, Mulder I, Lan A, Bridonneau C, Rochet V, Pisi A, De Paepe M, Brandi G, Eberl G, Snel J, Kelly D, Cerf-Bensussan N (2009) The key role of segmented filamentous bacteria in the coordinated maturation of gut helper T cell responses. Immunity 31:677–689

Gao K, Mu CL, Farzi A, Zhu WY (2020) Tryptophan metabolism: a link between the gut microbiota and brain. Adv Nutr 11:709–723

Gareau MG, Jury J, Macqueen G, Sherman PM, Perdue MH (2007) Probiotic treatment of rat pups normalises corticosterone release and ameliorates colonic dysfunction induced by maternal separation. Gut 56:1522–1528

Geda YE (2012) Mild cognitive impairment in older adults. Curr Psychiat Rep 14:320–327
Goyal D, Ali SA, Singh RK (2021) Emerging role of gut microbiota in modulation of neuroinflammation and neurodegeneration with emphasis on Alzheimer's disease. Prog Neuro-Psychoph 106:110112
Grinan-Ferre C, Bellver-Sanchis A, Izquierdo V, Corpas R, Roig-Soriano J, Chillon M, Andres-Lacueva C, Somogyvari M, Soti C, Sanfeliu C, Pallas M (2021) The pleiotropic neuroprotective effects of resveratrol in cognitive decline and Alzheimer's disease pathology: from antioxidant to epigenetic therapy. Ageing Res Rev 67:101271
Guo M, Peng J, Huang X, Xiao L, Huang F, Zuo Z (2021) Gut microbiome features of chinese patients newly diagnosed with Alzheimer's disease or mild cognitive impairment. J Alzheimers Dis 80:299–310
Gupta S, Allen-Vercoe E, Petrof EO (2016) Fecal microbiota transplantation in perspective. Therap Adv Gastroenterol 9:229–239
Gwak MG, Chang SY (2021) Gut-brain connection: microbiome, gut barrier and environmental sensors. . Immune Netw 21:E20
Hao W, Wu J, Yuan N, Gong L, Huang J, Ma Q, Zhu H, Gan H, Da X, Deng L, Li X, Chen J (2021a) Xiaoyaosan improves antibiotic-induced depressive-like and anxiety-like behavior in mice through modulating the gut microbiota and regulating the Nlrp3 inflammasome in the colon. Front Pharmacol 12:619103
Hao WZ, Ma QY, Tao G, Huang JQ, Chen JX (2021b) Oral coniferyl ferulate attenuated depression symptoms in mice via reshaping gut microbiota and microbial metabolism. Food Funct 12:12550–12564
Haran JP, Bhattarai SK, Foley SE, Dutta P, Ward DV, Bucci V, Mccormick BA (2019) Alzheimer's Disease microbiome is associated with dysregulation of the anti-inflammatory p-glycoprotein pathway. Mbio 10:E00632-E719
Hazan S (2020) Rapid improvement in Alzheimer's disease symptoms following fecal microbiota transplantation: a case report. J Int Med Res 48:300060520925930
He Y, Li B, Sun D, Chen S (2020) Gut microbiota: implications in Alzheimer's disease. J Clin Med 9:2042
Heidarzadeh-Rad N, Gokmen-Ozel H, Kazemi A, Almasi N, Djafarian K (2020) effects of a psychobiotic supplement on serum brain-derived neurotrophic factor levels in depressive patients: a post hoc analysis of a randomized clinical trial. J Neurogastroenterol Motil 26:486–495
Heym N, Heasman BC, Hunter K, Blanco SR, Wang GY, Siegert R, Cleare A, Gibson GR, Kumari V, Sumich AL (2019) The role of microbiota and inflammation in self-judgement and empathy: implications for understanding the brain-gut-microbiome axis in depression. Psychopharmacology 236:1459–1470
Huang HL, Chen HT, Luo QL, Xu HM, He J, Li YQ, Zhou YL, Yao F, Nie YQ, Zhou YJ (2019) Relief of irritable bowel syndrome by fecal microbiota transplantation is associated with changes in diversity and composition of the gut microbiota. J Dig Dis 20:401–408
Huang X, Li L, Ling Z, Gao L, Chen H, Duan X (2021a) Gut Microbiome diversity mediates the association between right dorsolateral prefrontal cortex and anxiety level. Brain Imaging Behav 16:397–405
Huang Y, Shen Z, He W (2021b) Identification of gut microbiome signatures in patients with post-stroke cognitive impairment and affective disorder. Front Aging Neurosci 13:706765
Huang Y, Shi X, Li Z, Shen Y, Shi X, Wang L, Li G, Yuan Y, Wang J, Zhang Y, Zhao L, Zhang M, Kang Y, Liang Y (2018) Possible association of firmicutes in the gut microbiota of patients with major depressive disorder. Neuropsychiatr Dis Treat 14:3329–3337
Hwang YH, Park S, Paik JW, Chae SW, Kim DH, Jeong DG, Ha E, Kim M, Hong G, Park SH, Jung SJ, Lee SM, Na KH, Kim J, Chung YC (2019) Efficacy and safety of lactobacillus plantarum c29-fermented soybean (Dw2009) in individuals with mild cognitive impairment: A 12-week, multi-center, randomized, double-blind placebo-controlled clinical trial. Nutrients 11:305

Jiang H, Ling Z, Zhang Y, Mao H, Ma Z, Yin Y, Wang W, Tang W, Tan Z, Shi J, Li L, Ruan B (2015) Altered fecal microbiota composition in patients with major depressive disorder. Brain Behav Immun 48:186–194

Jiang HY, Zhang X, Yu ZH, Zhang Z, Deng M, Zhao JH, Ruan B (2018) Altered gut microbiota profile in patients with generalized anxiety disorder. J Psychiatr Res 104:130–136

Kazemi A, Noorbala AA, Azam K, Eskandari MH, Djafarian K (2019) Effect of probiotic and prebiotic vs placebo on psychological outcomes in patients with major depressive disorder: a randomized clinical trial. Clin Nutr 38:522–528

Kedinger M, Freund JN (2006) Gut epithelium. In Encyclopedic reference of genomics and proteomics in molecular medicine. Springer, Berlin, Heidelberg.

Kelly JR, Kennedy PJ, Cryan JF, Dinan TG, Clarke G, Hyland NP (2015) Breaking down the barriers: the gut microbiome, intestinal permeability and stress-related psychiatric disorders. Front Cell Neurosci 9:392

Khan MZ, He L (2017) The role of polyunsaturated fatty acids and Gpr40 receptor in brain. Neuropharmacology 113:639–651

Kobayashi Y, Kuhara T, Oki M, Xiao JZ (2019) Effects of bifidobacterium breve a1 on the cognitive function of older adults with memory complaints: a randomised, double-blind, placebo-controlled trial. Benef Microbes 10:511–520

Kuo SM (2013) The interplay between fiber and the intestinal microbiome in the inflammatory response. Adv Nutr 4:16–28

Lai WT, Zhao J, Xu SX, Deng WF, Xu D, Wang MB, He FS, Liu YH, Guo YY, Ye SW, Yang QF, Zhang YL, Wang S, Li MZ, Yang YJ, Liu TB, Tan ZM, Xie XH, Rong H (2021) Shotgun metagenomics reveals both taxonomic and tryptophan pathway differences of gut microbiota in bipolar disorder with current major depressive episode patients. J Affect Disord 278:311–319

Leblhuber F, Ehrlich D, Steiner K, Geisler S, Fuchs D, Lanser L, Kurz K (2021) The immunopathogenesis of Alzheimer's disease is related to the composition of gut microbiota. Nutrients 13:361

Leblhuber F, Steiner K, Schuetz B, Fuchs D, Gostner JM (2018) Probiotic supplementation in patients with Alzheimer's dementia - an explorative intervention study. Curr Alzheimer Res 15:1106–1113

Lew LC, Hor YY, Yusoff NAA, Choi SB, Yusoff MSB, Roslan NS, Ahmad A, Mohammad JAM, Abdullah M, Zakaria N, Wahid N, Sun Z, Kwok LY, Zhang H, Liong MT (2019) Probiotic lactobacillus plantarum p8 alleviated stress and anxiety while enhancing memory and cognition in stressed adults: a randomised, double-blind, placebo-controlled study. Clin Nutr 38:2053–2064

Li B, He Y, Ma J, Huang P, Du J, Cao L, Wang Y, Xiao Q, Tang H, Chen S (2019) Mild cognitive impairment has similar alterations as alzheimer's disease in gut microbiota. Alzheimers Dement 15:1357–1366

Li C, Huang B, Zhang YW (2021a) Chinese herbal medicine for the treatment of depression: effects on the neuroendocrine-immune network. Pharmaceuticals (basel) 14:65

Li C, Wang N, Zheng G, Yang L (2021b) Oral administration of resveratrol-selenium-peptide nanocomposites alleviates alzheimer's disease-like pathogenesis by inhibiting abeta aggregation and regulating gut microbiota. Acs Appl Mater Interfaces 13:46406–46420

Li M, Van Esch B, Wagenaar GTM, Garssen J, Folkerts G, Henricks PAJ (2018) Pro- and anti-inflammatory effects of short chain fatty acids on immune and endothelial cells. Eur J Pharmacol 831:52–59

Li S, Song J, Ke P, Kong L, Lei B, Zhou J, Huang Y, Li H, Li G, Chen J, Li X, Xiang Z, Ning Y, Wu F, Wu K (2021c) The gut microbiome is associated with brain structure and function in schizophrenia. Sci Rep 11:9743

Li S, Zhuo M, Huang X, Huang Y, Zhou J, Xiong D, Li J, Liu Y, Pan Z, Li H, Chen J, Li X, Xiang Z, Wu F, Wu K (2020) Altered gut microbiota associated with symptom severity in schizophrenia. PeerJ 8:E9574

Lin P, Ding B, Feng C, Yin S, Zhang T, Qi X, Lv H, Guo X, Dong K, Zhu Y, Li Q (2017) Prevotella and klebsiella proportions in fecal microbial communities are potential characteristic parameters for patients with major depressive disorder. J Affect Disord 207:300–304

Ling Y, Gu Q, Zhang J, Gong T, Weng X, Liu J, Sun J (2020) Structural change of gut microbiota in patients with post-stroke comorbid cognitive impairment and depression and its correlation with clinical features. J Alzheimers Dis 77:1595–1608

Liskiewicz P, Kaczmarczyk M, Misiak B, Wronski M, Baba-Kubis A, Skonieczna-Zydecka K, Marlicz W, Bienkowski P, Misera A, Pelka-Wysiecka J, Kucharska-Mazur J, Konopka A, Loniewski I, Samochowiec J (2021) Analysis of gut microbiota and intestinal integrity markers of inpatients with major depressive disorder. Prog Neuropsychopharmacol Biol Psychiatry 106:110076

Liu J, Liu X, Xiong XQ, Yang T, Cui T, Hou NL, Lai X, Liu S, Guo M, Liang XH, Cheng Q, Chen J, Li TY (2017) Effect of vitamin A supplementation on gut microbiota in children with autism spectrum disorders - a pilot study. Bmc Microbiol 17:204

Liu P, Jia XZ, Chen Y, Yu Y, Zhang K, Lin YJ, Wang BH, Peng GP (2021a) Gut microbiota interacts with intrinsic brain activity of patients with amnestic mild cognitive impairment. Cns Neurosci Ther 27:163–173

Liu P, Wu L, Peng G, Han Y, Tang R, Ge J, Zhang L, Jia L, Yue S, Zhou K, Li L, Luo B, Wang B (2019a) Altered microbiomes distinguish Alzheimer's disease from amnestic mild cognitive impairment and health in a Chinese cohort. Brain Behav Immun 80:633–643

Liu S, Gao J, Liu K, Zhang HL (2021b) Microbiota-gut-brain axis and Alzheimer's disease: implications of the blood-brain barrier as an intervention target. Mech Ageing Dev 199:111560

Liu X, Cao S, Zhang X (2015) Modulation of gut microbiota-brain axis by probiotics, prebiotics, and diet. J Agric Food Chem 63:7885–7895

Liu X, Teng T, Li X, Fan L, Xiang Y, Jiang Y, Du K, Zhang Y, Zhou X, Xie P (2021c) Impact of inosine on chronic unpredictable mild stress-induced depressive and anxiety-like behaviors with the alteration of gut microbiota. Front Cell Infect Microbiol 11:697640

Liu Y, Du T, Zhang W, Lu W, Peng Z, Huang S, Sun X, Zhu X, Chen C, Qian L, Wen L, Xu P, Zhang Y (2019b) Modified huang-lian-jie-du decoction ameliorates abeta synaptotoxicity in a murine model of Alzheimer's disease. Oxid Med Cell Longev 2019:8340192

Liu Z, Li L, Ma S, Ye J, Zhang H, Li Y, Sair AT, Pan J, Liu X, Li X, Yan S, Liu X (2020) High-dietary fiber intake alleviates antenatal obesity-induced postpartum depression: roles of gut microbiota and microbial metabolite short-chain fatty acid involved. J Agric Food Chem 68:13697–13710

Logsdon AF, Erickson MA, Rhea EM, Salameh TS, Banks WA (2018) Gut reactions: how the blood-brain barrier connects the microbiome and the brain. Exp Biol Med (maywood) 243:159–165

Lyketsos CG, Carrillo MC, Ryan JM, Khachaturian AS, Trzepacz P, Amatniek J, Cedarbaum J, Brashear R, Miller DS (2011) Neuropsychiatric symptoms in Alzheimer's disease. Alzheimers Dement 7:532–539

Ma X, Asif H, Dai L, He Y, Zheng W, Wang D, Ren H, Tang J, Li C, Jin K, Li Z, Chen X (2020) Alteration of the gut microbiome in first-episode drug-naive and chronic medicated schizophrenia correlate with regional brain volumes. J Psychiatr Res 123:136–144

Mahmoudiandehkordi S, Arnold M, Nho K, Ahmad S, Jia W, Xie G, Louie G, Kueider-Paisley A, Moseley MA, Thompson JW, Williams SJ, L., Tenenbaum, J. D., Blach, C., Baillie, R., Han, X., Bhattacharyya, S., Toledo, J. B., Schafferer, S., Klein, S., Koal, T., Risacher, S. L., Kling, M. A., Motsinger-Reif, A., Rotroff, D. M., Jack, J., Hankemeier, T., Bennett, D. A., De Jager, P. L., Trojanowski, J. Q., Shaw, L. M., Weiner, M. W., Doraiswamy, P. M., Van Duijn, C. M., Saykin, A. J., Kastenmuller, G., Kaddurah-Daouk, R., Alzheimer's Disease Neuroimaging, I. & The Alzheimer Disease Metabolomics, C. (2019) Altered Bile acid profile associates with cognitive impairment in alzheimer's disease-an emerging role for gut microbiome. Alzheimers Dement 15:76–92

Makris AP, Karianaki M, Tsamis KI, Paschou SA (2021) The role of the gut-brain axis in depression: endocrine, neural, and immune pathways. Hormones (athens) 20:1–12

Martinez C, Gonzalez-Castro A, Vicario M, Santos J (2012) Cellular and molecular basis of intestinal barrier dysfunction in the irritable bowel syndrome. Gut Liver 6:305–315
Miao Y, Li X, Yuan XX, Zhang LY, Pang LJ, Zhang XY, Wang SY, Song XQ (2021) Effect of the correlation between gut microbiota and folic acid in first-episode schizophrenia. Zhonghua Yi Xue Za Zhi 101:3012–3017
Miranda PM, De Palma G, Serkis V, Lu J, Louis-Auguste MP, Mccarville JL, Verdu EF, Collins SM, Bercik P (2018) High salt diet exacerbates colitis in mice by decreasing lactobacillus levels and butyrate production. Microbiome 6:57
Misiak B, Loniewski I, Marlicz W, Frydecka D, Szulc A, Rudzki L, Samochowiec J (2020) The hpa axis dysregulation in severe mental illness: can we shift the blame to gut microbiota? Prog Neuro-Psychoph 102:109951
Miyaoka T, Kanayama M, Wake R, Hashioka S, Hayashida M, Nagahama M, Okazaki S, Yamashita S, Miura S, Miki H, Matsuda H, Koike M, Izuhara M, Araki T, Tsuchie K, Azis IA, Arauchi R, Abdullah RA, Oh-Nishi A, Horiguchi J (2018) Clostridium butyricum miyairi 588 as adjunctive therapy for treatment-resistant major depressive disorder: a prospective open-label trial. Clin Neuropharmacol 41:151–155
Mizuno S, Masaoka T, Naganuma M, Kishimoto T, Kitazawa M, Kurokawa S, Nakashima M, Takeshita K, Suda W, Mimura M, Hattori M, Kanai T (2017) Bifidobacterium-rich fecal donor may be a positive predictor for successful fecal microbiota transplantation in patients with irritable bowel syndrome. Digestion 96:29–38
Nagpal R, Neth BJ, Wang S, Craft S, Yadav H (2019) Modified mediterranean-ketogenic diet modulates gut microbiome and short-chain fatty acids in association with Alzheimer's disease markers in subjects with mild cognitive impairment. EBioMedicine 47:529–542
Naseribafrouei A, Hestad K, Avershina E, Sekelja M, Linlokken A, Wilson R, Rudi K (2014) Correlation between the human fecal microbiota and depression. Neurogastroenterol Motil 26:1155–1162
Neufeld KM, Kang N, Bienenstock J, Foster JA (2011) Reduced anxiety-like behavior and central neurochemical change in germ-free mice. Neurogastroenterol Motil 23(255–64):E119
Nguyen TT, Kosciolek T, Daly RE, Vazquez-Baeza Y, Swafford A, Knight R, Jeste DV (2021) Gut microbiome in schizophrenia: altered functional pathways related to immune modulation and atherosclerotic risk. Brain Behav Immun 91:245–256
Nguyen TT, Kosciolek T, Maldonado Y, Daly RE, Martin AS, Mcdonald D, Knight R, Jeste DV (2019) Differences in gut microbiome composition between persons with chronic schizophrenia and healthy comparison subjects. Schizophr Res 204:23–29
Okubo R, Koga M, Katsumata N, Odamaki T, Matsuyama S, Oka M, Narita H, Hashimoto N, Kusumi I, Xiao J, Matsuoka YJ (2019) Effect of bifidobacterium breve A-1 on anxiety and depressive symptoms in schizophrenia: a proof-of-concept study. J Affect Disord 245:377–385
Oleskin AV, Shenderov BA (2016) Neuromodulatory effects and targets of the scfas and gasotransmitters produced by the human symbiotic microbiota. Microb Ecol Health Dis 27:30971
Oleskin AV, Shenderov BA (2019) Probiotics and psychobiotics: the role of microbial neurochemicals. Probiotics Antimicrob Proteins 11:1071–1085
Olsson C, Holmgren S (2011) Autonomic control of gut motility: a comparative view. Auton Neurosci 165:80–101
Ooi KM, Vacy K, Boon WC (2021) Fatty acids and beyond: age and alzheimer's disease related changes in lipids reveal the neuro-nutraceutical potential of lipids in cognition. Neurochem Int 149:105143
Painold A, Morkl S, Kashofer K, Halwachs B, Dalkner N, Bengesser S, Birner A, Fellendorf F, Platzer M, Queissner R, Schutze G, Schwarz MJ, Moll N, Holzer P, Holl AK, Kapfhammer HP, Gorkiewicz G, Reininghaus EZ (2019) A step ahead: exploring the gut microbiota in inpatients with bipolar disorder during a depressive episode. Bipolar Disord 21:40–49
Paoli A (2014) Ketogenic diet for obesity: friend or foe? Int J Environ Res Public Health 11:2092–2107

Pinto-Sanchez MI, Hall GB, Ghajar K, Nardelli A, Bolino C, Lau JT, Martin FP, Cominetti O, Welsh C, Rieder A, Traynor J, Gregory C, De Palma G, Pigrau M, Ford AC, Macri J, Berger B, Bergonzelli G, Surette MG, Collins SM, Moayyedi P, Bercik P (2017) Probiotic bifidobacterium longum ncc3001 reduces depression scores and alters brain activity: a pilot study in patients with irritable bowel syndrome. Gastroenterology 153(448–459):E8

Rao J, Qiao Y, Xie R, Lin L, Jiang J, Wang C, Li G (2021) Fecal microbiota transplantation ameliorates stress-induced depression-like behaviors associated with the inhibition of glial and Nlrp3 inflammasome in rat brain. J Psychiatr Res 137:147–157

Ren T, Gao Y, Qiu Y, Jiang S, Zhang Q, Zhang J, Wang L, Zhang Y, Wang L, Nie K (2020) Gut microbiota altered in mild cognitive impairment compared with normal cognition in sporadic parkinson's disease. Front Neurol 11:137

Rong H, Xie XH, Zhao J, Lai WT, Wang MB, Xu D, Liu YH, Guo YY, Xu SX, Deng WF, Yang QF, Xiao L, Zhang YL, He FS, Wang S, Liu TB (2019) Similarly in depression, nuances of gut microbiota: evidences from a shotgun metagenomics sequencing study on major depressive disorder versus bipolar disorder with current major depressive episode patients. J Psychiatr Res 113:90–99

Round JL, Mazmanian SK (2010) Inducible Foxp3+ regulatory T-cell development by a commensal bacterium of the intestinal microbiota. P Natl Acad Sci Usa 107:12204–12209

Rudzki L, Ostrowska L, Pawlak D, Malus A, Pawlak K, Waszkiewicz N, Szulc A (2019) Probiotic lactobacillus plantarum 299v decreases kynurenine concentration and improves cognitive functions in patients with major depression: a double-blind, randomized, placebo controlled study. Psychoneuroendocrino 100:213–222

Ruggiero M (2016) Fecal microbiota transplantation and the brain microbiota in neurological diseases. Clin Endosc 49:579

Rusek M, Pluta R, Ulamek-Koziol M, Czuczwar SJ (2019) Ketogenic diet in Alzheimer's disease. Int J Mol Sci 20:3892

Sampaio LP (2016) Ketogenic diet for epilepsy treatment. Arq Neuropsiquiatr 74:842–848

Severance EG, Gressitt KL, Stallings CR, Katsafanas E, Schweinfurth LA, Savage CLG, Adamos MB, Sweeney KM, Origoni AE, Khushalani S, Dickerson FB, Yolken RH (2017) Probiotic normalization of candida albicans in schizophrenia: a randomized, placebo-controlled, longitudinal pilot study. Brain Behav Immun 62:41–45

Shen T, Yue Y, He T, Huang C, Qu B, Lv W, Lai HY (2021) The association between the gut microbiota and parkinson's disease. A Meta-Analysis. Front Aging Neurosci 13:636545

Shen Y, Xu J, Li Z, Huang Y, Yuan Y, Wang J, Zhang M, Hu S, Liang Y (2018) Analysis of gut microbiota diversity and auxiliary diagnosis as a biomarker in patients with schizophrenia: a cross-sectional study. Schizophr Res 197:470–477

Siegel MP, Conklin SM (2020) Acute intake of B. longum probiotic does not reduce stress, anxiety, or depression in young adults: a pilot study. Brain Behav Immun Health 2:100029

Skonieczna-Zydecka K, Grochans E, Maciejewska D, Szkup M, Schneider-Matyka D, Jurczak A, Loniewski I, Kaczmarczyk M, Marlicz W, Czerwinska-Rogowska M, Pelka-Wysiecka J, Dec K, Stachowska E (2018) Faecal short chain fatty acids profile is changed in polish depressive women. Nutrients 10:1939

Slykerman, R. F., Hood, F., Wickens, K., Thompson, J. M. D., Barthow, C., Murphy, R., Kang, J., Rowden, J., Stone, P., Crane, J., Stanley, T., Abels, P., Purdie, G., Maude, R., Mitchell, E. A. & Probiotic In Pregnancy Study, G (2017) Effect of lactobacillus rhamnosus hn001 in pregnancy on postpartum symptoms of depression and anxiety: a randomised double-blind placebo-controlled trial. EBioMedicine 24:159–165

Sommer F, Backhed F (2013) The gut microbiota-masters of host development and physiology. Nat Rev Microbiol 11:227–238

Strandwitz P (2018) Neurotransmitter modulation by the gut microbiota. Brain Res 1693:128–133

Sun J, Xu J, Ling Y, Wang F, Gong T, Yang C, Ye S, Ye K, Wei D, Song Z, Chen D, Liu J (2019) Fecal microbiota transplantation alleviated Alzheimer's disease-like pathogenesis in App/Ps1 transgenic mice. Transl Psychiat 9:189

Takiishi T, Fenero CIM, Camara NOS (2017) Intestinal barrier and gut microbiota: shaping our immune responses throughout life. Tissue Barriers 5:E1373208

Tamtaji OR, Heidari-Soureshjani R, Mirhosseini N, Kouchaki E, Bahmani F, Aghadavod E, Tajabadi-Ebrahimi M, Asemi Z (2019) Probiotic and selenium co-supplementation, and the effects on clinical, metabolic and genetic status in Alzheimer's disease: a randomized, double-blind, controlled trial. Clin Nutr 38:2569–2575

Thayer JF, Sternberg EM (2010) Neural aspects of immunomodulation: focus on the vagus nerve. Brain Behav Immun 24:1223–1228

Ton AMM, Campagnaro BP, Alves GA, Aires R, Coco LZ, Arpini CM, Guerra EOT, Campos-Toimil M, Meyrelles SS, Pereira TMC, Vasquez EC (2020) Oxidative stress and dementia in Alzheimer's patients: effects of synbiotic supplementation. Oxid Med Cell Longev 2020:2638703

Tran TTT, Corsini S, Kellingray L, Hegarty C, Le Gall G, Narbad A, Muller M, Tejera N, O'toole, P. W., Minihane, A. M. & Vauzour, D. (2019) Apoe genotype influences the gut microbiome structure and function in humans and mice: relevance for alzheimer's disease pathophysiology. Faseb J 33:8221–8231

Valles-Colomer M, Falony G, Darzi Y, Tigchelaar EF, Wang J, Tito RY, Schiweck C, Kurilshikov A, Joossens M, Wijmenga C, Claes S, Van Oudenhove L, Zhernakova A, Vieira-Silva S, Raes J (2019) The neuroactive potential of the human gut microbiota in quality of life and depression. Nat Microbiol 4:623–632

Vighi G, Marcucci F, Sensi L, Di Cara G, Frati F (2008) Allergy and the gastrointestinal system. Clin Exp Immunol 153(Suppl 1):3–6

Vinolo MA, Rodrigues HG, Nachbar RT, Curi R (2011) Regulation Of Inflammation By Short Chain Fatty Acids. Nutrients 3:858–876

Vogt NM, Kerby RL, Dill-Mcfarland KA, Harding SJ, Merluzzi AP, Johnson SC, Carlsson CM, Asthana S, Zetterberg H, Blennow K, Bendlin BB, Rey FE (2017) Gut Microbiome Alterations In Alzheimer's Disease. Sci Rep 7:13537

Vogt NM, Romano KA, Darst BF, Engelman CD, Johnson SC, Carlsson CM, Asthana S, Blennow K, Zetterberg H, Bendlin BB, Rey FE (2018) The gut microbiota-derived metabolite trimethylamine n-oxide is elevated In Alzheimer's disease. Alzheimers Res Ther 10:124

Walker WA (2017) The importance of appropriate initial bacterial colonization of the intestine in newborn, child, and adult health. Pediatr Res 82:387–395

Wang H, Braun C, Murphy EF, Enck P (2019a) Bifidobacterium longum 1714 strain modulates brain activity of healthy volunteers during social stress. Am J Gastroenterol 114:1152–1162

Wang M, Cao J, Gong C, Amakye WK, Yao M, Ren J (2021a) Exploring the microbiota-Alzheimer's disease linkage using short-term antibiotic treatment followed by fecal microbiota transplantation. Brain Behav Immun 96:227–238

Wang X, Sun G, Feng T, Zhang J, Huang X, Wang T, Xie Z, Chu X, Yang J, Wang H, Chang S, Gong Y, Ruan L, Zhang G, Yan S, Lian W, Du C, Yang D, Zhang Q, Lin F, Liu J, Zhang H, Ge C, Xiao S, Ding J, Geng M (2019b) Sodium oligomannate therapeutically remodels gut microbiota and suppresses gut bacterial amino acids-shaped neuroinflammation to inhibit alzheimer's disease progression. Cell Res 29:787–803

Wang Z, Chen WH, Li SX, He ZM, Zhu WL, Ji YB, Wang Z, Zhu XM, Yuan K, Bao YP, Shi L, Meng SQ, Xue YX, Xie W, Shi J, Yan W, Wei H, Lu L, Han Y (2021b) Gut microbiota modulates the inflammatory response and cognitive impairment induced by sleep deprivation. Mol Psychiatr 26:6277–6292

Waxenbaum JA, Reddy V, Varacallo M (2021) Anatomy, autonomic nervous system. In Statpearls. Treasure Island (Fl).

Xi J, Ding D, Zhu H, Wang R, Su F, Wu W, Xiao Z, Liang X, Zhao Q, Hong Z, Fu H, Xiao Q (2021) Disturbed microbial ecology in Alzheimer's disease: evidence from the gut microbiota and fecal metabolome. Bmc Microbiol 21:226

Xiao J, Katsumata N, Bernier F, Ohno K, Yamauchi Y, Odamaki T, Yoshikawa K, Ito K, Kaneko T (2020) Probiotic bifidobacterium breve in improving cognitive functions of older

adults with suspected mild cognitive impairment: a randomized, double-blind, placebo-controlled trial. J Alzheimers Dis 77:139–147
Xu R, Wu B, Liang J, He F, Gu W, Li K, Luo Y, Chen J, Gao Y, Wu Z, Wang Y, Zhou W, Wang M (2020) Altered gut microbiota and mucosal immunity in patients with schizophrenia. Brain Behav Immun 85:120–127
Yamamura R, Okubo R, Katsumata N, Odamaki T, Hashimoto N, Kusumi I, Xiao J, Matsuoka YJ (2021) Lipid and energy metabolism of the gut microbiota is associated with the response to probiotic bifidobacterium breve strain for anxiety and depressive symptoms in schizophrenia. J Pers Med 11:987
Yang J, Fu X, Liao X, Li Y (2020) Effects of gut microbial-based treatments on gut microbiota, behavioral symptoms, and gastrointestinal symptoms in children with autism spectrum disorder: a systematic review. Psychiatry Res 293:113471
Yuan X, Kang Y, Zhuo C, Huang XF, Song X (2019) The gut microbiota promotes the pathogenesis of schizophrenia via multiple pathways. Biochem Biophys Res Commun 512:373–380
Yuan X, Wang Y, Li X, Jiang J, Kang Y, Pang L, Zhang P, Li A, Lv L, Andreassen OA, Fan X, Hu S, Song X (2021) Gut microbial biomarkers for the treatment response in first-episode, drug-naive schizophrenia: a 24-week follow-up study. Transl Psychiat 11:422
Zhang Q, Yun Y, An H, Zhao W, Ma T, Wang Z, Yang F (2021) Gut microbiome composition associated with major depressive disorder and sleep quality. Front Psychiatry 12:645045
Zhao L, Xiong Q, Stary CM, Mahgoub OK, Ye Y, Gu L, Xiong X, Zhu S (2018) Bidirectional Gut-brain-microbiota axis as a potential link between inflammatory bowel disease and ischemic stroke. J Neuroinflammation 15:339
Zheng P, Zeng B, Zhou C, Liu M, Fang Z, Xu X, Zeng L, Chen J, Fan S, Du X, Zhang X, Yang D, Yang Y, Meng H, Li W, Melgiri ND, Licinio J, Wei H, Xie P (2016) Gut microbiome remodeling induces depressive-like behaviors through a pathway mediated by the host's metabolism. Mol Psychiatr 21:786–796
Zhou Y, Wang Y, Quan M, Zhao H, Jia J (2021) Gut Microbiota changes and their correlation with cognitive and neuropsychiatric symptoms in Alzheimer's disease. J Alzheimers Dis 81:583–595
Zhu C, Zheng M, Ali U, Xia Q, Wang Z, Chenlong Y, L., Chen, Y., Yan, J., Wang, K., Chen, J. & Zhang, X. (2021) Association between abundance of haemophilus in the gut microbiota and negative symptoms of schizophrenia. Front Psychiatry 12:685910
Zhu F, Ju Y, Wang W, Wang Q, Guo R, Ma Q, Sun Q, Fan Y, Xie Y, Yang Z, Jie Z, Zhao B, Xiao L, Yang L, Zhang T, Feng J, Guo L, He X, Chen Y, Chen C, Gao C, Xu X, Yang H, Wang J, Dang Y, Madsen L, Brix S, Kristiansen K, Jia H, Ma X (2020) Metagenome-wide association of gut microbiome features for schizophrenia. Nat Commun 11:1612
Zhuang Z, Yang R, Wang W, Qi L, Huang T (2020) Associations between gut microbiota and Alzheimer's disease, major depressive disorder and schizophrenia. J Neuroinflam 17:288
Zhuang ZQ, Shen LL, Li WW, Fu X, Zeng F, Gui L, Lu Y, Cai M, Zhu C, Tan YL, Zheng P, Li HY, Zhu J, Zhou HD, Bu XL, Wang YJ (2018) Gut microbiota is altered in patients with Alzheimer's disease. J Alzheimers Dis 63:1337–1346
Zmora N, Suez J, Elinav E (2019) You are what you eat: diet, health and the gut microbiota. Nat Rev Gastroenterol Hepatol 16:35–56

Chapter 15
"Aging Gut Microbiota and Colorectal Cancer Pathways Correlations"

Roberto Catanzaro, Morena Sciuto, Nima Fattahi, Atoosa Saidpour, and Francesco Marotta

Abstract Colorectal cancer (CRC) is the third most common cancer worldwide. Its pathogenesis is very complex and is still under study. The involvement of the intestinal microbiota has been known for many years, in particular, in the elderly subject who is more exposed to factors determining the prevalence of pathogenic and carcinogenic bacterial and fungal species than the young and the adult. These include diet, medications, lifestyle habits, hospitalization, etc. An intestinal dysbiosis is thus established which can favor inflammation, the breakdown of tight junctions, and an abnormal immune response that favor the genesis and proliferation of CRC.

Keywords Gut Microbiota · Colorectal Cancer · Aging · Dysbiosis · Candida Albicans · Colorectal Adenoma

---

R. Catanzaro (✉)
Department of Clinical and Experimental Medicine, Gastroenterology Section– "Gaspare Rodolico" Policlinico Hospital, University of Catania, Catania, Italy
e-mail: rcatanza@unict.it

M. Sciuto
Department of Clinical and Experimental Medicine, Graduate School of Geriatrics, University of Catania, Catania, Italy
e-mail: morena.sciuto@virgilio.it

N. Fattahi
Gastroenterology Department, Duke University, Durham, NC 27708, United States

A. Saidpour
National Nutrition and Food Technology Research Institute, Department of Clinical Nutrition & Dietetics, Faculty of Nutrition Sciences and Food Technology, Shahid Beheshti University of Medical Sciences, Tehran, Iran

F. Marotta (✉)
ReGenera R&D International for Aging Intervention, Milano, Italy
e-mail: fmarchimede@libero.it

F. Marotta (ed.), *Gut Microbiota in Aging and Chronic Diseases*, Healthy Ageing and Longevity 17, https://doi.org/10.1007/978-3-031-14023-5_15

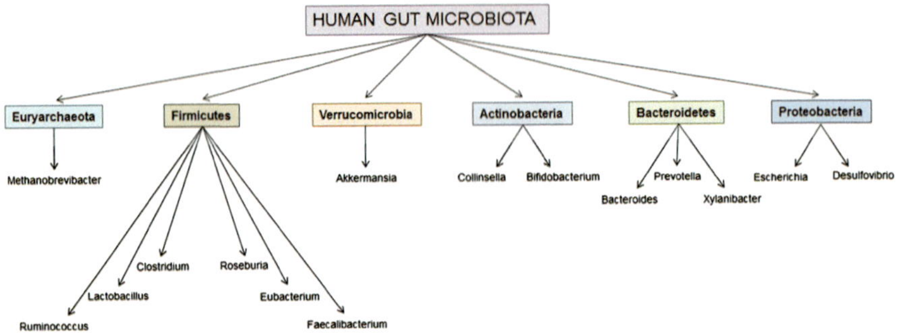

Fig. 15.1 The six main phyla of the human intestinal microbiota and the respective species most represented in the human intestine

15.1 Introduction

The human intestine is made up of about ten trillion different symbionts (bacteria, viruses, fungi, and protozoa), which make up the so-called "microbiota". These microorganisms encode over three million genes that produce thousands of metabolites. Furthermore, the components of gut microbiota (GM) interact with the host's immune system in ways that influence physiological homeostasis and disease development. The normal human GM includes six main phyla: *Bacteroidetes, Firmicutes, Actinobacteria, Verrucomicrobia, Euryarchaeota, and Proteobacteria* (Fig. 15.1) (Fong et al. 2020; Bliss and Whiteside 2018).

15.2 Features of Gut Microbiota

GM is a dynamic reality that has adapted to colonize the intestinal environment. Some specific characteristics are essential to survive in this environment. For example, the ability to evade bacteriophages and the host's immune system (Adak and Khan 2019). Furthermore, microbes would have to withstand physical and chemical stresses in the gut in order to proliferate. The relationship between host and microbes is commensal (one partner benefits while the other appears unaffected) rather than mutualistic (both partners benefited) (Ley et al. 2006).

Human GM is classified into three discrete enterotypes: *Bacteroides*, *Prevotella*, and *Ruminococcus*. Enterotype 1 is characterized by the dominance of *Bacteroides* with saccharolytic and proteolytic activities; it is also involved in the synthesis of biotin, riboflavin, pantothenate, and ascorbate (Arumugam et al. 2011). Enterotype 2 is dominant over *Prevotella* and it acts as a degrader of mucin glycoprotein and is involved in the synthesis of thiamine and folate. Enterotype 3 is characterized by the dominance of *Ruminococcus* capable of degrading mucin and transporting sugars on the membrane. However, the concept of enterotypes is debated due to the observation of a high degree of variation between individuals and some data showing more variability within the same individual related to diet (Knights et al. 2014). In fact, the intake of animal proteins and fats for a long period favors the *Bacteroides* enterotype while the diet rich in carbohydrates favors *Prevotella* (Wu et al. 2011).

GM is constantly evolving along with its host. It is acquired at birth and develops parallel to the host, playing an important role in the body from adulthood until death (Adak and Khan 2019).

15.3 Changes in Microbiota During Growth and in the Elderly

Colonization of the various parts of the body, including the intestine, is a dynamic process that begins at the moment of birth and that undergoes numerous changes during the development of each individual (Palmer et al. 2007).

The type of bacteria that colonize the newborn depends primarily on the mode of delivery: vaginal or by caesarean section. In fact, in subjects born by caesarean section, intestinal colonization by the microbiota begins later. The species found are, in particular, lower numbers of *Bifidobacteria* and prevalence of *Escherichia coli* and *Clostridium difficile*. Instead in the case of infants born by vaginal birth, there prevail *Lactobacilli* and *Prevotella spp.* GM colonization is also delayed in babies born by caesarean section compared to those born by vaginal delivery (Penders et al. 2006; Lakshminarayanan et al. 2014).

Another factor that influences the colonization by the microbiota is represented by the feeding of the infant. Breast-fed subjects are mainly colonized by *E. coli, Streptococci,* and *Bifidobacteria*. On the other hand, babies fed with artificial milk are colonized by *Enterobacteria, Clostridia,* and *Bacteroidetes* (Milani et al. 2017).

Further differences concern the different areas in which the children live and raise. In fact, the Western diet involves a high amount of *Firmicutes* and *Proteobacteria* and a low share of *Bacteroidetes* and *Actinobacteria* (Lakshminarayanan et al. 2014). Differences in the composition of the microbiota also exist in the case in which one lives in the city or in the rural suburbs (Boyajian et al. 2021).

A recent study on South Korean adults and seniors found significant differences in the gut microbiota between individuals who lived in the city and peers residing in the villages. For example, the *Firmicutes* genus was more abundant in village elders, while *Bacteroidetes* prevailed in urbanized elders (Kim et al. 2019).

Over the course of life, the intestinal microbiota undergoes various changes related to age, diet, environment, and disease. Various studies have proven such changes in the microbiota of different model organisms. In flies, alterations in the microbiota cause dysfunction of the intestinal barrier. Transplantation experiments from aged donor flies to young recipient flies were also conducted and they induced dramatic changes in the microbiota and reduced recipient longevity (Bana and Cabreiro 2019). Age-related changes, such as increased intestinal permeability and inflammation, were also observed in mice. Various studies have found an abundance of the *Bacteroidaceae* family in older mice compared to juvenile mice, while the families *Lactobacillaceae*, *Prevotellaceae*, *Rikenellaceae,* and *Lachnospiraceae* were more prevalent in young mice than in old mice. Aged microbiota in mice could affect the bioavailability of B vitamins such as cobalamin (B_{12}) and biotin (B_7) and other metabolites, such as creatine. Furthermore, alterations in DNA repair mechanisms and carbohydrate metabolism have also been observed (Langille et al. 2014; Thevaranjan et al. 2017).

GM changes due to age-related events also occur in humans. These modifications depend on various conditions, such as nutrition, lifestyle, teething, and stress (Weiss and Hennet 2017). Aging also involves changes in the microbiota that are influenced by the alterations that normal physiological processes undergo in this phase of life: reduction of gastrointestinal motility, delayed gastric emptying, reduced intestinal blood flow, constipation, alterations in the immune response, and impoverishment of the diet (Rinninella et al. 2019). In addition, hospitalization, reduced physical activity, and recurrent infections with consequent use of antibiotics are more frequent in the elderly (Claesson et al. 2012).

Modifications of the microbiota generally consist in a reduction in the diversity of bacterial species. In particular, the levels of *Bifidobacteria* and *Bacteroides* are reduced, while those of *Enterobacteriaceae* and *Clostridia* increase (Claesson et al. 2011; Woodmansey 2007). Furthermore, an Italian study compared the microbiota of young and centenarian subjects. In the latter, they found a rearrangement in the *Firmicutes* population and an increase in facultative anaerobes. This alteration predisposes to the development of an inflammatory state, with an increase in serum IL-6 and IL-8. This is associated with a reduction of *Faecalibacterium prauznitzii* and other species with anti-inflammatory action (Biagi et al. 2010). According to Collino et al. (2013), the centenarians are also characterized by a reduced concentration of sphingomyelin, diacylphosphatidylcholine, and eicosanoids. These metabolic changes are associated with changes in the GM as well.

As in the mouse, age-related changes in GM are also related to increased intestinal permeability and inflammation in old men. Furthermore, the regenerative capacity of the tissues tends to decrease with age. In fact, aging hinders the renewal and differentiation capacity of the stem cells of the gastrointestinal tract. It has been known for decades that the onset of various diseases, especially those with a chronic course, is related to lifestyle, diet, pleasure habits, the geographical region in which one lives, as well as having been breastfed or less. All these factors also determine the different composition of the intestinal microbiota, which is therefore probably an important factor in facilitating or preventing the onset of certain diseases. Changes in the microbiota also seem to be related to age-related conditions that alter the physiology of the host. For example, diseases such as Alzheimer's disease or osteoporosis induce changes in the human gut microbiota (Bana and Cabreiro 2019).

Another interesting aspect is that the composition of the GM is significantly correlated with frailty and nutritional status. Furthermore, the microbiota of elderly guests in long-term care facilities seems to be subject to greater diversification than that of peers residing in the community. This correlates with an increase in frailty (Bischoff 2016). Indeed, such changes in the microbiota promote inflammation, also leading to opportunistic infections (mainly from *Proteobacteria*) (Rampelli et al. 2013). Aging leads to the development of a pro-inflammatory environment in the colon, but also in the innate immune system (Franceschi et al. 2000). Several studies suggest that longevity may underline a balance between inflammatory and anti-inflammatory bacterial species (Biagi et al. 2010).

Another feature of aging is the loss of protein homeostasis (proteostasis). This condition causes protein aggregation in various tissues. Proteostasis is determined by the presence of molecular chaperones and proteolytic systems. The former guarantee the correct folding of proteins; the latter are responsible for the short-term degradation of proteins. Any defect in this proteostasis network can accelerate aging and the progression of aging-related diseases (López-Otín et al. 2013).

An exogenous factor that affects the composition of the intestinal microbiota is the environment in which the elderly person lives. In fact, important microbiota differences have been found between elderly people living in long-term care facilities and elderly living in the community. These changes are also represented by the loss of species associated with a healthy microbiome, which are generally found in the young (Jackson et al. 2016).

15.4 Age, Gut Microbiota, and CRC

The normal human gastrointestinal flora is made up of beneficial and harmful microbes. The first are involved in digestive processes and in the production of substances that help prevent the onset of diseases. Instead, pathogenic microbes produce toxins and substances that can also have a carcinogenic effect (Cheng et al. 2020). When an imbalance is generated between beneficial and pathogenic bacteria, we speak of dysbiosis, particularly when harmful microbes prevail. In fact, several studies have suggested the causal relationship between intestinal microbial dysbiosis and the pathogenesis of some diseases, including the development of digestive and hepatobiliary neoplasms, especially colorectal cancer (CRC) (Fig. 15.2) (Lee et al. 2021). When dysbiosis occurs, bacterial species generally tend to predominate and are normally present in small quantities (Petersen and Round 2014). The CRC is the third most widespread cancer worldwide and is associated with high morbidity and mortality. In addition, the incidence rates have increased in individuals < 50 years and important etiopathogenetic factors are alcohol, smoking, obesity, diabetes, and physical inactivity (Bray et al. 2018). Nutrition can also influence the development of CRC or not. A diet particularly rich in processed foods, animal fats, and red meats and low in fiber and fruit instead represents an important risk factor for this neoplasm (Lee et al. 2021).

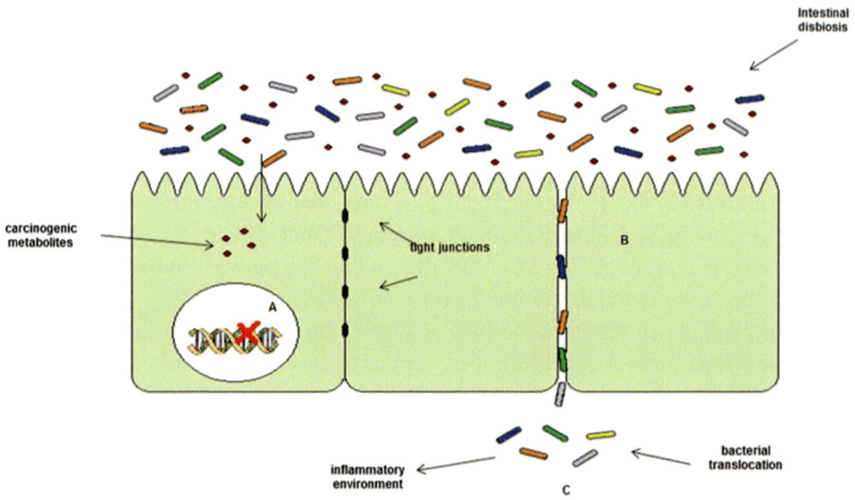

Fig. 15.2 Harmful effect of dysbiosis on cells and the extracellular environment. Intestinal dysbiosis involves the production of carcinogenic metabolites (hydroxyl radicals, polyketide-peptide genotoxin, colibactin, etc.) which penetrate inside the cells. Here (A) they cause DNA damage (genomic instability, mutations). Furthermore, some pathogens cause the breaking of tight junctions (B); this involves the translocation of bacteria into the extracellular tissue environment (C). Here the bacteria cause the establishment of an inflammatory environment (through the increase of IL-2, IL-6, IL-17, and TNF-α)

Bacteria—In addition to the aforementioned exogenous factors, endogenous factors have also been identified. These include the aforementioned intestinal dysbiosis, which in these cases involves the expression of some bacterial species including, the most studied to date, *Fusobacterium nucleatum* (Fong et al. 2020). *F. nucleatum* is a Gram-negative anaerobe. It is found physiologically in the oral cavity, and in the healthy subject, in general, it cannot be found in other parts of the body. Its role in the pathogenesis of colorectal cancer was initially regarded with skepticism as this microbe is normally part of the bacterial flora of the oral cavity. However, it turned out that, in some pathological conditions, it is possible to find *F. nucleatum* in other parts of the body, as in the case of the CRC (Han 2015). In particular, it appears to be involved in promoting genomic instability, cell mutations, and the uncontrolled proliferation of abnormal cells. Furthermore, it could also be responsible for the genesis of a pro-inflammatory tumor environment. All these processes are facilitated by the ability of this bacterium to evade the host's immune response (Brennan and Garrett 2019). In particular, *F. nucleatum* causes an upregulation and evasion of nuclear factor kappa B (NF-κB) in the lysis of tumor cells mediated by natural killer (NK) cells. Upregulation of NF-κB contributes to an increase in pro-inflammatory cytokines (IL-2, IL-6, IL-17) and TNF-α. Furthermore, *F. nucleatum* also acts as an immunosuppressant against $CD4^+$ T lymphocytes, thus counteracting the elimination of tumor cells by the immune system (Bashir et al. 2016). *F. nucleatum* enters host cells via FadA, a surface adhesion molecule capable of binding E-cadherin present on the surface of the epithelial cell. In this way, it determines the activation of the signaling pathways involved in the upregulation of the expression of oncogenes. Furthermore, the synergy between *F. nucleatum* and *Streptococcus* and *Campylobacter spp.* seems to enhance inflammation and tumorigenesis (Allen-Vercoe and Jobin 2014). This microbe can also affect the prognosis of people with CRC. It has been associated with reduced survival in these patients and promotes chemoresistance. It emerged that the survival of subjects with a high proportion of *F. nucleatum* is lower than those who do not have or have minimal quantities of this bacterium (Mima et al. 2016; Yu et al. 2017). Among other things, it was also proved that Fusobacterium can be detected in the feces of colon cancer patients. This means microbiome analyzes on fecal samples could be used as a non-invasive screening tool for CRC (LaCourse et al. 2021).

Peptostreptococcus anaerobius is a Gram-positive anaerobic coccus that colonizes the gastrointestinal and vaginal tracts (Legaria et al. 2021). The role played by *P. anaerobius* is not so certain. However, *Long* et al. have shown how this bacterium is able to interact more easily with cancer cells than healthy cells. From this interaction, it would seem to derive a cascade of signals capable of stimulating cell proliferation (Long et al. 2019).

Instead, the action of enterotoxigenic *Bacteroides fragilis* (ETBF) in the intestine is better defined. It is an obligate anaerobic Gram-negative bacillus and it produces an enterotoxin that acts on colonic epithelial tight junctions. The resulting effect is a response from the epithelium which increases the permeability of the mucosal barrier (Chung et al. 2018). ETBF is also capable of producing a biofilm that facilitates its damaging action on DNA (LaCourse et al. 2021).

Streptococcus bovis is a Gram-positive coccus. A probable contribution to carcinogenesis appears to be the induction of colonic inflammation. In fact, inflammation stimulates cell proliferation and also causes DNA damage (Krishnan and Eslick 2014). Furthermore, some animal studies have shown that *S. bovis* is capable of promoting the development and progression of preneoplastic lesions (Ellmerich et al. 2000). In particular, the intestinal mucosa of rats colonized by *S. bovis* has been shown to have increased levels of interleukin-8 (IL-8), a chemokine that stimulates oxidative stress and induces the formation of carcinogenic compounds. Furthermore, it promotes angiogenesis and the proliferation of cancer cells (Xie 2011).

Enterococcus faecalis is a Gram-positive bacteria, a member of *Firmicutes*. It has been associated with the development of CRC due to its ability to damage the DNA of colonic cells through the production of hydroxyl radicals (De Almeida et al. 2018).

Escherichia coli is a Gram-negative commensal bacterium of the *Enterobacteriaceae* family. It is associated with numerous pathological conditions of the digestive tract, including CRC. In fact, several studies have found a higher rate of colonization by *E. coli* in subjects with CRC than in healthy subjects (Jang et al. 2017). *E. coli* is able to synthesize a polyketide-peptide genotoxin, colibactin.

It appears to promote cellular senescence and facilitate the proliferation of cancer cells (Cheng et al. 2020). Colibactin also induces incomplete DNA repair and chromosomal instability. Colibactin-producing *E. coli* strains induce a cytopathic response in primary colonic epithelial cell lines in vitro, which suggests a possible involvement in the genesis and development of the CRC (Fiorentini et al. 2020).

On the other hand, bacteria such as *Lachnospiraceae* spp., *Bifidobacterium animalis,* and *Streptococcus thermophilus* appear to be deficient or absent in patients with CRC. Therefore, it is possible to hypothesize that these bacteria exert a protective effect against CRC (Dai et al. 2018).

Fungi—However, it is important to emphasize that not only the action of a single bacterial species can be harmful to the mucous membrane of the colon. The coexistence of certain microbial species acting synergistically can also cause an increased risk of developing CRC (Fig. 15.3). In addition to bacteria, certain species of fungi can also be involved in the genesis of neoplasms in humans. The fungal phyla residing in the digestive tract are *Basidiomycota*, *Glomeromycota,* and *Ascomycota*. Some, such as *Saccharomyces* and *Cladosporium*, physiologically colonize the gastrointestinal tract. Others, on the other hand, are pathogenic (Qin et al. 2021). Dysbiosis was detected in subjects with CRC with an increased ratio of *Basidiomycota/Ascomycota* (Gao et al. 2017). Among the pathogenic fungi, there is worth mentioning *Phoma*, which was found to be the predominant genus in intestinal biopsy samples of subjects with adenomas (Luan et al. 2015). *Phoma* is a genus of fungi belonging to the phylum *Ascomycota*. To date, more than 220 species have been distinguished. Some are conidia-forming species, and others

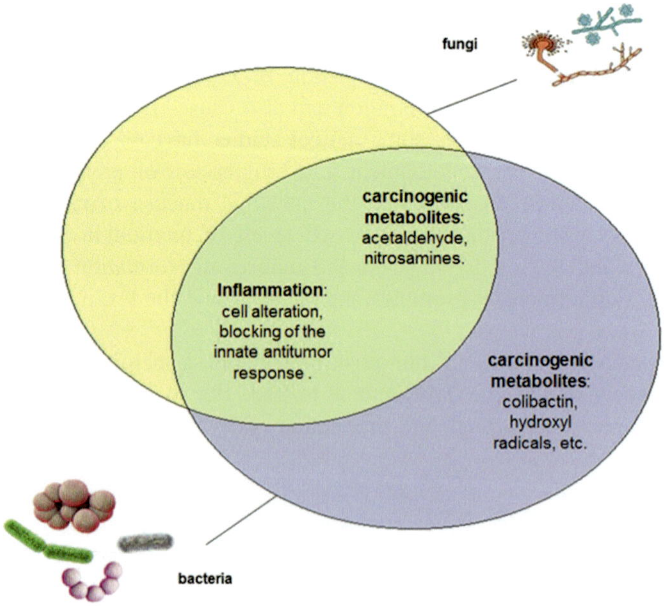

Fig. 15.3 Carcinogenic mechanisms shared by fungi and bacteria

produce pycnidia or chlamydospores (Bennett et al. 2018). *Phoma* species are ubiquitous; found in soil, organic matter, plants, and water sources. Some of these species are phytopathogens, while others are opportunistic invasive pathogens of animals and, in particular, of humans. In fact, they have been associated with the onset of lung masses, as well as in the formation of colonic adenomas (Aveskamp et al. 2008).

Candida spp., an opportunistic fungal pathogen that rarely colonizes the gastrointestinal tracts of healthy subjects, was also discovered in biopsy specimens of colon adenomas. Starý et al. (2020) found a strong over-representation of *Candida albicans* in cases of CRC or adenoma. *Candida spp.* is able to form a biofilm that facilitates its ability to infect the host by evading the immune system. Furthermore, these fungi can produce high levels of acetaldehyde which are carcinogenic. In fact, acetaldehyde increases intracellular reactive oxygen species (ROS) and Ca^{++} concentrations. This results in mitochondrial dysfunction leading to cytotoxicity and rupture of tight epithelial junctions (Luan et al. 2015; Douglas 2003). It has been shown that *C. albicans* can not only induce inflammation in the tissue but it can also attenuate the immune responses. In recent years, it has emerged that *C. albicans* could also promote CRC and metastases through a pro-inflammatory response, mediated by an increase in cytokines and the expression of adhesion molecules. Furthermore, when *C. albicans* induces an immune response, there is an increase in Th17 cells that produce IL-17, implicated in the resistance against *C. albicans*. However, Th17 cells also produce other interleukins, such as IL-23 which, however, promote angiogenesis as well as the genesis and growth of the tumor. Furthermore, IL-23 antagonizes IL-12 and IFN-γ, both of which are critical in antitumor immune responses. Thus, *C. albicans* can also have an indirect carcinogenic effect (Ramirez-Garcia et al. 2016; Langowski et al. 2006).

In addition, more fungal diversity was found in the late stage of the CRC compared to the early stage of the CRC. In particular, a greater amount of opportunistic fungi *Malassezia* and *Trichosporon* was noted as the largest contributor to CRC progression (Zhang et al. 2021).

Viruses–Alongside the GM, it is also important to mention the presence of the gut virome. In fact, the human intestine is physiologically also made up of RNA or DNA viruses. They can be in eukaryotic form or, for the most part, in prokaryotic form as bacteriophages (Ogilvie and Jones 2015). The latter, under normal conditions, behave like symbionts and do not harm the host. However, in certain situations, an alteration in the behavior of these microbes can occur. For example, in Crohn's disease, there is an increase in the bacteriophage population which also causes a reduction in bacterial diversity (Dahiya and Renuka 2017). It has also been hypothesized that the diet may have an effect on the gut virome and this was supported by the fact that the alterations in the gut virome are superimposable in individuals following a similar diet. Hence, similarly to the gut microbiota, the composition of the viral population in the human gut could also be determined by diet (Minot et al. 2011; Kim and Bae 2016).

Alongside the gut virome, there are those pathogenic viruses that, in physiological conditions, are not found in the human body. Some of them are potentially oncogenic as they are capable of implementing various strategies to favor the development and progression of the tumor. In fact, they can create the conditions for the uncontrolled replication of colic cells (e.g., induction of angiogenesis and inhibition of DNA repair mechanisms), as well as maximize viral production by producing more proviruses (Hanahan and Weinberg 2011). Among these, a role is played by Human Papilloma Virus (HPV), of which there are over 100 types of HPV, with types 16 and 18 being the most prone to cervical cancer. In fact, a study conducted by Buyru et al. (2006) pointed out that CRC due to HPV 16 and HPV 18 accounts for 51.8% of cases. In this study, 55% of HPV-infected patients had codon 12 mutations of the protooncogene K-Ras. The latter, if presenting a mutation in codons 12 or 13, loses the ability to bind to the GTPase activating protein and remains constantly activated and this favors the development of cancer. Furthermore, HPV proteins E6 and E7 inhibit the tumor suppressor proteins p53 and pRb, which can induce the development of CRC. However, it should be noted that in order to develop CRC, other conditions that can cause the disease must coexist. The presence of HPV alone is not sufficient to cause tumorigenesis, but it itself can determine the onset of conditions (p53 and K-Ras mutations) that contribute to the development of CRC (Buyru et al. 2006; Ambrosio et al. 2019).

Another association found is that between lower gastrointestinal tract neoplasms and the John Cunningham virus.

(JC virus). It belongs to the *Poliomaviridae* family and 70–80% of adult humans have antibodies specific to this virus. Generally, JC virus infection occurs in childhood and then the virus remains latent in various locations, including the intestinal mucosa (Sinagra et al. 2014). In conjunction with pathologies determining a state of immunosuppression, the virus can reactivate and induce harmful effects on various cells. This occurs through the expression of various proteins. In the colon, for example, large T-antigen protein blocks p53 and pRb making cells immortal and promoting tumor development. This protein is also capable of stimulating CRC metastasis. Agnoprotein is also capable of altering the cell cycle and inducing genomic instability (Artemev et al. 2022). JC virus is also able to integrate its own DNA into that of the host cell, but this has only been observed in cancer cells, less frequently in adenomatous cells and never in healthy cells. Therefore, it is evident that the process of integrating the viral DNA into that of the host could be the basis of chromosomal damage and malignant transformation (Coelho et al. 2010).

An ambiguous role is instead played by Epstein Barr Virus (EBV). In a study by Tafvizi et al. (2015), DNA fragments of the virus were found in 38% of patients with colorectal cancer and in 50% of patients with non-malignant colon lesions. In particular, the virus has been found most often in moderately differentiated adenocarcinomas. In another study, Fiorina et al. (2014) obtained similar results, with the difference that, in the samples of their patients, the virus was present not in the tumor cells but in the lymphocyte aggregates present in the tumor itself. In particular, EBV produces oncoproteins (LMP1, LMP2A, and EBNA1) which favor the development and progression of carcinoma through the initiation of the epithelial-mesenchymal transition (EMT) (Oleynikova et al. 2021). These proteins act by means of microRNAs (miRNAs) and cause cellular alterations in the host. In particular, miR-BAR T19-3p seems to attack WIF1, a major colon cancer gene, while miR-BAR T1 acts on PSA T1, a gene that promotes the development and growth of cancer cells (Mjelle et al. 2019).

Parasites—Finally, another important role is played by parasites. Already over a century ago, the association between CRC and schistosomiasis was noted. Several evidence associate chronic *Schistosoma japonicum* infestation with CRC (Hamid 2019). A case control reported that patients with chronic schistosomiasis have three times increased risk of developing CRC compared to those who have never been exposed to the infestation. Furthermore, the longer the infestation lasts, the greater the risk of developing CRC appears to be (Qiu et al. 2005). From a pathophysiological point of view, in the colon, the eggs of the schistosome infiltrate the mucosa and submucosa. Here, they determine a focal inflammation that leads to the formation of microulcerations of the mucosa, microabscesses, and the formation of granulomas. This favors the progressive onset of fibrosis, hyperplasia, polyposis, and pseudopolyposis (Gryseels 2012). Besides the induction of these processes through egg infiltration, Schistosoma is also a producer of toxins. It expresses inducible nitric oxide synthase (iNOS) which is involved in p53 mutation and tumor angiogenesis (Cianchi et al. 2003). *S. japonicum* infestation also affects CRC prognosis. Indeed, the 5-year survival of patients with CRC associated with Schistosoma infestation was lower than the 5-year survival of patients without the infestation (Hamid 2019).

15.5 Diet, Gut Microbiota, Xenobiotics, and CRC

It is now increasingly evident that when foods change the intestinal microbial composition and its diversity, the balance between beneficial and harmful microbes is altered and this can favor the development of cancer (De Almeida et al. 2019). As for complex carbohydrates, it has long been known that the presence of fiber in the diet reduces the risk of CRC. In particular, this effect is attributed to short-chain fatty acids (SCFA), such as acetate, propionate, and, above all, butyrate. The latter appears to mitigate the harmful effect of the high-fat diet, resulting in an improved intestinal barrier (Wu et al. 2018; Fang et al. 2019). These substances act both at the cellular level, for example by stimulating the proliferation of normal colonocytes and hindering that of cancer cells, and on the microbiota. In fact, various studies have shown the ability of dietary fibers to modify the intestinal microbiota. This is also demonstrated by the different composition of the microbiota in subjects who live in different parts of the world and who, therefore, have different eating habits (Loke et al. 2020). Indeed, in the long run, the prevalent consumption of fibers leads to an abundance of *Prevotella*, while a low-fiber diet, such as the Western one, determines the prevalence of *Bacteroides* (De Filippis et al. 2016). However, the fermentation of dietary fibers to produce butyrate is in turn mainly mediated by the gut microbiota, in particular *Faecalibacterium prausnitzii*, *Eubacterium rectale*,

Roseburia faecis, etc. (Baxter et al. 2019). Therefore, not only does butyrate have a beneficial effect on the intestinal flora, but also the intestinal microbiome can influence the production of butyrate. On the other hand, the association between the microbiota and simple carbohydrates has been poorly studied. A certain correlation between food and CRC is that of excessive consumption of processed meat and, to a lesser extent, red meat (Loke et al. 2020). These effects occur through mechanisms mediated by various substances called xenobiotics that are produced following the ingestion of these foods:

- N-nitrous compounds (NOC). Their main exogenous source includes processed, smoked, or cured meats. NOCs can also be produced in the intestine by the bacterial flora, starting from amino acids. However, NOC-producing bacterial species are found in low concentrations in healthy subjects. The growth of these bacteria is, in fact, favored by the excessive intake of nitrates and nitrites in the diet. This microbial dysbiosis can contribute to colon carcinogenesis (Bouvard et al. 2015; Fahrer and Kaina 2013; Kobayashi 2018).
- Heterocyclic amines (HCAs). They are obtained from the conversion of creatine, amino acids, and sugars contained in meats cooked at high temperatures. These substances could alter the metabolic activities of the microbiota which, in the long term, can disrupt intestinal homeostasis (Helmus et al. 2013; Defois et al. 2018).
- Trimethylamine-N-oxide (TMAO). It is a metabolite produced by the intestinal flora following the ingestion of saltwater fish, eggs, dairy products, and red meats (Wang et al. 2019). When an alteration of the normal intestinal flora occurs, an altered production of this metabolite can also be generated, with an increased risk of developing colic damage (Chan et al. 2019).
- Polycyclic aromatic hydrocarbons (PAHs) are a group of organic compounds formed by two or more benzene rings linked together and arranged in various configurations (Kim et al. 2013). They can be found in the atmosphere but also in foods (oils, vegetables, etc.) that have been subjected to special processing or cooking, such as frying and roasting,. Among these, benzo(a)pyrene (BaP) is the one considered carcinogenic by the IARC. However, it is not possible to define a minimum amount to ingest that can be considered carcinogenic, nor can the exact amount of PAH in individual foods be determined. It is only known that the foods with the greatest amount of these substances are smoked ones and grilled meats (Alomirah et al. 2011).

The GM can interact with food xenobiotics in two ways: xenobiotics influence the composition and activity of microbiota and, in turn, the microbiota can also metabolize and transform xenobiotics. For example, oral exposure to benzo(a) pyrene can cause moderate inflammation of the ileal mucosa and colon and induce alterations in the composition of the GM. In these cases, in fact, there are increases in certain bacterial species, such as *Bacteroidaceae*, and reductions in others, such as *Lactobacillaceae* (Nogacka et al. 2019). In other cases, changes in the metabolome and transcriptome were observed and not in the composition of the GM (Defois et al. 2017).

GM can also metabolize xenobiotics by modifying their toxicity. In some cases, the effect can be in favor of the host, so the GM can favor the excretion of toxic substances or can metabolize them making them less harmful. In other cases, GM determines an enhancement of their toxicity or even transforms them into mutagenic substances (Nogacka et al. 2019).

A change in the toxicity of xenobiotics can also be determined by the interactions between GM and host. One example is the exacerbation of xenobiotic toxicity via the enterohepatic cycle. In the liver, xenobiotics are often conjugated to glucuronic acid, stored in the gallbladder, and released back into the intestine with the bile during digestive processes. Here, the conjugated xenobiotic is broken down from glucuronic acid by bacterial glucuronidases. The xenobiotic is then transformed back into a toxic substance (Humblot et al. 2007).

Another way by which the toxicity of the xenobiotic in the intestine is the alteration of the expression of some host genes by the GM. An example is the hepatic enzyme CYP1A1 (aryl-4 monooxygenase), belonging to the cytochrome P450 family, involved in phase I of hepatic detoxification. It is involved in the detoxification of benzo(a)pyrene and depends on the membrane receptor TLR2, which in turn is activated by bacterial lipoproteins. In fact, various studies suggest that the intestinal microbiota can modulate the xenobiotic metabolism of the host through TLR2 signaling (Carmody and Turnbaugh 2014; Do et al. 2012).

Finally, several studies have shown how a high-fat diet can lead to intestinal dysbiosis, with an increase in *Firmicutes* and a reduction in *Bacteroidetes*. This seems to predispose to the onset of various diseases, including tumors. TMAO or N8-acetylspermidine were detected at high levels in the case of aged microbiota gut, as well as the metabolites of polyamines (putrescine, spermidine, and spermine), which are involved in cell proliferation and differentiation and are also found in patients with CRC (Yoshimoto et al. 2021). In addition to producing these polyamines directly, some intestinal bacteria that are abundant in elderly subjects (such as enterotoxigenic *B. fragilis*) cause an increased synthesis of polyamines by host cells (Louis et al. 2014).

In old age, since changes in the composition of the microbiota occur, there are also changes in the bacterial metabolites produced. SCFA, the major bacterial metabolites in the colon, are lower in the elderly than in young adults. A possible cause is attributable to the fact that in young people the bacterial metabolism is mainly saccharolytic, while in the elderly it is mainly putrefactive (Salazar et al. 2017). Furthermore, with the progress of age, the diet also changes in most cases, both from a quantitative and a qualitative point of view. This is due to multiple factors. In the elderly, in fact, the appetite tends to significantly reduce with consequent adaptation by the organism and a reduction in the demand for calories. Loss of appetite is often due to impaired production of the hormones that regulate the sense of hunger, such as ghrelin or cholecystokinin (Amarya et al. 2015; Remond et al. 2015). This dysregulation not only determines a reduced appetite, but also causes an alteration in the type of food sought. This entails a high risk of malnutrition, a condition that often characterizes the elderly.

Alongside these factors, there are many others, such as loneliness, depression, and polypharmacy. All this also influences the eating style and, consequently, the composition of the GM (Soderstrom et al. 2013).

15.6 Conclusion

In conclusion, it is possible to state that among the various causes of CRC there is also the alteration of the GM. This condition, which can also occur in young people, is more frequent in the elderly. In fact, as we age, events occur that cause GM changes: infections, use of antibiotics, diet changes, etc. Several bacteria have been implicated in the etiopathogenesis of CRC: *F. nucleatum, P. anaerobius, B. fragilis, S. bovis, E. faecalis, E. coli, etc.* Alongside bacterial species, viruses and fungi capable of inducing carcinogenesis have also been identified. All these microbes can induce direct alterations in the DNA of the host cells. They can cause, for example, p53 or pRb mutations. But they can also modify toxic metabolites from certain foods, enhancing their damaging effect. The quantitative and qualitative alterations of the diet that occur in the elderly facilitate intestinal dysbiosis. Therefore, the elderly person, compared to the young, has an increased risk of developing CRC.

References

Adak A, Khan MR (2019) An insight into gut microbiota and its functionalities. Cell Mol Life Sci 76:473–493. https://doi.org/10.1007/s00018-018-2943-4

Allen-Vercoe E, Jobin C (2014) Fusobacterium and Enterobacteriaceae: important players for CRC? Immunol Lett 162:54–61. https://doi.org/10.1016/j.imlet.2014.05.014

Alomirah H, Al-Zenki S, Al-Hooti S, Zaghloul S, Sawaya W, Ahmed N et al (2011) Concentrations and dietary exposure to polycyclic aromatic hydrocarbons (PAHs) from grilled and smoked foods. Food Control 22:2028–2035. https://doi.org/10.1016/j.foodcont.2011.05.024

Amarya S, Sinhg H, Sabharwalh M (2015) Changes during ageind and their association with malnutrition. J Clin Gerontol Geriatr 6:78–84. https://doi.org/10.1016/j.jcgg.2015.05.003

Ambrosio MR, Vernillo R, De Carolis S, Carducci A, Mundo L, Ginori A et al (2019) Putative role of circulating human papillomavirus DNA in the development of primary squamous cell carcinoma of the middle rectum: a case report. Front Oncol 9:93. https://doi.org/10.3389/fonc.2019.00093

Artemev A, Naik S, Pougno A, Honnavar P, Shanbhag NM (2022) The association of microbiome dysbiosis with colorectal cancer. Cureus 14:e22156. https://doi.org/10.7759/cureus.22156

Arumugam M, Raes J, Pelletier E, Le Paslier D, Yamada T, Mende DR et al (2011) Enterotypes of the human gut microbiome. Nature 473:174–180. https://doi.org/10.1038/nature09944

Aveskamp MM, de Gruyter J, Crous PW (2008) Biology and recent developments in the systematics of Phoma, a complex genus of major quarantine significance. Fungal Divers 31:1–18

Bana B, Cabreiro F (2019) The microbiome and aging. Annu Rev Genet 53:239–261. https://doi.org/10.1146/annurev-genet-112618-043650

Bashir A, Miskeen AY, Hazari YM, Asrafuzzaman S, Fazili KM (2016) Fusobacterium nucleatum, inflammation, and immunity: the fire within human gut. Tumour Biol 37:2805–2810. https://doi.org/10.1007/s13277-015-4724-0

Baxter NT, Schmidt AW, Venkataraman A, Kim KS, Waldron C, Schmidt TM (2019) Dynamics of human gut microbiota and short-chain fatty acids in response to dietary interventions with three fermentable fibers. mBio10;10:e02566-e02518. https://doi.org/10.1128/mBio.02566-18

Bennett A, Ponder MM, Garcia-Diaz J (2018) Phoma infections: classification, potential food sources, and its clinical impact. Microorganisms 6:58. https://doi.org/10.3390/microorganisms6030058

Biagi E, Nylund L, Candela M, Ostan R, Bucci L, Pini E et al (2010) Through ageing, and beyond: gut microbiota and inflammatory status in seniors and centenarians. PLoS ONE 5:e10667. https://doi.org/10.1371/journal.pone.0010667

Bischoff SC (2016) Microbiota and aging. Curr Opin Clin Nutr Metab Care 19:26–30. https://doi.org/10.1097/MCO.0000000000000242

Bliss ES, Whiteside E (2018) The gut-brain axis, the human gut microbiota and their integration in the development of obesity. Front Physiol 9:900. https://doi.org/10.3389/fphys.2018.00900

Bouvard V, Loomis D, Guyton KZ, Grosse Y, Ghissassi FE, Benbrahim-Tallaa L et al (2015) Carcinogenicity of consumption of red and processed meat. Lancet Oncol 16:1599–1600. https://doi.org/10.1016/s1470-2045(15)00444-1

Boyajian JL, Ghebretatios M, Schaly S, Islam P, Prakash S (2021) Microbiome and human aging: probiotic and prebiotic potentials in longevity, skin health cell senescence. Nutrients 13:4550. https://doi.org/10.3390/nu13124550

Bray F, Ferlay J, Soerjomataram I, Siegel RL, Torre LA, Jemal A (2018) Global cancer statistics 2018: GLOBOCAN estimates of incidence and mortality worldwide for 36 cancers in 185 countries. CA Cancer J Clin 68:394–424. https://doi.org/10.3322/caac.21492

Brennan CA, Garrett WS (2019) Fusobacterium nucleatum-symbiont, opportunist and oncobacterium. Nat Rev Microbiol 17:156–166. https://doi.org/10.1038/s41579-018-0129-6

Buyru N, Tezol A, Dalay N (2006) Coexistence of K-ras mutations and HPV infection in colon cancer. BMC Cancer 6:115. https://doi.org/10.1186/1471-2407-6-115

Carmody RN, Turnbaugh PJ (2014) Host-microbial interactions in the metabolism of therapeutic and diet-derived xenobiotics. J Clin Investig 124:4173–4181. https://doi.org/10.1172/JCI72335

Chan CWH, Law BMH, Waye MMY, Chan JYW, So WKW, Chow KM (2019) Trimethylamine-N-oxide as one hypothetical link for the relationship between intestinal microbiota and cancer - where we are and where shall we go? J Cancer 10:5874–5882. https://doi.org/10.7150/jca.31737

Cheng WT, Kantilal HK, Davamani F (2020) The mechanism of bacteroides fragilis toxin contributes to colon cancer formation. Malays J Med Sci 27:9–21. https://doi.org/10.21315/mjms2020.27.4.2

Chung L, Orberg ET, Geis AL, Chan JL, Fu K, DeStefano Shields CE et al (2018) Bacteroides fragilis toxin coordinates a procarcinogenic inflammatory cascade via targeting of colonic epithelial cells. Cell Host Microbe 23:203–14.e5. https://doi.org/10.1016/j.chom.2018.01.007

Cianchi F, Cortesini C, Fantappiè O, Messerini L, Schiavone N, Vannacci A et al (2003) Inducible nitric oxide synthase expression in human colorectal cancer: correlation with tumour angiogenesis. Am J Pathol 162:793–801

Claesson MJ, Cusack S, O'Sullivan O, Greene-Diniz R, de Weerd H, Flannery E et al (2011) Composition, variability, and temporal stability of the intestinal microbiota of the elderly. Proc Natl Acad Sci USA 108:4586–4591. https://doi.org/10.1073/pnas.1000097107

Claesson MJ, Jeffery IB, Conde S, Power SE, O'Connor EM, Cusack S et al (2012) Gut microbiota composition correlates with diet and health in the elderly. Nature 488:178–184. https://doi.org/10.1038/nature11319

Coelho TR, Almeida L, Lazo PA (2010) JC virus in the pathogenesis of colorectal cancer, an etiological agent or another component in a multistep process? Virol J 7:42. https://doi.org/10.1186/1743-422X-7-42

Collino S, Montoliu I, Martin FP, Scherer M, Mari D, Salvioli S et al (2013) Metabolic signatures of extreme longevity in northern Italian centenarians reveal a complex remodeling of lipids, amino acids, and gut microbiota metabolism. PLoS ONE 8:e56564. https://doi.org/10.1371/journal.pone.0056564

Dai Z, Coker OO, Nakatsu G, Wu WKK, Zhao L, Chen Z et al (2018) Multi-cohort analysis of colorectal cancer metagenome identified altered bacteria across populations and universal bacterial markers. Microbiome 6:70. https://doi.org/10.1186/s40168-018-0451-2

Dahiya DK, Renuka, (2017) The gut virome: a neglected actor in colon cancer pathogenesis. Future Microbiol 12:1345–1348. https://doi.org/10.2217/fmb-2017-0159

De Almeida CV, Taddei A, Amedei A (2018) The controversial role of Enterococcus faecalis in colorectal cancer. Therap Adv Gastroenterol 11:1756284818783606. https://doi.org/10.1177/1756284818783606

De Almeida CV, de Camargo MR, Russo E, Amedei A (2019) Role of diet and gut microbiota on colorectal cancer immunomodulation. World J Gastroenterol 25:151–162. https://doi.org/10.3748/wjg.v25.i2.151

De Filippis F, Pellegrini N, Vannini L, Jeffery IB, La Storia A, Laghi L et al (2016) High-level adherence to a mediterranean diet beneficially impacts the gut microbiota and associated metabolome. Gut 65:1812–1821. https://doi.org/10.1136/gutjnl-2015-309957

Defois C, Ratel J, Denis S, Batut B, Beugnot R, Peyretaillade E et al (2017) Environmental pollutant benzo[a]pyrene impacts the volatile metabolome and transcriptome of the human gut microbiota. Front Microbiol 8:1562. https://doi.org/10.3389/fmicb.2017.01562

Defois C, Ratel J, Garrait G, Denis S, Le Goff O, Talvas J et al (2018) Food chemicals disrupt human gut microbiota activity and impact intestinal homeostasis as revealed by in vitro systems. Sci Rep 8:11006. https://doi.org/10.1038/s41598-018-29376-9

Do KN, Fink LN, Jensen TE, Gautier L, Parlesak A (2012) TLR2 Controls intestinal carcinogen detoxication by CYP1A1. PLoS ONE 7:e32309. https://doi.org/10.1371/journal.pone.0032309

Douglas LJ (2003) Candida biofilms and their role in infection. Trends Microbiol 11:30–36. https://doi.org/10.1016/s0966-842x(02)00002-1

Ellmerich S, Schöller M, Duranton B, Gossé F, Galluser M, Klein JP et al (2000) Promotion of intestinal carcinogenesis by Streptococcus bovis. Carcinogenesis 21:753–756. https://doi.org/10.1093/carcin/21.4.753

Fahrer J, Kaina B (2013) O6 -methylguanine-DNA methyltransferase in the defense against N-nitroso compounds and colorectal cancer. Carcinogenesis 34:2435–2442. https://doi.org/10.1093/carcin/bgt275

Fang W, Xue H, Chen X, Chen K, Ling W (2019) Supplementation with sodium butyrate modulates the composition of the gut microbiota and ameliorates high-fat diet-induced obesity in mice. J Nutr 149:747–754. https://doi.org/10.1093/jn/nxy324

Fiorentini C, Carlini F, Germinario EAP, Maroccia Z, Travaglione S, Fabbri A (2020) Gut Microbiota and Colon Cancer: A Role for Bacterial Protein Toxins? Int J Mol Sci 21:6201. https://doi.org/10.3390/ijms21176201

Fiorina L, Ricotti M, Vanoli A, Luinetti O, Dallera E, Riboni R et al (2014) Systematic analysis of human oncogenic viruses in colon cancer revealed EBV latency in lymphoid infiltrates. Infect Agent Cancer 9:18. https://doi.org/10.1186/1750-9378-9-18

Fong W, Li Q, Yu J (2020) Gut microbiota modulation: a novel strategy for prevention and treatment of colorectal cancer. Oncogene 39:4925–4943. https://doi.org/10.1038/s41388-020-1341-1

Franceschi C, Bonafè M, Valensin S, Olivieri F, De Luca M, et al (2000) Inflamm-aging. An evolutionary perspective on immunosenescence. Ann NY Acad Sci;908:244–54. doi: https://doi.org/10.1111/j.1749-6632.2000.tb06651.x

Gao R, Kong C, Li H, Huang L, Qu X, Qin N et al (2017) Dysbiosis signature of mycobiota in colon polyp and colorectal cancer. Eur J Clin Microbiol Infect Dis 36:2457–2468. https://doi.org/10.1007/s10096-017-3085-6

Gryseels B (2012) Schistosomiasis. Infect Dis Clin North Am 26:383–397. https://doi.org/10.1016/j.idc.2012.03.004

Hamid HKS (2019) Schistosoma japonicum-Associated Colorectal Cancer: A Review. Am J Trop Med Hyg 100:501–505. https://doi.org/10.4269/ajtmh.18-0807

Han YW (2015) Fusobacterium nucleatum: a commensal-turned pathogen. Curr Opin Microbiol 23:141–147. https://doi.org/10.1016/j.mib.2014.11.013

Hanahan D, Weinberg RA (2011) Hallmarks of cancer: the next generation. Cell 144:646–674. https://doi.org/10.1016/j.cell.2011.02.013

Helmus DS, Thompson CL, Zelenskiy S, Tucker TC, Li L (2013) Red meat-derived heterocyclic amines increase risk of colon cancer: A populationbased case-control study. Nutr Cancer 65:1141–1150. https://doi.org/10.1080/01635581.2013.834945

Humblot C, Murkovic M, Rigottier-Gois L, Bensaada M, Bouclet A, Andrieux C et al (2007) Glucuronidase in human intestinal microbiota is necessary for the colonic genotoxicity of the food-borne carcinogen 2-amino-3-methylimidazo[4,5-f]quinoline in rats. Carcinogenesis 28:2419–2425. https://doi.org/10.1093/carcin/bgm170

Jackson MA, Jeffery IB, Beaumont M, Bell JT, Clark AG, Ley RE et al (2016) Signatures of early frailty in the gut microbiota. Genome Med 8:8. https://doi.org/10.1186/s13073-016-0262-7

Jang J, Hur HG, Sadowsky MJ, Byappanahalli MN, Yan T, Ishii S (2017) Environmental Escherichia coli: ecology and public health implications-a review. J Appl Microbiol 123:570–581. https://doi.org/10.1111/jam.13468

Kim BS, Choi CW, Shin H, Jin AP, Bae JS, Han M et al (2019) Comparison of the Gut Microbiota of Centenarians in Longevity Villages of South Korea with Those of Other Age Groups. J. Microbiol Biotechnol 29:429–440. https://doi.org/10.4014/jmb.1811.11023

Kim KH, Jahan SA, Kabir E, Brown RJ (2013) A review of airborne polycyclic aromatic hydrocarbons (PAHs) and their human health effects. Environ Int 60:71–80. https://doi.org/10.1016/j.envint.2013.07.019

Kim MS, Bae JW (2016) Spatial disturbances in altered mucosal and luminal gut viromes of diet-induced obese mice. Environ Microbiol 18:1498–1510. https://doi.org/10.1111/1462-2920.13182

Knights D, Ward TL, McKinlay CE, Miller H, Gonzalez A, McDonald D et al (2014) Rethinking "enterotypes." Cell Host Microbe 16:433–437. https://doi.org/10.1016/j.chom.2014.09.013

Krishnan S, Eslick GD (2014) Streptococcus bovis infection and colorectal neoplasia: a meta-analysis. Colorectal Dis 16:672–680. https://doi.org/10.1111/codi.12662

Kobayashi J (2018) Effect of diet and gut environment on the gastrointestinal formation of N-nitroso compounds: A review. Nitric. Oxide 73:66–73. https://doi.org/10.1016/j.niox.2017.06.001

LaCourse KD, Johnston CD, Bullman S (2021) The relationship between gastrointestinal cancers and the microbiota. Lancet Gastroenterol Hepatol 6:498–509. https://doi.org/10.1016/S2468-1253(20)30362-9

Lakshminarayanan B, Stanton C, O'Toole PW, Ross RP (2014) Compositional dynamics of the human intestinal microbiota with aging: implications for health. J Nutr Health Aging 18:773–786. https://doi.org/10.1007/s12603-014-0549-6

Langille MG, Meehan CJ, Koenig JE, Dhanani AS, Rose RA, Howlett SE et al (2014) Microbial shifts in the aging mouse gut. Microbiome 2:50. https://doi.org/10.1186/s40168-014-0050-9

Langowski JL, Zhang X, Wu L, Wu L, Mattson JD, Chen T et al (2006) IL-23 promotes tumour incidence and growth. Nature 442:461–465. https://doi.org/10.1038/nature04808

Lee KA, Luong MK, Shaw H, Nathan P, Bataille V, Spector TD (2021) The gut microbiome: what the oncologist ought to know. Br J Cancer 125:1197–1209. https://doi.org/10.1038/s41416-021-01467-x

Legaria MC, Nastro M, Camporro J, Heger F, Barberis C, Stecher D, et al (2021) Peptostreptococcus anaerobius: Pathogenicity, identification, and antimicrobial susceptibility. Review of monobacterial infections and addition of a case of urinary tract infection directly identified from a urine sample by MALDI-TOF MS. Anaerobe;72:102461. doi: https://doi.org/10.1016/j.anaerobe.2021.102461.

Ley RE, Peterson DA, Gordon JI (2006) Ecological and evolutionary forces shaping microbial diversity in the human intestine. Cell 124:837–848. https://doi.org/10.1016/j.cell.2006.02.017

Loke YL, Chew MT, Ngeow YF, Lim WWD, Peh SC (2020) Colon Carcinogenesis: the interplay between diet and gut microbiota. Front Cell Infect Microbiol 10:603086. https://doi.org/10.3389/fcimb.2020.603086

Long X, Wong CC, Tong L, Chu ESH, Ho Szeto C, Go MYY et al (2019) Peptostreptococcus anaerobius promotes colorectal carcinogenesis and modulates tumour immunity. Nat Microbiol 4:2319–2330. https://doi.org/10.1038/s41564-019-0541-3

López-Otín C, Blasco MA, Partridge L, Serrano M, Kroemer G (2013) The hallmarks of aging. Cell 153:1194–1217. https://doi.org/10.1016/j.cell.2013.05.039

Louis P, Hold GL, Flint HJ (2014) The gut microbiota, bacterial metabolites and colorectal cancer. Nat Rev Microbiol 12:661–672. https://doi.org/10.1038/nrmicro3344

Luan C, Xie L, Yang X, Miao H, Lv N, Zhang R et al (2015) Dysbiosis of fungal microbiota in the intestinal mucosa of patients with colorectal adenomas. Sci Rep 5:7980. https://doi.org/10.1038/srep07980

Milani C, Duranti S, Bottacini F, Casey E, Turroni F, Mahony J et al (2017) The First Microbial Colonizers of the Human Gut: Composition, Activities, and Health Implications of the Infant Gut Microbiota. Microbiol Mol Biol Rev 81:e00036-e117. https://doi.org/10.1128/MMBR.00036-17

Mima K, Nishihara R, Qian ZR, Cao Y, Sukawa Y, Nowak JA et al (2016) Fusobacterium nucleatum in colorectal carcinoma tissue and patient prognosis. Gut 65:1973–1980. https://doi.org/10.1136/gutjnl-2015-310101

Minot S, Sinha R, Chen J, Li H, Keilbaugh SA, Wu GD et al (2011) The human gut virome: inter-individual variation and dynamic response to diet. Genome Res 21:1616–1625. https://doi.org/10.1101/gr.122705.111

Mjelle R, Sjursen W, Thommesen L, Sætrom P, Hofsli E (2019) Small RNA expression from viruses, bacteria and human miRNAs in colon cancer tissue and its association with microsatellite instability and tumor location. BMC Cancer 19:161. https://doi.org/10.1186/s12885-019-5330-0

Nogacka AM, Gómez-Martín M, Suárez A, González-Bernardo O, de Los Reyes-Gavilán CG, González S (2019) Xenobiotics Formed during Food Processing: Their Relation with the Intestinal Microbiota and Colorectal Cancer. Int J Mol Sci 20:2051. https://doi.org/10.3390/ijms20082051

Ogilvie LA, Jones BV (2015) The human gut virome: a multifaceted majority. Front Microbiol 6:918. https://doi.org/10.3389/fmicb.2015.00918

Oleynikova NA, Danilova NV, Grimuta MO, Malkov PG (2021) Epstein-Barr Virus in the Development of Colorectal Cancer (Review). Sovrem Tekhnologii Med 13:82–91. https://doi.org/10.17691/stm2021.13.4.09

Palmer C, Bik EM, DiGiulio DB, Relman DA, Brown PO (2007) Development of the human infant intestinal microbiota. PLosBiol 5:e177. https://doi.org/10.1371/journal.pbio.0050177

Penders J, Thijs C, Vink C, Stelma FF, Snijders B, Kummeling I et al (2006) Factors influencing the composition of the intestinal microbiota in early infancy. Pediatrics 118:511–521. https://doi.org/10.1542/peds.2005-2824

Petersen C, Round JL (2014) Defining dysbiosis and its influence on host immunity and disease. Cell Microbiol 16:1024–1033. https://doi.org/10.1111/cmi.12308

Qin X, Gu Y, Liu T, Wang C, Zhong W, Wang B et al (2021) Gut mycobiome: A promising target for colorectal cancer. Biochim Biophys Acta Rev Cancer 1875:188489. https://doi.org/10.1016/j.bbcan.2020.188489

Qiu DC, Hubbard AE, Zhong B, Zhang Y, Spear RC (2005) A matched, case-control study of the association between Schistosoma japonicum and liver and colon cancers, in rural China. Ann Trop Med Parasitol 99:47–52. https://doi.org/10.1179/136485905X19883

Ramirez-Garcia A, Rementeria A, Aguirre-Urizar JM, Moragues MD, Antoran A, Pellon A et al (2016) Candida albicans and cancer: Can this yeast induce cancer development or progression? Crit Rev Microbiol 42:181–193. https://doi.org/10.3109/1040841X.2014.913004

Rampelli S, Candela M, Turroni S, Biagi E, Collino S, Franceschi C et al (2013) Functional metagenomic profiling of intestinal microbiome in extreme ageing. Aging 5:902–912. https://doi.org/10.18632/aging.100623

Remond D, Shahar DR, Gille D, Pinto P, Kachal J, Peyron MA et al (2015) Understanding the gastrointestinal tract of the elderly to develop dietary solutions that prevent malnutrition. Oncotarget 6:13858–13898. https://doi.org/10.18632/oncotarget.4030

Rinninella E, Raoul P, Cintoni M, Franceschi F, Miggiano GAD, Gasbarrini A et al (2019) What is the Healthy Gut Microbiota Composition? A Changing Ecosystem across Age, Environment, Diet, and Diseases. Microorganisms 7:14. https://doi.org/10.3390/microorganisms7010014

Salazar N, Valdés-Varela L, González S, Gueimonde M, de Los Reyes-Gavilán CG (2017) Nutrition and the gut microbiome in the elderly. Gut Microbes 8:82–97. https://doi.org/10.1080/19490976.2016.1256525

Sinagra E, Raimondo D, Gallo E, Stella M, Cottone M, Orlando A et al (2014) Could JC virus provoke metastasis in colon cancer? World J Gastroenterol 20:15745–15749. https://doi.org/10.3748/wjg.v20.i42.15745

Soderstrom L, Thors Adolfsson E, Rosenblad A, Frid H, Saletti A, Bergkvist L (2013) Mealtime habits and meal provision are associated with malnutrition among elderly patients admitted to hospital. Clin Nutr 32:281–288. https://doi.org/10.1016/j.clnu.2012.07.013

Starý L, Mezerová K, Vysloužil K, Zbořil P, Skalický P, Stašek M et al (2020) Candida albicans culture from a rectal swab can be associated with newly diagnosed colorectal cancer. Folia Microbiol (praha) 65:989–994. https://doi.org/10.1007/s12223-020-00807-3

Tafvizi F, Fard ZT, Assareh R (2015) Epstein-Barr virus DNA in colorectal carcinoma in Iranian patients. Pol J Pathol 66:154–160. https://doi.org/10.5114/pjp.2015.53012

Thevaranjan N, Puchta A, Schulz C, Naidoo A, Szamosi JC, Verschoor CP et al (2017) Age-associated microbial dysbiosis promotes intestinal permeability, systemic inflammation, and macrophage dysfunction. Cell Host Microbe 21:455-466.e4. https://doi.org/10.1016/j.chom.2017.03.002

Wang Z, Bergeron N, Levison BS, Li XS, Chiu S, Jia X et al (2019) Impact of chronic dietary red meat, white meat, or non-meat protein on trimethylamine N-oxide metabolism and renal excretion in healthy men and women. Eur Heart J 40:583–594. https://doi.org/10.1093/eurheartj/ehy799

Weiss GA, Hennet T (2017) Mechanisms and consequences of intestinal dysbiosis. Cell Mol Life Sci 74:2959–2977. https://doi.org/10.1007/s00018-017-2509-x

Woodmansey EJ (2007) Intestinal bacteria and ageing. J Appl Microbiol 102:1178–1186. https://doi.org/10.1111/j.1365-2672.2007.03400.x

Wu GD, Chen J, Hoffmann C, Bittinger K, Chen Y-Y, Keilbaugh SA et al (2011) Linking long-term dietary patterns with gut microbial enterotypes. Science 334:105–108. https://doi.org/10.1126/science.1208344

Wu X, Wu Y, He L, Wu L, Wang X, Liu Z (2018) Effects of the intestinal microbial metabolite butyrate on the development of colorectal cancer. J Cancer 9:2510–2517. https://doi.org/10.7150/jca.25324

Yoshimoto S, Mitsuyama E, Yoshida K, Odamaki T, Xiao JZ (2021) Enriched metabolites that potentially promote age-associated diseases in subjects with an elderly-type gut microbiota. Gut Microbes 13:1–11. https://doi.org/10.1080/19490976.2020.1865705

Yu T, Guo F, Yu Y, Sun T, Ma D, Han J et al (2017) Fusobacterium nucleatum promotes chemoresistance to colorectal cancer by modulating autophagy. Cell 170:548–63.e516. https://doi.org/10.1016/j.cell.2017.07.008

Xie K (2011) Interleukin-8 and human cancer biology. Cytokine Growth Factor Rev 12:375–391. https://doi.org/10.1016/s1359-6101(01)00016-8

Zhang L, Zhan H, Xu W, Yan S, Ng SC (2021) The role of gut mycobiome in health and diseases. Therap Adv Gastroenterol 14:17562848211047130. https://doi.org/10.1177/17562848211047130

Part IV
Interventions on Gut Microbiota

Chapter 16
Gerobiotics: Probiotics for Healthy Aging

Ying-Chieh Tsai, Sabrina Wang, Li-Hao Cheng, One-Jang Jeng, and Francesco Marotta

Abstract Aging, despite its ubiquity, and its underlying mechanisms are still not well understood. A collection of biological processes known as hallmarks of aging offer to explain this unavoidable phenomenon, while the interdisciplinary field of geroscience strives to define its physiological mechanisms at both a molecular and cellular level. Geroscience hypothesizes that comprehensively targeting these hallmarks can prevent aging-related pathogenesis, therefore, improving the healthspan of most people. There is also growing awareness that probiotics can be used prophylactically to maintain health and improve life quality among the elderly. Rapidly advancing geroscience research includes a focus on anti-aging probiotics. If extensive preclinical and clinical research confirms a particular probiotic's efficacy to truly attenuate the fundamental mechanisms of aging, reduce physiological aging processes, and thereby extend the healthspan of the host, then this strain along with its derived postbiotics and para-probiotics could be designated as "gerobiotics". This chapter highlights the significance of the novel field of gerobiotics, epitomizing biomarkers of aging and recommending a path by which to discover gerobiotic strains and develop their ensuing products. Finally, we will provide several examples of potential gerobiotics. By exploiting these unique strains, we could formulate interventional strategies that slow or even reverse expressions of aging, improving healthspan and longevity for future generations.

Y.-C. Tsai (✉)
Institute of Biochemistry and Molecular Biology, National Yang Ming Chiao Tung University, Taipei 11221, Taiwan
e-mail: tsaiyc66@gmail.com

S. Wang
Institute of Anatomy and Cell Biology, National Yang Ming Chiao Tung University, Taipei 11221, Taiwan

L.-H. Cheng · O.-J. Jeng
Bened Biomedical Co., Ltd., Taipei 10448, Taiwan

F. Marotta
ReGenera R&D International for Aging Intervention and Longevity & Vitality in Medical Science Commission, FEMTEC World Federation, Milan, Italy

F. Marotta (ed.), *Gut Microbiota in Aging and Chronic Diseases*, Healthy Ageing and Longevity 17, https://doi.org/10.1007/978-3-031-14023-5_16

Keywords Gerobiotics · Geroscience · Biomarkers · *Lactobacillus paracasei* PS23

16.1 Introduction

Today's global population is, on average, older than ever before, with life expectancy reaching a mean of 73.4 years in 2019 (World Health 2019). This is an increase of more than six years since 2000, when the average person could expect to live for 66.8 years. Declining mortality is a major contributor to this rapid aging, with those over 65 years of age making up 8.7% of the population (World Health 2019). This "Silver Tsunami," as the trend is often referred to, is a potential threat to public healthcare systems around the world. The National Council on Aging estimates four in five seniors have one chronic disease, while 77% suffer from at least two (National Council on Aging). Heart disease, stroke, cancer, and diabetes are the most seen ailments, altogether contributing to about two-thirds of deaths. Mental health conditions, including dementia and depression, also severely impact health of the elderly.

It's often discussed whether aging itself leads to an increase in chronic illness or if these diseases are what accelerate the aging process. Geroscience is a new field of study exploring the basic physiological mechanisms of conditions and diseases that contribute to senescence (Kennedy et al. 2014; Yabluchanskiy et al. 2018; Seals et al. 2016; Sierra 2016). In the past, combating aging mainly focused on the healthcare of those suffering from age-related ailments. Geroscience promotes targeting the underlying biological mechanisms of aging, which will simultaneously delay the onset of both disease and senescence to extend not only lifespan but also healthspan. An important concept is to begin active aging prevention as early as possible, even before obvious signs have begun to appear. In the past, it was typically only after symptoms of elevated blood pressure, blood sugar, or cholesterol levels began to manifest that individuals—usually in their 50 s and 60 s—became cognizant of needing to pay special care to age-related health issues. Declining physical strength was the rule rather than the exception. Geroscience, in contrast, emphasizes undertaking aging-prevention measures as early as in one's 40 s or even 30 s.

As long as a whole century ago, Elie Metchnikoff suggested that the onset of aging could potentially be delayed by improving the intestinal microbiome by consuming yogurt that contained beneficial bacteria (Mackowiak 2013). Metchnikoff consequently has been revered as the father of probiotics, which are not only seen today as one of the most influential aspects of human health but have also grown into a multi-billion-dollar industry. These beneficial microorganisms have also become an important biomedical research field of their own. According to Pubmed, more than 5000 journal articles on the topic of probiotics were published in 2021 alone.

This chapter will first discuss microbiome dysbiosis as a hallmark of aging. It will then introduce the newly proposed concept of gerobiotics (Tsai et al. 2021) and discuss how to effectively develop this novel class of probiotics, and then several examples of potential gerobiotics bacterial strains will be provided.

16.2 The Gut Microbiome as a Hallmark of Aging

In 2013, a milestone review by López-Otín et al. (2013) was published, in which nine hallmarks of aging were outlined. These hallmarks, the underlying biological mechanisms that advance the aging process, are divided into three categories. First are primary hallmarks, which cause direct cellular damage and are comprised of genomic instability, telomere attrition, epigenetic alterations, and loss of proteostasis. Next are antagonistic hallmarks that are responses to the primary damage: deregulated nutrient sensing, mitochondrial dysfunction, and cellular senescence. Finally, integrated hallmarks occur as a result of the two previous groups. They include stem cell exhaustion as well as altered intercellular communication and are responsible for the aging of a complete organism.

Each hallmark was selected based on the following criteria: (1) manifests during normal aging, (2) accelerates aging if aggravated experimentally, and (3) retards the normal aging process (i.e., increases healthspan) during experimental amelioration. Though the scientific world generally recognizes these nine hallmarks, the proposal is not without critics. For example, Gems and de Magalhaes (2021) argue the hallmarks are somewhat arbitrary, failing to serve as a framework that explains the actual causes of aging and the range of diseases it can bring.

Mounting evidence points toward the role the gut microbiome plays in longevity, and consequently, microbiome dysbiosis has been proposed in several studies as an additional hallmark (Carter 2021; Guerville et al. 2020; Bana and Cabreiro 2019). However, to truly be considered at the same level, microbiome dysbiosis must be assessed by the same aforementioned three criteria put forth by López-Otín et al. (2013). First of all, do dysbiosis manifest during the normal aging process? Both clinical and preclinical evidence indeed indicates this is the case. Furthermore, dysbiosis of the microbiome triggers a chain of inflammatory and pathological events, all highly correlated with aging-associated morbidities (Ragonnaud and Biragyn 2021; Kim and Benayoun 2020; Kim and Jazwinski 2018). Next, do experiments in which the microbiota is either aggravated or ameliorated result in respective acceleration or delay of the aging process, thus affecting healthy lifespan duration? Several animal model studies have shown that age-related microbiota can play a role in the lifespan of the host. Recolonizing the gut of a middle-aged African turquoise killifish with microbiota from a young fish extended the older fish's lifespan, delayed its behavioral decline, and resulted in its own gut microbiota resembling that of a more youthful fish (Smith et al. 2017). In another experiment, fecal microbiota was transplanted from wild-type mice into progeroid model mice. The transplant recipients had greater α diversity, more probiotic genera overall, and higher populations of short-chain fatty acid-producing genera than the control group (Bñrcena et al. 2019). These results indicate modulation of gut microbiota may slow the aging process, lengthen lifespan, and even rejuvenate certain aspects of health (Chen et al. 2020). The evidence suggests microbiome dysbiosis truly is qualified as an aging hallmark and does play a causal role in the aging process. Into which of the aging hallmark categories (primary, antagonistic, or integrated) should

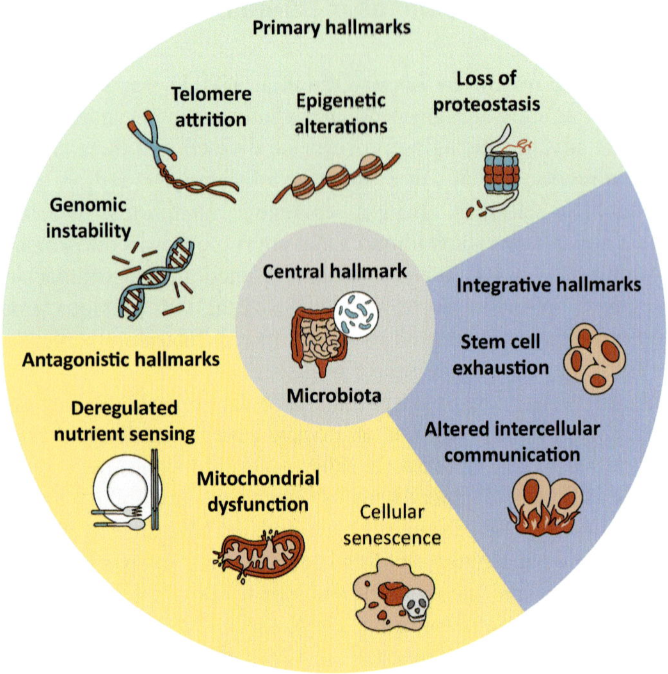

Fig. 16.1 The hallmarks of aging. The relationship between hallmarks of aging and microbiota

microbiota dysbiosis be grouped? The microbiota is well known for its intricate roles in regulating many physiological functions of the host and is intimately linked to the other aging hallmarks (Bana and Cabreiro 2019). We thus suggest that microbiota dysbiosis should be positioned as a central hallmark of aging (Fig. 16.1).

16.3 Gerobiotics: A New Term for a Novel Concept

In 2014, the International Scientific Association for Probiotics and Prebiotics (ISAPP) released a consensus statement defining the term "probiotic" (Hill et al. 2014). The paper stated that among the mechanisms of action by which probiotics confer benefits to their hosts, several are widespread among nearly all strains. These include acid and SCFA production, resistance to colonization by harmful strains, normalization of perturbed microbiota, and competitive exclusion of pathogens. In contrast, some mechanisms are strain-specific, such as the production of specific bioactive molecules and unique immunological, neurological, or endocrinological effects. These functions can be extremely rare and only discovered or demonstrated through rigorously designed studies; therefore, these distinct strains are sometimes given specific designations, e.g., those with immunological effects are called immunobiotics (Clancy 2003) and those with neurological or mental health benefits are termed psychobiotic (Dinan et al. 2013).

"Gerobiotics" have been defined as "probiotic strains and their derived postbiotics and para-probiotics that are able to beneficially attenuate the fundamental mechanisms of aging, reduce physiological aging processes, and thereby expand the healthspan of the host" (Tsai et al. 2021). If providing benefits to the quality of life in aging was the only requisite to be a gerobiotic, most probiotics could safely assume the title. However, the scope of what qualifies as a gerobiotic should be narrower, specifically those strains capable of attenuating the underlying physiological mechanisms of the aging process, ultimately resulting in the host's extended healthspan. A strain that possesses specific immune-modulatory, anti-inflammatory, or infection-reducing properties is not necessarily a gerobiotic. To be classified as a gerobiotic, a strain must demonstrate lifespan expansion effects. Early on, these effects may be seen in *Drosophila melanogaster* or *Caenorhabditis elegans* models. Further evidence of actual aging marker improvement must be shown in rodent studies, possibly expressed through improved aged appearance, cognition, muscle strength, etc. Certain well-studied probiotic strains with anti-aging properties should be newly classified as gerobiotics. Here, we offer two main reasons for doing so.

16.3.1 Rapid Growth in the Field of Geroscience

Recent increasing interest in the basic mechanisms of aging and how they can be utilized to prevent and/or treat age-related chronic illness is the foundation of an emerging field known as geroscience. As a result, "aging-related" has been given its own designated code (XT9T) in the latest 2018 version of the WHO's International Classification of Diseases (ICD-11). This code can be used in conjunction with any other condition found in ICD-11 (e.g., aging-related gastrointestinal tract dystrophy, aging-related mental disorders) and supports the concept of pathological aging. It is an early, yet important, step on the path to recognizing aging as a disease. Were this the case, aging could then be an indication for investigating new drugs in clinical trials. Geroscience and the potential of related therapies have attracted wide attention. Pharmaceutical companies are racing to develop effective and safe geroprotectors (Moskalev et al. 2017; Anis et al. 2022), while scientists are reconsidering the anti-aging functions of metformin, rapamycin, and other long-used medications.

Probiotics have played a long-standing role in this area. Aside from supporting the overall health and wellness of elderly individuals, probiotics are also used to prevent many specific conditions associated with aging. Therefore, an entirely new term "gerobiotics" for specific strains that act directly on primary aging mechanisms will help facilitate their development.

16.3.2 A Reassessed Marketing Strategy for Gerobiotics

Consumers who purchase conventional probiotics for gut health have come to anticipate rather rapid effects. They expect improved gastrointestinal function in a matter of days. Immunobiotics and psychobiotics understandably take longer to work, relieving allergies and balancing mental health, respectively, over the course of weeks or even months. However, gerobiotics that aim to slow the aging process could take years or even longer before their anti-aging effects can be realized.

Therefore, a different strategy is warranted to market this new class of probiotics. A crucial initial step is simply promoting the use of the term gerobiotics. It communicates a distinction between these and other probiotics, conveying that these are the strains directly associated with aging processes. Accordingly, their effects will only be seen after persistent, extended use. Marketing the benefits of gerobiotics should emphasize improvements that are more easily perceived, such as memory and physical strength. As blood-based biomarker panels that measure key indicators of aging are being used more and more, gerobiotic research and development should incorporate biomarkers and help determine a blood panel that objectively measures how gerobiotic strains modulate mechanisms of aging. These types of science-based results will help build consumer confidence in the unique effectiveness of gerobiotics.

16.4 Biomarkers for Identifying and Evaluating Gerobiotic Candidates

The search for a gerobiotic strain and to prove its prolongevity effects is time-consuming and expensive, especially when using human clinical trials. Therefore, a biomarker panel that can be analyzed relatively easily and can extensively and truly reflect the aging process must be included in research designs. A good aging biomarker must translate from invertebrate to vertebrate models, as well as from preclinical to clinical studies, and it must clearly indicate changes that occurred in physiological mechanisms of aging. Furthermore, biomarkers used in clinical trials must be relatively simple to test, and sample requisition must be non-invasive (e.g., taken from blood, urine, or saliva), facilitating multiple tests that track aging progression over the course of a long-term trial. Many blood-based and tissue-based biomarkers have already been reported, and the most comprehensive list of aging-related indicators can be found in the Digital Ageing Atlas portal's database (http://ageing-map.org).

Table 16.1 illustrates four of the recently reported blood-based aging biomarker panels. Since they can be tested from a simple blood draw, they can conveniently be used in gerobiotic preclinical experiments and clinical trials. Their compiled data can also facilitate the promotion of future gerobiotic products on the market. The biomarkers that we feel are most appropriate for use in studying the efficacy and

Table 16.1 Potential aging biomarkers

		High priority	Medium priority
TAME trail (Justice et al. 2018)	Inflammation	**IL-6, CRP, TNFα**	Eotxin, CMV antibody
	Nutrient sign	**IGF-1, Insulin, IGFBPs**	
	Cardio vascular	**NT-proBNP**	Fibrinogen, AGE/rAGE, tPA Isoprostane
	Stress response	**GDF15**	TSH
	Others	**Cystatin C, DHEAS, Telomere Length, Adiponectin, Leptin**	
Cardoso et al. (2018)	Inflammation	CXCL10, CX3CL1, IL-6	Pentraxin, sVCAM/sICAM, **Defensin**
	Mitochondria	**GDF15, FDNC5**, Vimentin	**APP**, LDH
	Ca homeostasis	**Regucalcin**, Calreticulin	S100B
	Fibrosis	PLAU, AGT	TGF-β, PAI-1, TGM2 sRAGE, **HMGB1**, C3/Caq, ST2, **Agrin**
	NMJ & Neurons	**BDNF**, Progranulin	**Leptin, IGF1, Resistin, Adiponectin, Ghrelin, GH**
	Cytoskeleton	α-KL, FGF23, FGF21	GpnmB, Microparticle panel, Lactoferrin
	Others	**miRNA**, AHCY, KRT18	
Wagner et al. (2016)	Physical and anthropometric	Walking speed, Chair stand, Standing balance, Grip strength, BMI, Waist circumference, Muscle mass	
	Blood-based	**IL-6, TNF-α, CRP, HbA1C**, glu, **Adipokines**, TSH, **Vit D, NT-proBNP**	
	DNA based	**Telomere length**, DNA repair, DNA/chromosomal damage	
	Novel	**Bilirubin**, AGEs, **MTs**, DNA methylation, **miRNA**	
Guerville et al. (2020)	Genomic instability: Micronucleus assay	Telomere attrition: **Telomere length**	
	Epigenetic alterations: DNA methylation	Loss of proteostasis: **Clusterin**	
	Nutrient sensing: **Sirtuin 1**	Mitochondrial dysfunction: **GDF-15**, Apelin	
	Cellular Senescence: **P16INK4A**	Stem cell exhaustion: COP	
	Inter. Communication: Inflammasomes, IMM-AGE score		

The chosen biomarkers for gerobiotics development are in boldface. TSH: Thyroid hormones, AGEs: Advanced glycation end products, MTs: Metallothioneins, COP: Circulating osteogenic progenitors, NMJ: neuromuscular junction, Inter. Communication: intercellular communication, CMV: Cytomegalovirus, FNDC5: Fibronectin Type III Domain Containing 5, APP: Amyloid precursor protein, GH: Growth hormone, AHCY: Adenosylhomocysteinase, IGFBPs: insulin-like growth factor-binding protein, LDH: Lactate Dehydrogenase, BDNF: Brain-derived neurotrophic factor, HMGB1: High mobility group box 1, GDF-15: Growth and differentiation factor 15, TGM2: Transglutaminase 2, GpnmB: Glycoprotein Nmb, CRP: C-Reactive Protein, tPA: Tissue plasminogen activator developing gerobiotics are printed in bold. The first set is the group of biomarkers selected by the TAME (Targeting Aging with Metformin) trial. The TAME trial is a landmark clinical geroscience study managed by the American Federation for Aging Research (AFAR). Over the course of six years, it will continue to test 3000 subjects, aged 65–80, who have slowed gait speeds or age-related diseases (Justice et al. 2018). From among the trial's 258 candidate molecules associated with aging and aging-related diseases, nine were selected as high-priority and 11 as medium-priority blood-based markers. The nine high-priority blood biomarkers will be used to assess metformin's anti-aging effects. AFAR expects the TAME trial to provide a proof of concept for testing the geroscientific hypothesis in humans.

The other three sets of biomarkers in Table 16.1 are more suitable for smaller-scale clinical studies. Cardoso et al. identified 44 fundamental biomarker panels, among which 19 were considered high priority, 22 medium priority, and three low priority (Cardoso et al. 2018). Wagner et al. selected 9 blood-based, 3 DNA-based, and several relatively novel aging biomarkers (Wagner et al. 2016). Physical functions and anthropometric markers, such as walking speed, chair standing, standing balance, grip strength, BMI, waist circumference, and muscle mass, were included as useful sensors with which to measure aging. miRNAs were chosen by both Cardoso et al. and Wagner et al. as novel biomarkers. As revealed by a growing number of studies, miRNAs are potential aging and cellular senescence indicators (Kinser and Pincus 2020; Kumar et al. 2017; Teodori et al. 2019). Multiple miRNAs have been suggested as useful biomarkers of aging and aging-associated pathologies. These include miR-21, miR-34a (Chua and Tang 2019), miR-146a (Mensa et al. 2019), miR-92a-5p, miR-532-5p (Carini et al. 2021), miR-101-3p, miR-142-5p (Carini et al. 2022), and miR-486 (Liu et al. 2021). For each hallmark of aging, Guerville et al. designated one or two biomarkers that are closely related to mortality, frailty, or a particular aging-related disease (Guerville et al. 2020). For example, the micronucleus assay, often used to test genotoxicity, was selected to assess genomic instability. Circulating osteogenic progenitor (COP) cells, which have been suggested as a proxy for the mesenchymal stem cell population, were selected to measure the depletion of stem cells. Inflammasomes and the novel-method IMM-AGE score were designated to evaluate chronic inflammation. Once again, the four sets of biomarkers in Table 16.1 are blood-based, relatively easy to analyze at a reasonable cost, and associated with aging or aging-related diseases as demonstrated in animal or clinical trials. Those in bold are the markers we feel are best for use in the development of gerobiotics. Additionally, in the continual search for new aging-related biomarkers, Almeida et al. recently suggested serum lipids as an appropriate target (Almeida et al. 2021). More attention is being given to the application of multi-omics technologies, including lipidomics, in biomarker discovery (Rivero-Segura et al. 2020; Almeida et al. 2021).

We expect the number of aging biomarkers for gerobiotic discovery and research will continue to grow. Nevertheless, it is still unclear whether the aging-related biomarkers will show visible changes during short-term (e.g., three to six months) gerobiotic interventions. Successful preclinical identifying of aging-related biomarkers that respond to short-term gerobiotic interventions will greatly facilitate future investigation by allowing for use of the same biomarkers during clinical trials.

16.5 Development of Potential Gerobiotic Strains

As already mentioned, possessing solely immune-boosting or anti-inflammatory benefits does not sufficiently qualify a probiotic strain as a gerobiotic. To be designated as such, we recommend a three-step, systematic approach. First, functional-probiotic strain banks should be screened via appropriate cell or invertebrate models to determine strains with anti-aging potential. This step ideally will utilize a *D. melanogaster* or *C. elegans* model to show a strain's ability to effectively extend its host's lifespan. Second, it must also be proved in rodent models to ameliorate biological aging markers such as senescent appearance, cognition, and muscle strength. Preferably, the results shown in these animal model studies will identify which specific aging fundamental process is mitigated by each potential strain. Third, candidate strain gerobiotic function must be confirmed through attentively designed clinical studies. Evidence of objective improvement to aging, such as biomarkers like those listed in Table 16.1, is essential to gaining consumer confidence in gerobiotics. Data collected from preclinical and clinical trials must clearly convey the benefits of extended healthspan or reduction of aging-related chronic disease. Only carefully formulated, systematic preclinical and clinical research will yield the necessary biomarker results. Provided with this type of scientific data, consumers will be convinced of the benefits that can be obtained from the long-term use of these products. Here, we propose detailed strategies for gerobiotic development (Fig. 16.2) which are based on the abovementioned research steps.

(1) To facilitate the screening process, begin with currently available probiotic strains with specific functions. Their effects can be immunomodulatory, anti-inflammatory, anti-allergic, anti-infective, psychiatric, etc.
(2) High-throughput screening of a functional-probiotic bank can take advantage of in-vitro cell models developed especially for screening small-molecule geroprotectors (Auclair et al. 2020; Chen et al. 2019a; Xu et al. 2021).
(3) Despite being more complicated and time-consuming than cell models, invertebrate animal models (e.g., *C. elegans* or *D. melanogaster*) can be employed as a primary or secondary method to screen candidate strains.
(4) The core of any screening pipeline should be rodent models, and each potential strain should be evaluated by multiple designs. While most will use aged animals, even normal naïve mice or rats can be studied and changes in aging-related biomarkers analyzed, such as in the example of *Bifidobacterium breve* B-3 (Toda et al. 2020).

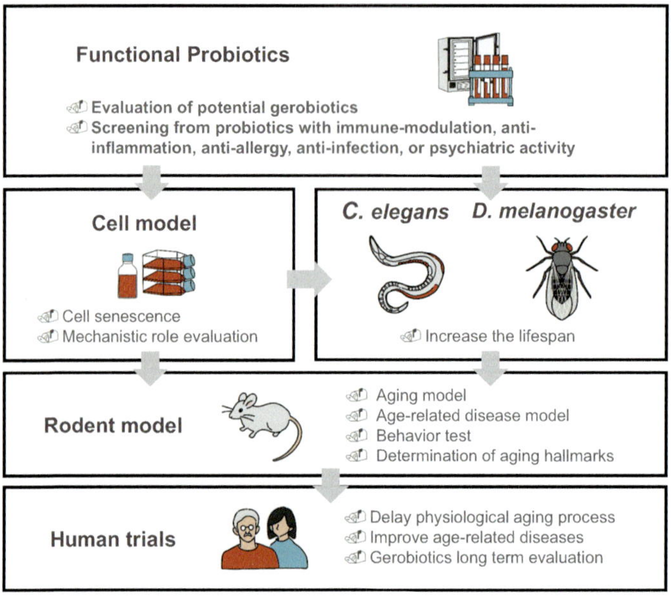

Fig. 16.2 Strategies for gerobiotic development

Focus may tend to be on biomarkers affected by microbes residing in the gastrointestinal tract. After all, that is where probiotics usually remain, from the time they are ingested orally until excreted fecally. However, markers that are seemingly unrelated to the digestive tract should also be considered. Numerous potential biomarkers, all described in published articles, are listed in Table 16.1. As many as possible must be analyzed to explore the maximum number of possible biological pathways involved in each candidate strain's anti-aging activities, thus greatly facilitating subsequent clinical study design. Apart from established aging-rodent models, aging-related disease models (e.g., Parkinson's disease, Alzheimer's disease, and frailty models) could also be used for further validating the candidate strains' efficacy. Although aging does not necessarily cause diseases and mitigating the disease symptom is not the main focus of gerobiotics, beneficial effects on aging-related disease are certainly a welcomed property of gerobiotics.

(5) Success or failure of human trials will ultimately determine a strain's potential. Observing positive changes in measures of muscle mass, grip strength, bone mineral density, or gait velocity would prove very convincing. Still, these changes are unlikely to be demonstrated in the short term, and within the context of a clinical study, it is impractical to expect a strain to be shown effective at significantly extending lifespan. Therefore, blood-based or other biological biomarkers, first selected according to animal models, are essential for developing potential gerobiotics.

16.6 Selected Probiotics as Potential Gerobiotics

The concept of gerobiotics might be new; however, the quest for "anti-aging" probiotics has been ongoing for more than a decade. Several specific probiotics that display beneficial health effects on age-related dysfunctions have been identified in preclinical and clinical studies [for review see (Tsai et al. 2021)].

One of the better studies of anti-aging probiotics is *Lactobacillus paracasei* PS23 (PS23). PS23 was isolated from healthy human feces. This strain has been reported as psychobiotics, in either the maternal separation early life stress mice model (Liao et al. 2019) or the chronic corticosterone-treated mice depression model (Wei et al. 2019). Both live and heat-killed PS23 effectively ameliorated the depression-and anxiety-like behaviors induced by maternal separation or chronic corticosterone treatment. The anti-aging properties of PS23 had been tested in the senescence-accelerated mouse prone 8 (SAMP8) model and D-galactose-induced aging mice model. SAMP8 mouse model is generated by selective breeding whereas D-galactose-induced aging model is based on the production of reactive oxygen species (ROS) generated by excessive intake of D-galactose. The ROS could facilitate cellular aging. Long-term PS 23 administration could modulate gut microbiota composition and improve gastrointestinal function in SAMP8 mice. The PS23-treated aging mice displayed longer intestine length, lower intestine permeability, higher phagocytosis, and lower pro-inflammatory cytokines and immunoglobulins in the intestinal mucosa, suggesting better intestine integrity and function (Chen et al. 2021). Furthermore, PS23 supplementation could reduce age-related loss of muscle mass and strength (Chen et al. 2019b; Cheng et al. 2021). In particular, the mitochondrial function in muscle was significantly improved in PS23 supplemented group, and the relative mRNA expressions of genes related to mitochondrial biogenesis and mitochondrial DNA copy number were increased (Chen et al. 2019b). Mitochondria dysfunction is one of the major contributing factors of sarcopenia. The improvement in mitochondrial function by PS23 certainly countered one of the hallmarks of aging which is mitochondrial dysfunction. In the aging population, sarcopenia frailty is associated with disability and mortality rate (Cooper et al. 2010). Improvement in muscle mass and strength would greatly improve healthy living in the elderly.

Mental health is another important issue of healthy living in the elderly. Preclinical studies had demonstrated that in addition to physical strength, PS23 could also improve memory function and reduce anxiety-like behaviors in aged mice. In both the SAMP8 model and D-galactose-induced aging model, PS23 supplementation alleviated aging-induced cognitive decline (Huang et al. 2018; Cheng et al. 2022). More specifically, PS23 supplementation improved the memory function in the Morris water maze test. This enhancement was associated with the increased brain-derived trophic factor (BDNF) and monoamine levels in the hippocampus (Huang et al. 2018; Cheng et al. 2022). The anti-oxidative enzymes superoxide dismutase (SOD) and glutathione peroxidase in the hippocampus of aged mice were also significantly increased following PS23 supplementation.

Further hippocampal RNA-seq analysis revealed that PS23 supplementation upregulated several genes involved in protection against inflammation, neural plasticity, and adaption to psychological stress (Cheng et al. 2022). In addition, the PS23 administration also reduced anxiety-like behaviors in aged mice (Huang et al. 2018; Cheng et al. 2022). All in all, the anti-oxidant, anti-inflammatory, and anxiolytic effects could all contribute to the cognitive enhancing effects of PS23. Thus, PS23 supplementation demonstrated great potential as treatment to alleviate emotional and cognitive deficits in the aging population. One unique feature of PS23 was that many beneficial effects were also observed in heat-killed PS23 supplementation, and this feature greatly prolonged the shelf-life for commercial distribution. Currently, there were two ongoing clinical trials examining the effects of PS23 supplementation on treating sarcopenia (ClinicalTrials, gov identifier: NCT0402087) and mild cognitive impairment (ClinicalTrials, gov identifier: NCT04971096). The sarcopenia trial has been completed and the manuscript is under preparation. The mild cognitive impairment trial is currently recruiting research subjects.

A number of other probiotics also displayed memory-enhancing effects in aged rodents, for example, *Lactobacillus brevis* OW38 (Jeong et al. 2016), *Lactobacillus pentosus var. plantarum* C29 (C29) (Jeong et al. 2015; Woo et al. 2014), *Lactobacillus plantarum* NDC 75,017 (Peng et al. 2014), and *L. plantarum* GKM3 (GKM3) (Lin et al. 2021). Among these probiotics, memory-enhancing effects of C29 had been well studied in several animal models (Jeong et al. 2015; Jung et al. 2012; Lee et al. 2018; Woo et al. 2014). Long-term C29 administration in D-galactose-induced aging mice could increase the expression of BDNF, doublecortin (DCX), and cAMP-response element-binding protein (CREB) and reduce inflammatory and senescence markers in the hippocampus (Woo et al. 2014). In aged F344 rat, C29 administration reduced aging-associated activation of Akt, mTOR, and FOX3a in both colon and the hippocampus (Jeong et al. 2015). More importantly, the memory-enhancing effect has been demonstrated in a multi-center, randomized, double-blind, placebo-controlled clinical study (Hwang et al. 2019). In a cohort of 100 elderly subjects with mild cognitive impairment (ages between 55 and 85, average 69), the C29-fermented soybean (DW2009) consumption group showed improvements in cognitive functions related to memory and attention in comparison to the placebo group. Furthermore, cognitive improvement was associated with increased serum BDNF levels. This clinical study demonstrates that DW2009 is safe for human consumption, and it exhibits cognition-enhancing properties in elderly individuals with mild cognitive impairment.

In addition to alleviating age-related physical and mental deficits, probiotics could also reduce yet another hallmark of aging, the telomere attrition. Administration of *L. plantarum* DR7 (DR7), *Lactobacillus fermentum* DR9 (DR9), and *Lactobacillus reuteri* 8513d could significantly reduce the shortening of telomere in D-galactose-induced aging rat (Lew et al. 2019). In addition to reducing aging-induced shorting of telomere, DR7 and DR9 also improved hyperlipidemia and liver steatosis in aging rats; demonstrating potentials on reducing risks of cardiovascular and liver diseases (Lew et al. 2020). DR7 also improved memory function and reduced anxiety-like behavior in the aging rat (Zaydi et al. 2020). Furthermore, a randomized, double-blind, placebo-controlled study on stressed adult humans had demonstrated that 12 weeks of DR7 consumption could alleviate stress and anxiety in stressed subjects (Chong et al. 2019); suggesting DR7 also processes the potential of reducing stress and anxiety in the elderly. Studies on DR9 revealed that DR9 supplementation could lower the gene expression of p53, a known senescent biomarker, in gastrocnemius muscle and tibia (Hor et al. 2019) and improved musculoskeletal health in D-galactose-induced aging rat (Hor et al. 2021).

Since it is believed that obesity can accelerate the aging process, anti-obesity probiotics may also play a role in anti-aging (Salvestrini et al. 2019). The strain *B. breve* B-3 (B3) has been shown to suppress the accumulation of body weight and improve serum cholesterol in a mouse model of high-fat-diet-induced obesity (Kondo et al. 2010). A subsequent human study demonstrated that 12 weeks of B3 consumption could reduce body fat in healthy pre-obese ($25 \leq$ BMI < 30) adults (Minami et al. 2018). Furthermore, in a randomized, double-blind, placebo-controlled trial on healthy elderly subjects, the consumption of B3 in combination with 3 other *Bifidobacteria* species could improve mental health and body weight in comparison to the placebo group (Inoue et al. 2018). GKM3 also exhibits anti-obesity properties in the high-fat diet-induced obese rats (Mohammadi et al. 2019). More importantly, this probiotic had been shown to promote the lifespan of the SAMP8 mice (Lin et al. 2021). Based on the abovementioned studies, several probiotics could be qualified as gerobiotics.

16.7 Conclusion

The benefits of low cost, ease of administration, a strong safety profile, and high user acceptance make probiotics an appealing option as a preventative intervention. Utilizing a geroscience-based approach will advance anti-aging probiotic development and help formulate effective marketing strategies. A new term, gerobiotics, has been devised for those strains that have been confirmed, through substantial preclinical and clinical studies, to target multiple hallmarks of aging. Gerobiotics demonstrate prolongevity effects and improve senescence-related physiological signs. Well-designed clinical trials must validate both the anti-aging effects and the underlying mechanistic pathways of candidate strains. Many intracellular signaling pathways are involved in aging, and we expect any single gerobiotic strain to act on only a limited number of pathways. The combined action of multiple gerobiotics with other geroprotectors may have additive or synergistic benefits. Solid preclinical and clinical evidence is key to convincing financial investors and assuring consumers of the promise of gerobiotics. Marketing strategies will rely on scientific biomarker data garnered from these studies. The definition of gerobiotics also includes postbiotics and para-probiotics derived from these strains, since we anticipate their shared role in healthspan extension.

References

Almeida I, Magalhaes S, Nunes A (2021) Lipids: biomarkers of healthy aging. Biogerontology 22:273–295

Anis E, Xie A, Brundin L, Brundin P (2022) Digesting recent findings: gut alpha-synuclein, microbiome changes in Parkinson's disease. Trends Endocrinol Metab 33:147–157

Auclair M, Guenantin AC, Fellahi S, Garcia M, Capeau J (2020) HIV antiretroviral drugs, dolutegravir, maraviroc and ritonavir-boosted atazanavir use different pathways to affect inflammation, senescence and insulin sensitivity in human coronary endothelial cells. PLoS ONE 15:E0226924

Bana B, Cabreiro F (2019) The microbiome and aging. Annu Rev Genet 53:239–261

Bñrcena C, Vald¢S-Mas R, Mayoral P, Garabaya C, Durand S, Rodrýguez F, Fernñndez-Garcýa MT, Salazar N, Nogacka AM, Garatachea N, Bossut N, Aprahamian F, Lucia A, Kroemer G, Freije JMP. Quirnature Med 25:1234–1242

Cardoso AL, Fernandes A, Aguilar-Pimentel JA, De Angelis MH, Guedes JR, Brito MA, Ortolano S, Pani G, Athanasopoulou S, Gonos ES, Schosserer M, Grillari J, Peterson P, Tuna BG, Dogan S, Meyer A, Van Os R, Trendelenburg AU (2018) Towards frailty biomarkers: candidates from genes and pathways regulated in aging and age-related diseases. Ageing Res Rev 47:214–277

Carini G, Mingardi J, Bolzetta F, Cester A, Bolner A, Nordera G, La Via L, Ieraci A, Russo I, Maggi S, Calza S, Popoli M, Veronese N, Musazzi L, Barbon A (2022) Mirnome profiling detects Mir-101-3p and Mir-142-5p as putative blood biomarkers of frailty syndrome. Genes 13

Carini G, Musazzi L, Bolzetta F, Cester A, Fiorentini C, Ieraci A, Maggi S, Popoli M, Veronese N, Barbon A (2021) The potential role of mirnas in cognitive frailty. Front Aging Neurosci 13:763110

Carter CS (2021) A "gut feeling" to create a 10th hallmark of aging. J Gerontol A Biol Sci Med Sci 76:1891–1894

Chen B, Sun Y, Zhang J, Zhu Q, Yang Y, Niu X, Deng Z, Li Q, Wang Y (2019a) Human embryonic stem cell-derived exosomes promote pressure ulcer healing in aged mice by rejuvenating senescent endothelial cells. Stem Cell Res Ther 10:142

Chen LH, Huang SY, Huang KC, Hsu CC, Yang KC, Li LA, Chan CH, Huang HY (2019b) Lactobacillus paracasei Ps23 decelerated age-related muscle loss by ensuring mitochondrial function in Samp8 mice. Aging (ALBANY Ny) 11:756–770

Chen Y, Zhang S, Zeng B, Zhao J, Yang M, Zhang M, Li Y, Ni Q, Wu, Li Y (2020) Transplant of microbiota from long-living people to mice reduces aging-related indices and transfers beneficial bacteria. Aging (Albany Ny) 12:4778–4793

Chen LH, Wang MF, Chang CC, Huang SY, Pan CH, Yeh YT, Huang CH, Chan CH, Huang HY (2021) Lacticaseibacillus paracasei Ps23 effectively modulates gut microbiota composition and improves gastrointestinal function in aged samp8 mice. Nutrients 13

Cheng L-H, Cheng S-H, Wu C-C, Huang C-L, Wen P-J, Chang M-Y, Tsai Y-C (2021) Lactobacillus paracasei Ps23 dietary supplementation alleviates muscle aging via ghrelin stimulation in D-galactose-induced aging mice. J Funct Foods 85:104651

Cheng LH, Chou PY, Hou AT, Huang CL, Shiu WL, Wang S (2022) Lactobacillus paracasei Ps23 improves cognitive deficits via modulating hippocampal gene expression and gut microbiota in D-galactose-induced aging mice. Food Funct

Chong HX, Yusoff NAA, Hor YY, Lew LC, Jaafar MH, Choi SB, Yusoff MSB, Wahid N, Abdullah M, Zakaria N, Ong KL, Park YH, Liong MT (2019) Lactobacillus plantarum Dr7 alleviates stress and anxiety in adults: a randomised, double-blind, placebo-controlled study. Benef Microbes 10:355–373

Chua CEL, Tang BL (2019) Mir-34a In neurophysiology and neuropathology. J Mol Neurosci 67:235–246

Clancy R (2003) Immunobiotics and the probiotic evolution. Fems Immunol Med Microbiol 38:9–12

Cooper R, Kuh D, Hardy R (2010) Objectively measured physical capability levels and mortality: systematic review and meta-analysis. BMJ 341:C4467

Dinan TG, Stanton C, Cryan JF (2013) Psychobiotics: a novel class of psychotropic. Biol Psychiatry 74:720–726

Gems D, De Magalhaes JP (2021) The hoverfly and the wasp: a critique of the hallmarks of aging as a paradigm. Ageing Res Rev 70:101407

Guerville F, De Souto Barreto P, Ader I, Andrieu S, Casteilla L, Dray C, Fazilleau N, Guyonnet S, Langin D, Liblau R, Parini A, Valet P, Vergnolle N, Rolland Y, Vellas B (2020) Revisiting the hallmarks of aging to identify markers of biological age. J Prev Alzheimers Dis 7:56–64

Hill C, Guarner F, Reid G, Gibson GR, Merenstein DJ, Pot B, Morelli L, Canani RB, Flint HJ, Salminen S, Calder PC, Sanders ME (2014) The international scientific association for probiotics and prebiotics consensus statement on the scope and appropriate use of the term probiotic. Nat Rev Gastroenterol Hepatol 11:506–514

Hor Y-Y, Ooi C-H, Khoo B-Y, Choi S-B, Seeni A, Shamsuddin S, Oon C-E, Ong K-L, Jeong W-S, Liong M-T (2019) Lactobacillus strains alleviated aging symptoms and aging-induced metabolic disorders in aged rats. J Med Food 22:1–13

Hor YY, Ooi CH, Lew LC, Jaafar MH, Lau AS, Lee BK, Azlan A, Choi SB, Azzam G, Liong MT (2021) The molecular mechanisms of probiotic strains in improving ageing bone and muscle of d-galactose-induced ageing rats. J Appl Microbiol 130:1307–1322

Huang SY, Chen LH, Wang MF, Hsu CC, Chan CH, Li JX, Huang HY (2018) Lactobacillus paracasei Ps23 delays progression of age-related cognitive decline in senescence accelerated mouse prone 8 (Samp8) mice. Nutrients 10

Hwang YH, Park S, Paik JW, Chae SW, Kim DH, Jeong DG, Ha E, Kim M, Hong G, Park SH, Jung SJ, Lee SM, Na K. H., Kim, J. & Chung, Y. C. 2019. Efficacy and safety of lactobacillus plantarum C29-fermented soybean (Dw2009) in individuals with mild cognitive impairment: a 12-week, multi-center, randomized, double-blind, placebo-controlled clinical trial. Nutrients 11

Inoue T, Kobayashi Y, Mori N, Sakagawa M, Xiao JZ, Moritani T, Sakane N, Nagai N (2018) Effect of combined bifidobacteria supplementation and resistance training on cognitive function, body composition and bowel habits of healthy elderly subjects. Benef Microbes 9:843–853

Jeong JJ, Kim KA, Hwang YJ, Han MJ, Kim DH (2016) Anti-inflammaging effects of lactobacillus brevis Ow38 in aged mice. Benef Microbes 7:707–718

Jeong JJ, Woo JY, Kim KA, Han MJ, Kim DH (2015) Lactobacillus pentosus var. plantarum C29 ameliorates age-dependent memory impairment in fischer 344 rats. Lett Appl Microbiol 60:307–314

Jung IH, Jung MA, Kim EJ, Han MJ, Kim DH (2012) Lactobacillus pentosus var. plantarum C29 protects scopolamine-induced memory deficit in mice. J Appl Microbiol 113:1498–1506

Justice JN, Ferrucci L, Newman AB, Aroda VR, Bahnson JL, Divers J, Espeland MA, Marcovina S, Pollak MN, Kritchevsky SB, Barzilai N, Kuchel GA (2018) A framework for selection of blood-based biomarkers for geroscience-guided clinical trials: report from the tame biomarkers workgroup. Geroscience 40:419–436

Kennedy BK, Berger SL, Brunet A, Campisi J, Cuervo AM, Epel ES, Franceschi C, Lithgow GJ, Morimoto RI, Pessin JE, Rando TA, Richardson A, Schadt EE, Wyss-Coray T, Sierra F (2014) Geroscience: linking aging to chronic disease. Cell 159:709–713

Kim M, Benayoun BA (2020) The microbiome: an emerging key player in aging and longevity. Transl Med Aging 4:103–116

Kim S, Jazwinski SM (2018) The Gut microbiota and healthy aging: a mini-review. Gerontology 64:513–520

Kinser HE, Pincus Z (2020) Micrornas as modulators of longevity and the aging process. Hum Genet 139:291–308

Kondo S, Xiao JZ, Satoh T, Odamaki T, Takahashi S, Sugahara H, Yaeshima T, Iwatsuki K, Kamei A, Abe K (2010) Antiobesity effects of bifidobacterium breve strain B-3 supplementation in a mouse model with high-fat diet-induced obesity. Biosci Biotechnol Biochem 74:1656–1661

Kumar S, Vijayan M, Bhatti JS, Reddy PH (2017) Micrornas as peripheral biomarkers in aging and age-related diseases. Prog Mol Biol Transl Sci 146:47–94

Lee HJ, Hwang YH, Kim DH (2018) Lactobacillus plantarum C29-fermented soybean (Dw2009) alleviates memory impairment in 5xfad transgenic mice by regulating microglia activation and gut microbiota composition. Mol Nutr Food Res 62:E1800359

Lew L-C, Hor Y-Y, Jaafar M-H, Lau A-S-Y, Lee B-K, Chuah L-O, Yap K-P, Azlan A, Azzam G, Choi S-B, Liong M-T (2020) Lactobacillus strains alleviated hyperlipidemia and liver steatosis in aging rats via activation of Ampk. Int J Mol Sci 21:5872

Lew LC, Hor YY, Jaafar MH, Lau ASY, Ong JS, Chuah LO, Yap KP, Azzam G, Azlan A, Liong MT (2019) Lactobacilli modulated Ampk activity and prevented telomere shortening in ageing rats. Benef Microbes 10:883–892

Liao JF, Hsu CC, Chou GT, Hsu JS, Liong MT, Tsai YC (2019) Lactobacillus paracasei Ps23 reduced early-life stress abnormalities in maternal separation mouse model. Benef Microbes 10:425–436

Lin SW, Tsai YS, Chen YL, Wang MF, Chen CC, Lin WH, Fang TJ (2021) Lactobacillus plantarum Gkm3 promotes longevity, memory retention, and reduces brain oxidation stress in Samp8 mice. Nutrients 13

Liu H-C, Han D-S, Hsu C-C, Wang J-S (2021) Circulating Microrna-486 and Microrna-146a serve as potential biomarkers of sarcopenia in the older adults. BMC Geriatr 21:86

Lopez-Otin C, Blasco MA, Partridge L, Serrano M, Kroemer G (2013) The hallmarks of aging. Cell 153:1194–1217

Mackowiak PA (2013) Recycling Metchnikoff: probiotics, the intestinal microbiome and the quest for long life. Front Public Health 1:52

Mensa E, Giuliani A, Matacchione G, Gurau F, Bonfigli AR, Romagnoli F, De Luca M, Sabbatinelli J, Olivieri F (2019) Circulating Mir-146a in healthy aging and type 2 diabetes: age- and gender-specific trajectories. Mech Ageing Dev 180:1–10

Minami J, Iwabuchi N, Tanaka M, Yamauchi K, Xiao JZ, Abe F, Sakane N (2018) Effects of bifidobacterium breve B-3 on body fat reductions in pre-obese adults: a randomized, double-blind, placebo-controlled trial. Biosci Microbiota Food Health 37:67–75

Mohammadi G, Dargahi L, Peymani A, Mirzanejad Y, Alizadeh SA, Naserpour T, Nassiri-Asl M (2019) The effects of probiotic formulation pretreatment (Lactobacillus helveticus R0052 and Bifidobacterium longum R0175) on a lipopolysaccharide rat model. J Am Coll Nutr 38:209–217

Moskalev A, Chernyagina E, Kudryavtseva A, Shaposhnikov M (2017) Geroprotectors: a unified concept and screening approaches. Aging Dis 8:354–363

National Council On Aging. Healthy Aging Facts [Online]. National council on aging

Peng X, Meng J, Chi T, Liu P, Man C, Liu S, Guo Y, Jiang Y (2014) Lactobacillus plantarum Ndc 75017 alleviates the learning and memory ability in aging rats by reducing mitochondrial dysfunction. Exp Ther Med 8:1841–1846

Ragonnaud E, Biragyn A (2021) Gut microbiota as the key controllers of "healthy" aging of elderly people. Immun Ageing 18:2

Rivero-Segura NA, Bello-Chavolla OY, Barrera-Vñzquez OS, Gutierrez-Robledo LM, Gomez-Verjan JC (2020) Promising biomarkers of human aging: in search of a multi-omics panel to understand the aging process from a multidimensional perspective. Ageing Res Rev 64:101164

Salvestrini V, Sell C, Lorenzini A (2019) Obesity may accelerate the aging process. Front Endocrinol (lausanne) 10:266

Seals DR, Justice JN, Larocca TJ (2016) Physiological geroscience: targeting function to increase healthspan and achieve optimal longevity. J Physiol 594:2001–2024

Sierra F (2016) The emergence of geroscience as an interdisciplinary approach to the enhancement of health span and life span. Cold Spring Harb Perspect Med 6:A025163

Smith P, Willemsen D, Popkes M, Metge F, Gandiwa E, Reichard M, Valenzano DR (2017) Regulation of life span by the gut microbiota in the short-lived African turquoise killifish. Elife 6

Teodori L, Petrignani I, Giuliani A, Prattichizzo F, Gurau F, Matacchione G, Olivieri F, Coppari S, Albertini MC (2019) Inflamm-aging micrornas may integrate signals from food and gut microbiota by modulating common signalling pathways. Mech Ageing Dev 182:111127

Toda K, Yamauchi Y, Tanaka A, Kuhara T, Odamaki T, Yoshimoto S, Xiao JZ (2020) Heat-killed bifidobacterium breve B-3 enhances muscle functions: possible involvement of increases in muscle mass and mitochondrial biogenesis. Nutrients 12

Tsai YC, Cheng LH, Liu YW, Jeng OJ, Lee YK (2021) Gerobiotics: probiotics targeting fundamental aging processes. Biosci Microbiota Food Health 40:1–11

Wagner KH, Cameron-Smith D, Wessner B, Franzke B (2016) Biomarkers of aging: from function to molecular biology. Nutrients 8

Wei CL, Wang S, Yen JT, Cheng YF, Liao CL, Hsu CC, Wu CC, Tsai YC (2019) Antidepressant-like activities of live and heat-killed lactobacillus paracasei Ps23 in chronic corticosterone-treated mice and possible mechanisms. Brain Res 1711:202–213

Woo JY, Gu W, Kim KA, Jang SE, Han MJ, Kim DH (2014) Lactobacillus pentosus var. plantarum C29 ameliorates memory impairment and inflammaging in a D-galactose-induced accelerated aging mouse model. Anaerobe 27:22–26

World Health Organization (2019) World health statistics overview 2019: monitoring health for the sdgs, sustainable development goals. World Health Organization, Geneva

Xu Q, Fu Q, Li Z, Liu H, Wang Y, Lin X, He R, Zhang X, Ju Z, Campisi J, Kirkland JL, Sun Y (2021) The flavonoid procyanidin C1 has senotherapeutic activity and increases lifespan in mice. Nat Metab 3:1706–1726

Yabluchanskiy A, Ungvari Z, Csiszar A, Tarantini S (2018) Advances and challenges in geroscience research: an update. Physiol Int 105:298–308

Zaydi AI, Lew LC, Hor YY, Jaafar MH, Chuah LO, Yap KP, Azlan A, Azzam G, Liong MT (2020) Lactobacillus plantarum Dr7 improved brain health in aging rats via the serotonin, inflammatory and apoptosis pathways. Benef Microbes 11:753–766

Chapter 17
Screening Potential Probiotics Against Obesity and Metabolism Abnormalities in the Elderly

Fang He, Xi Shen, Ruyue Cheng, and Francesco Marotta

Abstract With the progress in the nutrition and medicinal science, the life-expectancy of the peoples has significantly increased in the both of developed and developing countries, and the ageing is undergoing rather than expected. Ageing is also deeply associated with increased susceptibility to degenerative or infectious diseases. The industrialized societies are suffering from dynamically increasing cost and spending for health care and medical treatment of their senior citizens. The growing scientific evidences have well suggested that ageing can alter the intestinal microbiota qualitatively and quantitatively, generating a different microbe community with the age characterized structure. These ageing related gut microbiota will directly or indirectly abnormalize physiological metabolisms of middle and elderly people and have been considered as the risk to cause obesity and metabolic diseases. Therefore, screening new anti-obesity probiotics to maintain the health of the intestinal microbiota of middle-aged and elderly people will help them to prevent or alleviate various metabolic diseases related to obesity and might be a practical method to promote the middle-aged and elderly people. In this chapter, two testing models, the combination of macrophage and adipocyte and fecal culture, were suggested to evaluate /or screen the probiotics strains with anti-obesity effects. With these testing models, some selected probiotics strains such as *Lactobacillus* GG and *Bifidoacterium bifidum* TMC3115 exhibited its strain dependent abilities to activate macrophages by which to inhibit the adipogenic differentiation of preadipocytes, indicating that these may mediate the host immune response and affect their fat metabolism function. The results from fecal culture model indicate that the modification of the composition and function of intestinal microbiota might be one of the underlying mechanisms by which these bacteria could express the anti-obesity effects. Such anti-obesity effects have also been

F. He (✉) · X. Shen · R. Cheng
Department of Nutrition and Food hygiene, West China School of Public Health and West China Fourth Hospital, Sichuan University, Sichuan 610041, Chengdu, China
e-mail: nrb47389@nifty.com; hf18602880124@163.com

F. Marotta
ReGenera R&D International for Aging, Intervention and San Babila Clinic, Corso Matteotti, 1/A, 20121 Milan, Italy

F. Marotta (ed.), *Gut Microbiota in Aging and Chronic Diseases*, Healthy Ageing and Longevity 17, https://doi.org/10.1007/978-3-031-14023-5_17

demonstrated with animal and human studies, at least partly. These results indicate that there might be deep association between intestinal microbiota and health of the elderly and some intestinal microbes, might play important role in the development of obesity and metabolisms.

Keywords Tailored probiotics · Obesity · Metabolic syndrome · Elderly health care · Preventing chronic diseases

17.1 Introduction

The average life expectancy has increased over the last century with the significant progress in medical science and health care (Saunier and Dore 2002). The old (>65 years) and the "oldest" (>85 years) age group are the fastest growing sub-population in the world, especially in the industrialized societies which is also called as aged societies. WHO indicates there are currently about 580 million peoples in the world aged 60 or over, and this figure is expected to rise to over billion within next 20 years (Kalache 1999).

It has been well known that many physiological functions, such as immunity and gut function, in humans being usually decline progressively with age after the 30 years old (Saunier and Dore 2002). Among the major consequences of the people living longer might be an increased health problem and health concern. Epidemiological researches have shown that the highest mortality rate among middle-aged and elderly people might be cardiovascular disease. The main risk factors for ageing related cardiovascular disease are still considered as obesity and obesity related metabolic syndrome (Ortega et al. 2016). Therefore, the effective measures should be much sought to modify/ or stabilize the age related decline (or imbalance) in physiological function to protect the elderly from obesity.

It has well been known that the intestinal microbes could mediate many crucial events toward the protection or alteration of human health. This chapter would like to summarize the current knowledge and findings about the intestinal mictobiota in the elderly with focusing on obesity and metabolism of the elderly and the testing model to screening new microbes specific for using as probiotics against obesity and metabolism of the elderly.

17.2 Age Altered Respects of the Intestinal Microbiota

The gastrointestinal tract (GI tract) has been well known to servers as one of the biggest interfaces between the body and the external environment (Schneeman 2002). This GI tract is highly specialized organ system that allows us to consume food in discrete meals as well as very diverse array foodstuffs to meet nutrient needs. The organs of the GIT include the mouth, esophagus, stomach, small intestine and large intestine; in addition, the pancreas and liver secrete into the small intestine. The systems is connected to the vascular, lymphatic and nervous systems to facilitate regulation of the digestive response, delivery of absorbed compounds to organ of body and regulation of the food intake.

Different from other body organs, there are numerous endogenous microbes naturally colonize/reside the surface of GI tract through whole life of the host and consist a complex community inside of the host known as intestinal microbiota, a characteristic internal ecosystem. Normal aging is associated with significant changes in the function of the most organs and tissues, such as decreased taste thresholds, hypochlorhydria due to atrophyic gastritiss and decreased liver blood flow and size (Russell 1992). The GI tract is no exception, and there is increased evidence of the several impaired gastrointestinal function with ageing (Russell 1992; Gill et al. 2001; Schneeman 2002). In the gastrointestinal tract of the elderly, the age-related changes include decreased acid secretion by the gastric mucosa, and greater permeability of mucosal membranes which have been linked to increase in circulating antibodies to components of the intestinal microbiota in the elderly subjects. Therefore, the certain microbes which can take advantage of new ecological niches are assumed to become predominant inhabitants, leading a dranmic shift in the composition of the gut microflora upon age. In general, the aging may decrease or dimnish bifidobacteria, and significantly increase clostridia including *C. perfringens*, and slightly overgrow lactobacilli, streptococci and enterobacteriacea. The total bacterial count and anaerobes/aerobes in the intestinal microflora of the elderly are relatively lower rather than those of the healthy infants and adults.

A recent study has been conducted to investigate the possible association between gut microbes and immunity among healthy middle-aged and elderly individuals in southwest China, a geographically specific region with its own traditional diet habits (Shen et al. 2018). In this study, 148 healthy adults aged $\geq$ 50 years including many retired medical scientists and clinical doctors were radamicaly divided into two age groups: middle-aged group (50–59 years; n = 67, 54.13 $\pm$ 3.32) and elderly group ($\geq$ 60 years; n = 81, 64.70 $\pm$ 3.93). Their blood samples were collected to measure serum immune and other biochemical indices. The fecal samples were also collected for profiling fecal bacteria compositions using 16S rRNA gene sequencing. Same to those reported in previous studies to analyze human intestinal microbiota, *Firmicutes*, *Bacteriodetes*, *Proteobacteria*, *Verrumicrobia,* and *Actinobacteria* were still detected as the perdomoniat microbes in the tested fecal samples. *Bacteroidetes/Firmicutes* ratio was almost 0.35, which was relatively lower than those reported in other elderly people (Mariat et al. 2009). Low *Bacteroidetes/Firmicutes* ratio has been well suggested/or confirmed as one of the hallmarks of human gut microbiota, because this ratio has been observed to be lower in obese individuals in some human studies (Compare et al. 2016). Clinical examination indicated that the tested participants were clinically healthy and the most of them were keeping normal body weight. However, they are still at a potential risk of metabolic disorders as other elderly in light of their intestinal microbial composition based on the previous studies (Compare et al. 2016).

Furthermore, these results might raise a question why the tested elderly from southwest China might be a less obese and overweight, although their intestinal microbiota have become obesity type characteristically with low *Bacteroidetes/Firmicutes* ratio.

In this study, the possible association between the immunity and intestinal microbiota in the tested middle and elderly people has also been analyzed. In the middle-aged group, *Bacteroidetes* was positively correlated with serum immunoglobulin G (IgG) levels and the percent of CD8+T cells and negatively correlated with the CD4 + /CD8 + ratio; the presence of *Firmicutes* was negatively correlated with IgM levels; *Bacteroidetes/Firmicutes* ratio was positively correlated with IgG and IgM levels and Simpson index was negatively correlated with the percent of CD8+ T cells and positively correlated with CD4 +/CD8 + ratio. In the elderly group, the presence of *Verrucomicrobia* (identified as genus *Akkermansia*) was positively correlated with IgA levels and the percent of CD8+T cells and negatively correlated with the percent of CD4+T cells and CD4+/CD8 + ratio; the Chao index and observed species were positively correlated with IgA levels. These results indicate that ageing could significantly correlate with the composition of gut microbiota in terms of quantity and quality. Such changes in gut microbiota caused by ageing, characterized by decreased *Bacteroidetes* levels, might be associated with immunosenescence among healthy middle-aged and elderly people in southwest China. These results also indicate that some specific intestinal microbes resides the gut of the tested people which could interact with immune cells by which to influence/allaivate the altered energy metabolism from ageing related intestinal microbiota, therefore which protect the tested people from obesity.

In order to answer over question, the fecal samples collected from these healthy volunteers were also investigated for their fecal bifidobacterial composition of the tested 148 healthy volunteers in detailed by real time fluorescence quantitative PCR (qPCR) using 16S rRNA targeting designed gene and species specific primers (Wang et al. 2017). Bifidobacteria can be detected in the all of the tested fecal samples as key member of fecal microbiota with more than 10^8 cells/g fecal. The detected fecal bifidobacteria were quantitatively less than infant, children, and the youngs as reported in the previous studies. Compared to other studies, it was interesting that more species generally predominated in the infants were found theses tested subjects. Therefore, the possible association between the detected bifidobacteria and physiological function of the tested subjects were analyzed. *B. adolescentis* was negatively associated with body mass index (BMI), while *B. catenulatum* was positive to BMI. *B. breve* and high density lipoproteincholesterol (HDLC) were negatively correlated, but *B. infantis* and HDLC were positively correlated. *B. bifidum* was found to be positive associated with IgA. These results indicated that *Bifidobacterium* might affect the host physiologically and immunology in the species dependent manner. Keeping intestinal bifidobacteria in the ideal species composition might be one of effective ways to slow immune senescence and promote the health of the elderly.

Another study also investigate the possible relationship between the gut microbiota and hyperlipidemia, hyperglycemia in middle-aged and elderly people in Chengdu, Sichuan, China. Total 51 of middle-aged and elderly people participated in this study. The blood glucose and lipid, and the other physiological parameters of the subjects were analyzed, and their fecal microbiota was also profiled by sequencing technique. The fecal microbiota in the middle-aged and elderly were mainly *Firmicutes, Bacteroidetes, Proteobacteria, Fusobacteria, Actinobacteria, Verrucomicrobia.* The diversity of fecal microbiota was significantly lower in the subjects with higher blood glucose, while it did not exhibited relationship with changes in blood lipids. At the bacterial phyla: *Proteobacteria* and *Bacteroidetes* were respectively significantly higher in middle-aged and elderly women with higher blood glucose. At the bacterial genus: *Faecalibacterium* was significantly lower in the hyperglycemic subjects, its abundance and blood glucose levels were negatively correlated; *Prevotella Paraprevotella* were significantly lower in the high blood lipids group, and both of them were negatively correlated with blood lipid levels. The fecal microbiota in the tested middle-aged and elderly subjects varied with blood glucose and blood lipid concentration. Intestinal microbes might posse close relationship with blood glucose and lipid metabolism in middle-aged and elderly people.

A recent study conducted also in Chengdu China to profile the intestinal bifidobacteria in the middle-aged and elderly people with different blood glucose and blood lipid levels (Wang et al. 2022). Total 289 middle-aged and elderly people aged 45 were divided into healthy group, dyslipidemia group, hyperglycemia group, and double-high group (with both dyslipidemia and hyperglycemia). *Bifidobacterium bifidum* in the healthy group was negatively correlated with TG, and *B. breve* was negatively correlated with GLU, TC, and LDL-C, *B. dentate* was positively correlated with GLU, and *B. infantis* was negatively correlated with TC. *B. chain* and *B. infantis* in hyperglycemia group were positively correlated with HDL-C; *B. bifidum* in double hyperglycemia group was negatively correlated with HDL-C, *B. breve* was positively correlated with TG and HDL-C. These results indicated that intestinal bifidobacteria species might have a stronger species dependent correlation with blood glucose and blood lipid levels of host, and may influence glucose and lipid metabolism in different ways.

These studies point out that the changes in the gut microbiota caused by aging will directly or indirectly affect various physiological metabolisms of middle and elderly people, and have the risk of causing obesity and metabolic diseases. Therefore, screening out new anti-obesity probiotics to maintain the health of the intestinal microbiota of middle-aged and elderly people will help prevent or alleviate various metabolic diseases related to obesity, and might be a practical method to promote the middle-aged and elderly people.

17.3 Screening Potential Probiotics Against Obesity and Metabolism of the Elderly with Immune and Fat Cell Lines

In general, a body mass index ≥ 30 could be considered obese. Excessive growth of adipose tissue is among the important pathogenic/clinical aspects of obesity. Such excessive adipose growth might result from over enlargement of existing adipocytes (hypertrophy) and formation of new adipocytes (hyperplasia) through differentiation of stromal preadipocytes (adipogenesis).

Adipocytes have been well known to play positive important role to normalize adequate balance of energy of host animals by storing calories in lipids, mobilizing energy sources in response to hormonal stimulation and to modulate body functions by secreting various signal. Emerging scientific evidences suggest that adipose tissue may have much greater impacts to physiological and metabolic function of host animals, and adipocyte activity could greatly determinate/or influence the most obesity-associated pathogenic events. Therefore, the modification of hypertrophy and hyperplasia could effectively regulate overgrowth of adipose tissue, which may be a practical strategy for the management of obesity.

It is well documented that obesity and metabolic syndrome might have a deep association with the intestinal immune function of host animal. Furthermore, some selected probiotics can modulate the immune responses of the host animal by which to alter its lipid metabolism.

In light of these findings, a testing model was developed to test/or explore the possibilities of the microbes to modify the growth of adipose tissue through host animals immune-competent cells (Fig. 17.1). *Lactobacillus* GG (LGG) and *L. gasseri* TMC0356 (TMC0356) are well studied probiotics strains with many health promoting effects. These tow bacteria were used this testing model to investigate whether they could activate the macrophages by which to inhibit the development of dadipocyte as one of the key underlying mechanisms to express their anti-obesity effects in animal studies (Shi et al. 2013; Miyazawa et al. 2021).

In this study, the heat-killed LGG and TMC0356 were co-cultured with the murine macrophage-like cell line J774.1 (J774.1). The activation of J774.1 were evaluated by profile interleukin (IL)-6, IL-12, and tumor necrosis factor-α secreted from the J774.1 cells using enzyme-linked immunosorbent assay methods. The cell free conditioned medium from cultured J774.1 cells was then moved to the pre-adipocyte cell line 3T3-L1, a mouse embryonic fibroblast-adipose-like cell line. In order to monitor growth and differentiation of the tested 3T3-L1 cells, lipid accumulation and expression of peroxisome proliferator-activated receptor (PPAR)-γ mRNA were analyzed. The lipid accumulation in 3T3-L1 cells was significantly suppressed in a dose-dependent manner by the conditioned medium of LGG and TMC0356. Furthermore, PPAR-γ mRNA expression in 3T3-L1 cells was also significantly decreased ($P < 0.01$, $P < 0.05$, respectively).These results suggest that some microbes, such as LGG and TMC0356 may strain dependently activate macrophage of host animals by which to significantly suppress differentiation of

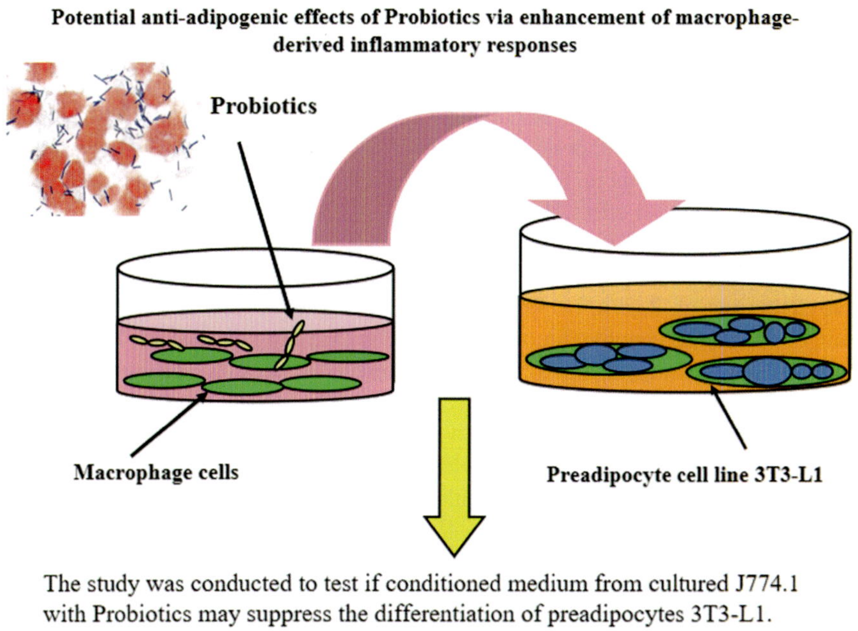

Fig. 17.1 Screening potential probiotics against obesity with immune and fat cell lines

preadipocytes. Therefore, this new testing model could screen new anti-obesity probiotics which can characteristically alter the immune responses of macrophages as one of the underlying mechanism to implicate adipose cells.

Streptococcus thermophiles is the most commonly used lactic acid bacteria in fermented milk production. However, a large number of *Streptococcus thermophiles* are considered to have weak acid-resistance and they could not survive and colonize in the human intestine as do other lactic acid bacteria, especially lactobacilli used as probiotics. Therefore, there are very few studies conducted to evaluate the health promoting effects of *Streptococcus thermophiles* and new approaches should be developed to find out/or demonstrate the health promoting effects of these bacteria.

Streptococcus thermophiles MN-ZLW-002 (MN-ZLW-002) was isolated from a traditional fermented milk collected from Gansu region of Gansu Province, China (Kang et al. 2012). We have tried to investigate the possibilities of MN-ZLW-002 against obesity with this testing model (Kang et al. 2021a, b). MN-ZLW-002 was co-cultured with RAW264.7 macrophages. The conditioned medium (CM) of MN-ZLW-002 was added to 3T3-L1 pre-adipocytes. Both of viable and heat-killed MN-ZLW-002 significantly stimulated RAW264.7 cells to produce significant interleukin (IL)-6 and tumor necrosis factor (TNF)-α and induced intense phosphorylation of P38, p44/42 mitogen-activated protein kinase (MAPK) (extracellular signal-regulated kinase (ERK)) and nuclear factor κB (NF-κB). Cytokine production reduced dramatically when heat-killed MN-ZLW-002 was treated with

Ribonuclease. As control, LGG induced less cytokine production and little signaling protein activation. MN-ZLW-002-stimulated RAW264.7-CM notably inhibited pre-adipocytes differentiation. These findings suggest that MN-ZLW-002 might suppress adipocytes differentiation and lipid accumulation though implicating the immune response, possibly via the release of cytokines, particularly TNF-α at least partly; Furthermore, the RNA of MN-ZLW-002 may be a key component in its immune activation effect.

In light of over results, 12-week animal study was done to investigate the possibilities of MN-ZLW-002 against HFD caused obesity observed in cell line study (Kang et al. 2021a, b). The orally fed MN-ZLW-002 was found to significantly reduced the weight gain of mice caused by the high-fat diet (HFD) at weeks 3–8; and suppressed fasting blood glucose levels at week 4 and week 8; down regulated serum total triglyceride level at week 12. MN-ZLW-002 also reduced serum IL-1β and chemokine ligand 3 levels in the HFD-fed mice. These results well demonstrated that MN-ZLW-002 could partly improve metabolism in the early stage of HFD intervention though regulating the immune response of host animal.

Bifidobacterium bifidum TMC3115 (TMC3115) originally isolated from healthy infants. In the previous research with cell line and animals studies, this bacterium has been found to have a strain dependent immune regulation effect and can improve various immunity and anti-allergy disorders (He et al. 2002; Cheng et al. 2019). Recent studies also found that this bacterium also has some benefits on nerves development (Cheng et al. 2020; Wang et al. 2020). However, the possible anti-obesity effect of this bacterium remains to be further studied. Therefore, this test model was used to test the possibilities of TMC3115 to express anti-obesity and metabolic disease-modifying effects of TMC3115 (Luo et al. 2018).

TMC3115 was co-cultured with mouse macrophage cell line J774.1 (J774.1) and the conditioned medium after co-cultured J774.1 was used to suppress the adipogenic differentiation of 3T3-L1 (3T3-L1). As results, TMC3115 was found to significantly activate J774.1 cells and stimulated the secretion of IL-6 and TNF-α from J774.1 cells. Furthermore, the TMC3115 conditioned J774.1 medium was observed to significantly inhibite the adipogenic differentiation of 3T3-L1 preadipocytes (Fig. 17.2). These results suggest that TMC3115 might indirectly inhibits the adipogenic differentiation though its strain dependent ability to activate macrophages and inducing their secretion of cytokines.

Following the cell line study, three weeks TMC3115 intervention was conducted with 47 subjects with mild hyperglycemia and dyslipidemia aged 45 to 75 years to investige the anti-obesity effects of TMC3115 (Wang et al. 2018). TMC3115 significantly decreased plasma TC and LDL-C levels of the tested subjects after intervention. The orally fed TMC3115 were also significantly improved frequencies of defaecation and faecal odour. TMC3115 administration also implicated fecal microbiota characteristically by increasing 1 *Firmicutes*, *Bacteroides* and *Actinobacteria* and decreasing *Proteobacteria* and *Fusobacteria*. These results indicate that TMC3115 might exhibit beneficial effects on the serum cholesterol metabolism of subjects with dyslipidaemia through modulation of their intestinal microbiota besides the immune modifiaction.

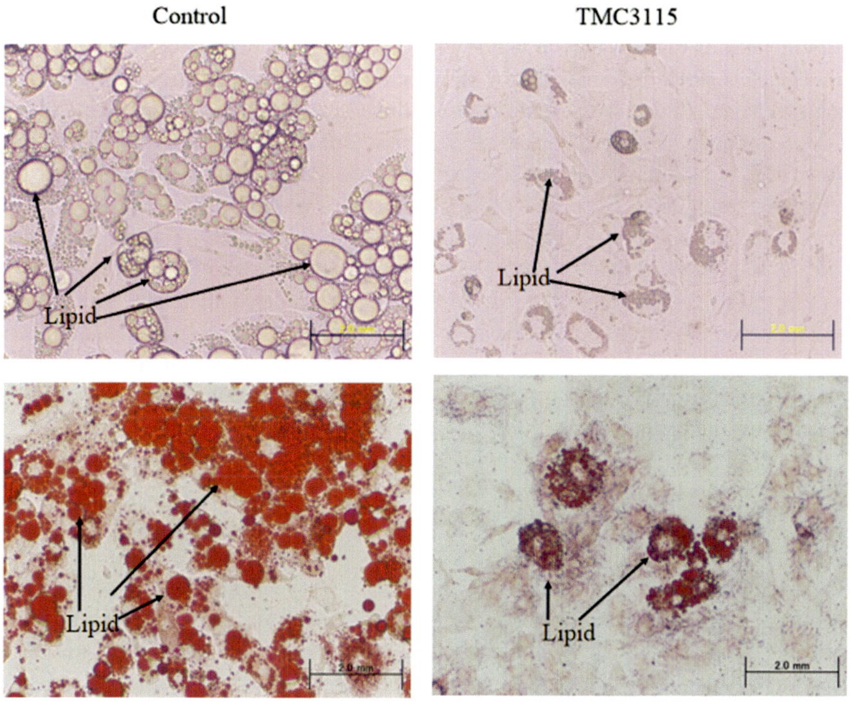

Fig. 17.2 TMC3115 activated macrophages and inducing their secretion of cytokines, By which to inhibit the adipogenic differentiation of preadipocytes

17.4 Screnting Potential Probiotics Against Obesity and Metabolism of Elderly with Fecal Culture

As well known, the intestinal microbes could play important roles to protect host animal from various infection and physiological disorders. Probiotics have been received a deal of attention as one of the practical ways to modulate/or stabilize intestinal microbiota. The health benefits of probiotic to the intestinal microbiota have been studied in different in vitro models. However, such in vitro models with over-simplified tests have not considering the influence of the accompanying *in site* microbiota or only focusing on microbiota composition without considering functional traits. Therefore, the fecal culture model, new in vitro model, was suggested to determine the impact of selected probiotics strains on the intestinal microbiota in term of the composition but also of functionality (Nogacka et al. 2021).

In this fecal culture model, six bacteria including LGG and TMC3115 were compared for their responses to intestinal microbiota from normal-weight (NW, BMI < 25 kg/m^2, 40 ± 9 years old) and severely obese subjects (OW, BMI < 25 kg/m^2, 44 ± 9 years old). All participants were asked to follow an unrestricted diet and had not taken antibiotics during the previous 6 months. Fecal samples were thawed at 37 °C under anaerobic conditions. As results, the tested probiotic strains did not induce major changes on the microbiota composition with well detected almost exclusively for the species added. Some of the tested bacteria, such as TMC3115 and LGG significantly reduced gas production from fecal culture of obese subjects. Furthermore, all of *Bifidobacterium* strains including TMC3115 and LGG characteristically modified/or restored the function of the intestinal microbiota and cells from the obese subjects similar to those obtained from normal-weight subjects in strain dependent way, at least partly. These results also indicate that the modification of the composition and function of intestinal microbiota from OB subjects might be one of the underlying mechanisms by which these bacteria could express the anti-obesity effects. The results also suggest that effects of probiotics may be influence/or modify by different intestinal microbiota for example those from OB, and it is important to take into account the mediation of the surrounding microbiota in the final effect on the host.

17.5 Conclusion

In this chapter, two testing models, the combination of macrophage and adipocyte and fecal culture, were suggested to evaluate /or screen the probiotics strains with anti-obesity effects. With these testing models, some selected probiotics strains such as *Lactobacillus* GG and *B. bifidum* TMC3115 exhibited its strain dependent abilities to activate macrophages by which to inhibit the adipogenic differentiation of preadipocytes, indicating that these may mediate the host immune response and affect their fat metabolism function. The results also indicate that the modification of the composition and function of intestinal microbiota might be one of the underlying mechanisms by which these bacteria could express the anti-obesity effects. Such anti-obesity effects have also been demonstrated with animal and human studies, at least partly. These results indicate that there might be deep association between intestinal microbiota and health of the elderly and some intestinal microbes might play important role in the development of obesity and metabolisms.

References

Cheng R, Guo J, Pu F, Wan C, Shi L, Li H, Yang Y, Huang C, Li M, He F (2019) Loading ceftriaxone, vancomycin, and *Bifidobacteria bifidum* TMC3115 to neonatal mice could differently and consequently affect intestinal microbiota and immunity in adulthood. Sci Rep 9 (1):3254

Cheng R, Xu T, Zhang Y, Wang F, Zhao L, Jiang Y, He F (2020) *Lactobacillus rhamnosus* GG and *Bifidobacterium bifidum* TMC3115 can affect development of hippocampal neurons cultured in vitro in a strain-dependent manner. Probiotics Antimicrob Proteins 12(2):589–599

Compare D, Rocco A, Zamparelli MS, Nardone G (2016) The gut bacteria-driven obesity development. Dig Dis 34(3):221–229

Gill HS, Darragh AJ, Cross ML (2001) Optimizing immunity and gut function in the healthy. J Nutr Health Aging 5(2):80–91

He F, Morita H, Ouwehand AC, Hosoda M, Hiramatsu M, Kurisaki J, Isolauri E, Benno Y, Salminen S (2002) Stimulation of the secretion of pro-inflammatory cytokines by Bifidobacterium strains. Microbiol Immunol 46(11):781–785

Kalache A (1999) Active ageing makes the differentce. Bull WHO 77:299

Kang X, Ling N, Sun G, Zhou Q, Zhang L, Sheng QJ (2012) Complete genome sequence of *Streptococcus thermophilus* strain MN-ZLW-002. Bacteriol 194(16):4428–4429

Kang X, Liang H, Luo Y, Li Z, He F, Han X, Zhang L (2021a) *Streptococcus thermophilus* MN-ZLW-002 Can Inhibit Pre-adipocyte Differentiation through Macrophage Activation. Biol Pharm Bull 44(3):316–324

Kang X, Liang H, Luo Y, Li Z, He F, Han X, Zhang L (2021b) Anti-adipogenesis and metabolism-regulating effects of heat-inactivated *Streptococcus thermophilus* MN-ZLW-002. Lett Appl Microbiol 72(6):677–687

Luo ZH, Lang CH, Wan Q, Ge L, Li M, Li L, Yang UH, He F (2018) Effect of *Bifidobacterium bifidum* TMC3115 to regulate immunity and lipometabolism. Food Sci Technol 43(09):13–18

Mariat D, Firmesse O, Levenez F, Guimarães V, Sokol H, Doré J, Corthier G, Furet JP (2009) The *Firmicutes/Bacteroidetes* ratio of the human microbiota changes with age. BMC Microbiol 9:123

Miyazawa K, He F, Yoda K, Hiramatsu M (2021) Potent effects of, and mechanisms for, modification of crosstalk between macrophages and adipocytes by lactobacilli. Microbiol Immunol 56(12):847–854

Nogacka AM, de Los Reyes-Gavilán CG, Arboleya S, Ruas-Madiedo P, Martínez-Faedo C, Suarez A, He F, Harata G, Endo A, Salazar N, Gueimonde M (2021) In vitro selection of probiotics for microbiota modulation in normal-weight and severely obese individuals: focus on gas production and interaction with intestinal epithelial cells. Front Microbiol 12:630572

Ortega FB, Lavie CJ, Blair SN (2016) Obesity and cardiovascular disease. Circ Res 118(11):1752–1770

Rusell RM (1992) Changes in gastrointestinal function attributed to aging. Am J Clin Nutr 55:1230–1207

Saunier K, Dore J (2002) Gastrointestinal tract and the elderly: functional foods, gut microflora and healthy ageing. Digest Liver Dis 34(supple 2):19–24

Schneeman BO (2002) Gastrointestinal physiology and function. Br J Nutr 88(2):159–163

Shen X, Miao J, Wan Q, Wang S, Li M, Pu F, Wang G, Qian W, Yu Q, Marotta F, He F (2018) Possible correlation between gut microbiota and immunity among healthy middle-aged and elderly people in southwest China. Gut Pathogens 10. https://doi.org/10.1186/s13099-018-0231-3

Shi L, Li M, Miyazawa K, Li Y, Hiramatsu M, Xu J, Gong C, Jing X, He F, Huang C (2013) Effects of heat-inactivated *Lactobacillus gasseri* TMC0356 on metabolic characteristics and immunity of rats with the metabolic syndrome. Br J Nutr 28;109(2):263–72

Wang SY, Yu XH, Li M, Li SS, Miao JJ, Yu Q, He FA (2017) Preliminary analysis of bifidobacterial species dependent influence on the health of the elderly. Sichuan Da Xue Xue Bao Yi Xue Ban 48(5):745–749

Wang K, Yu X, Li Y, Guo Y, Ge L, Pu F, Ma X, Cui W, Marrota F, He F, Li M (2018) *Bifidobacterium bifidum* TMC3115 can characteristically influence glucose and lipid profile and intestinal microbiota in the middle-aged and elderly. Probiotics Antimicrob Proteinshttps://doi.org/10.1007/s12602-018-9441-8

Wang K, Cui J, Lu QX, Chen SQ, Peng TY, Li M, Guo Y, Jiang L, He F, Yu XH, Wu X, Liu XR (2022) Preliminary study on association between intestine microbiota and blood glucose, blood lipid metabolism in middle-aged and elderly people in Chengdu. Sichuan Da Xue Xue Bao Yi Xue Ban 53(2):297–302

Wang F, Xu T, Zhang Y, Zheng T, He Y, He F, Jiang YF (2020) Long-term combined administration of *Bifidobacterium bifidum* TMC3115 and Lactobacillus plantarum 45 alleviates spatial memory impairment and gut dysbiosis in APP/PS1 mice. FEMS Microbiol Lett 1;367 (7):fnaa048

Chapter 18
Microbiome and Probiotics for Ischemic Stroke

Loo Keat Wei

Abstract Ischemic stroke is a deliberating disease, and the case is increasing as populations aged. This multifactorial disease is affected by various risk factor, not limited to hypertension, hyperlipidemia, genetics inheritance, epigenetics factors as well as microbiome. Mircobiome influences the gut-brain axis through phyla Bacteroidetes and Firmicutes followed by Proteobacteria and others in both clinical studies and in vivo studies. Genus *Anaerostipes*, *Bilophila*, *Butyricimonas*, *Enterococcus*, *Escherichia*, *Lactobacillus*, Morganella, Oscillospira, *Parabacteroides*, *Prevotella*, *Ruminococcus*, and *Subdoligranulum*, *Streptococcus* tend to be the predominant microbes for ischemic stroke cases, which also affect the neurological outcomes. Likewise, various strains of probiotics including *Bifidobacterium breve*, *Clostridium* butyricum, *Lactobacillus casei*, *Lactobacillus bulgaricus*, and *Lactobacillus acidophilus* have been widely used in various clinical trials for ischemic stroke. Of which, these strains may accelerate antioxidant enzymes production, improve gut microflora, attenuate neurological deficits and improvise the ischemic stroke outcomes.

Keywords Microbiome · Ischemic stroke · Probiotics · Prebiotics · Firmicutes · Bacteroidetes · Lactobaccilus · Escherichia

18.1 Introduction

Ischemic stroke is classified as a multifactorial disease that has claimed more than a dozen of life. This disease can be subclassified into five subtypes, where all of which exhibits different prognosis and diagnosis. Past studies confirmed that this

---

L. K. Wei (✉)
Department of Biological Science, Faculty of Science, Universiti Tunku Abdul Rahman, Bandar Barat, 31900 Kampar, Perak, Malaysia
e-mail: wynnelkw@gmail.com; lookw@utar.edu.my

F. Marotta (ed.), *Gut Microbiota in Aging and Chronic Diseases*, Healthy Ageing and Longevity 17, https://doi.org/10.1007/978-3-031-14023-5_18

non-communicable disease is contributed by epigenetics factors (Wei et al. 2015), family history (Chung et al. 2016), genetic polymorphisms (Au et al. 2015, 2017; Wei et al. 2017) and poor lifestyles (Joundi et al. 2021), uncontrolled diets (Spence 2018), etc. Recently, Lee et al. (2021) and others (Stanley et al. 2016; Singh et al. 2016) confirmed that microbiome played an important role in ischemic stroke through the gut-brain axis.

18.2 Microbiome and Ischemic Stroke

There are many kinds of definitions for microbiome, e.g. microbiome refers to the microbial community which resided in the reasonably well-defined habitat with distinct physio-chemical properties (Whipps et al. 1988). Subsequently, Lederberg and Mccray (2001) describe microbiome as a community of commensal, pathogenic and symbiotic microbes within a body space or other environment. Later, Marchesi and Ravel (2015) define microbiome as genomes, gene expression, and proteomes of microbes in an abiotic or a biotic environment.

It is suggested that the symbiotic relationship as well as spatial and temporal characteristics of microbe-microbe and microbe-host interactions contribute to human microbiome (Berg et al. 2020). For example, human microbiome entails of both potentially harmful and helpful microbes, i.e. archaeal, bacterial, fungal, microbial, viral, and taxa, which works synergistically in an environment. Most of these microbes exhibit beneficial effects, while there are a small number of them are pathogenic to human. Interestingly, these microbes which weighted about 2% of the adult body mass can coexist in a healthy individual's body without exerting any problems (Ursell et al. 2012).

The METAgenomics of the Human Intestinal Tract Project (2021) and The Human Microbiome Project (2021) which recruited more than 2000 healthy participants world-widely since 2006 suggested various body sites for the microbial analysis. These body sites are not limited to brain, eyes, mouth, skin, lungs, heart, gut, placenta, and urogenital tracts. Likewise, human blood is a good residential area for microbial growth. Depending on one's health status, human blood consists of 54% plasma, 45% erythrocytes, 0.7% lymphocytes, and numerous thrombocytes (Alberts et al. 2003). Thus, human blood-microbiome initiative initiates future research (Castillo et al. 2019).

Notwithstanding, the bidirectional gut-brain axis refers to the two-way communication between the central and the enteric nervous system. It connects cognitive and emotional centres of the brain with peripheral intestinal functions (Carabotti et al. 2015). Of which, the connection between gut microbiota and brain may contribute to the pathophysiology of ischemic stroke (Lee et al. 2021). As such, the communication between brain and gut through hypothalamic–pituitary–adrenal axis with the immune system at the autonomic nervous system is initiated in response to stressful stimuli. Integration of neuroimmune and neuroendocrine pathways promotes gut tissue inflammation, while deteriorating intestinal homeostasis (Battaglini et al. 2020).

Both local and systemic inflammatory responses are accelerated following an ischemic stroke event. The monocytes enter the ischemic brain and differentiate into tissue macrophages within 24 h following an ischemic stroke onset. These monocytes which could exert neuroprotective effects, reaching their maximum number at about three to five days of an onset (Battaglini et al. 2020). On the other hand, two-facets neutrophils exhibit both neuroprotective and neuroinflammatory effects on an ischemic brain. Neutrophils accelerate reactive oxidative species and metalloproteinase production and release, respectively. Meanwhile, neutrophils can be remodeled toward an N2 phenotype within the inflamed brain tissue by initiating the peroxisome proliferator-activated receptor-γ nuclear receptor (Cuartero et al. 2013). Hence, the neurological outcome following an ischemic stroke event is highly affected by the gut microbiota composition (Table 14.1).

In fact, gut microbiota composition changes gradually over a period, especially following a stressful event of disease onset. Clinical-based study with a proper study design such as a prospective, retrospective or follow-up study may allow researchers to puzzle how gut microbiota affects the neurological outcome. However, >88% of the clinical-based microbiome-ischemic stroke studies available to date still focused on the case–control design (Haak et al. 2021; Tan et al. 2021; Li et al. 2020, 2019; Huang et al. 2019; Wang et al. 2018; Yamashiro et al. 2017; Yin et al. 2015). Of which, fecal samples remained as the preferred biospecimens for the case–control studies (Haak et al. 2021; Tan et al. 2021; Li et al. 2020, 2019; Huang et al. 2019; Wang et al. 2018; Yamashiro et al. 2017; Yin et al. 2015). Meanwhile, saliva samples are used as the biospecimen for the prospective observational study (Boaden et al. 2017).

On the other hand, most of the in vivomicrobiome studies related to ischemic stroke were focused on animal models, but not limited to mice (Benakis et al. 2016, Singh et al. 2016, Crapser et al. 2016, Jandzinski 2015, Spychala et al. 2018, Singh et al. 2018, Stanley et al. 2018). The commonest mice model for ischemic stroke analysis is C57BL/6. For example, wild-type C57BL/6, KikGR33, Trdc-eGFP, $Il10^{-/-}$ and $Il17a^{-/-}$ as well as Il17a-eGFP have been used to examine the association between neuroprotective effects and altered abundances of microbial compositions (Benakis et al. 2016). In addition, the neuroinflammatory responses following stroke dysbiosis are determined by using germ-free C57BL/6 J and GF $Rag1^{-/-}$ female mice, and wild-type C57BL/6 J and $Rag1^{-/-}$ male mice (Singh et al. 2016, 2018).

18.3 Key Findings for Clinical-Based and Animal-Based Studies

When examining both clinical-based and animal-based studies for ischemic stroke, it has been discovered that phyla Bacteroidetes and Firmicutes were dominant over phyla Actinobacteria, Deferribacteraceae, Proteobacteria, Spirochaetes, Tenericutes and Verrucomicrobia (Stanley et al. 2016; Benakis et al. 2016; Singh et al. 2016; Crapser et al. 2016; Jandzinski 2015; Haak et al. 2021; Tan et al. 2021; Li et al. 2020; Huang et al. 2019; Li et al. 2019; Wang et al. 2018; Yamashiro et al. 2017; Yin et al. 2015; Xia et al. 2019; Spychala et al. 2018; Singh et al. 2018; Stanley et al. 2018; Boaden et al. 2017).

An increment of Firmicutes/Bacteroidetes ratio indicates a progressive aging event, which may also contribute to ischemic stroke pathophysiology (Lee et al. 2021). For example, phylum Firmicutes with genera *Lactobacillus*, *Prevotella* and *Streptococcus* predominate in the biospecimens of ischemic stroke cases, which acts as the sovereign for ischemic stroke risk (Li et al. 2020, 2019; Huang et al. 2019; Yamashiro et al. 2017). Of which, *Streptococcus species*, including *S. infantis,* have contributed to 70% of the oral microbiota for ischemic stroke cases (Boaden et al. 2017; Huang et al. 2019). Likewise, genera *Bilophila, Butyricimonas*, *Oscillospira,* and *Parabacteroides* are significantly enriched in the fecal samples for ischemic stroke patients (Xia et al. 2019). In addition, fecal samples also contain abundance of *P. copri* (Huang et al. 2019) as well as *Lactobacillus* and *Lactococcus* (Li et al. 2020) than healthy controls. On the contrary, Yamashiro et al. (2017) reported that decrease in *L. sakei* and increase in *L. ruminis* are the predictor for ischemic stroke risk.

It is important to take note that aerobic *Enterococcus* spp. and *Escherichia coli* as well as facultatively anaerobic *Morganella morganii*, and obligate anaerobic *Anaerostipes*, *Ruminococcus*, and *Subdoligranulum* are observed in the biospecimens of ischemic stroke patients, not limited to blood, urine or sputum (Stanley et al. 2016; Haak et al. 2021). Oxygen concentration level in the ischemic penumbra and ischemic core affect the neurological outcome. Thus, when examining the relationship between gut microbiota and neurological outcomes, it has been observed that higher abundance of *Gammaproteobacteria* and relatively lower abundance of *Bacteroidia*, *Prevotella* and *Faecalibacterium* are linked to cases with neurological deterioration (Yin et al. 2015; Wang et al. 2018; Haak et al. 2021). *Ruminococcaceae_UCG_005,* the short chain fatty acids producer for acetate, butyrate and propionate, which exhibited relatively higher abundances in cases, also contributed to severe neurological outcomes (Li et al. 2020, 2019). In addition, ischemic stroke cases with poor neurological outcomes tend to have increased abundance of genera *Akkermansia*, *Enterobacteriacea*, *Lactobacillaceae,* and *Porphyromonadaceae* (Tan et al. 2021).

Like the clinical studies, phyla Firmicutes and Bacteroidetes predominate in the gut and fecal of acute middle cerebral artery occlusion mice (Singh et al. 2016; Jandzinski 2015). When determining the microbial composition following a series of antibiotic courses, it has been discovered that phyla Firmicutes and Bacteroidetes outweigh in antibiotic-resistant mice, while phylum Proteobacteria prevail in antibiotic-sensitive mice (Benakis et al. 2016). Different genus of bacteria tends to take possession of different organs of mice following an ischemic event. For example, genera *Staphylococcus* and *Enterococcus* colonize peripheral organs (Crapser et al. 2016), genera *Actinomyces*, *Brevundimonas*, *Clostridium*, *Eshcherichia*, *Lactobacilus*, and *Shigella* outweigh in lungs (Stanley et al. 2016), and genera *Akkermansia*, *Alistipes*, *Anaerotruncus*, *Clostridium*, *Parabacteroides*, *Roseburia* prevail in gastrointestinal tract (Stanley et al. 2018).

18.4 Probiotics for Ischemic Stroke

There are limited publications on probiotics and ischemic stroke per se. Probiotics have been regarded as a tool that could modify gut microbiome. It is a dietary supplement that contain a wide range of viable microbes that provides short-term and long-term health benefits to the host when administrated in sufficient amounts. These beneficial microbes, not limited to *Bifidobacterium* and *Lactobacillus*, are facilitating and enhancing the propagation of favorable indigenous gastrointestinal microflora. Moreover, probiotics also helps in attenuate antimicrobial-associated diarrhea, astriction, and dysbacteriosis in both animals and humans (Sun et al. 2016). Due to the positive effects exerted by the probiotics, these beneficial microbes have been used in various types of ischemic stroke treatments. An in vivo study which involved 30 male BLC57 mice revealed that the infract size in the posterior middle cerebral artery region was shrunken by 52% following the probiotics pre-treatment (Akhoundzadeh et al. 2018). For example, probiotics pre-treatment may decrease the contents of oxidative stress biomarker and proinflammatory cytokine within the brain tissue significantly. Of which, the malondialdehyde levels of mice which have been accelerated following the breakdown of antioxidant enzyme formation, are being restored to its baseline levels following the probiotics administration (Akhoundzadeh et al. 2018). One of the prominent examples, i.e. *Clostridium butyricum* can produce significant amount of butyrate in the gut lumen which could also exert neuroprotective effects against cerebral ischemia/reperfusion injury via anti-oxidation mechanism (Sun et al. 2016). As for the anti-oxidative effects, *C. butyricum* decreases the neurological deficit score in mice by accelerating antioxidant enzymes production to scavenge the free radicals, while decelerating the reactive oxygen species production (Sun et al. 2016).

It has been known that extreme oxidative stress levels can contribute to excessive proinflammatory cytokine production, thereby, enhancing the proinflammatory signalling pathway. Akhoundzadeh et al. (2018) suggested that probiotics pre-treatment on mice through *Bifidobacterium breve*, *Lactobacillus casei*, *Lactobacillus bulgaricus*, and *Lactobacillus acidophilus* may attenuate the levels of proinflammatory cytokine tumor necrosis factor-alpha within their neuronal cells.

These probiotics exert the neuroprotective effects by enhancing brain-derived neurotrophic factor (BDNF) levels while decelerating the apoptosis pathway (Akhoundzadeh et al. 2018). Post-hoc analysis of a double-blind randomized controlled trial of 28 patients with low-to-moderate depression who receiving probiotic vs 25 patients with low-to-moderate depression who receiving prebiotic and placebo suggested that a course of eight-week supplementation with *Bifidobacterium longum* R0175 and *Lactobacillus helveticus* R0052 might increase serum BDNF levels, and thus, attenuate the depression symptoms (Messaoudi et al. 2011). BDNF protects neuronal cells from 3-NP-induced apoptotic cascades, by counter-regulating the Bax/Bcl-2. By which, *C. butyricum* may attenuate the cerebral ischemia/reperfusion by enhancing butyrate levels in neuronal cells (Sun et al. 2016).

Recent meta-analysis on the efficacy of probiotics for ischemic stroke suggested that enteral nutrition coupled with probiotics may improvise the nutritional status of patients and modulate the intestinal microflora (Chen et al. 2021). This meta-analysis encompasses 26 randomized controlled trials highlighted that enteral nutrition coupled with probiotics group exhibited lesser incidence of gastrointestinal complications (OR 0.29, 95% CI 0.24–0.36, $p < 0.05$) and infection (OR 0.27, 95% CI 0.21–0.36, $p < 0.05$) as well as shorter length of hospitaliization (MD– 8.70, 95% CI−13.24 to −4.16, $p < 0.05$) and dysbacteriosis rate (OR 0.17, 95% CI 0.07–0.41, $p < 0.05$) as compared to early enteral nutrition group. When comparing the biochemical profiles between early enteral nutrition group vs enteral nutrition coupled with probiotics group it has been discovered that the latter exhibited higher albumin, pre-albumin, total protein, hemoglobin as well as immunoglobulins A and G levels (MD 3.38, 95% CI 2.74–4.02, $p < 0.05$ and MD 32.20, 95% CI 24.42–39.98, $p < 0.05$, MD 4.91, 95% CI 3.20–6.62, $p < 0.05$, MD 9.62, 95% CI 7.92–11.32, $p < 0.05$, MD 0.23, 95% CI 0.12–0.34, $p < 0.05$, MD 0.33, 95% CI 0.21–0.45, $p < 0.05$, respectively) (Chen et al. 2021). All of these biochemical profiles indicate probiotics may increase the surface area of small intestinal villi and enhance nutrients absorption. Hence, it is obvious that probiotics exhibits different kinds of beneficial effects on ischemic stroke treatment.

18.5 Conclusion

Various strains of probiotics supplementation such as *B. breve*, *C.* butyricum, *L. casei*, *L. bulgaricus*, and *L. acidophilus* may be beneficial for ischemic stroke patients. These strains may accelerate antioxidant enzymes production, decelerate eactive oxygen species production, balance oxidative stress levels, improve gut microflora, and attenuate neurological deficits. Nonetheless, a synbiotics treatment i.e. by combining probiotics and prebiotics strains is suggested for future research, which is to minimize the side effects related to probiotics, while to exert long-term neuroprotective effects against ischemic stroke.

Acknowledgements This research was supported by Ministry of Higher Education (MOHE), Malaysia through Fundamental Research Grant Scheme FRGS/1/2021/SKK0/UTAR/02/4 and Universiti Tunku Abdul Rahman Research Fund (UTARRF) IPSR/RMC/UTARRF/2021-C1/L07.

Conflict of Interest Statement None declared.

References

Akhoundzadeh K, Vakili A, Shadnoush M, Sadeghzadeh J (2018) Effects of the oral ingestion of probiotics on brain damage in a transient model of focal cerebral ischemia in mice. Iranian J Med Sci 43(1):32

Alberts B, Johnson A, Lewis J, Raff M, Roberts K, Walter P (2003) Molecular biology of the cell. Scand J Rheumatol 32(2):125–125

Au A, Griffiths LR, Cheng KK, Kooi CW, Irene L, Wei LK (2015) The influence of OLR1 and PCSK9 gene polymorphisms on ischemic stroke: evidence from a meta-analysis. Sci Rep 5 (1):1–11

Au A, Griffiths LR, Irene L, Kooi CW, Wei LK (2017) The impact of APOA5, APOB, APOC3 and ABCA1 gene polymorphisms on ischemic stroke: evidence from a meta-analysis. Atherosclerosis 265:60–70

Battaglini D, Pimentel-Coelho PM, Robba C, Dos Santos CC, Cruz FF, Pelosi P, Rocco PRM (2020) Gut microbiota in acute ischemic stroke: from pathophysiology to therapeutic implications. Front Neurol 11:598

Benakis C, Brea D, Caballero S, Faraco G, Moore J, Murphy M, Sita G, Racchumi G, Ling L, Pamer EG, Iadecola C (2016) Commensal microbiota affects ischemic stroke outcome by regulating intestinal γδ T cells. Nat Med 22(5):516–523

Berg G, Rybakova D, Fischer D, Cernava T, Vergès MCC, Charles T, Chen X, Cocolin L, Eversole K, Corral GH, Kazou M (2020) Microbiome definition re-visited: old concepts and new challenges. Microbiome 8(1):1–22

Boaden E, Lyons M, Singhrao SK, Dickinson H, Leathley M, Lightbody CE, McLoughlin A, Khan Z, Crean S, Smith C, Higham S (2017) Oral flora in acute stroke patients: A prospective exploratory observational study. Gerodontology 34(3):343–356

Carabotti M, Scirocco A, Maselli MA, Severi C (2015) The gut-brain axis: interactions between enteric microbiota, central and enteric nervous systems. Ann Gastroenterol Q Publ Hellenic Soc Gastroenterol 28(2):203

Castillo DJ, Rifkin RF, Cowan DA, Potgieter M (2019) The healthy human blood microbiome: fact or fiction? Front Cell Infect Microbiol 9:148

Chen X, Hu Y, Yuan X, Yang J, Li K (2021) Effect of early enteral nutrition combined with probiotics in patients with stroke: a meta-analysis of randomized controlled trials. Eur J Clin Nutr:1–12

Chung JW, Kim BJ, Han MK, Kang K, Park JM, Park SS, Park TH, Cho YJ, Hong KS, Lee KB, Kim JG (2016) Family history and risk of recurrent stroke. Stroke 47(8):1990–1996

Crapser J, Ritzel R, Verma R, Venna VR, Liu F, Chauhan A, Koellhoffer E, Patel A, Ricker A, Maas K, Graf J (2016) Ischemic stroke induces gut permeability and enhances bacterial translocation leading to sepsis in aged mice. Aging (albany NY) 8(5):1049

Cuartero MI, Ballesteros I, Moraga A, Nombela F, Vivancos J, Hamilton JA, Corbí ÁL, Lizasoain I, Moro MA (2013) N2 neutrophils, novel players in brain inflammation after stroke: modulation by the PPARγ agonist rosiglitazone. Stroke 44(12):3498–3508

Haak BW, Westendorp WF, van Engelen TS, Brands X, Brouwer MC, Vermeij JD, Hugenholtz F, Verhoeven A, Derks RJ, Giera M, Nederkoorn PJ (2021) Disruptions of anaerobic gut bacteria

are associated with stroke and post-stroke infection: a prospective case–control study. Transl Stroke Res 12(4):581–592
Huang L, Wang T, Wu Q, Dong X, Shen F, Liu D, Qin X, Yan L, Wan Q (2019) Analysis of microbiota in elderly patients with Acute Cerebral Infarction. PeerJ 7:e6928
Jandzinski M (2015) Manipulation of the microbiome and its impact on functional recovery following ischemic stroke
Joundi RA, Patten SB, Williams JV, Smith EE (2021) Association between excess leisure sedentary time and risk of stroke in young individuals. Stroke 52(11):3562–3568
Lederberg J, McCray AT (2001) Ome SweetOmics–A genealogical treasury of words. The Scientist 15(7):8–8
Lee YT, Mohd Ismail NI, Wei LK (2021) Microbiome and ischemic stroke: a systematic review. PloS One 16(1):e0245038
Li N, Wang X, Sun C, Wu X, Lu M, Si Y, Ye X, Wang T, Yu X, Zhao X, Wei N (2019) Change of intestinal microbiota in cerebral ischemic stroke patients. BMC Microbiol 19(1):1–8
Li H, Zhang X, Pan D, Liu Y, Yan X, Tang Y, Tao M, Gong L, Zhang T, Woods CR, Du Y (2020) Dysbiosis characteristics of gut microbiota in cerebral infarction patients. Transl Neurosci 11 (1):124–133
Marchesi JR, Ravel J (2015) The vocabulary of microbiome research: a proposal
Messaoudi M, Lalonde R, Violle N, Javelot H, Desor D, Nejdi A, Bisson JF, Rougeot C, Pichelin M, Cazaubiel M, Cazaubiel JM (2011) Assessment of psychotropic-like properties of a probiotic formulation (Lactobacillus helveticus R0052 and Bifidobacterium longum R0175) in rats and human subjects. Br J Nutr 105(5):755–764
METAgenomics of the Human Intestinal Tract Project (2021) https://www.gutmicrobiotaforhealth.com/metahit/
Singh V, Roth S, Llovera G, Sadler R, Garzetti D, Stecher B, Dichgans M, Liesz A (2016) Microbiota dysbiosis controls the neuroinflammatory response after stroke. J Neurosci 36 (28):7428–7440
Singh V, Sadler R, Heindl S, Llovera G, Roth S, Benakis C, Liesz A (2018) The gut microbiome primes a cerebroprotective immune response after stroke. J Cereb Blood Flow Metab 38 (8):1293–1298
Spence JD (2018) Diet for stroke prevention. Stroke Vascular Neurol 3(2)
Spychala MS, Venna VR, Jandzinski M, Doran SJ, Durgan DJ, Ganesh BP, Ajami NJ, Putluri N, Graf J, Bryan RM, McCullough LD (2018) Age-related changes in the gut microbiota influence systemic inflammation and stroke outcome. Ann Neurol 84(1):23–36
Stanley D, Mason LJ, Mackin KE, Srikhanta YN, Lyras D, Prakash MD, Nurgali K, Venegas A, Hill MD, Moore RJ, Wong CH (2016) Translocation and dissemination of commensal bacteria in post-stroke infection. Nat Med 22(11):1277–1284
Stanley D, Moore RJ, Wong CH (2018) An insight into intestinal mucosal microbiota disruption after stroke. Sci Rep 8(1):1–12
Sun J, Ling Z, Wang F, Chen W, Li H, Jin J, Zhang H, Pang M, Yu J, Liu J (2016) Clostridium butyricum pretreatment attenuates cerebral ischemia/reperfusion injury in mice via anti-oxidation and anti-apoptosis. Neurosci Lett 613:30–35
Tan C, Wu Q, Wang H, Gao X, Xu R, Cui Z, Zhu J, Zeng X, Zhou H, He Y, Yin J (2021) Dysbiosis of gut microbiota and short-chain fatty acids in acute ischemic stroke and the subsequent risk for poor functional outcomes. J Parenter Enter Nutr 45(3):518–529
The Human Microbiome Project (2021) https://hmpdacc.org/
Ursell LK, Metcalf JL, Parfrey LW, Knight R (2012) Defining the human microbiome. Nutr Rev 70(suppl_1):S38–S44
Wang W, Li X, Yao X, Cheng X, Zhu Y (2018) The characteristics analysis of intestinal microecology on cerebral infarction patients and its correlation with apolipoprotein E. Medicine 97(41)
Wei KL, Sutherland H, Au A, Camilleri E, Haupt LM, Gan SH, Griffiths LR (2015) A potential epigenetic marker mediating serum folate and vitamin B12 levels contributes to the risk of ischemic stroke. BioMed Res Int

Wei LK, Au A, Menon S, Griffiths LR, Kooi CW, Irene L, Zhao J, Lee C, Alekseevna AM, Hassan MRA, Aziz ZA (2017) Polymorphisms of MTHFR, eNOS, ACE, AGT, ApoE, PON1, PDE4D, and ischemic stroke: meta-analysis. J Stroke Cerebrovasc Dis 26(11):2482–2493

Whipps J, Lewis K, Cooke R (1988) Mycoparasitism and plant disease control. In: Burge M (ed) Fungi biol control syst. Manchester University Press, pp 161–187

Xia GH, You C, Gao XX, Zeng XL, Zhu JJ, Xu KY, Tan CH, Xu RT, Wu QH, Zhou HW, He Y (2019) Stroke dysbiosis index (SDI) in gut microbiome are associated with brain injury and prognosis of stroke. Front Neurol 10:397

Yamashiro K, Tanaka R, Urabe T, Ueno Y, Yamashiro Y, Nomoto K, Takahashi T, Tsuji H, Asahara T, Hattori N (2017) Gut dysbiosis is associated with metabolism and systemic inflammation in patients with ischemic stroke. PLoS ONE 12(2):e0171521

Yin J, Liao SX, He Y, Wang S, Xia GH, Liu FT, Zhu JJ, You C, Chen Q, Zhou L, Pan SY (2015) Dysbiosis of gut microbiota with reduced trimethylamine-N-oxide level in patients with large-artery atherosclerotic stroke or transient ischemic attack. J Am Heart Assoc 4(11): e002699

Chapter 19
Physical Exercise and Gut Microbiota in the Aging Process: From Osteoporosis to Preventive Bone Health Strategy

Cong Huang

Abstract The incidence of osteoporosis is increasing globally and becoming a serious public health problem for the elderly. Physical activity engagements could promote bone health and prevent osteoporosis via cytokines, RANKL/RANK/OPG pathway, or Wnt-β-Catenin pathway. Some of these regulatory pathways may be influenced by physical activity-induced changes in gut microbiota composition and its metabolites. In light of the evidence presented in this chapter, it may be possible to demonstrate the existence of the "exercise-gut-bone" axis. Further studies are warranted to confirm this hypothesis and clarify the mechanisms and effects of various physical activity elements (type, time, frequency and intensity) on gut microbiota composition and bone metabolism.

Keywords Physical activity · Gut microbiota · Bone metabolism · Osteoporosis · Aging

19.1 Introduction

Human aging has been a topic of great interest. Aging is a complex, multifaceted process that affects every organ and tissue (Schumacher et al. 2021). This physiological process refers to the structural and functional decline of systemic tissues and organs with the human body growth and age. Notably, loss of bone mass or bone density is an important manifestation of the aging process. In this regard, bone mass begins to gradually decrease toward the end of the third decade of life. As a result of this phenomenon, the incidence of osteoporosis is increasing, particularly in postmenopausal women. Osteoporosis and its fractures not only significantly impair health and quality of life but also create a serious economic burden on health care systems. By 2025, the economic burden of osteoporotic fracture is projected to

C. Huang (✉)
Department of Sports Science, Zhejiang University, 866 Yuhangtang Road,
Hangzhou 310058, China
e-mail: cohuang@zju.edu.cn

F. Marotta (ed.), *Gut Microbiota in Aging and Chronic Diseases*, Healthy Ageing and Longevity 17, https://doi.org/10.1007/978-3-031-14023-5_19

be 25.3 billion dollars in the US (Kemmak et al. 2020). Therefore, the development of enhanced strategies for osteoporosis prevention, bone health improvement, and active aging promotion has become an area of active research interest.

Physical activity is a validated anti-bone aging method. In fact, the two bone cell types, osteoclasts and osteoblasts, are influenced by physical activity (Kadow-Romacker et al. 2009). Previous review showed that 2 to 3 times of weighted exercises per week could maintain or increase bone mineral density at the lumbar spine and hip in postmenopausal women with osteopenia and osteoporosis (Zehnacker and Bemis-Dougherty 2007). However, studies of how physical activity improves bone health are required to further understand this observation.

In recent years, gut microbiota has been shown to have an impact on bone health (Ohlsson and Sjögren 2015; Santos et al. 2017; Zhang et al. 2018). Notably, due to the rapid development of microbial research technology, gut microbiota has become one of the most popular biological research topics. These advances enable further understanding of the relationships of gut microbiota with bone metabolism. In this regard, interactions between gut microbiota and bone health may be mediated by nutrition, immunity, endocrine, and metabolites mechanisms.

In this chapter, we will review recent discoveries of the associations among physical activity, gut microbiota, and bone health. Additionally, we will summarize the potential mechanisms by which physical activity and gut microbiota influence bone health.

19.2 Aging and Bone Health

Bone is a dynamic organ with constant remodeling through life. This remodeling process is coordinated by the activity of two cell types, osteoclasts and osteoblasts. Bone resorption is carried out by osteoclasts, while new bone formation is carried out by osteoblasts (Quach and Britton 2017). Bone remodeling consists of five stages: activation of osteoclasts, resorption, reversal, formation, and quiescence (Parfitt 1984). Aging is accompanied by the loss of bone mass and strength, increasing the risk of osteoporosis (Demontiero et al. 2012). Osteoporosis is characterized by decreased bone mass and deteriorated bone tissue microarchitecture. This systemic bone metabolic disease may lead to enhanced skeletal fragility and subsequent increase in fracture risk (Organization 1994). Peak bone mass is generally thought to occur by the end of the second or early in the third decade of life (Baxter-Jones et al. 2011), when calcium storage has reached its maximum and begun to decline (Santos et al. 2017). Particularly in postmenopausal women, the rate and amount of bone loss increase dramatically and greatly exceeding those in men (Hendrickx et al. 2015) (Fig. 19.1).

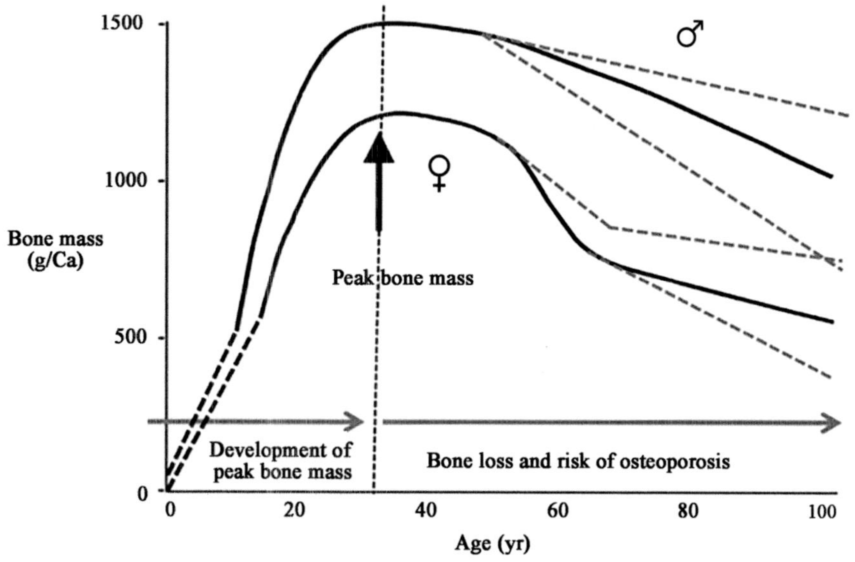

Fig. 19.1 Age and Bone Mass (From: Harvey N, Dennison E, Cooper C. Osteoporosis: a life course approach. JBMR, 2014, 29(9): 1917–1925.)

19.3 Physical Activity and Bone Health

Physical activity habits are critical to preventing bone loss and maintaining bone strength. Physical activity is defined as any bodily movement produced by skeletal muscles and results in energy expenditure (Caspersen et al. 1985). This term includes the full range of human movements from competitive sport and exercise to daily life hobbies or activities. Exercise represents a type of structured physical activity with the aim of improving or maintaining one or more components of physical fitness (Caspersen et al. 1985). Exercise makes bones and muscles stronger.

When a person exercises regularly, their bones will adapt by building more bones and becoming denser. There are many different types of daily exercise with health benefits. Weight-bearing and strength-training exercises are considered the two most effective ways to strengthen bones (Mosti et al. 2013). Weight-bearing exercise refers to on-foot physical activities that force bones and muscles to work against gravity. This type of exercise could help prevent bone loss and strengthen bone after young adulthood. Examples of weight-bearing exercises include rope jumping, dancing, stair climbing, brisk walking, and hiking. Strength-training exercise refers to resistance training on bones and muscles. Although resistance training focuses on increasing muscle mass, it also creates bone stress and strengthens bone-building capacity. Common types of strength-training exercises include weightlifting and resistance band work-out.

Many cross-sectional (Nguyen et al. 2000; Muraki et al. 2007), prospective or longitudinal (Heesch et al. 2008), and randomized controlled trials (RCTs) (Allison et al. 2013; Cheung and Giangregorio 2012; Harding et al. 2020) have implicated exercise in the regulation of bone health in elderly individuals. For example, a cross-sectional study of older Japanese women showed that the body mineral density was higher in participants with physical activity habits (Muraki et al. 2007). Additionally, an interventional study in 65-to-80-year-old male participants indicated that 12-month unilateral exercise increased femoral neck bone mineral density, bone mineral content, and geometry (Allison et al. 2013). Other researchers also found that an 8-month improvement in trochanteric bone mineral density was greater in the high-intensity progressive resistance and impact training group than the non-training group (Harding et al. 2020). These findings confirm that physical exercise plays a very important role in improving bone health in the elderly.

There are several potential mechanisms, such as cytokines, RANKL/RANK/OPG pathway, and Wnt-β-Catenin pathway, which may explain the role of physical activity on bone health. Firstly, physical exercise may regulate bone resorption by altering cytokine levels. Previous studies have shown that both anti-inflammatory and pro-inflammatory cytokines contribute to promoting the proliferation and activity of osteoclasts, leading to increased bone resorption (Takayanagi 2007; Nakashima and Takayanagi 2009; Pacifici 2010). Specifically, tumor necrosis factor-α (TNF-α), and interleukin (IL)-1 and IL-17 are important osteoclast activators, whereas IL-12, IL-18, IL-33, and interferons are powerful suppressors of osteoclast differentiation and inhibit bone loss (Schett 2011). Furthermore, the RANKL/RANK/OPG pathway may be an important regulatory pathway of osteoclast maturation (He et al. 2017). The main components of this signaling system are receptor activator of NF-kB (RANK), receptor activator of NF-kB ligand (RANKL), and osteoprotegerin (OPG). RANKL promotes osteoclast differentiation and activity by binding to its receptor RANK on the surface of osteoclast progenitors. On the other hand, OPG prevents RANKL from binding to RANK and thus inhibits RANK activation, preventing osteoclast recruitment and differentiation (Yuan et al. 2016). It has been shown that exercise increases OPG expression and decreases RANKL expression, both of which help to inhibit osteoclast differentiation and activity (Kish et al. 2015; Bergström et al. 2012). In animal studies, resistance exercise could increase OPG expression (Notomi et al. 2014). Moreover, exercise-elicited mechanical loading also promotes bone formation, in which the Wnt-β-Catenin pathway might play an important role. The theory of mechanical loading leads to increased bone mass accretion which was first developed by Frost (Frost 2003). According to this theory, the internal biological system of the bone causes bone formation in response to high mechanical strains, thereby strengthening the bone. Notably, exercise might regulate osteoblastic differentiation of mesenchymal precursors through mechanical loading-activated Wnt signaling (Yuan et al. 2016). In this context, osteocytes maintain bone homeostasis by sensing mechanical loads and transmitting information to osteoblasts and osteoclasts. Leal and colleagues reported that 8 weeks of power training could significantly upregulate Wnt1 and β-catenin protein expression in physically active young men (Leal et al. 2011). Likewise, an interventional trial has indicated that resistance exercise could increase Wnt4 and β-catenin levels in young men (Spillane et al. 2015). These observations suggest that exercise may positively impact bone health by upregulating the expression of Wnt molecules.

19.4 Physical Activity and Gut Microbiota

Microbiota represents a large number of bacteria, archaea, viruses, and unicellular eukaryotes in the human body (Sekirov et al. 2010). For example, the intestinal tract is a nutrient-rich environment packed with up to 100 trillion microbiotas (Ley et al. 2006). The gut microbiome composition evolves from early childhood to old age. While the gut microbiota is stable in healthy adults, changes in the composition and number of these microbes can occur in the elderly (Mariat et al. 2009). The change of gut microbiota can reflect the health status of the body, whereas its composition is easily affected by the external environment and host factors, such as physical activity (Queipo-Ortuño et al. 2013).

Observational evidence has indicated that physical activity is closely related to the diversity and composition of gut microbiota in humans. In young adults, people who regularly exercise show a greater abundance of individual microbial alpha diversity, especially the *Firmicutes* and *Clostridiales* (McFadzean 2014). In the studies comparing athletes to the healthy males (Clarke et al. 2014; Han et al. 2020) or sedentary (Kulecka et al. 2020) controls, higher physical activity was associated with more abundance of alpha diversity. Moreover, physically inactive young adults are more likely to have a lower immune function and a higher abundance of pro-inflammatory gut bacteria than elite athletes and physically active young adults (Xu et al. 2022). There are also differences in alpha richness and diversity, assessed by the Shannon and Simpson diversity indices, between higher-level athletes and lower-level athletes (Liang et al. 2019). In older adults, the alpha diversity of gut microbiota increased with age. Notably, the alterations in microbial composition and function induced by aging were reshaped by regular exercise (Zhu et al. 2020). Furthermore, compared to healthy older adults, the older athletes had a more homogeneous microbiota within the group and a higher abundance of *Faecalibacterium prausnitzii*, which has been shown to have beneficial health properties (Fart et al. 2020).

As the exploration of exercise intervention and gut microbiota has become a research topic of high interest in exercise science and rehabilitation medicine, several RCTs have been conducted in recent years (Table 19.1). Dupuit et al. indicated that 12 weeks of combined high-intensity interval and resistance training could significantly alter microbiota composition in overweight or obese post-menopausal women (Dupuit et al. 2021). Similarly, a combined resistance and aerobic intervention could change both beta diversity and composition of gut microbiota such as *Firmicutes* and *Bacteroidetes*, in physically inactive elderly women (Zhong et al. 2021). Based on the current evidence from the RCTs in Table 19.1,

Table 19.1 RCTs on physical activity and gut microbiota

Author, publish year	Characteristics of objects	Exercise type	Period and frequency	Main results
Quiroga et al. (2020)	N = 39, obese children, 7 ~ 12 years old	Combine resistant and aerobic/control	12 weeks; 2 times a week for 45 min	(phyla) ↓ *Proteobacteria*; (class) * *Clostridia*, *Flavobacteriia*, *Actinobacteria* and *Gammaproteobacteria*; (genus) ↑ *Blautia*, *Dialister*, and *Roseburia*
Resende et al. (2021)	N = 24, healthy men, (C) 25.5 ± 4.66 (E) 25.58 ± 5.07 years old	Aerobic/control	10 weeks; 3 times a week for 50 min	–α diversity; - β diversity; –(phyla) High individual differences (genus) ↑ *Streptococcus*; ↓ an unclassified genus belonging to *Clostridiales* order
Mahdieh et al. (2021)	N = 16, female, 19 ~ 30 years old	Aerobic/control	10 weeks; 3 times a week for 30 ~ 45 min	(genus) ↑ *Bifidobacterium*; –*Lactobacillus*
Cronin et al. (2018)	N = 17, adults with Inflammatory Bowel Disease, 18 ~ 40 years old	Combine resistant and aerobic/control	8 weeks; 3 times a week	–α diversity; - β diversity; –Metabolic pathways
Moitinho-Silva et al. (2021)	N = 42, healthy adults without exercise, 20 ~ 45 years old	Aerobic/strength/control	6 weeks; 3 times a week for 30 min +	–α diversity; - β diversity; –bacteria abundance
Mokhtarzade et al. (2021)	N = 42, adults with Multiple Sclerosis, (E) 35.06 ± 8.18 (C) 36.38 ± 9.13 years old	Combine resistant and aerobic/control	6 months; 5 times a week (3 aerobic and 2 resistant exercises)	(genus) ↑ *Prevotella*; ↓ *Akkermansia muciniphila*; –*Faecalibacterium prausnitzii*, *Bacteroides*
Kern et al. (2020)	N = 88, overweight or obese adults, around 36 years old	Bike/moderate exercise (50% of VO_{2peak}-reserve)/vigorous exercise (70% of VO_{2peak}-reserve)/control	6 months; weekly energy expenditure of exercise: female 1600 kcal, male 2100 kcal	* α diversity; - β diversity; –(genus) and (phyla)
Warbeck et al. (2021)	N = 23, adults with Celiac Disease, (E) 42.0 ± 12.3 (C) 46.2 ± 10.2 years old	HIIT + lifestyle/control	12 weeks 2 times a week for 60 min	–α diversity; * β diversity; (genus) (C) ↑ *Roseburia intestinalis*, *Klebsiella*, and *Adlercreutzia*; (E) ↑ *Defluviitaleaceae_UCG_011* and *Parabacteroides*;

(continued)

Table 19.1 (continued)

Author, publish year	Characteristics of objects	Exercise type	Period and frequency	Main results
				↓ *Lachnoclostridium* and *Allistipe*; ↑ *Ruminococcaceae_UCG_002. bacterium*, *Dialister*, and *Streptococcus*
Dupuit et al. (2021)	N = 17, postmenopausal women with overweight or obesity, (C) 60.9 ± 4.8 (E) 58.8 ± 5.3 years old	HIIT + resistant/control	12 weeks; 3 times a week	–α diversity; * β diversity; –(phyla) and (family);
Zhong et al. (2021)	N = 12, elderly female, 60 ~ 75 years old	Combine resistant and aerobic/control	8 weeks; 4 times a week for 60 min	–α diversity; - β diversity; (phyla) (E) ↓ *Firmicutes*; (C) ↑ *Bacteroidetes*; (C) ↓ *Actinobacteria*; (class) ↓ *Clostridia*, ↑ *Betaproteobacteria*; (Order) (C) ↓ *Bifidobacteriales*, (E) ↑ (C) ↓ *Burkholderiales*; (family) ↑ *Acidaminococcaceae*, (C) ↑ *Bacteroidaceae*, * *Bifidobacteriaceae*, * *Eubacteriaceae*; (Genus) ↑ *Phascolarctobacterium* and *Mitsuokella*, (C) ↑ *Bacteroides* and *Parabacteroides*; * *Bifidobacterium*, *Gemmiger*, and *Parabacteroides*
Taniguchi et al. (2018)	N = 31, elderly men, 62 ~ 76 years old	Aerobic/control	5 weeks; 3 times a week for 30 ~ 45 min	–α diversity; - β diversity; (genus) ↑ *Oscillospira* (was no increase after adjusting diet); (species) ↓ *Clostridium difficile* (significant after adjusting diet)

*: change; -: no change; ↑: increase; ↓: decrease; (C): control group; (E): exercise group; VO_{2peak}: peak oxygen consumption; HIIT: high-intensity interval training

most study subjects are young and/or middle-aged adults. A few studies have focused on the older population. The main consequences of the alterations in gut microbiome composition observed in these studies were not consistent because of the variations in the detected and reported gut bacteria. Therefore, rigorous analyses of gut microbiota in future studies will contribute to the accumulation of relevant research evidence. In the light of exercise intervention approach, more researchers prefer aerobic exercise independently or combined with resistance exercise. Recent studies by Warbeck (Warbeck et al. 2021) and Dupuit (Dupuit et al. 2021) also examined the effect of high-intensity interval training on gut microbiota in adults with Celiac disease or postmenopausal elderly women with overweight or obesity. It is worth noting that both studies found this type of exercise intervention can significantly alter gut microbiome beta diversity. Nevertheless, the impact of different exercise frequencies, intensity times, and types on gut microbiota are still largely unexplored. Additional experimental trials are required to reveal how these physical activity elements influence the human gut microbiota.

Although physical activity can affect gut microbiota directly or indirectly, knowledge of potential mechanisms of this phenomenon is limited. During exercise, blood flow is concentrated in skeletal muscle, heart, and lungs, causing ischemia and hypoxia in visceral tissues, such as the intestine, resulting in loss of membrane integrity of intestinal epithelium (Shepherd and Vanhoutte 1978; Clausen 1977; van Wijck et al. 2012; Derikx et al. 2008). At the end of ischemia and hypoxia, intestinal reperfusion occurs. Cellular signals are then released to trigger an inflammatory response that clears out damaged cells (Carden and Granger 2000; Jaeschke 2003; Arumugam et al. 2004). Apoptosis of mature intestinal epithelial cells and recovery of epithelial cell integrity occur during this process (Derikx et al. 2008). The process of intestinal ischemia and reperfusion speeds up cell renewal, which may be a mechanism for gut microbiome alterations. A hot environment created by exercise and the loss of body fluids from regulating body temperature could also accelerate this process (Vargas and Marino 2016; Pires et al. 2017; Zuhl et al. 2014; Lambert et al. 2008). In addition, exercise can change the transit time of gut (Oettle 1991), and subsequently modulate its microbiota (Müller et al. 2020). Furthermore, the study found that the hypothalamic–pituitary–adrenal (HPA) axis could be regulated by exercise (Field 2011), which in turn can influence the gut microbiome composition (Bermon et al. 2015). In this regard, HPA axis activation can release a variety of hormones under physical or psychological stress conditions, leading to an imbalance of gut microbiota in the body (Galley and Bailey 2014). Lastly, the bile acids have been linked to physical activity and gut microbiota. As a host factor, bile acids can modulate gut microbiome composition in rats (Islam et al. 2011; Zheng et al. 2017). Additionally, total plasma bile acids were significantly decreased after resistant training (Morville et al. 2018). As a result, bile acids may be an underlying mechanism by which exercise regulates gut microbiota and improves body metabolism.

19.5 Gut Microbiota and Bone Health

The development of microbial detection technology has advanced the understanding of the important role of gut microbiota and its metabolites in human health. More emphasis has recently been placed on the role of gut microbiota in bone metabolisms.

Earlier studies found that germ-free mice had higher bone mass, indicating an involvement of gut microbiota in this process (Sjögren et al. 2012). Antibiotic tests in germ-free mice supported this evidence. Age and gender were also contributing factors to the effects of gut microbiota on bone health (Cho et al. 2012; Cox et al. 2014). In addition, ovariectomized mice with *Lactobacillus acidophilus* showed reduced bone resorption markers and inhibition of osteoclast formation (Ohlsson et al. 2014). In contrast, with the use of *Lactobacillus reuteri*, a reduction in the number of T lymphocytes and minimized bone loss was observed (Collins et al. 2019). Gut microbial dysregulation has recently been suggested as one of the risk factors for human bone disease (Tu et al. 2021). The overgrowth of gut microbiota may lead to the development of metabolic bone disease. Stefano et al. found that in patients with overgrowth of gut microbiota, lumbar spine and femur bone mineral density values are significantly lower than those of healthy subjects (Di Stefano et al. 2001). Increased abundance of some gut microbial metabolites, such as short chain fatty acids (SCFAs) (Lucas et al. 2018), trimethylamine n-oxide (TMAO) (Lin et al. 2020), may also be associated with osteoporosis. Gut microbiota-derived metabolites can also be directly or indirectly involved in bone health regulation.

To date, the underlying mechanism of the connection between gut microbiota and bone health is still elusive. Based on the available evidence, this part summarizes four potential mechanisms that may explain this relationship: (1) intestinal barrier and nutrition, (2) immune system function, (3) endocrine system function, and (4) metabolites of gut microbiota (Fig. 19.2).

19.5.1 Intestinal barrier and nutrition

The frontline of intestinal mucosal barrier is maintained by a single layer of specialized epithelial cells that are linked together by tight junction proteins to regulate the permeability of the intestinal barrier (Ulluwishewa et al. 2011). The gut microbiota and its metabolites are involved in maintaining intestinal epithelial barrier function. Disruption of the microbiota balance could weaken the tight junction proteins between intestinal epithelial cells, reducing barrier function (Ulluwishewa et al. 2011). Subsequent translocation of microbiota and their bacterial byproducts such as lipopolysaccharide may contribute to systemic inflammation (Emanuele et al. 2010). This process could enable the gut microbiota to influence distant organs like bone (Proal et al. 2017) as inflammatory changes at the bone

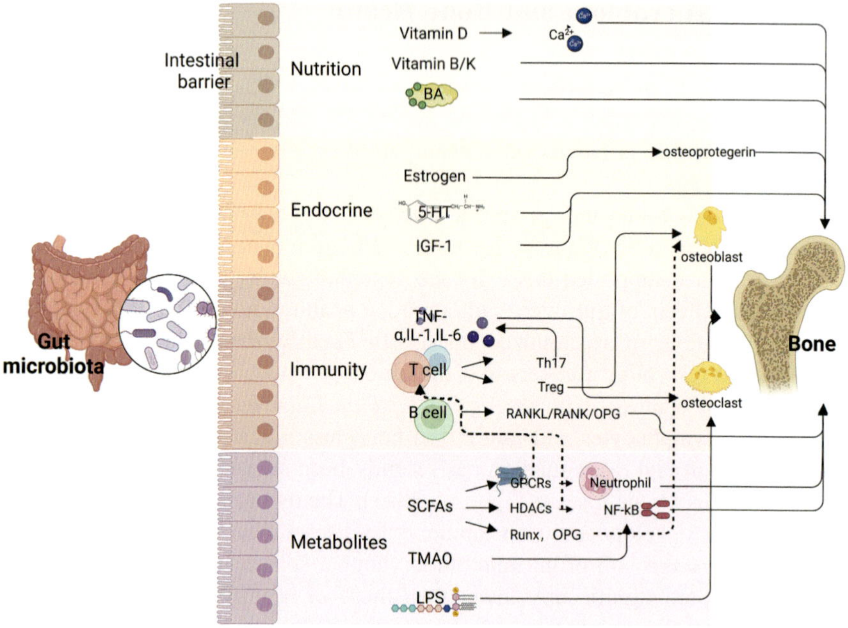

Fig. 19.2 The potential mechanisms of the involvement of gut microbiota in bone health

microenvironment result in both excessive bone loss and bone formation (Adamopoulos 2018). Moreover, the intestinal mucosal barrier also controls the absorption of nutrients such as vitamin D, B, K, and bile acids, which are important for bone formation. For instance, vitamin D has certain benefits on bone health (Ooi et al. 2013). When calcium supply is adequate, absorbed vitamin D and its intestinal metabolites can improve calcium balance and promote mineral deposition in the bone matrix. When calcium deficiency occurs, vitamin D enhances bone resorption while inhibiting bone mineralization, thereby maintaining calcium homeostasis at the cost of reducing bone mass. Therefore, one possible mechanism by which the gut microbiota controls bone health is through modulation of the intestinal mucosal barrier and nutrient absorption.

Immune system function

Changes in gut microbiota may act on bone through the immune system. As early as the 1970s (Horton et al. 1972; Mundy et al. 1974), the close relationship between the immune system and bone has been recognized due to pioneering work on soluble immune cell-derived osteoclast-activating factors. In fact, a separate discipline of Osteoimmunology was formed. Gut microbiota may regulate osteoclasts and osteoblasts by regulating host immune cells (B and T cells), and immune cytokines (TNF-α, IL-1, and IL-6) (Lorenzo et al. 2008). B cells are inextricably linked to the bone. These cells can play a bidirectional regulatory role in the skeletal system (Pietschmann et al. 2016). The interaction between B cells and bone may revolve around the RANKL/RANK/OPG pathway (Titanji 2017), which was discussed earlier in the physical activity and bone health section. Li and colleagues found that B-cell-deficient mice were osteoporotic and lack bone marrow OPG, indicating that B cells play a critical role in maintaining bone homeostasis and peak bone mass (Li et al. 2007). B cells are also thought to be the source of RANKL and OPG, two important regulatory factors of bone homeostasis (Bucay et al. 1998). On the other hand, a subset of T cells, called regulatory T cells, can not only control autoimmunity but also directly or indirectly stimulate bone formation (Nutsch et al. 2016). Additionally, Sato et al. found that T helper (Th) 17 cells, an osteoclastogenic subset, have profound relevance to bone damage (Sato et al. 2006). Th17 may induce osteoclast differentiation by stimulating TNF-α and IL-6 production (Sato et al. 2006). These findings also predict that gut microbiota may improve bone health through the immune system.

Endocrine system function

Endocrine hormones act on various organs of the human body and are involved in the development of many diseases. Gut microbiota can control the secretion of hormones and neurotransmitters such as estrogens (Flores et al. 2012; Chiang and Pan 2013), 5-hydroxytryptamine (5-HT) (Kennedy et al. 2017) and insulin-like growth factor type 1 (IGF-1) (Yakar et al. 2010), thereby influencing bone health. Estrogen is directly related to the richness and diversity of gut microbiota, especially *Clostridia* taxa and several *Ruminococcaceae* family members, which regulate the enterohepatic circulation of estrogen (Flores et al. 2012). On the other hand, estrogen can promote osteoprotegerin secretion, reduce osteoclast differentiation, and maintain bone metabolic homeostasis (Zhang et al. 2014). In mice, estrogen deficiency may increase the number of osteoclasts on cancellous and cortical bone, potentially leading to bone loss (Manolagas et al. 2013). Therefore, estrogen deficiency is considered an important risk factor for osteoporosis in postmenopausal women. 5-HT is another potential factor mediating signal transduction between gut microbiota and bone health. This molecule exerts a complex but significant physiologic impact on bone remodeling (Tu et al. 2021). Lowering 5-HT levels with a synthetic inhibitor prevented ovariectomized-induced bone loss in mice (Yadav et al. 2010). Sjogren et al. indicated that germ-free mice had decreased 5-HT levels and increased trabecular bone mass (Sjögren et al. 2012). Lastly, IGF-1 is a key regulator of healthy mammalian growth. Compared to germ-free mice, mice with intact gut microbiota (conventionally reared mice) had higher levels of circulating IGF-1 (Schwarzer et al. 2016). Meanwhile, the femoral bone mineral density, cortical bone thickness, and trabecular bone density were all decreased in germ-free mice, suggesting a critical role for IGF-1 in the promotion of bone development by gut microbiome (Schwarzer et al. 2016). Collectively, these findings suggest that the gut microbiota may participate in bone formation and resorption by regulating the levels of hormones and neurotransmitters within the body.

Metabolites of gut microbiota

Microbial byproducts not only aid in nutrient digestion and absorption, but also regulate bone health (Chen et al. 2017). SCFAs are the main products of dietary fiber fermentation in the intestinal tract, including acetic acid, butyric acid, and propionic acid, which are important energy sources for intestinal epithelial cells (Gaudier et al. 2004). Evidence showed that SCFAs can act as inhibitors of histone deacetylases and ligands of G-protein coupled receptors. These metabolites also play a role in the immune system (Rooks and Garrett 2016). In this regard, SCFAs can inactivate NF-kB signaling to regulate pro-inflammatory responses and influence of T cells by inhibiting HDACs. They could also regulate neutrophil chemotaxis and the expansion and suppressive function of regulatory T cells through G-protein coupled receptors (Rooks and Garrett 2016). Some SCFAs (e.g., butyrate) may influence the Runx and OPG signaling pathways, which play important roles in bone formation and mineralization (Katono et al. 2008). In addition to SCFAs, TMAO, a gut microbial metabolite, could accelerate bone loss and osteoporosis by regulating NF-κB-dependent bone marrow mesenchymal stem cell function (Lin et al. 2020). Moreover, animal studies showed that the lipopolysaccharides, potent activators of monocytes, have a local effect on bone resorption, suggesting that this substance influences osteoclast development and function (Orcel et al. 1993).

19.6 Summary

In this chapter, we describe the relationship between physical exercise, gut microbiota, and bone health. In general, physical activity plays an important role in improving bone mass and bone mineral intensity. Therefore, certain people such as elders could participate in regular physical exercise to maintain bone health and prevent osteoporosis. On the other hand, given the important role of the gut microbiota in bone metabolisms, additional attention has been placed on how bone health is regulated by gut microbiota. Pre-clinical studies have already provided important mechanistic evidence of the influence of the human gut microbiota on bone. These studies indicated that gut microbiota and microbial products can modulate processes of bone gain and bone loss by regulating osteoblasts and osteoclasts. Based on these discoveries, it may be possible to demonstrate the existence of the "exercise-gut-bone" axis. In other words, the beneficial effects of exercise on bone may be correlated with changes in gut microbiome composition. Additionally, more RCTs of elderly subjects are necessary to reveal the mechanisms and effects of various physical exercise methods on gut microbiome composition and bone metabolism.

References

Adamopoulos IE (2018) Inflammation in bone physiology and pathology. Curr Opin Rheumatol 30:59

Allison SJ, Folland JP, Rennie WJ, Summers GD, Brooke-Wavell K (2013) High impact exercise increased femoral neck bone mineral density in older men: a randomised unilateral intervention. Bone 53:321–328

Arumugam TV, Shiels IA, Woodruff TM, Granger DN, Taylor SM (2004) The role of the complement system in ischemia-reperfusion injury. Shock 21:401–409

Baxter-Jones AD, Faulkner RA, Forwood MR, Mirwald RL, Bailey DA (2011) Bone mineral accrual from 8 to 30 years of age: an estimation of peak bone mass. J Bone Mineral Res 26:1729–1739

Bergström I, Parini P, Gustafsson SA, Andersson G, Brinck J (2012) Physical training increases osteoprotegerin in postmenopausal women. J Bone Mineral Metab 30:202–207

Bermon S, Petriz B, Kajeniene A, Prestes J, Castell L, Franco OL (2015) The microbiota: an exercise immunology perspective. Exercise Immunol Rev 21:70–79

Bucay N, Sarosi I, Dunstan CR, Morony S, Tarpley J, Capparelli C, Scully S, Tan HL, Xu W, Lacey DL (1998) Osteoprotegerin-deficient mice develop early onset osteoporosis and arterial calcification. Genes Dev 12:1260–1268

Carden DL, Granger DN (2000) Pathophysiology of ischaemia-reperfusion injury. J Pathol 190:255–266

Caspersen CJ, Powell KE, Christenson GM (1985) Physical activity, exercise, and physical fitness: definitions and distinctions for health-related research. Public Health Rep 100:126

Chen Y-C, Greenbaum J, Shen H, Deng H-W (2017) Association between gut microbiota and bone health: potential mechanisms and prospective. J Clin Endocrinol Metab 102:3635–3646

Cheung AM, Giangregorio L (2012) Mechanical stimuli and bone health: what is the evidence? Curr Opin Rheumatol 24:561–566

Chiang S-S, Pan T-M (2013) Beneficial effects of phytoestrogens and their metabolites produced by intestinal microflora on bone health. Appl Microbiol Biotechnol 97:1489–1500

Cho I, Yamanishi S, Cox L, Methé BA, Zavadil J, Li K, Gao Z, Mahana D, Raju K, Teitler I (2012) Antibiotics In early life alter the murine colonic microbiome and adiposity. Nature 488:621–626

Clarke SF, Murphy EF, O'sullivan O, Lucey AJ, Humphreys M, Hogan A, Hayes P, O'reilly M, Jeffery IB, Wood-Martin R (2014) Exercise and associated dietary extremes impact on gut microbial diversity. Gut 63:1913–1920

Clausen JP (1977) Effect of physical training on cardiovascular adjustments to exercise in man. Physiol Rev 57:779–815

Collins FL, Rios-Arce ND, Schepper JD, Jones AD, Schaefer L, Britton RA, Mccabe LR, Parameswaran N (2019) Beneficial effects of lactobacillus Reuteri 6475 on bone density in male mice is dependent on lymphocytes. Sci Rep 9:1–17

Cox LM, Yamanishi S, Sohn J, Alekseyenko AV, Leung JM, Cho I, Kim SG, Li H, Gao Z, Mahana D (2014) Altering the intestinal Microbiota during a critical developmental window has lasting metabolic consequences. Cell 158:705–721

Cronin O, Barton W, Skuse P, Penney NC, Garcia-Perez I, Murphy EF, Woods T, Nugent H, Fanning A, Melgar S (2018) A prospective metagenomic and metabolomic analysis of the impact of exercise and/or whey protein supplementation on the gut microbiome of sedentary adults. Msystems 3:E00044–18

Demontiero O, Vidal C, Duque G (2012) Aging and bone loss: new insights for the clinician. Therap Adv Musculosk Dis 4:61–76

Derikx JP, Matthijsen RA, De Bruïne AP, Van Bijnen AA, Heineman E, Van Dam RM, Dejong CH, Buurman WA (2008) Rapid reversal of human intestinal ischemia-reperfusion induced damage by shedding of injured enterocytes and reepithelialisation. PLoS ONE 3: E3428

Di Stefano M, Veneto G, Malservisi S, Corazza GR (2001) Small intestine bacterial overgrowth and metabolic bone disease. Dig Dis Sci 46:1077–1082

Dupuit M, Rance M, Morel C, Bouillon P, Boscaro A, Boisseau N, Vincent M, Vazeille E, Barnich N, Chassaing B (2021) Impact of concurrent training on body composition and gut microbiota in postmenopausal women with overweight or obesity. Med Sci Sport Exer 54:517–529

Emanuele E, Orsi P, Boso M, Broglia D, Brondino N, Barale F, Di Nemi SU, Politi P (2010) Low-grade endotoxemia in patients with severe autism. Neurosci Lett 471:162–165

Fart F, Rajan SK, Wall R, Rangel I, Lindqvist CM (2020) Differences in gut Microbiome composition between senior orienteering athletes and community-dwelling older adults. Nutrients 12:2610

Field T (2011) Tai Chi research review. Complement Ther Clin Pract 17:141–146

Flores R, Shi J, Fuhrman B, Xu X, Veenstra TD, Gail MH, Gajer P, Ravel J, Goedert JJ (2012) Fecal microbial determinants of Fecal and systemic Estrogens and Estrogen metabolites: a cross-sectional study. J Transl Med 10:1–11

Frost HM (2003) Bone's mechanostat: a 2003 update. Anato Record Part A Disc Molec Cell Evolut Biology off Public Am Assoc Anato 275:1081–1101

Galley JD, Bailey MT (2014) Impact of stressor exposure on the interplay between commensal microbiota and host inflammation. Gut Microbes 5:390–396

Gaudier E, Jarry A, Blottiere H, De Coppet P, Buisine M, Aubert J, Laboisse C, Cherbut C, Hoebler C (2004) Butyrate specifically modulates muc gene expression in intestinal epithelial goblet cells deprived of glucose. Am J Physiol Gastroint Liver Physiol 287:G1168–G1174

Han M, Yang K, Yang P, Zhong C, Chen C, Wang S, Lu Q, Ning K (2020) Stratification of athletes' gut microbiota: the multifaceted hubs associated with dietary factors, physical characteristics and performance. Gut Microbes 12:1842991

Harding AT, Weeks BK, Lambert C, Watson SL, Weis LJ, Beck BR (2020) A comparison of bone-targeted exercise strategies to reduce fracture risk in middle-aged and older men with osteopenia and osteoporosis: Liftmor-M semi-randomized controlled trial. J Bone Mineral Res 35:1404–1414

He X-F, Zhang L, Zhang C-H, Zhao C-R, Li H, Zhang L-F, Tian G-F, Guo M-F, Dai Z, Sui F-G (2017) Berberine alleviates oxidative stress in rats with osteoporosis through receptor activator of Nf-Kb/receptor activator of Nf-Kb ligand/osteoprotegerin (Rank/Rankl/Opg) pathway. Bosn J Basic Med Sci 17:295

Heesch KC, Byles JE, Brown WJ (2008) Prospective association between physical activity and falls in community-dwelling older women. J Epidemiol Community Health 62:421–426

Hendrickx G, Boudin E, Van Hul W (2015) A look behind the scenes: the risk and pathogenesis of primary osteoporosis. Nat Rev Rheumatol 11:462–474

Horton JE, Raisz LG, Simmons HA, Oppenheim JJ, Mergenhagen SE (1972) Bone resorbing activity in supernatant fluid from cultured human peripheral blood leukocytes. Science 177:793–795

Islam KS, Fukiya S, Hagio M, Fujii N, Ishizuka S, Ooka T, Ogura Y, Hayashi T, Yokota A (2011) Bile acid is a host factor that regulates the composition of the cecal microbiota in rats. Gastroenterology 141:1773–1781

Jaeschke H (2003) Molecular mechanisms of hepatic ischemia-reperfusion injury and preconditioning. Am J Physiol Gastroint Liver Physiol 284:G15–G26

Kadow-Romacker A, Hoffmann J, Duda G, Wildemann B, Schmidmaier G (2009) Effect of mechanical stimulation on osteoblast-and osteoclast-like cells In Vitro. Cells Tissues Organs 190:61–68

Katono T, Kawato T, Tanabe N, Suzuki N, Iida T, Morozumi A, Ochiai K, Maeno M (2008) Sodium butyrate stimulates mineralized nodule formation and Osteoprotegerin expression by human osteoblasts. Arch Oral Biol 53:903–909

Kemmak AR, Rezapour A, Jahangiri R, Nikjoo S, Farabi H, Soleimanpour S (2020) Economic burden of osteoporosis in the world: a systematic review. Med J Islam Repub Iran 34:154

Kennedy PJ, Cryan JF, Dinan TG, Clarke G (2017) Kynurenine pathway metabolism and the microbiota-gut-brain axis. Neuropharmacology 112:399–412

Kern T, Blond MB, Hansen TH, Rosenkilde M, Quist JS, Gram AS, Ekstrøm CT, Hansen T, Stallknecht B (2020) Structured exercise alters the gut microbiota in humans with overweight and obesity—a randomized controlled trial. Int J Obes 44:125–135

Kish K, Mezil Y, Ward WE, Klentrou P, Falk B (2015) Effects of plyometric exercise session on markers of bone turnover in boys and young men. Eur J Appl Physiol 115:2115–2124

Kulecka M, Fraczek B, Mikula M, Zeber-Lubecka N, Karczmarski J, Paziewska A, Ambrozkiewicz F, Jagusztyn-Krynicka K, Cieszczyk P, Ostrowski J (2020) The composition and richness of the gut microbiota differentiate the top polish endurance athletes from sedentary controls. Gut Microbes 11:1374–1384

Lambert GP, Lang J, Bull A, Pfeifer P, Eckerson J, Moore G, Lanspa S, O'brien J (2008) Fluid restriction during running increases Gi permeability. Intern J Sports Med 29:194–198

Leal ML, Lamas L, Aoki MS, Ugrinowitsch C, Ramos MSC, Tricoli V, Moriscot AS (2011) Effect of different resistance-training regimens on the Wnt-signaling pathway. Eur J Appl Physiol 111:2535–2545

Ley RE, Peterson DA, Gordon JI (2006) Ecological and evolutionary forces shaping microbial diversity in the human intestine. Cell 124:837–848

Li Y, Toraldo G, Li A, Yang X, Zhang H, Qian W-P, Weitzmann MN (2007) B cells and t cells are critical for the preservation of bone homeostasis and attainment of peak bone mass In Vivo. Blood 109:3839–3848

Liang R, Zhang S, Peng X, Yang W, Xu Y, Wu P, Chen J, Cai Y, Zhou J (2019) Characteristics of the gut microbiota in professional martial arts athletes: a comparison between different competition levels. PLoS ONE 14:E0226240

Lin H, Liu T, Li X, Gao X, Wu T, Li P (2020) The role of gut microbiota metabolite trimethylamine N-Oxide in functional impairment of bone marrow mesenchymal stem cells in osteoporosis disease. Ann Transl Med 8:1009–1009

Lorenzo J, Horowitz M, Choi Y (2008) Osteoimmunology: interactions of the bone and immune system. Endocr Rev 29:403–440

Lucas S, Omata Y, Hofmann J, Bottcher M, Iljazovic A, Sarter K, Albrecht O, Schulz O, Krishnacoumar B, Krönke G (2018) Short-chain fatty acids regulate systemic bone mass and protect from pathological bone loss. Nat Commun 9:1–10

Mahdieh MS, Maryam J, Bita B, Neda F, Motahare M, Mahboobeh B, Lebris SQ, Kalani Behrooz S (2021) A pilot study on the relationship between lactobacillus, bifidibactrium counts and inflammatory factors following exercise training. Arch Physiol Biochem, 1–10

Manolagas SC, O'brien CA, Almeida M (2013) The role of estrogen and androgen receptors in bone health and disease. Nat Rev Endocrinol 9:699

Mariat D, Firmesse O, Levenez F, Guimarães V, Sokol H, Doré J, Corthier G, Furet J (2009) The firmicutes/bacteroidetes ratio of the human microbiota changes with age. BMC Microbiol 9:1–6

Mcfadzean R (2014) Exercise can help modulate human gut microbiota. Undergraduate Honors Theses, 155

Moitinho-Silva L, Wegener M, May S, Schrinner F, Akhtar A, Boysen TJ, Schaeffer E, Hansen C, Schmidt T, Rühlemann MC (2021) Short-term physical exercise impacts on the human holobiont obtained by a randomised intervention study. BMC Microbiol 21:1–14

Mokhtarzade M, Shamsi MM, Abolhasani M, Bakhshi B, Sahraian MA, Quinn LS, Negaresh R (2021) Home-based exercise training influences gut bacterial levels in multiple sclerosis. Compl Ther Clin Pract 45:101463

Morville T, Sahl RE, Trammell SA, Svenningsen JS, Gillum MP, Helge JW, Clemmensen C (2018) Divergent effects of resistance and endurance exercise on plasma bile acids, Fgf19, And Fgf21 In humans. Jci Insight 3:e122737

Mosti MP, Kaehler N, Stunes AK, Hoff J, Syversen U (2013) Maximal strength training in postmenopausal women with osteoporosis or osteopenia. J Stren Condit Res 27:2879–2886

Müller M, Hermes GD, Canfora EE, Smidt H, Masclee AA, Zoetendal EG, Blaak EE (2020) Distal colonic transit is linked to gut microbiota diversity and microbial fermentation in humans with slow colonic transit. Am J Physiol Gastroint Liver Physiol 318:G361-G369
Mundy GR, Raisz LG, Cooper RA, Schechter GP, Salmon SE (1974) Evidence for the secretion of an osteoclast stimulating factor in myeloma. N Engl J Med 291:1041–1046
Muraki S, Yamamoto S, Ishibashi H, Oka H, Yoshimura N, Kawaguchi H, Nakamura K (2007) Diet and lifestyle associated with increased bone mineral density: cross-sectional study of Japanese elderly women at an osteoporosis outpatient clinic. J Orthop Sci 12:317–320
Nakashima T, Takayanagi H (2009) Osteoimmunology: crosstalk between the immune and bone systems. J Clin Immunol 29:555–567
Nguyen T, Center J, Eisman J (2000) Osteoporosis in elderly men and women: effects of dietary calcium, physical activity, and body mass index. J Bone Mineral Res 15:322–331
Notomi T, Karasaki I, Okazaki Y, Okimoto N, Kato Y, Ohura K, Noda M, Nakamura T, Suzuki M (2014) Insulinogenic sucrose+ amino acid mixture ingestion immediately after resistance exercise has an anabolic effect on bone compared with non-insulinogenic fructose+ amino acid mixture in growing rats. Bone 65:42–48
Nutsch K, Chai JN, Ai TL, Russler-Germain E, Feehley T, Nagler CR, Hsieh C-S (2016) Rapid and efficient generation of regulatory t cells to commensal antigens in the periphery. Cell Rep 17:206–220
Oettle G (1991) Effect of moderate exercise on bowel habit. Gut 32:941–944
Ohlsson C, Sjögren K (2015) Effects of the gut microbiota on bone mass. Trends Endocrinol Metab 26:69–74
Ohlsson C, Engdahl C, Fåk F, Andersson A, Windahl SH, Farman HH, Movérare-Skrtic S, Islander U, Sjögren K (2014) Probiotics protect mice from ovariectomy-induced cortical bone loss. PLoS ONE 9:E92368
Ooi JH, Li Y, Rogers CJ, Cantorna MT (2013) Vitamin D regulates the gut microbiome and protects mice from dextran sodium sulfate-induced colitis. J Nutr 143:1679–1686
Orcel P, Feuga M, Bielakoff J, De Vernejoul M (1993) Local bone injections of Lps and M-Csf increase bone resorption by different pathways in Vivo In rats. Am J Physiol Endocrinol Metabol 264:E391–E397
World Health Organization (1994) Assessment of fracture risk and its application to screening for postmenopausal osteoporosis: report of a who study group [Meeting Held In Rome From 22 To 25 June 1992]
Pacifici R (2010) The immune system and bone. Archives Biochem Biophy 503:41–53
Parfitt A (1984) The cellular basis of bone remodeling: the quantum concept reexamined in light of recent advances in the cell biology of bone. Calcif Tissue Int 36:S37–S45
Pietschmann P, Mechtcheriakova D, Meshcheryakova A, Föger-Samwald U, Ellinger I (2016) Immunology of osteoporosis: a mini-review. Gerontology 62:128–137
Pires W, Veneroso CE, Wanner SP, Pacheco DA, Vaz GC, Amorim FT, Tonoli C, Soares DD, Coimbra CC (2017) Association between exercise-induced hyperthermia and intestinal permeability: a systematic review. Sports Med 47:1389–1403
Proal AD, Lindseth IA, Marshall TG (2017) Microbe-microbe and host-microbe interactions drive microbiome dysbiosis and inflammatory processes. Discov Med 23:51–60
Quach D, Britton RA (2017) Gut microbiota and bone health. In: McCabe LR, Parameswaran N. (eds) Understanding the gut-bone signaling axis: advances in experimental medicine and biology 1033:47–58
Queipo-Ortuño MI, Seoane LM, Murri M, Pardo M, Gomez-Zumaquero JM, Cardona F, Casanueva F, Tinahones FJ (2013) Gut microbiota composition in male rat models under different nutritional status and physical activity and its association with serum Leptin and ghrelin levels. PLoS ONE 8:E65465
Quiroga R, Nistal E, Estébanez B, Porras D, Juárez-Fernández M, Martínez-Flórez S, García-Mediavilla MV, De Paz JA, González-Gallego J, Sánchez-Campos S (2020) Exercise training modulates the gut microbiota profile and impairs inflammatory signaling pathways in obese children. Exp Mol Med 52:1048–1061

Resende AS, Leite GS, Lancha Junior AH (2021) Changes in the gut bacteria composition of healthy men with the same nutritional profile undergoing 10-week aerobic exercise training: a randomized controlled trial. Nutrients 13:2839

Rooks MG, Garrett WS (2016) Gut microbiota, metabolites and host immunity. Nat Rev Immunol 16:341–352

Santos L, Elliott-Sale KJ, Sale C (2017) Exercise and bone health across the lifespan. Biogerontology 18:931–946

Sato K, Suematsu A, Okamoto K, Yamaguchi A, Morishita Y, Kadono Y, Tanaka S, Kodama T, Akira S, Iwakura Y (2006) Th17 functions as an osteoclastogenic helper T cell subset that links T cell activation and bone destruction. J Exp Med 203:2673–2682

Schett G (2011) Effects of inflammatory and anti-inflammatory cytokines on the bone. Eur J Clin Invest 41:1361–1366

Schumacher B, Pothof J, Vijg J, Hoeijmakers JH (2021) The Central role of Dna damage in the ageing process. Nature 592:695–703

Schwarzer M, Makki K, Storelli G, Machuca-Gayet I, Srutkova D, Hermanova P, Martino ME, Balmand S, Hudcovic T, Heddi A (2016) Lactobacillus plantarum strain maintains growth of infant mice during chronic undernutrition. Science 351:854–857

Sekirov I, Russell SL, Antunes LCM, Finlay BBJPR (2010) Gut microbiota in health and disease. Physiol Rev 90:959–904

Shepherd J, Vanhoutte P (1978) Role of the venous system in circulatory control. Mayo Clinic Proc, 247–255

Sjögren K, Engdahl C, Henning P, Lerner UH, Tremaroli V, Lagerquist MK, Bäckhed F, Ohlsson C (2012) The gut microbiota regulates bone mass in mice. J Bone Mineral Res 27:1357–1367

Spillane M, Schwarz N, Willoughby DS (2015) Upper-body resistance exercise augments vastus lateralis androgen receptor-dna binding and canonical Wnt/Β-catenin signaling compared to lower-body resistance exercise in resistance-trained men without an acute increase in serum testosterone. Steroids 98:63–71

Takayanagi H (2007) Osteoimmunology: shared mechanisms and crosstalk between the immune and bone systems. Nat Rev Immunol 7:292–304

Taniguchi H, Tanisawa K, Sun X, Kubo T, Hoshino Y, Hosokawa M, Takeyama H, Higuchi M (2018) Effects of short-term endurance exercise on gut microbiota in elderly men. Physiol Rep 6:E13935

Titanji K (2017) Beyond antibodies: B cells and the Opg/Rank-Rankl pathway in health, non-Hiv disease and Hiv-induced bone loss. Front Immunol 8:1851

Tu Y, Yang R, Xu X, Zhou X (2021) The microbiota-gut-bone axis and bone health. J Leukoc Biol 110:525–537

Ulluwishewa D, Anderson RC, Mcnabb WC, Moughan PJ, Wells JM, Roy NC (2011) Regulation of tight junction permeability by intestinal bacteria and dietary components. J Nutr 141:769–776

Vargas N, Marino F (2016) Heat stress, gastrointestinal permeability and interleukin-6 signaling—implications for exercise performance and fatigue. Temperature 3:240–251

Van Wijck K, Lenaerts K, Grootjans J, Wijnands KA, Poeze M, Van Loon LJ, Dejong CH, Buurman WA (2012) Physiology and pathophysiology of splanchnic hypoperfusion and intestinal injury during exercise: strategies for evaluation and prevention. Am J Physiol Gastroint Liver Physiol 303:G155–G168

Warbeck C, Dowd AJ, Kronlund L, Parmar C, Daun JT, Wytsma-Fisher K, Millet GY, Schick A, Reimer RA, Fung T (2021) Feasibility and effects on the gut microbiota of a 12-week high-intensity interval training plus lifestyle education intervention on inactive adults with celiac disease. Appl Physiol Nutr Metab 46:325–336

Xu Y, Zhong F, Zheng X, Lai H-Y, Wu C, Huang C (2022) Discrepancy of gut microbiota composition among elite athletes and young adults with different physical activity independent of dietary status: a matching study. Front Nutr, 9:843076

Yadav VK, Balaji S, Suresh PS, Liu XS, Lu X, Li Z, Guo XE, Mann JJ, Balapure AK, Gershon MD (2010) Pharmacological inhibition of gut-derived serotonin synthesis is a potential bone anabolic treatment for osteoporosis. Nat Med 16:308–312

Yakar S, Courtland HW, Clemmons D (2010) Igf-1 and bone: new discoveries from mouse models. J Bone Mineral Res 25:2543–2552

Yuan Y, Chen X, Zhang L, Wu J, Guo J, Zou D, Chen B, Sun Z, Shen C, Zou J (2016) The roles of exercise in bone remodeling and in prevention and treatment of osteoporosis. Prog Biophy Molec Biol 122:122–130

Zehnacker CH, Bemis-Dougherty A (2007) Effect of weighted exercises on bone mineral density in post menopausal women a systematic review. J Geriat Phy Therapy 30:79–88

Zhang C, Peng J, Wu S, Jin Y, Xia F, Wang C, Liu K, Sun H, Liu M (2014) Dioscin promotes osteoblastic proliferation and differentiation Via Lrp5 and Er pathway in mouse and human osteoblast-like cell lines. J Biomed Sci 21:1–13

Zhang J, Lu Y, Wang Y, Ren X, Han J (2018) The impact of the intestinal microbiome on bone health. Intract Rare Dis Res 7:148–155

Zheng X, Huang F, Zhao A, Lei S, Zhang Y, Xie G, Chen T, Qu C, Rajani C, Dong B (2017) Bile acid is a significant host factor shaping the gut microbiome of diet-induced obese mice. BMC Biol 15:1–15

Zhong F, Wen X, Yang M, Lai H-Y, Momma H, Cheng L, Sun X, Nagatomi R, Huang C (2021) Effect of an 8-week exercise training on gut microbiota in physically inactive older women. Int J Sports Med 42:610–623

Zhu Q, Jiang S, Du G (2020) Effects of exercise frequency on the gut microbiota in elderly individuals. Microbiologyopen 9:E1053

Zuhl MN, Lanphere KR, Kravitz L, Mermier CM, Schneider S, Dokladny K, Moseley PL (2014) Effects of oral glutamine supplementation on exercise-induced gastrointestinal permeability and tight junction protein expression. J Appl Physiol 116:183–191

Chapter 20
Gut Microbiota Peculiarities in Aged HIV-Infected Individuals: Molecular Understanding and Therapeutic Perspectives

Ramachandran Vignesh, Pitchaipillai S. Ganesh, Sathish Sankar, Indranil Chattopadhyay, Yean K. Yong, Marie Larsson, and Esaki M. Shankar

Abstract Aging is a progressive physiological process that involves alterations in the population of normal gut microbiota. HIV infection in conjunction with aging complicates the quality of immune responses paving way for increased epithelial permeability and translocation of commensals and pathogen-derived substances into the systemic circulation. Enteric dysbiosis is linked with other comorbidities of aging including cardiovascular diseases and dementia in the HIV-infected population. The diversity of gut microbiota declines with age, and HIV infection compromises the effective functioning of the immune system complicating the metabolic as well as regulatory programming of the host's physiological machinery.

R. Vignesh and P. S. Ganesh contributed equally to the manuscript.

R. Vignesh
Preclinical Department, Royal College of Medicine Perak (UniKL RCMP), Universiti Kuala Lumpur, Ipoh, Malaysia

P. S. Ganesh · S. Sankar
Department of Microbiology, Saveetha Dental College and Hospital, Saveetha Institute of Medical and Technical Sciences, Velappanchavadi, Chennai 600 077, India

I. Chattopadhyay
Cell and Cancer Biology, Department of Life Sciences, Central University of Tamil Nadu, Neelakudi 610 005, India

Y. K. Yong
Laboratory Centre, Xiamen University Malaysia, 43900 Sepang, Malaysia

M. Larsson
Molecular Medicine and Virology, Department of Biomedical and Clinical Sciences, Linköping University, 581 85 Linköping, Sweden

E. M. Shankar (✉)
Infection Biology, Department of Life Sciences, Central University of Tamil Nadu, Thiruvarur 610 005, India
e-mail: shankarem@cutn.ac.in

F. Marotta (ed.), *Gut Microbiota in Aging and Chronic Diseases*, Healthy Ageing and Longevity 17, https://doi.org/10.1007/978-3-031-14023-5_20

Given the scenario of the global HIV/AIDS and the COVID-19 pandemics, it is also likely that alterations in gut microbiota potentially could contribute to post-acute COVID-19 syndrome as well as accelerate the rate of HIV disease progression. The current chapter discusses the role of normal gut microbiota in the aged HIV-infected population and their functional implications in normal cellular and molecular mechanisms in the host.

Keywords Aging · Dysbiosis · HIV infection · Microbiota · SARS-CoV-2

20.1 Introduction

The human gastrointestinal (GI) tract harbors a wide array of complex microbial community, which plays a pivotal role against the pathogenic establishment (biological barriers), aids in several key cellular and metabolic functions, and also maintains the overall health of individuals (Bull and Plummer 2014; Grace et al. 2020). Of the diverse array of microbial communities present (Matijašić et al. 2020), a vast majority belong to one of the following bacterial groups: Bacteroidetes, Firmicutes, Actinobacteria, Proteobacteria, Fusobacteria, Cyanobacteria, and Verrucomicrobia (Rinninella et al. 2019). The Bacteroidetes and Firmicutes phyla (obligate anaerobes) predominantly dominate the gut microbiota pool forming >90% of the total flora (Hong et al. 2011). The complex and dynamic microbial community is host-specific, evolves throughout life (Vemuri et al. 2018), and is influenced by both exogenous and endogenous alterations (Rowland et al. 2018). Similarly, intestinal microbes also play a critical role in maintaining gut immune homeostasis (Rakoff-Nahoum et al. 2004). Pathogenic invasion of the GI tract can result in dysbiosis eventually leading to immune activation as well as dysregulation (Pickard et al. 2017) paving way for the onset of metabolic disorders such as type 1 diabetes (T1D) (Vaarala, 2012), inflammatory bowel disease (IBD) (Goethel et al. 2018), rheumatic disorders (Zhong et al. 2018), cardiovascular diseases (CVD) (Jie et al. 2017), and atherosclerosis.

Acquired immunodeficiency syndrome (AIDS) is a life-threatening infectious disease caused by the human immunodeficiency virus (HIV). By weakening the immune system following progressive attrition of CD4+ T cells in the host, HIV paves way for pathogenic invasion and renders individuals susceptible to a broad array of opportunistic infections and neoplasms (Mildvan et al. 1982; Marchetti et al. 2013). In recent years, numerous studies have suggested that HIV infection could cause perturbation in the population of the normal gut microbiota (Machiavelli et al. 2019). HIV upon entry into the host replicates in the gut-associated lymphoid tissue (GALT) causing massive attrition of mucosal effector CD4+ T cells, including Th17 cells, as well as dendritic cells and macrophages (Okoye and Picker 2013). Such immune perturbation occurring especially in the

HIV-infected elderly population results in systemic invasion by commensals (as well as their byproducts) (microbial translocation) leading to sepsis and certain other systemic complications (Brenchley and Douek 2008).

20.2 Impact of Aging on the Dynamics of Gut Microbiota

Maintenance of good health and well-being is influenced heavily by nutritional status. There are many predisposing factors that significantly contribute to the pathophysiology of aging (Leslie and Hankey 2015). Primary aging refers to basic age-related changes immaterial to the disease process and the environment, whereas secondary aging refers to changes that are influenced by the disease process and the environment (Vemuri et al. 2018; Tosato et al. 2007; Harman 2003). The age-related changes especially involve GI complications starting with dysphagia, malabsorption, and gastroesophageal reflux disease (GERD) to more complicated and chronic diseases (Dumic et al. 2019). During the process of aging, the ongoing GI changes disturb assimilation and digestion, which, in turn, reduces nutrient supply eventually leading to malnutrition (Rémond et al. 2015). As a consequence, the stomach capacity shrinks, causing hormonal imbalance of enterohormones, cholecystokinin, and ghrelin (Leeuwendaal et al. 2021). The GI tract, apart from the maintenance of nutrient absorption, also plays a critical role in the activation of innate mucosal immune responses and protects the host from inflammation and other diseases (Wu and Wu 2012). Aged mucosal surfaces become less functional (immunosenescence) and become increasingly permissive to infectious agents (Sato et al. 2015). Changes in the GI functions are inevitable in the aging process that includes gastric acid secretion, gastric motility, structure and motility of the small intestine, pancreatic secretions, and most importantly, the microbiota (Dumic et al. 2019). The major physiological changes in the GI tract are dependent on gut microbiota as it correlates with the host's proinflammatory status, disease incidence, and malabsorption (Hills et al. 2019). It is, therefore, imperative to weigh gut microbiota in parallel with the aging physiology given their age-related alterations (Fig. 20.1).

The gut microbiota supports the growth of beneficial microbial communities by providing them with necessary short-chain fatty acids (SCFAs) such as acetate, propionate, butyrate, and several gases (Parada Venegas et al. 2019). These SCFAs also protect the colon by maintaining homeostasis, mucosal epithelial and immune cell integrity, oxygen balance, gut hormones, metabolic regulations, and prevent gut dysbiosis (Parada Venegas et al. 2019; Ríos-Covián et al. 2016; Van der Hee and Wells 2021). Certain metabolites released by the gut microbiota, for instance, trimethylamine and indole propionic acid alleviate the risk of the development of atherosclerosis and cardiac dysfunctions (Wang et al. 2011).

Individuals suffering from certain chronic infectious conditions often report an imbalance in gut microbiota. Alterations in the functional development and maintenance of the immune system are closely related to dysbiosis (Zheng et al. 2020).

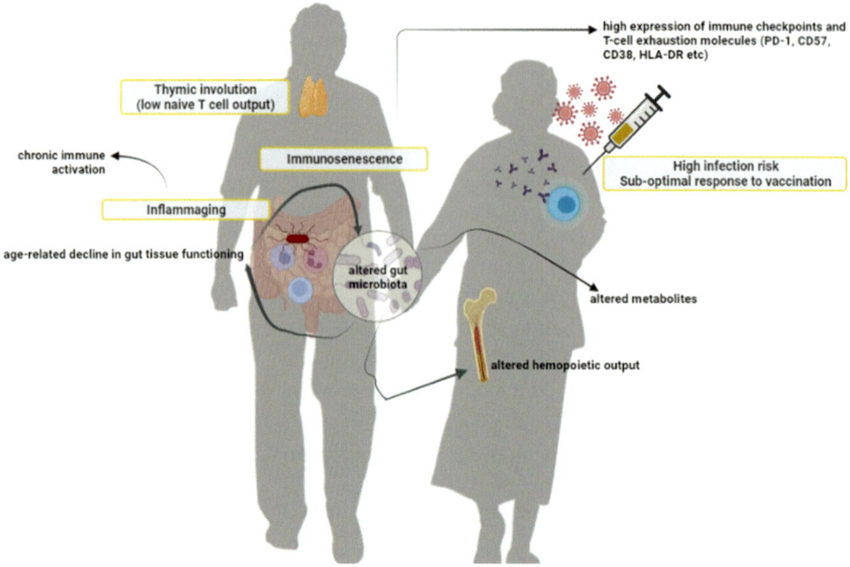

Fig. 20.1 Effects of age-associated physiological changes impacted by the gut microbiota. Aging negatively impacts gut tissue functions such as poor muscular functions leading to sub-optimal bowel movements and peristalsis. Gut microbial changes can also harness inflammaging and chronic immune activation stemming from systemic low-grade inflammation. Poor levels of metabolome and microbial presence also results in immunosenescence entailing in increased risk of infection and sub-optimal responses to vaccination as well as poor sustenance of vaccine-generated humoral immune responses (antibody decay)

For instance, tuberculosis leads to alterations in the gut flora (Shi et al. 2021). Furthermore, changes in the spectrum of microbial flora, viz., depletion or enrichment of the specific classes of microbes has been reported in certain chronic infectious manifestations (Hu et al. 2019). Evidence also suggests that gut microbiota favors the rapid resolution of an infection by selecting the necessary immune cell signaling pathway to alleviate the magnitude of inflammation as well as tissue damage/injury. Probiotics and fecal microbiota transplantation (FMT) have been shown to play a beneficial role in the prevention and control of several infectious diseases (Fehily et al. 2021). In the elderly, gradual decline in the functioning of the immune system, infectious diseases, and the related consequences of age-related dysbiosis concertedly could leave a huge adverse impact on the healthy microbiome (Bosco and Noti 2021). More studies on age-related changes in the gut microbial community could help design appropriate clinical interventions to improve overall health and well-being.

20.3 HIV Infection and Gut Microbiota

In spite of successful antiretroviral therapy (ART), complete eradication of HIV infection remains a wild goose chase, due to the persistence of HIV reservoirs in different anatomical sites and antiviral resistance. Apart from replication-competent viruses, defective proviruses also contribute to inflammation and T cell activation during viral persistence or latency (Dufour et al. 2020). Recent studies provide evidence for the role of gut microbiota in HIV-associated inflammation and immune activation in latent HIV infection.

Studies on the spectrum of gut microbiota in HIV-infected patients reveal variation in specific microbial communities. Significant differences in the bacterial communities between men having sex with men (MSM) and those practicing heterosexual activities have been observed indicating the impact of sexual practices on the altered homeostasis of the gut microbiota (Vujkovic-Cvijin et al. 2020; Valeri and Endres 2021).

HIV-associated gut mucosal damage results in the destruction of the epithelial barrier and immune aggravation (Nazli et al. 2010). Damage to the epithelial barrier results in increased permeability and translocation of commensals and pathogens (microbial translocation resulting from 'leaky gut') and their virulence factors (Fig. 20.2). This enhances the progression of HIV infection complicating ART (Márquez et al. 2016). A recent study on SIV in a primate model showed significant alterations in the gut microbiome during acute and chronic HIV infection. The differences in ART-naïve and ART-treated patients indicate the impact of ART emphasizing the need for gut microbiota restoration to be one of the few prioritized therapeutic modalities (Crakes and Jiang 2019). The association has been reported in children as well (Nguyen et al. 2021). Apart from the gut, microbiota from other anatomical sites also indicate a paradigm shift among the infected versus uninfected as well as between ART-naïve and -treated (Dubourg et al. 2017).

20.4 Do Age-Associated Comorbidities Impact the Dynamics of Microbiota Among People Living with HIV?

Immunosenescence is imminent in old age owing to the aging-associated chronic inflammation and dysregulation of immune activation (Shankar et al. 2015). Reports indicate a common occurrence of impaired immune activity and lowered intestinal function leading to compromised digestion and absorption of nutrients among the old population than in the younger groups (Odamaki et al. 2016). In addition, involution of the thymus as a result of aging can lead to compromised T cell functionality, reduced memory T cell population, and impeded production of IL-2 (Asher et al. 2016). There are striking similarities observed in the gut in both aging and HIV infection wherein there is a loss of resident T cells contributing to

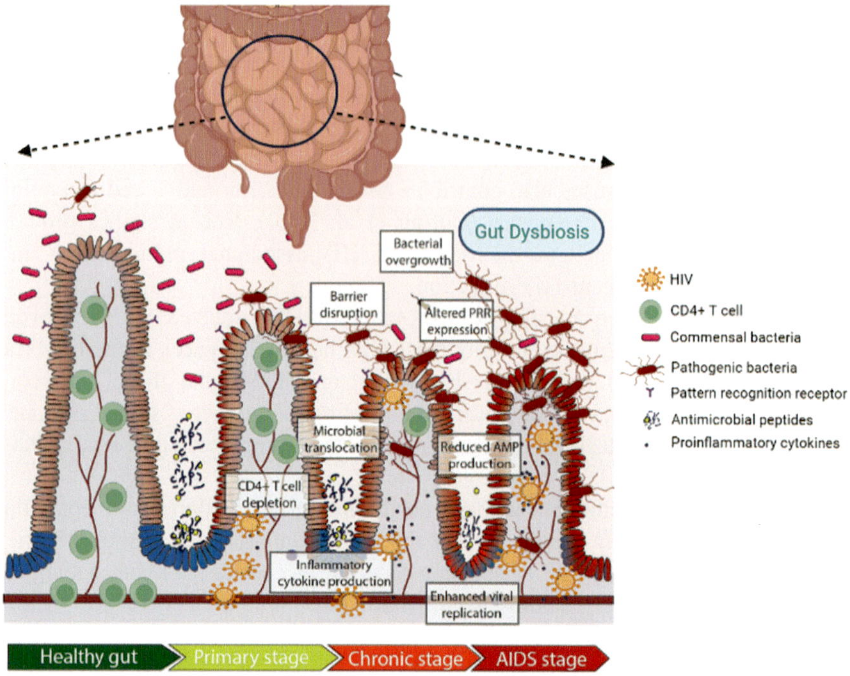

Fig. 20.2 The role of alterations in gut microbiota during different stages of HIV infection. The healthy gut with an intact mucosal epithelial lining together with a thick and thin mucosal lining harboring a rich population of microbiota is impermeable to pathogenic bacteria. During the initial stages of infection, the predominantly CCR5-tropic HIV specifically targets the gut-resident memory CD4+ T cells leading to their attrition that precedes the onset of a 'leaky gut', which further get intensified during the chronic stage of HIV disease. The terminal AIDS stage has a compromised gut immunity characterized by altered synthesis of antimicrobial peptides, pathogenic and commensal overgrowth (opportunistic pathogens) that serves as a prelude to immune activation and immune senescence

dysbiosis of microbiota, intestinal permeability, and systemic inflammation. This finding is supported by the partial reversal of dysbiosis upon the increase in CD4 + T cell counts in response to antiviral therapy in primates (Guillén et al. 2019). Both HIV infection and aging have been known to be associated with a chronic low-level inflammatory state that could augment the risk of other comorbidities (Peterson and Baker 2019).

Though the advent of antiretroviral therapy (ART) has significantly impacted the morbidity and mortality associated with opportunistic infections, elderly PLWH are prone to the earlier onset and high rates of age-associated comorbidities such as CVD, neurocognitive disorders, diabetes, liver disease, kidney disease, reduced bone mineral density, and non-AIDS-defining malignancies (Masters et al. 1999). Also, there is a disproportionately higher incidence of frailty, reduced physical function, and functional disabilities among older PLWH occurring at younger ages and associated with increased measures of inflammation (Greene et al. 2015;

Butterfield et al. 2020; Woldesemayat 2020). Aging in PLWH is known to exacerbate chronic inflammation, immune exhaustion, microbial translocation, and dysbiosis, thus dysfunctional immunometabolism could be one of the factors that mediate age-related comorbidities.

Microbial dysbiosis or the alterations in microbial communities are studied to be associated with increased inflammation and age-associated comorbidities in PLWH (Mutlu et al. 2014). During HIV infection, the loss of gut integrity and ensuing microbial translocation can trigger persistent inflammation even among virally suppressed PLWH (Dillon and Wilson 2020). Studies have extensively documented the alterations of enteric microbial communities in the elderly populations wherein their dysbiotic profiles predominantly featured an abundance of facultative anaerobes like Enterobacteriaceae species and a lower abundance of probiotics and those anaerobes with immune-regulatory and/or inflammatory potential (Buford 2017). Interestingly, enteric dysbiosis is linked with other comorbidities of aging including CVDs and dementia (Dillon and Wilson 2020).

The enteric microbiota of PLWH is reported to be different when compared to those without HIV infection. Dysbiotic profiles of microbiota are observed among both naïve and ART-treated PLWH, often characterized by a greater abundance of *Prevotella*, higher abundance of *Proteobacteria* (Table 20.1), and lower abundance of obligate anaerobes including butyrate-producing bacteria (Vujkovic-Cvijin and Somsouk 2019). Though ART has been shown to exert several favorable outcomes among the HIV-infected population, it could contribute to dysbiosis (Deusch et al. 2018; Pinto-Cardoso et al. 2018).

There are similarities between the profiles of enteric microbiota in the elderly population and among PLWH. Among the elderly population, HIV infection and age are independently associated with distinct alterations in the microbial communities (Liu et al. 2019). The role of aging in impacting inflammation and immune activation has been demonstrated by elderly PLWH having significantly higher levels of markers like sCD163, sCD14, and soluble TNF when compared with healthy controls (Liu et al. 2019). Moreover, there is a decrease in the levels of IL-1Rα, IL-2, and IL-5 inhibitors in long-term controlled and aged HIV-infected populations. Older populations with long-term controlled HIV infection have significant alterations in their microbial communities partially due to the loss of compartmentalization. The effect of aging on the gut microbial communities and associations between microbiota and inflammatory markers differed based on the HIV serostatus (Rhoades et al. 2019). In addition to the age-associated factors, increasing need for administration of drugs like antibiotics and malnutrition among the elderly PLWH can play a key role in alterations in the composition of gut microbiota (Rondanelli et al. 2015).

CVD represents a leading cause of morbidity and mortality in PLWH (Miller et al. 2014). The role of inflammation in the development of atherosclerosis and other cardiovascular events has been well established. An increase in inflammation has been linked to a heightened risk of CVD among PLWH (Siedner et al. 2016). There has been an association between increased incidence of myocardial infarction and stroke with the increased inflammation in PLWH (Tenorio et al. 2014).

Table 20.1 Summary of seminal literature on gut dysbiosis in the HIV-infected compared to healthy individuals

S. No	Age group	Study groups	HIV-infected and altered gut microbiome	Healthy controls and gut microbiome	References
1	Age >35	HIV positive (N = 18) and HIV negative (N = 14)	Acinetobacter, Blautia, Coprococcus	Prevotella, Bacteroides	Dillon et al. (2014)
2	Age >40	HIV cases N = 21 and Control cases (N-16)	Proteobacteria Gammaproteobacteria, Enterobacteriales, Enterobacteriaceae, Erysipelotrichi, Erysipelotrichales, Erysipelotrichaceae Phylum Firmicutes, Genus—Barnesiella Phylum Bacteroidetes	Rikenellaceae and Alistipes Bacteroidetes	Dinh et al. (2015)
3	Age group >44 (Infected) Control >37	HIV-infected individuals N = 129 (82.7%) and control N = 27 (17.3%)	Prevotella Prevotella, Alloprevotella, Succinvibrio, Dorea, Desulfovibrio, Phascolarctobacterium and Bacteroidales	Prevotella and Bacteroides	Noguera-Julian et al. (2016)
4	Age ratio 28.91 (6.43) and 34.10 (6.36)	HIV positive (N = 45), HIV negative (N = 200)	Peptoniphilus, Porphyromonas and Prevotella and depletion of Bacteroides, Brachyspira and Faecalibacterium	Bacteriodes	Cook et al. (2019)

(continued)

Table 20.1 (continued)

S. No	Age group	Study groups	HIV-infected and altered gut microbiome	Healthy controls and gut microbiome	References
5	Age Mean age 45	HIV positive (N = 21) and control (N = 22) human subjects of all ages	Increase in Brachyspira, Campylobacter, Catenibacterium, Escherichia and other unclassified Enterobacteriaceae, unclassified Fusobacteriaceae, Mogibacterium, Prevotella, and Ralstonia	Akkermansia, Bacteroides, Blautia, Coprococcus, Dialister, Dorea, Faecalibacterium, Lachnospira, Roseburia, Ruminococcus, Odoribacter, Oscillospira, Barnesiellaceae, Lachnospiraceae, Peptostreptococcaceae, Rikenellaceae, and Ruminococcaceae	Mutlu et al. (2014)
6	Pediatric individuals	aged 1 to 19; HIV positive (n = 87) and control (n = 82)	*Prevotella* and *Clostridia Akkermansia muciniphila* and *Faecclibacterium prausnitzii*	High Prevotella and low Bacteroides and various lineages of *Clostridia*, regardless of HIV-infection status	Abange et al. (2021)
7	Age >47	HIV-infected individuals (N = 16) and healthy control (N = 16)	Sutterella, Rhizobium, Delftia, Anaerofilum and Oscillospira genera	Blautia and Anaerostipes	Vesterbacka et al. (2017)
8	Age >35	Adult infected individuals and control	*P.aeruginosa* and *C.albicans* and decreased abundance of Lactobacilli and Bifidobacteria (ranging from [9.1 ± 1.0] × 107 to [8.9 ± 1.0] × 107 cells/g [wet weight] of feces)	*P.aeruginosa* and *C. albicans* and decreased abundances of *Lactobacilli* and *Bifidobacteria* (~104 cells/g [wet weight] of feces)	Gori et al. (2008)

(continued)

Table 20.1 (continued)

S. No	Age group	Study groups	HIV-infected and altered gut microbiome	Healthy controls and gut microbiome	References
9	Adults	28 IR patients, 30 INR patients and 36 healthy control	*Lachnoclostridium, Megasphaera, Escherichia-Shigella, Veillonella, Streptococcus, Fusobacterium, and Ruminococcus gnavus*	*Faecalibacterium, Eubacterium rectale, Alistipes, Subdoligranulum, Bifidobacterium, Roseburia, Ruminococcaceae* and *Parasutterella*	Xie et al. (2021)
10	Age HIV-infected age group 31 ± 4.36 and Negative control age group (37 ± 12.26)	HIV-infected patients (N = 3) and Seronegative (N = 13)	*Erysipelotrichaceae (Eubacterium biforme, Catenibacterium mitsuoki), Prevotellaceae (Prevotella copri* and *Prevotella stercorea), Veillonellaceae (Dialister propionicifaciens, Mitsuokella jalaludinii, Allisonella histaminiformans, Megasphaera elsdenii,* and *Acidaminococcus fermentans), Ruminococcaceae (Ruminococcus callidus),* and *Desulfovibrionaceae (Desulfovibrio piger*	Bacteroidaceae (Bacteroides), Rikenellaceae, (Alistipes), and Porphyromonadaceae (Parabacteroides)	Lozupone et al. (2013)

Lipopolysaccharide (LPS), the product of microbial translocation has been implicated as a risk factor for increased CVD among PLWH. Primate studies have demonstrated the role of LPS in immune activation and CVD by its action of driving immune activation of monocytes and subsequent coagulopathy (Schechter et al. 2017). Additionally, among the PLWH, the levels of gut microbial-related choline metabolite trimethylamine-N-oxide (TMAO) were associated with CVD markers and atherosclerotic CVD (Srinivasa et al. 2015; Shan et al. 2018). While investigating the microbial communities in the elderly population with HIV and age-associated comorbidities, no specific bacterial taxa were found to be associated with coronary heart disease in elderly PLWH. However, differences in the co-occurrence of bacterial clusters between PLWH with and without coronary heart diseases indicate a potential link between the gut microbiota and coronary heart disease in PLWH (Kehmann et al. 2019).

There have been reports on the increasing prevalence of diabetes mellitus (DM) in PLWH compared to those without HIV (Butterfield et al. 2020). The proinflammatory state associated with the alterations in gut microbiota is common in both HIV and DM. HIV patients with DM have a reduction in the relative proportion of several bacteria known to produce butyrate, linked to anti-inflammation and improvement of insulin sensitivity. Also, these bacteria were inversely proportional to the kynurenine/tryptophan ratio, an indicator of tryptophan catabolism, involved in bacterial translocation and inflammation (Estrada and Gonzalez 2018). The occurrence of inflammatory disorders such as DM, ulcerative colitis, and Crohn's disease are associated with the lower abundance of beneficial bacterium *Akkermansia muciniphila* (*A. muciniphila*) in the gut microbiota of PLWH (Ouyang et al. 2020a, b).

Increased measures of inflammation in PLWH have been linked to frailty characterized by physical shrinking, low activity, weakness, and fatigue in the elderly population (Hearps et al. 2012). A recent exploratory study investigating the possible links between the abundance of gut microbiota and physical functional abilities of elderly PLWH impact their revealed mixed associations. The findings were different in the uninfected population, thereby suggesting a unique gut microbiome-physical function axis in older people with HIV (Dillon et al. 2021) (Fig. 20.2).

20.5 HIV Infection, Gut Microbiota, and the Elderly Brain

When compared to HIV uninfected population, PLWH experience an increased rate of neurocognitive decline, and HIV-associated alternations in gut microbiota are also more pronounced (Rich et al. 2020; Seider et al. 2014). This dysbiosis is implicated in the neurocognitive changes in PLWH. Nearly half of all PLWH are reported to undergo HIV-associated neurocognitive decline (HAND) (Sacktor et al. 2016). In addition to aging, elevated levels of amyloid beta are a risk factor for dementia and are associated with long-term HIV infection. Having at least one apolipoprotein E4 allele has been associated with late-onset of Alzheimer's disease in PLWH probably via tau-mediated mechanisms of neurodegeneration (Montufar et al. 2017; Yamazaki et al. 2019).

The gut microbiome has been implicated as one potential source of the persistent neuroinflammation observed in PLWH. HIV-associated dementia (HAD) is associated with an increase in the relative abundance of Bacteroides and lower levels of Firmicutes. Gut microbiome-induced increase in the levels of LPS (endotoxin) leading to immune activation is the mechanism behind the observed neurocognitive dysfunction (Perez-Santiago et al. 2017). HAD in PLWH is also associated with viral replication, monocyte activation, T cell activation, and other markers of inflammation (McGuire et al. 2015). The human microbiome, predominantly of the gut, has been implicated to impact the health of the brain through several mechanisms collectively known by the microbiome–gut–brain axis. The mechanisms proposed under this concept include:

(a) ***Gut microbiota and the link with neurogenesis***

New neurons are generated by the differentiation of neural stem cells and the maturation of neural progenitor cells known by the process of neurogenesis that continues to be persistent in the brain throughout adulthood (Cosacak et al. 2020). Based on the arterial role of the hippocampus in maintaining cognition and memory and that it is the location where the new neurons are produced, it is considered that perturbation in neurogenesis could lead to impaired neurological conditions (Liu et al. 2021). Studies have documented accelerated age-dependent decline of neurogenesis by excessive calorie intake, while an inverse effect was noted with physical exercise and restriction of calories. In this process of modulation of neurogenic processes, the role of gut microbiota and other intrinsic factors are implicated (Cavallucci et al. 2020). The concept of the gut microbiota's role in the onset and modulation of neurogenesis was evidenced by the findings of augmented hippocampal neurogenesis in germ-free mice than in conventionally raised mice (Ogbonnaya et al. 2015). Thus, concurrent development and maturation of the microbiome, the GI tract, and hippocampal neurogenesis during the early stages of development is considered one of the mechanisms of the microbiome–gut–brain axis.

(b) ***Dysbiosis and increased intestinal permeability induces bacterial translocation causing inflammation in the central nervous system***

Accumulating evidence point toward the role of alterations in gut microbial composition in the development of various neurodegenerative diseases. Fecal microbial patterns observed in Alzheimer's patients were different from healthy individuals and the abundance of the bacterium *Helicobacter pylori* is positively associated with cognitive impairment in these patients (Park et al. 2017). Likewise, patients with Parkinson's disease exhibited an increase in the proinflammatory group of bacteria and a lower abundance of butyrate-producing, anti-inflammatory bacterial genera implicating the role of gut microbiota in the modulation of this condition. Elevated levels of D-lactic acid, a product of microbial fermentation of carbohydrates were observed in patients with myalgic encephalomyelitis and neurocognitive dysfunction, thereby indicating increased intestinal permeability and bacterial translocation (Sheedy et al. 2009).

(c) ***Direct or indirect effects of metabolites, neurotransmitters, and other metabolites of gut bacteria on the brain***

Short-chain fatty acids (SCFAs), the products of bacterial fermentation are one of the microbial-derived metabolites with known immune-modulatory and anti-inflammatory properties (Galland 2014; Sivaprakasam et al. 2016). Since these SCFAs can pass through the blood–brain barrier, they can exert regulatory effects both in the peripheral and CNS. They are known to contribute to improving cognitive performance in animal model studies investigating neurodegenerative diseases. Neurotransmitters such as serotonin and tryptophan have a critical role as signaling molecules in both the CNS and GI system. Elevated hippocampal levels of serotonin and a high concentration of tryptophan in the plasma of germ-free mice are observed compared to conventionally raised mice. This points to the link between the changes in microbiota and the neurotransmission in the CNS and the importance of the gut–brain axis in the maintenance of homeostasis (Clarke et al. 2013). Additionally, the age-associated comorbidities among the elderly PLWH are significant risk factors for each other. While CVD and diabetes mellitus are studied as risk factors for HAND, the decline in neurocognition is a risk factor for frailty among PLWH (Zamudio-Rodriguez et al. 2018; Sacktor 2018). CVD and diabetes co-occurring with multi-morbidities in older PLWH are associated with a significantly increased risk of becoming frail within 30 months (Falasca et al. 2015).

20.6 Aging: Gut Microbiota Interactions in COVID-19 and HIV/AIDS—Generic Extrapolation or Tip of the Iceberg?

Coronavirus disease 2019 (COVID-19), caused by SARS coronavirus 2 (SARS-CoV-2), emerged in December 2019, and is the cause of a devastating pandemic. To date, the pandemic has resulted in more than 497 million infections and over 6.2 million deaths (https://www.worldometers.info/coronavirus/) and is surging exponentially by leaps and bounds. COVID-19 shows an extremely diverse clinical course ranging from asymptomatic to mild to severe viral pneumonia and respiratory distress syndrome (Huang et al. 2020). Although ~80% of COVID-19 patients reportedly present with mild flu-like symptoms and can recover spontaneously; up to a quarter of patients could develop more severe symptoms that require different degree of medical support ranging from hospitalization to use of ventilation and oxygen support. Approximately, 5% of COVID-19 cases will require intensive care, and aged individuals and those with preexisting comorbidities such as T2D, hypertension, obesity, and CVD are at increased risk of developing severe COVID-19 (Clarke et al. 2020).

While the risk of SARS-CoV-2 infection and disease severity has been widely assessed among the general population, little is known about SARS-CoV-2 infection in PLWH. More importantly, vis-à-vis the current aging context with gut microbiota, the mechanism largely remains unknown. Because HIV infects CD4 + T cells, the central orchestrator of systemic immune responses, PLWH appears to remain naturally vulnerable to SARS-CoV-2 infection as they are relatively more likely to become immunosuppressed or immunocompromised (Chang et al. 2013; Vizcarra et al. 2020). The clinical findings available in the literature thus far suggest that the disease presentation and prognosis of COVID-19 among PLWH appear to be consistent with those without HIV infection. By comparing 55 PLWH with SARS-CoV-2 infection with 69 PLWH without SARS-CoV-2 infection, one study investigated the factors associated with mortality rate. The study found that there were no significant differences between the two groups in terms of gender, time with HIV infection, nadir CD4+ T cell counts, current CD4+ and CD8+ T cell counts, and type of ART. The only two factors that were significantly associated with death were age and the number of comorbidities (Maggiolo et al. 2021), which was consistent with the general population without HIV infection (Byrd et al. 2020).

The use of ART has led to a significant increase in the number of aged PLHW over the years. Nonetheless, owing to chronic inflammation, the aging process in PLWH is relatively accelerated and individuals experience greater age advancement than their HIV-negative counterparts (Meir-Shafrir and Pollack 2012). These PLWH are expected to have close to normal lifespans, although they are more likely to develop age-related comorbidities (De Francesco et al. 2019) including CVD, T2D, dyslipidemia, metabolic syndrome, and renal impairment that are associated with more severe COVID-19 disease. Having said that the distinction across the COVID-19-HIV axis is seemingly narrow, and given the paucity of data available, especially in the aged COVID-19-HIV positive population, one can only speculate that alterations in gut microbiota would still occur in the aforesaid population, and increased rates of mortality among the elderly in COVID-19 likely hinting at the possible higher abundance of inflammatory bacteria, this could still be extrapolated in the aged COVID-19 plus HIV-infected individuals (Chattopadhyay and Shankar 2021). Further, mounting evidence suggests that COVID-19 likely imparts an adverse impact on PLWH (Eisinger et al. 2021). The unswerving finding is that COVID-19 severity in PLWH is strongly associated with underlying comorbid conditions that likely could harness gut dysbiosis (Fig. 20.3). Hence, based on the data available in the literature, the impact of gut microbiota on the SARS-CoV-2-HIV-aging axis remains the tip of the iceberg. Nonetheless, it is reasonable to claim that in the context of eubiosis wherein PLWH (under ART) encounters SARS-CoV-2 due to optimal immune responses, the ensuing COVID-19 manifestations are relatively mild and immaterial of HIV disease progression. In addition, because dysbiosis plus HIV can cause mucosal barrier disruption, the ensuing immune system largely fails to contain immune dysregulation

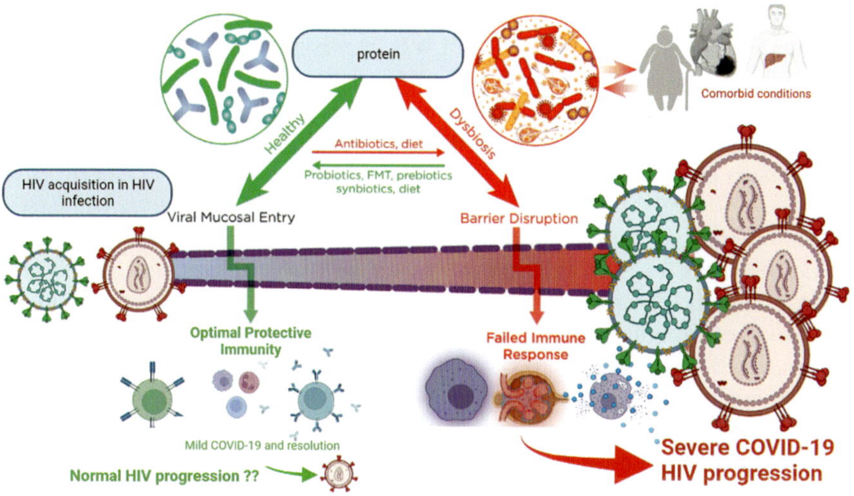

Fig. 20.3 The proposed role of gut microbiota on COVID-19/HIV disease progression. Mounting evidence suggests that COVID-19 severity in PLWH is strongly associated with underlying comorbid conditions that likely could harness gut dysbiosis. In the context of eubiosis in PLWH (under ART) encountering SARS-CoV-2, due to optimal immune responses the ensuing COVID-19 manifestations are relatively mild. On the contrary, because dysbiosis can cause mucosal barrier disruption, the ensuing immune system largely fails to contain immune dysregulation leading to onset of severe COVID-19 also accelerating the rate of HIV disease progression

leading to the onset of severe COVID-19 also impacting the HIV progression rate (Fig. 20.3). Considering the accelerated aging and age-related comorbidities that PLWH might have caution is warranted for PLWH, especially for those with low CD4+ T cell count, advanced disease, high viral load, those who are not on ART, and those with multiple comorbidities.

20.7 Aging, HIV and Gut Microbiota: Therapeutic Implications

Considering the role of microbiota in the pathogenesis of HIV related chronic inflammation in the elderly PLWH, improving gut health by targeted therapies could minimize aging and HIV-related comorbidities. Probiotic administration could be an effective therapeutic option in PLWH by altering the intestinal microbiota and thereby reconditioning and protecting the gut surface. It could interfere with the deleterious effects of HIV infection via several mechanisms including inhibition of epithelial invasion, lowering of inflammation, improving the integrity of the intestinal barrier, and prevention of microbial translocation (Falasca et al. 2015; Surendran Nair et al. 2017; d'Ettorre et al. 2015). Studies have demonstrated improvement in gut microbiota concentration, reduced inflammation, and improved NK cell activity among PLWH by prebiotic mixture supplementation with antiretroviral therapy (Gori et al. 2011; Kim et al. 2016). Supplementing probiotics in PLWH on ART showed a significant reduction in inflammation and microbial translocation and it also could help reduce the risk factors for comorbidities such as CVDs (Falasca et al. 2015; d'Ettorre et al. 2015; D'Angelo et al. 2017).

Consumption of extra virgin olive oil (EVOO) in PLWH over 50 years of age has demonstrated a significant decrease in total cholesterol, along with a significant increase in alpha diversity of intestinal microbiota of men and a decrease in the abundance of proinflammatory bacterial genera in the gut (Olalla et al. 2019). Studies have investigated the modulation of gut dysbiosis based on the association between microbiota and diabetes mellitus, one of the aging-associated comorbidities in PLWH. Improved insulin sensitivity in patients was observed with an increased abundance of SCFAs producing bacteria compared to the control group receiving autologous fecal microbiota transplantation (Kootte et al. 2017; Vrieze et al. 2012). There are also several reports that highlighted the beneficial effects of the gut-protective bacteria *A. muciniphila* in improving insulin sensitivity in mice (Everard et al. 2013; Depommier et al. 2019). It is known that in the gut of PLWH, there is a low abundance of butyrate-producing beneficial bacteria, and the administration of metformin has improved this abundance in the gut. Since a low abundance of butyrate-producing bacteria is associated with microbial translocation and immune activation in PLWH, metformin could aid in decreasing the inflammation by enhancing the intestinal barrier function (Dillon et al. 2017).

Probiotic supplementation among PLWH also can yield neurological benefits by reducing neuroinflammation and neurocognitive impairment. The PLWH who received probiotic supplementation had significantly decreased levels of neopterin microRNA-29a-c levels in cerebrospinal fluid than the control group and fared better in neurocognitive and neuropsychological tests (Ceccarelli et al. 2017a, b).

FMT could be an alternative therapeutic strategy given the dysbiosis and increased gut permeability in elderly PLWH. Primate studies have demonstrated the role of FMT in significantly increasing the number of peripheral Th17 and Th22 cells and reducing CD4+ T cell activation with the transplant being well tolerated (Ouyang et al. 2020a, b). FMT via colonoscopy in ART-suppressed PLWH resulted in partial microbial engraftment and an increase in *Faecalibacterium*, associated with anti-inflammatory effects was observed. Repeated oral capsular FMT significantly decreased the levels of I-FABP, a gut damage marker and the mild engraftment of the donor's microbiota persisted for up to 36 weeks after the transplantation (Serrano-Villar et al. 2021). However, considering the current threat of the COVID-19 pandemic, there remains the possibility of SARS-CoV-2 being detected in the stool specimen of donors and extreme caution to be exercised in such fecal transplant studies. The oral fecal microbiota transplantation feasibility study in Alzheimer's disease had to be terminated upon detection of SARS-CoV-2 in fecal material to minimize the risk to participants (https://clinicaltrials.gov/ct2/show/NCT03998423).

20.8 Concluding Remarks

The diversity of gut microbiota declines with age, and HIV infection compromises the effective functioning of the immune system complicating the metabolic and regulatory programming of the normal physiological machinery. Aging together with HIV infection also could weaken the gut barrier offering an opportunity for the gut flora to invade the systemic compartment increasing the chances of hospitalization and death, which especially holds significance as witnessed during the COVID-19 pandemic. Despite the voluminous quantum of research conducted on the role of gut microbiota in the HIV-infected population, the development of an optimal therapy to alleviate dysbiosis among the aged subset of PLWH still remain a gray area of investigation.

References

Abange WB, Martin C, Nanfack AJ, Yatchou LG, Nusbacher N, Nguedia CA, Kamga HG, Fokam J, Kennedy SP, Ndjolo A, Lozupone C, Nkenfou CN (2021) Alteration of the gut fecal microbiome in children living with HIV on antiretroviral therapy in Yaounde. Cameroon Sci Rep 11:7666. https://doi.org/10.1038/s41598-021-87368-8

Asher I, Guri KM, Elbirt D, Bezalel SR, Maldarelli F, Mor O, Grossman Z, Sthoeger ZM (2016) Characteristics and outcome of patients diagnosed with HIV at older age. Medicine (baltimore) 95:e2327. https://doi.org/10.1097/MD.0000000000002327

Bosco N, Noti M (2021) The aging gut microbiome and its impact on host immunity. Genes Immun 22:289–303. https://doi.org/10.1038/s41435-021-00126-8

Brenchley JM, Douek DC (2008) HIV infection and the gastrointestinal immune system. Mucosal Immunol 1:23–30. https://doi.org/10.1038/mi.2007.1

Buford TW (2017) (Dis)Trust your gut: the gut microbiome in age-related inflammation, health, and disease. Microbiome 5:80. https://doi.org/10.1186/s40168-017-0296-0

Bull MJ, Plummer NT (2014) Part 1: The human gut microbiome in health and disease. Integr Med (Encinitas, Calif.) 13:17–22

Butterfield TR, Landay AL, Anzinger JJ (2020) Dysfunctional immunometabolism in HIV infection: contributing factors and implications for age-related comorbid diseases. Curr HIV/AIDS Rep 17:125–137. https://doi.org/10.1007/s11904-020-00484-4

Byrd KM, Beckwith CG, Garland JM, Johnson JE, Aung S, Cu-Uvin S, Farmakiotis D, Flanigan T, Gillani FS, Macias-Gil R, Mileno M, Ramratnam B, Rybak NR, Sanchez M, Tashima K, Mylonakis E, Kantor R (2020) SARS-CoV-2 and HIV coinfection: clinical experience from Rhode Island, United States. J Int AIDS Soc 23(7):e25573. https://doi.org/10.1002/jia2.25573

Cavallucci V, Fidaleo M, Pani G (2020) Nutrients and neurogenesis: the emerging role of autophagy and gut microbiota. Curr Opin Pharmacol 50:46–52. https://doi.org/10.1016/j.coph.2019.11.004

Ceccarelli G, Brenchley JM, Cavallari EN, Scheri GC, Fratino M, Pinacchio C, Schietroma I, Fard SN, Scagnolari C, Mezzaroma I, Vullo V, d'Ettorre G (2017a) Impact of high-dose multi-strain probiotic supplementation on neurocognitive performance and central nervous system immune activation of HIV-1 infected individuals. Nutrients 9:1269. https://doi.org/10.3390/nu9111269

Ceccarelli G, Fratino M, Selvaggi C, Giustini N, Serafino S, Schietroma I, Corano Scheri G, Pavone P, Passavanti G, Alunni Fegatelli D, Mezzaroma I, Antonelli G, Vullo V, Scagnolari C, d'Ettorre G (2017b) A pilot study on the effects of probiotic supplementation on neuropsychological performance and microRNA-29a-c levels in antiretroviral-treated HIV-1-infected patients. Brain Behav 7(8):e00756. https://doi.org/10.1002/brb3.756

Chang CC, Crane M, Zhou J, Mina M, Post JJ, Cameron BA, Lloyd AR, Jaworowski A, French MA, Lewin SR (2013) HIV and co-infections. Immunol Rev 254(1):114–142. https://doi.org/10.1111/imr.12063

Chattopadhyay I, Shankar EM (2021) SARS-CoV-2-Indigenous microbiota nexus: does gut microbiota contribute to inflammation and disease severity in COVID-19? Front Cell Infect Microbiol 11:590874. https://doi.org/10.3389/fcimb.2021.590874

Clark A, Jit M, Warren-Gash C, Guthrie B, Wang HHX, Mercer SW, Sanderson C, McKee M, Troeger C, Ong KL, Checchi F, Perel P, Joseph S, Gibbs HP, Banerjee A, Eggo RM (2020) Centre for the Mathematical Modelling of Infectious Diseases COVID-19 working group. Global, regional, and national estimates of the population at increased risk of severe COVID-19 due to underlying health conditions in 2020: a modelling study. Lancet Glob Health 8(8):e1003-e1017 https://doi.org/10.1016/S2214-109X(20)30264-3

Clarke G, Grenham S, Scully P, Fitzgerald P, Moloney RD, Shanahan F, Dinan TG, Cryan JF (2013) The microbiome-gut-brain axis during early life regulates the hippocampal serotonergic system in a sex-dependent manner. Mol Psychiatry 18:666–673. https://doi.org/10.1038/mp.2012.77

Cosacak MI, Bhattarai P, Kizil C (2020) Alzheimer's disease, neural stem cells and neurogenesis: cellular phase at single-cell level. Neural Regen Res 15:824–827. https://doi.org/10.4103/1673-5374.268896

COVID-19 Coronavirus pandemic (2022). https://www.worldometers.info/coronavirus/

Crakes KR, Jiang G (2019) Gut microbiome alterations during HIV/SIV infection: Implications for HIV cure. Front Microbiol 10:1104. https://doi.org/10.3389/fmicb.2019.01104

D'Angelo C, Reale M, Costantini E (2017) Microbiota and probiotics in health and HIV infection. Nutrients 9:615. https://doi.org/10.3390/nu9060615

d'Ettorre G, Ceccarelli G, Giustini N, Serafino S, Calantone N, De Girolamo G, Bianchi L, Bellelli V, Ascoli-Bartoli T, Marcellini S, Turriziani O, Brenchley JM, Vullo V (2015) Probiotics reduce inflammation in antiretroviral treated, HIV-infected individuals: results of the "Probio-HIV" clinical trial. PLoS ONE 10:e0137200. https://doi.org/10.1371/journal.pone.0137200

De Francesco D, Wit FW, Bürkle A, Oehlke S, Kootstra NA, Winston A, Franceschi C, Garagnani P, Pirazzini C, Libert C, Grune T, Weber D, Jansen EHJM, Sabin CA, Reiss P (2019) the Co-morBidity in Relation to AIDS (COBRA) Collaboration. Do people living with HIV experience greater age advancement than their HIV-negative counterparts? AIDS 33 (2):259–268. https://doi.org/10.1097/QAD.0000000000002063

Depommier C, Everard A, Druart C, Plovier H, Van Hul M, Vieira-Silva S, Falony G, Raes J, Maiter D, Delzenne NM, de Barsy M, Loumaye A, Hermans MP, Thissen JP, de Vos WM, Cani PD (2019) Supplementation with *Akkermansia muciniphila* in overweight and obese human volunteers: a proof-of-concept exploratory study. Nat Med 25:1096–1103. https://doi.org/10.1038/s41591-019-0495-2

Deusch S, Serrano-Villar S, Rojo D, Martínez-Martínez M, Bargiela R, Vázquez-Castellanos JF, Sainz T, Barbas C, Moya A, Moreno S, Gosalbes MJ, Estrada V, Seifert J, Ferrer M (2018) Effects of HIV, antiretroviral therapy and prebiotics on the active fraction of the gut microbiota. AIDS (lond Engl) 32:1229–1237. https://doi.org/10.1097/QAD.0000000000001831

Dillon SM, Abdo MH, Wilson MP, Liu J, Jankowski CM, Robertson CE, Tuncil Y, Hamaker B, Frank DN, MaWhinney S, Wilson CC, Erlandson KM (2021) A unique gut microbiome–physical function axis exists in older people with HIV: an exploratory study. AIDS Res Hum Retrovirue 37:542–550. https://doi.org/10.1089/aid.2020.0283

Dillon SM, Kibbie J, Lee EJ, Guo K, Santiago ML, Austin GL, Gianella S, Landay AL, Donovan AM, Frank DN, McCARTER MD, Wilson CC (2017) Low abundance of colonic butyrate-producing bacteria in HIV infection is associated with microbial translocation and

immune activation. AIDS (lond Engl) 31:511–521. https://doi.org/10.1097/QAD.0000000000001366
Dillon SM, Lee EJ, Kotter CV, Austin GL, Dong Z, Hecht DK, Gianella S, Siewe B, Smith DM, Landay AL, Robertson CE, Frank DN, Wilson CC (2014) An altered intestinal mucosal microbiome in HIV-1 infection is associated with mucosal and systemic immune activation and endotoxemia. Mucosal Immunol 7:983–994. https://doi.org/10.1038/mi.2013.116
Dillon SM, Wilson CC (2020) What is the collective effect of aging and HIV on the gut microbiome? Curr Opin HIV AIDS 15:94–100. https://doi.org/10.1097/COH.0000000000000611
Dinh DM, Volpe GE, Duffalo C, Bhalchandra S, Tai AK, Kane AV, Wanke CA, Ward HD (2015) Intestinal microbiota, microbial translocation, and systemic inflammation in chronic HIV infection. J Infect Dis 211:19–27. https://doi.org/10.1093/infdis/jiu409
Dubourg G, Surenaud M, Lévy Y, Hüe S, Raoult D (2017) Microbiome of HIV-infected people. Microb Pathog 106:85–93. https://doi.org/10.1016/j.micpath.2016.05.015
Dufour C, Gantner P, Fromentin R, Chomont N (2020) The multifaceted nature of HIV latency. J Clin Invest 130:3381–3390. https://doi.org/10.1172/JCI136227
Dumic I, Nordin T, Jecmenica M, Stojkovic Lalosevic M, Milosavljevic T, Milovanovic T (2019) Gastrointestinal tract disorders in older age. Can J Gastroenterol Hepatol 6757524. https://doi.org/10.1155/2019/6757524
Eisinger RW, Lerner AM, Fauci AS (2021) Human immunodeficiency virus/AIDS in the era of Coronavirus Disease 2019: a juxtaposition of 2 pandemics. J Infect Dis 224:1455–1461. https://doi.org/10.1093/infdis/jiab114
Estrada V, Gonzalez N (2018) Gut microbiota in diabetes and HIV: Inflammation is the link. eBioMedicine 38:17–18. https://doi.org/10.1016/j.ebiom.2018.11.019
Everard A, Belzer C, Geurts L, Ouwerkerk JP, Druart C, Bindels LB, Guiot Y, Derrien M, Muccioli GG, Delzenne NM, de Vos WM, Cani PD (2013) Cross-talk between *Akkermansia muciniphila* and intestinal epithelium controls diet-induced obesity. Proc Natl Acad Sci USA 110:9066–9071. https://doi.org/10.1073/pnas.1219451110
Falasca K, Vecchiet J, Ucciferri C, Di Nicola M, D'Angelo C, Reale M (2015) Effect of probiotic supplement on cytokine levels in HIV infected individuals: a preliminary study. Nutrients 7:8335–8347. https://doi.org/10.3390/nu7105396
Fehily SR, Basnayake C, Wright EK, Kamm MA (2021) The gut microbiota and gut disease. Intern Med J 51:1594–1604. https://doi.org/10.1111/imj.15520
Galland L (2014) The gut microbiome and the brain. J Med Food 17:1261–1272. https://doi.org/10.1089/jmf.2014.7000
Goethel A, Croitoru K, Philpott DJ (2018) The interplay between microbes and the immune response in inflammatory bowel disease. J Physiol 596:3869–3882. https://doi.org/10.1113/JP275396
Gori A, Rizzardini G, Van't Land B, Amor KB, van Schaik J, Torti C, Quirino T, Tincati C, Bandera A, Knol J, Benlhassan-Chahour K, Trabattoni D, Bray D, Vriesema A, Welling G, Garssen J, Clerici M (2011) Specific prebiotics modulate gut microbiota and immune activation in HAART-naive HIV-infected adults: results of the "COPA" pilot randomized trial. Mucosal Immunol 4:554–563. https://doi.org/10.1038/mi.2011.15
Gori A, Tincati C, Rizzardini G, Torti C, Quirino T, Haarman M, Ben Amor K, van Schaik J, Vriesema A, Knol J, Marchetti G, Welling G, Clerici M (2008) Early impairment of gut function and gut flora supporting a role for alteration of gastrointestinal mucosa in human immunodeficiency virus pathogenesis. J Clin Microbiol 46:757–758. https://doi.org/10.1128/JCM.01729-07
Greene M, Covinsky KE, Valcour V, Miao Y, Madamba J, Lampiris H, Cenzer IS, Martin J, Deeks SG (2015) Geriatric syndromes in older HIV-infected adults. J Acquir Immune Defic Syndr 69:161–167. https://doi.org/10.1097/QAI.0000000000000556
Guillén Y, Noguera-Julian M, Rivera J, Casadellà M, Zevin AS, Rocafort M, Parera M, Rodríguez C, Arumí M, Carrillo J, Mothe B, Estany C, Coll J, Bravo I, Herrero C, Saz J, Sirera G, Torrella A, Navarro J, Crespo M, Negredo E, Brander C, Blanco J, Calle ML, Klatt NR,

Clotet B, Paredes R (2019) Low nadir CD4+ T-cell counts predict gut dysbiosis in HIV-1 infection. Mucosal Immunol 12:232–246. https://doi.org/10.1038/s41385-018-0083-7
Harman D (2003) The free radical theory of aging. Antioxid Redox Signal 5:557–561. https://doi.org/10.1089/152308603770310202
Hearps AC, Maisa A, Cheng W-J, Angelovich TA, Lichtfuss GF, Palmer CS, Landay AL, Jaworowski A, Crowe SM (2012) HIV infection induces age-related changes to monocytes and innate immune activation in young men that persist despite combination antiretroviral therapy. AIDS (lond Engl) 26:843–853. https://doi.org/10.1097/QAD.0b013e328351f756
Hensley-McBain T, Zevin AS, Manuzak J, Smith E, Gile J, Miller C, Agricola B, Katze M, Reeves RK, Kraft CS, Langevin S, Klatt NR (2016) Effects of fecal microbial transplantation on microbiome and immunity in simian immunodeficiency virus-infected macaques. J Virol 90:4981–4989. https://doi.org/10.1128/JVI.00099-16
Hills RD Jr, Pontefract BA, Mishcon HR, Black CA, Sutton SC, Theberge CR (2019) Gut microbiome: profound implications for diet and disease. Nutrients 11:1613. https://doi.org/10.3390/nu11071613
Hong PY, Croix JA, Greenberg E, Gaskins HR, Mackie RI (2011) Pyrosequencing-based analysis of the mucosal microbiota in healthy individuals reveals ubiquitous bacterial groups and micro-heterogeneity. PLoS ONE 6:e25042. https://doi.org/10.1371/journal.pone.0025042
Hu Y, Feng Y, Wu J, Liu F, Zhang Z, Hao Y, Liang S, Li B, Li J, Lv N, Xu Y, Zhu B, Sun Z (2019) The gut microbiome signatures discriminate healthy from pulmonary tuberculosis patients. Front Cell Infect Microbiol 9:90. https://doi.org/10.3389/fcimb.2019.00090
Huang C, Wang Y, Li X, Ren L, Zhao J, Hu Y, Zhang L, Fan G, Xu J, Gu X, Cheng Z, Yu T, Xia J, Wei Y, Wu W, Xie X, Yin W, Li H, Liu M, Xiao Y, Gao H, Guo L, Xie J, Wang G, Jiang R, Gao Z, Jin Q, Wang J, Cao B (2020) Clinical features of patients infected with 2019 novel coronavirus in Wuhan, China. Lancet 395(10223):497–506. https://doi.org/10.1016/S0140-6736(20)30183-5
Jie Z, Xia H, Zhong SL, Feng Q, Li S, Liang S, Zhong H, Liu Z, Gao Y, Zhao H, Zhang D, Su Z, Fang Z, Lan Z, Li J, Xiao L, Li J, Li R, Li X, Li F, Ren H, Huang Y, Peng Y, Li G, Bo W, Bo D, Chen ZY, Geng QS, Zhang ZW, Yang H, Wang J, Wang J, Zhang X, Madsen L, Brix S, Ning G, Xu X, Liu X, Hou Y, Jia H, He K, Kristiansen K (2017) The gut microbiome in atherosclerotic cardiovascular disease. Nat Commun 8:845. https://doi.org/10.1038/s41467-017-00900-1
Kehrmann J, Menzel J, Saeedghalati M, Obeid R, Schulze C, Holzendorf V, Farahpour F, Reinsch N, Klein-Hitpass L, Streeck H, Hoffmann D, Buer J, Esser S, HIV-HEART Study Group (2019) Gut microbiota in human immunodeficiency virus-infected individuals linked to coronary heart disease. J Infect Dis 219:497–508. https://doi.org/10.1093/infdis/jiy524
Kim CJ, Walmsley SL, Raboud JM, Kovacs C, Coburn B, Rousseau R, Reinhard R, Rosenes R, Kaul R (2016) Can probiotics reduce inflammation and enhance gut immune health in people living with HIV: Study designs for the probiotic Visbiome for inflammation and translocation (PROOV IT) pilot trials. HIV Clin Trials 17:147–157. https://doi.org/10.1080/15284336.2016.1184827
Kootte RS, Levin E, Salojärvi J, Smits LP, Hartstra AV, Udayappan SD, Hermes G, Bouter KE, Koopen AM, Holst JJ, Knop FK, Blaak EE, Zhao J, Smidt H, Harms AC, Hankemeijer T, Bergman JJGHM, Romijn HA, Schaap FG, Olde Damink SWM, Ackermans MT, Dallinga-Thie GM, Zoetendal E, de Vos WM, Serlie MJ, Stroes ESG, Groen AK, Nieuwdorp M (2017) Improvement of insulin sensitivity after lean donor feces in metabolic syndrome is driven by baseline intestinal microbiota composition. Cell Metab 26:611-619.e6. https://doi.org/10.1016/j.cmet.2017.09.008
Leeuwendaal NK, Cryan JF, Schellekens H (2021) Gut peptides and the microbiome: focus on ghrelin. Curr Opin Endocrinol Diabet Obes 28:243–252. https://doi.org/10.1097/MED.0000000000000616
Leslie W, Hankey C (2015) Aging, nutritional status and health. Healthcare (basel, Switzerland) 3:648–658. https://doi.org/10.3390/healthcare3030648

Liu C, Yang S-Y, Wang L, Zhou F (2021) The gut microbiome: implications for neurogenesis and neurological diseases. Neural Regen Res 17:53–58. https://doi.org/10.4103/1673-5374.315227

Liu J, Johnson R, Dillon S, Kroehl M, Frank DN, Tuncil YE, Zhang X, Ir D, Robertson CE, Seifert S, Higgins J, Hamaker B, Wilson CC, Erlandson KM (2019) Among older adults, age-related changes in the stool microbiome differ by HIV-1 serostatus. EBioMedicine 40:583–594. https://doi.org/10.1016/j.ebiom.2019.01.033

Lorenz DR, Mukerji SS, Misra V, Uno H, Gelman BB, Moore DJ, Singer EJ, Morgello S, Gabuzda D (2021) Predictors of transition to frailty in middle-aged and older people with HIV: a prospective cohort study. J Acquir Immune Defic Syndr 88:518–527. https://doi.org/10.1097/QAI.0000000000002810

Lozupone CA, Li M, Campbell TB, Flores SC, Linderman D, Gebert MJ, Knight R, Fontenot AP, Palmer BE (2013) Alterations in the gut microbiota associated with HIV-1 infection. Cell Host Microbe 14:329–339. https://doi.org/10.1016/j.chom.2013.08.006

Machiavelli A, Duarte RTD, Pires MMS, Zárate-Bladés CR, Pinto AR (2019) The impact of *in utero* HIV exposure on gut microbiota, inflammation, and microbial translocation. Gut Microbes 10:599–614. https://doi.org/10.1080/19490976.2018.1560768

Maggiolo F, Zoboli F, Arosio M, Valenti D, Guarneri D, Sangiorgio L, Ripamonti D, Callegaro A (2021) SARS-CoV-2 infection in persons living with HIV: a single center prospective cohort. J Med Virol 93(2):1145–1149. https://doi.org/10.1002/jmv.26352

Marchetti G, Tincati C, Silvestri G (2013) Microbial translocation in the pathogenesis of HIV infection and AIDS. Clin Microbiol Rev 26:2–18. https://doi.org/10.1128/CMR.00050-12

Márquez M, Gutiérrez F, del Álamo C, Girón-González JA (2016) Gut epithelial barrier dysfunction in human immunodeficiency virus-hepatitis C virus coinfected patients: Influence on innate and acquired immunity. World J Gastroenterol 22:1433–1448. https://doi.org/10.3748/wjg.v22.i4.1433

Masters MC, Landay AL, Robbins PD, Tchkonia T, Kirkland JL, Kuchel GA, Niedernhofer LJ (1999) Palella FJ (2022) Chronic HIV infection and aging: application of a geroscience-guided approach. J Acquir Immune Defic Syndr 89:S34–S46. https://doi.org/10.1097/QAI.0000000000002858

Matijašić M, Meštrović T, Paljetak HČ, Perić M, Barešić A, Verbanac D (2020) Gut microbiota beyond bacteria-Mycobiome, virome, archaeome, and eukaryotic parasites in IBD. Int J Mol Sci 21:2668. https://doi.org/10.3390/ijms21082668

McGuire JL, Gill AJ, Douglas SD, Kolson DL, CNS HIV Anti-Retroviral Therapy Effects Research (CHARTER) group (2015) Central and peripheral markers of neurodegeneration and monocyte activation in HIV-associated neurocognitive disorders. J Neurovirol 21:439–448. https://doi.org/10.1007/s13365-015-0333-3

Meir-Shafrir K, Pollack S (2012) Accelerated aging in HIV patients. Rambam Maimonides Med J 3(4):e0025. https://doi.org/10.5041/RMMJ.10089

Mildvan D, Mathur U, Enlow RW, Romain PL, Winchester RJ, Colp C, Singman H, Adelsberg BR, Spigland I (1982) Opportunistic infections and immune deficiency in homosexual men. Ann Intern Med 96:700–704. https://doi.org/10.7326/0003-4819-96-6-700

Miller CJ, Baker JV, Bormann AM, Erlandson KM, Huppler Hullsiek K, Justice AC, Neuhaus J, Paredes R, Petoumenos K, Wentworth D, Winston A, Wolfson J, Neaton JD (2014) Adjudicated morbidity and mortality outcomes by age among individuals with HIV infection on suppressive antiretroviral therapy. PLoS ONE 9:e95061. https://doi.org/10.1371/journal.pone.0095061

Montufar S, Calero C, Vinueza R, Correa P, Carrera-Gonzalez A, Villegas F, Moreta G, Paredes R (2017) Association between the APOE ε4 allele and late-onset Alzheimer's disease in an Ecuadorian Mestizo population. Int J Alzheimers Dis 2017:e1059678. https://doi.org/10.1155/2017/1059678

Mutlu EA, Keshavarzian A, Losurdo J, Swanson G, Siewe B, Forsyth C, French A, Demarais P, Sun Y, Koenig L, Cox S, Engen P, Chakradeo P, Abbasi R, Gorenz A, Burns C, Landay A (2014) A compositional look at the human gastrointestinal microbiome and immune activation parameters in HIV infected subjects. PLoS Pathog 10:e1003829. https://doi.org/10.1371/journal.ppat.1003829

Nazli A, Chan O, Dobson-Belaire WN, Ouellet M, Tremblay MJ, Gray-Owen SD, Arsenault AL, Kaushic C (2010) Exposure to HIV-1 directly impairs mucosal epithelial barrier integrity allowing microbial translocation. PLoS Pathog 6:e1000852. https://doi.org/10.1371/journal.ppat.1000852

Noguera-Julian M, Rocafort M, Guillén Y, Rivera J, Casadellà M, Nowak P, Hildebrand F, Zeller G, Parera M, Bellido R, Rodríguez C, Carrillo J, Mothe B, Coll J, Bravo I, Estany C, Herrero C, Saz J, Sirera G, Torrela A, Navarro J, Crespo M, Brander C, Negredo E, Blanco J, Guarner F, Calle ML, Bork P, Sönnerborg A, Clotet B, Paredes R (2016) Gut microbiota linked to sexual preference and HIV infection. EBioMedicine 5:135–146. https://doi.org/10.1016/j.ebiom.2016.01.032

Odamaki T, Kato K, Sugahara H, Hashikura N, Takahashi S, Xiao JZ, Abe F, Osawa R (2016) Age-related changes in gut microbiota composition from newborn to centenarian: a cross-sectional study. BMC Microbiol 16:90. https://doi.org/10.1186/s12866-016-0708-5

Ogbonnaya ES, Clarke G, Shanahan F, Dinan TG, Cryan JF, O'Leary OF (2015) Adult hippocampal neurogenesis is regulated by the microbiome. Biol Psychiatry 78:e7-9. https://doi.org/10.1016/j.biopsych.2014.12.023

Ogunrinola GA, Oyewale JO, Oshamika OO, Olasehinde GI (2020) The human microbiome and its impacts on health. Int J Microbiol 2020:1–7. https://doi.org/10.1155/2020/8045646

Okoye AA, Picker LJ (2013) CD4 (+) T-cell depletion in HIV infection: mechanisms of immunological failure. Immunol Rev 254:54–64. https://doi.org/10.1111/imr.12066

Olalla J, García de Lomas JM, Chueca N, Pérez-Stachowski X, De Salazar A, Del Arco A, Plaza-Díaz J, De la Torre J, Prada JL, García-Alegría J, Fernández-Sánchez F, García F (2019) Effect of daily consumption of extra virgin olive oil on the lipid profile and microbiota of HIV-infected patients over 50 years of age. Medicine (baltimore) 98:e17528. https://doi.org/10.1097/MD.0000000000017528

Ouyang J, Isnard S, Lin J, Fombuena B, Peng X, Nair Parvathy S, Chen Y, Silverman MS, Routy JP (2020a) Treating from the inside out: relevance of fecal microbiota transplantation to counteract gut damage in GVHD and HIV infection. Front Med (lausanne) 7:421. https://doi.org/10.3389/fmed.2020.00421

Ouyang J, Lin J, Isnard S et al (2020b) The bacterium *Akkermansia muciniphila*: A sentinel for gut permeability and its relevance to HIV-related inflammation. Front Immunol 11:645. https://doi.org/10.3389/fimmu.2020.00645

Parada Venegas D, De la Fuente MK, Landskron G, González MJ, Quera R, Dijkstra G, Harmsen HJM, Faber KN, Hermoso MA (2019) Short chain fatty acids (SCFAs)-mediated gut epithelial and immune regulation and its relevance for inflammatory bowel diseases. Front Immunol 10:277. https://doi.org/10.3389/fimmu.2019.00277

Park A-M, Omura S, Fujita M, Sato F, Tsunoda I (2017) *Helicobacter pylori* and gut microbiota in multiple sclerosis versus Alzheimer's disease: 10 pitfalls of microbiome studies. Clin Exp Neuroimmunol 8:215–232. https://doi.org/10.1111/cen3.12401

Pérez-Santiago J, Gianella S, Bharti A, Cooksonv D, Heaton RK, Grant I, Letendre S, Peterson SN (2017) The human gut microbiome and HIV-associated neurocognitive disorders. In: Proceedings of the conference on retroviruses and opportunistic infections (CROI), Seattle, WA, USA. pp 13–16. https://www.croiconference.org/abstract/human-gut-microbiome-and-hiv-associated-neurocognitive-disorders/

Peterson TE, Baker JV (2019) Assessing inflammation and its role in comorbidities among persons living with HIV. Curr Opin Infect Dis 32:8–15. https://doi.org/10.1097/QCO.0000000000000510

Pickard JM, Zeng MY, Caruso R, Núñez G (2017) Gut microbiota: role in pathogen colonization, immune responses, and inflammatory disease. Immun Rev 279:70–89. https://doi.org/10.1111/imr.12567

Pinto-Cardoso S, Klatt NR, Reyes-Terán G (2018) Impact of antiretroviral drugs on the microbiome: unknown answers to important questions. Curr Opin HIV AIDS 13:53–60. https://doi.org/10.1097/COH.0000000000000428

Rakoff-Nahoum S, Paglino J, Eslami-Varzaneh F, Edberg S, Medzhitov R (2004) Recognition of commensal microflora by Toll-like receptors is required for intestinal homeostasis. Cell 118:229–241. https://doi.org/10.1016/j.cell.2004.07.002

Rémond D, Shahar DR, Gille D, Pinto P, Kachal J, Peyron MA, Dos Santos CN, Walther B, Bordoni A, Dupont D, Tomás-Cobos L, Vergères G (2015) Understanding the gastrointestinal tract of the elderly to develop dietary solutions that prevent malnutrition. Oncotarget 6:13858–13898. https://doi.org/10.18632/oncotarget.4030

Rhoades N, Mendoza N, Jankeel A, Sureshchandra S, Alvarez AD, Doratt B, Heidari O, Hagan R, Brown B, Scheibel S, Marbley T, Taylor J, Messaoudi I (2019) Altered immunity and microbial dysbiosis in aged individuals with long-term controlled HIV infection. Front Immunol 10:463. https://doi.org/10.3389/fimmu.2019.00463

Rich S, Klann E, Bryant V, Richards V, Wijayabahu A, Bryant K, Mai V, Cook R (2020) A review of potential microbiome-gut-brain axis mediated neurocognitive conditions in persons living with HIV. Brain Behav Immun Health 9:100168. https://doi.org/10.1016/j.bbih.2020.100168

Rinninella E, Raoul P, Cintoni M, Franceschi F, Miggiano G, Gasbarrini A, Mele MC (2019) What is the healthy gut microbiota composition? A changing ecosystem across age, environment, diet, and diseases. Microorganisms 7:14. https://doi.org/10.3390/microorganisms7010014

Ríos-Covián D, Ruas-Madiedo P, Margolles A, Gueimonde M, de los Reyes-Gavilán CG, Salazar N (2016) Intestinal short chain fatty acids and their link with diet and human health. Front Microbiol 7:185. https://doi.org/10.3389/fmicb.2016.00185

Rondanelli M, Giacosa A, Faliva MA, Perna S, Allieri F, Castellazzi AM (2015) Review on microbiota and effectiveness of probiotics use in older. World J Clin Cases 3:156–162. https://doi.org/10.12998/wjcc.v3.i2.156

Rowland I, Gibson G, Heinken A, Scott K, Swann J, Thiele I, Tuohy K (2018) Gut microbiota functions: metabolism of nutrients and other food components. Eur J Nutr 57:1–24. https://doi.org/10.1007/s00394-017-1445-8

Sacktor N (2018) Changing clinical phenotypes of HIV-associated neurocognitive disorders. J Neurovirol 24:141–145. https://doi.org/10.1007/s13365-017-0556-6

Sacktor N, Skolasky RL, Seaberg E, Devlin KN, de la Monte SM, Chasman JD, Yan P, Tashima KT, Navia B, Cohen RA (2016) Prevalence of HIV-associated neurocognitive disorders in the multicenter AIDS Cohort study. Neurology 86:334–340. https://doi.org/10.1212/WNL.0000000000002277

Sato S, Kiyono H, Fujihashi K (2015) Mucosal immunosenescence in the gastrointestinal tract: a mini-review. Gerontology 61:336–342. https://doi.org/10.1159/000368897

Schechter ME, Andrade BB, He T, Tosh KW, Policicchio BB, Singh A, Raehtz KD, Sheikh V, Ma D, Brocca-Cofano E, Apetrei C, Tracy R, Ribeiro RM, Sher A, Francischetti IMB, Pandrea I, Sereti I (2017) Inflammatory monocytes expressing tissue factor drive SIV and HIV-coagulopathy. Sci Transl Med 9:eaam5441. https://doi.org/10.1126/scitranslmed.aam5441

Seider TR, Luo X, Gongvatana A, Devlin KN, de la Monte SM, Chasman JD, Yan P, Tashima KT, Navia B, Cohen RA (2014) Verbal memory declines more rapidly with age in HIV infected versus uninfected adults. J Clin Exp Neuropsychol 36:356–367. https://doi.org/10.1080/13803395.2014.892061

Serrano-Villar S, Talavera-Rodríguez A, Gosalbes MJ, Madrid N, Pérez-Molina JA, Elliott RJ, Navia B, Lanza VF, Vallejo A, Osman M, Dronda F, Budree S, Zamora J, Gutiérrez C, Manzano M, Vivancos MJ, Ron R, Martínez-Sanz J, Herrera S, Ansa U, Moya A, Moreno S (2021) Fecal microbiota transplantation in HIV: a pilot placebo-controlled study. Nat Commun 12(1):1139. https://doi.org/10.1038/s41467-021-21472-1

Shan Z, Clish CB, Hua S, Scott JM, Hanna DB, Burk RD, Haberlen SA, Shah SJ, Margolick JB, Sears CL, Post WS, Landay AL, Lazar JM, Hodis HN, Anastos K, Kaplan RC, Qi Q (2018) Gut microbial-related choline metabolite trimethylamine-n-oxide is associated with progression of carotid artery atherosclerosis in HIV infection. J Infect Dis 218:1474–1479. https://doi.org/10.1093/infdis/jiy356

Shankar EM, Velu V, Kamarulzaman A, Larsson M (2015) Mechanistic insights on immunosenescence and chronic immune activation in HIV-tuberculosis co-infection. World J Virol 4:17–24. https://doi.org/10.5501/wjv.v4.i1.17

Sheedy JR, Wettenhall REH, Scanlon D, Gooley PR, Lewis DP, McGregor N, Stapleton DI, Butt HL, Meirleir KL (2009) Increased D-lactic acid intestinal bacteria in patients with chronic fatigue syndrome. Vivo Athens Greece 23:621–628

Shi W, Hu Y, Ning Z, Xia F, Wu M, Hu YO, Xu B (2021) Alterations of gut microbiota in patients with active pulmonary tuberculosis in China: a pilot study. Int J Infect Dis 111:313–321. https://doi.org/10.1016/j.ijid.2021.08.064

Siedner MJ, Kim J-H, Nakku RS, Bibangambah P, Hemphill L, Triant VA, Haberer JE, Martin JN, Mocello AR, Boum Y 2nd, Kwon DS, Tracy RP, Burdo T, Huang Y, Cao H, Okello S, Bangsberg DR, Hunt PW (2016) Persistent immune activation and carotid atherosclerosis in HIV-infected ugandans receiving antiretroviral therapy. J Infect Dis 213:370–378. https://doi.org/10.1093/infdis/jiv450

Sivaprakasam S, Prasad PD, Singh N (2016) Benefits of short-chain fatty acids and their receptors in inflammation and carcinogenesis. Pharmacol Ther 164:144–151. https://doi.org/10.1016/j.pharmthera.2016.04.007

Srinivasa S, Fitch KV, Lo J, Kadar H, Knight R, Wong K, Abbara S, Gauguier D, Capeau J, Boccara F, Grinspoon SK (2015) Plaque burden in HIV-infected patients is associated with serum intestinal microbiota-generated trimethylamine. AIDS (lond Engl) 29:443–452. https://doi.org/10.1097/QAD.0000000000000565

Surendran Nair M, Amalaradjou MA, Venkitanarayanan K (2017) Antivirulence properties of probiotics in combating microbial pathogenesis. Adv Appl Microbiol 98:1–29. https://doi.org/10.1016/bs.aambs.2016.12.001

Tenorio AR, Zheng Y, Bosch RJ, Krishnan S, Rodriguez B, Hunt PW, Plants J, Seth A, Wilson CC, Deeks SG, Lederman MM, Landay AL (2014) Soluble markers of inflammation and coagulation but not T-cell activation predict non-AIDS-defining morbid events during suppressive antiretroviral treatment. J Infect Dis 210:1248–1259. https://doi.org/10.1093/infdis/jiu254

Tosato M, Zamboni V, Ferrini A, Cesari M (2007) The aging process and potential interventions to extend life expectancy. Clin Interven Aging 2:401–412

Vaarala O (2012) Gut microbiota and type 1 diabetes. Rev Diabet Stud 9:251–259. https://doi.org/10.1900/RDS.2012.9.251

Valeri F, Endres K (2021) How biological sex of the host shapes its gut microbiota. Front Neuroendocrinol 61:100912. https://doi.org/10.1016/j.yfrne.2021.100912

Vaiserman AM, Koliada AK, Marotta F (2017) Gut microbiota: a player in aging and a target for anti-aging intervention. Ageing Res Rev 35:36–45

Van der Hee B, Wells JM (2021) Microbial regulation of host physiology by short-chain fatty acids. Trends Microbiol 29:700–712. https://doi.org/10.1016/j.tim.2021.02.001

Vemuri R, Gundamaraju R, Shastri MD, Shukla SD, Kalpurath K, Ball M, Tristram S, Shankar EM, Ahuja K, Eri R (2018) Gut microbial changes, interactions, and their implications on human lifecycle: an ageing perspective. Biomed Res Int 2018:4178607. https://doi.org/10.1155/2018/4178607

Vemuri R, Shankar EM, Chieppa M, Eri R, Kavanagh K (2020) Beyond just bacteria: functional biomes in the gut ecosystem including virome, mycobiome, archaeome and helminths. Microorganisms 8(4):483. https://doi.org/10.3390/microorganisms8040483

Vesterbacka J, Rivera J, Noyan K, Parera M, Neogi U, Calle M, Paredes R, Sönnerborg A, Noguera-Julian M, Nowak P (2017) Richer gut microbiota with distinct metabolic profile in HIV infected elite controllers. Sci Rep 7:6269. https://doi.org/10.1038/s41598-017-06675-1

Vizcarra P, Pérez-Elías MJ, Quereda C, Moreno A, Vivancos MJ, Dronda F, Casado JL, COVID-19 ID Team (2020) Description of COVID-19 in HIV-infected individuals: a single-centre, prospective cohort. Lancet HIV 7(8):e554–e564 https://doi.org/10.1016/S2352-3018(20)30164-8

Vrieze A, Van Nood E, Holleman F, Salojärvi J, Kootte RS, Bartelsman JF, Dallinga-Thie GM, Ackermans MT, Serlie MJ, Oozeer R, Derrien M, Druesne A, Van Hylckama Vlieg JE, Bloks VW, Groen AK, Heilig HG, Zoetendal EG, Stroes ES, de Vos WM, Hoekstra JB, Nieuwdorp M (2012) Transfer of intestinal microbiota from lean donors increases insulin sensitivity in individuals with metabolic syndrome. Gastroenterology 143:913-916.e7. https://doi.org/10.1053/j.gastro.2012.06.031

Vujkovic-Cvijin I, Sklar J, Jiang L, Natarajan L, Knight R, Belkaid Y (2020) Host variables confound gut microbiota studies of human disease. Nature 587:448–454. https://doi.org/10.1038/s41586-020-2881-9

Vujkovic-Cvijin I, Somsouk M (2019) HIV and the gut microbiota: composition, consequences, and avenues for amelioration. Curr HIV/AIDS Rep 16:204–213. https://doi.org/10.1007/s11904-019-00441-w

Wang Z, Klipfell E, Bennett BJ, Koeth R, Levison BS, DuGar B, Feldstein AE, Britt EB, Fu X, Chung YM, Wu Y, Schauer P, Smith JD, Allayee H, Tang WHW, DiDonato JA, Lusis AJ, Hazen SL (2011) Gut flora metabolism of phosphatidylcholine promotes cardiovascular disease. Nature 472:57–63. https://doi.org/10.1038/nature09922

Woldesemayat EM (2020) Chronic diseases multimorbidity among adult people living with HIV at Hawassa university comprehensive specialized hospital. Southern Ethiopia. Int J Chronic Dis 2020:2190395. https://doi.org/10.1155/2020/2190395

Wu HJ, Wu E (2012) The role of gut microbiota in immune homeostasis and autoimmunity. Gut Microbes 3:4–14. https://doi.org/10.4161/gmic.19320

Xie Y, Sun J, Wei L, Jiang H, Hu C, Yang J, Huang Y, Ruan B, Zhu B (2021) Altered gut microbiota correlate with different immune responses to HAART in HIV-infected individuals. BMC Microbiol 21:11. https://doi.org/10.1186/s12866-020-02074-1

Yamazaki Y, Zhao N, Caulfield TR, Liu CC, Bu G (2019) Apolipoprotein E and Alzheimer disease: pathobiology and targeting strategies. Nat Rev Neurol 15:501–518. https://doi.org/10.1038/s41582-019-0228-7

Zamudio-Rodríguez A, Belaunzarán-Zamudio PF, Sierra-Madero JG, Cuellar-Rodríguez J, Crabtree-Ramírez BE, Alcala-Zermeno JL, Amieva H, Avila-Funes JA (2018) Association between frailty and HIV-associated neurodegenerative disorders among older adults living with HIV. AIDS Res Hum Retroviruses 34:449–455. https://doi.org/10.1089/AID.2017.0100

Zheng D, Liwinski T, Elinav E (2020) Interaction between microbiota and immunity in health and disease. Cell Res 30:492–506. https://doi.org/10.1038/s41422-020-0332-7

Zhong D, Wu C, Zeng X, Wang Q (2018) The role of gut microbiota in the pathogenesis of rheumatic diseases. Clin Rheumatol 37:25–34. https://doi.org/10.1007/s10067-017-3821-4

Index

F. Marotta (ed.), *Gut Microbiota in Aging and Chronic Diseases*, Healthy Ageing and Longevity 17, https://doi.org/10.1007/978-3-031-14023-5

GPSR Compliance

The European Union's (EU) General Product Safety Regulation (GPSR) is a set of rules that requires consumer products to be safe and our obligations to ensure this.

If you have any concerns about our products, you can contact us on ProductSafety@springernature.com

In case Publisher is established outside the EU, the EU authorized representative is:

Springer Nature Customer Service Center GmbH
Europaplatz 3
69115 Heidelberg, Germany

Batch number: 10756373

Printed by Printforce, the Netherlands